T0140858

Novel Lanthanide-Based Luminescent Probes for Biological and Lighting Applications

Zur Erlangung des akademischen Grades eines

DOKTORS DER NATURWISSENSCHAFTEN

(Dr. rer. nat.)

von der KIT-Fakultät für Chemie und Biowissenschaften

des Karlsruher Instituts für Technologie (KIT)

genehmigte

DISSERTATION

von

M. Sc. Alena S. Kalyakina

aus Obninsk, Russland

KIT-Dekan: Prof. Dr. Reinhard Fischer

Referent: Prof. Dr. Stefan Bräse

Korreferent: Prof. Dr. Peter Roesky

Tag der mündlichen Prüfung: 19.07.2018

Gedruckt mit Unterstützung des Deutschen Akademischen Austauschdienstes

Band 76
Beiträge zur organischen Synthese
Hrsg.: Stefan Bräse

Prof. Dr. Stefan Bräse
Institut für Organische Chemie
Karlsruher Institut für Technologie (KIT)
Fritz-Haber-Weg 6
D-76131 Karlsruhe

Bibliographic information published by the Deutsche Nationalbibliothek

The Deutsche Nationalbibliothek lists this publication in the Deutsche Nationalbibliografie; detailed bibliographic data are available in the Internet at http://dnb.d-nb.de

ISBN 978-3-8325-4810-0
ISSN 1862-5681

Logos Verlag Berlin GmbH
Comeniushof, Gubener Str. 47,
10243 Berlin
Tel.: +49 030 42 85 10 90
Fax: +49 030 42 85 10 92
INTERNET: http://www.logos-verlag.de

Fortune favors the prepared mind

– Louis Pasteur

Die vorliegende Arbeit wurde im Zeitraum vom 1. August 2015 bis 6. Juni 2018 am Institut für Organische Chemie und dem Institut für Toxikologie und Genetik der Fakultät für Chemie und Biowissenschaften am Karlsruher Institut für Technologie (KIT) unter der Leitung von Prof. Dr. STEFAN BRÄSE durchgeführt. Ein Teil der praktischen Arbeit mit dem Titel "Ln(III) complexes for FRET-assay" wurde zwischen dem 11. Februar 2018 und dem 05. Mai 2018 am Harvard Medical School - Massachusetts General Hospital (Boston, USA) unter der Leitung von Prof. Dr. RALPH MAZITSCHEK durchgeführt.

The present work was carried out at the Karlsruhe Institute of Technology, Faculty of Chemistry and Biosciences, Institute of Organic Chemistry in the period from August 1st 2015 to June 6th 2018 under supervision of Prof. Dr. STEFAN BRÄSE. Part of the practical work called "Ln(III) complexes for FRET-assay" was carried out between February, 11th 2018 and May, 5th 2018 at Harvard Medical School - Massachusetts General Hospital (Boston, USA) under the supervision of Prof. Dr. RALPH MAZITSCHEK.

Hiermit versichere ich, die vorliegende Arbeit selbstständig verfasst und keine anderen als die angegebenen Quellen und Hilfsmittel verwendet sowie Zitate kenntlich gemacht zu haben. Die Dissertation wurde bisher an keiner anderen Hochschule oder Universität eingereicht.

Table of contents

Abstract

The potential of lanthanide (Ln) coordination compounds in biological and lighting applications can hardly be overestimated due to their fascinating optical properties. However, the sensitivity of these probes is limited by the low efficiency of the ligand-to-lanthanide energy transfer or by quenching *via* high-energy bond vibrations (O-H, C-H). Therefore, new Ln-based probes with improved functional properties are of great interest.

In the present thesis, several strategies to improve the properties of Ln complexes have been established. The first strategy applies a novel approach to ligand fluorination in order to improve the luminescence, solubility and stability of Ln benzoates on the example of 85 Ln complexes with 9 different fluorinated benzoates. The structure-property relationship (SPR) was determined, analyzing 24 crystal structures and 38 powder diffraction patterns. This study was expanded on Eu ternary complexes, where the favorable geometry together with the minimized vibrational quenching results in exceptionally high performance with the quantum yield up to 90%, being the highest reported value for Eu complexes so far. To improve the luminescent properties of near infrared (NIR)-emitting Ln complexes, a bulky anthracene-based ligand with a higher π-conjugation was used, allowing to obtain Yb-based NIR-emitters with the quantum yield of 2.5%. The host-free organic light-emitting diode (OLED), fabricated using this compound, possessed solely an Yb^{3+} f-f emission band at 1000 nm, with the electroluminescence quantum yield of 0.21%. Besides, in this work a Tb(III) complex with an octadentate macrocyclic ligand was synthesized (Lumi4Tb®), exhibiting incredibly high luminescence intensity even in a nano-molar concentration range in aqueous buffers. The synthetic route to obtain this compound was optimized, proposing more reliable approach to linker incorporation.

Furthermore, substituted lanthanide benzoates were further modified and conjugated with molecular transporters (peptoids), using different conjugation methods, including copper(I)-catalyzed alkyne-azide cycloaddition (CuAAC) or strain-promoted alkyne-azide cycloaddition (SPAAC). The best results were obtained, when the conjugation was performed through an 1,10-phenanthroline-based or terpyridine-based auxiliary ligand, exhibiting intensive Eu-based luminescence, as well as aqueous solubility, being a very rare combination. To increase the coordination ability of Eu(III) ions, cyclic peptoids as chelating macrocycles were used, featuring heterocycle-based scaffolds in the peptoid backbone and carboxylic functionality in the side chains. These complexes, combining both, luminescent and transport properties, may find various applications for targeted luminescence.

Kurzzusammenfassung / Abstract in German

Das Potenzial von Koordinationsverbindungen mit Lanthanoiden (Ln), sowohl in biologischen als auch in optischen Anwendungsgebieten, kann aufgrund ihrer faszinierenden optischen Eigenschaften kaum überschätzt werden. Die Empfindlichkeit dieser Sonden ist jedoch durch die geringe Effizienz des Liganden-zu-Lanthanoid-Energietransfers oder durch Quenchen über hochenergetische Bindungsschwingungen (O-H, C-H) begrenzt. Daher sind neue, Ln-basierte Sonden mit verbesserten funktionellen Eigenschaften von großem Interesse.

In der vorliegenden Arbeit wurden verschiedene Strategien zur Verbesserung der Eigenschaften von Ln-Komplexen entwickelt. Die erste Strategie beinhaltet den neuartigen Ansatz der Ligandenfluorierung zur Verbesserung der Lumineszenz, Löslichkeit und Stabilität von Ln-Benzoaten am Beispiel von 85 Ln-Komplexen mit neun verschiedenen, fluorierten Benzoaten. Die Struktur-Eigenschafts-Beziehung wurde bestimmt, indem 24 Kristallstrukturen und 38 Pulvermuster analysiert wurden. Diese Studie wurde an Eu-ternären Komplexen erweitert, wobei deren günstige Geometrie, zusammen mit ihrer minimierten Schwingungslöschung zu einer außergewöhnlich hohen Lumineszenzeffizienz mit einer Quantenausbeute von bis zu 90% führte. Dies ist der bisher höchste berichtete Wert für Eu-Komplexe. Um die lumineszierenden Eigenschaften von im NIR-emittierenden Ln-Komplexen zu verbessern, wurde ein voluminöser Anthracen-Ligand mit höherer π-Konjugation verwendet, um Yb-basierte NIR-Emitter mit einer Quantenausbeute von 2.5% zu erhalten. Die hostfreie OLED, die unter Verwendung dieser Verbindung hergestellt wurde, besaß nur eine Yb^{3+} f-f-Emissionsbande bei 1000 nm mit einer Elektrolumineszenzquantenausbeute von 0.21%. Außerdem wurde in dieser Arbeit ein Tb(III)-Komplex mit einem oktadentaten, makrocyclischen Liganden (Lumi4Tb®) synthetisiert, der selbst im nanomolaren Konzentrationsbereich in wässrigen Puffern eine unglaublich hohe Lumineszenzintensität aufweist. Der Syntheseweg, um diese Verbindung zu erhalten, wurde optimiert, indem eine zuverlässigere Linkereinführung postuliert wurde.

Darüber hinaus wurden substituierte Lanthanoidbenzoate mit molekularen Transportern (Peptoiden) unter Verwendung verschiedener Konjugationsmethoden, einschließlich der Kupfer(I)-katalysierten Alkin-Azid-Cycloaddition (CuAAC) oder der Alkin-Azid-Cycloaddition mit gepannten Cycloalkinen (SPAAC), modifiziert und konjugiert. Die besten Ergebnisse wurden erhalten, wenn die Konjugation durch einen Hilfsliganden auf 1,10-Phenanthrolin- oder Terpyridin-Basis durchgeführt wurde, wobei die seltene Kombination aus intensiver, Eu-basierte Lumineszenz sowie Löslichkeit in wässrigem Medium erhalten wurde. Um die Koordinationsfähigkeit von Eu(III)-Ionen zu erhöhen, wurden cyclische Peptoide als chelatbildende Makrocyclen verwendet, die Heterocyclus-basierte Gerüste im Peptoid-Rückgrat und Carboxylfunktionen in den Seitenketten enthielten. Diese Komplexe, die sowohl Lumineszenz- als auch Transporteigenschaften kombinieren, können für verschiedene Anwendungen, derer es einer gezielten Lumineszenz bedarf, eingesetzt werden.

I. Introduction

I.1 Exciting compounds: the features of lanthanide-based luminescent probes

The fascinating features of lanthanide complexes make them very promising for various applications. At the very beginning, they were found to be of use as NMR shift agents, simplifying the spectral analysis of chiral organic compounds.[1,2] This was later expanded to structural biology, where lanthanide-based probes are proved to be very effective for the determination of protein structures by NMR spectroscopy and X-ray crystallography.[3,4] Subsequently, lanthanide complexes turned out to be almost irreplaceable as shift agents for magnetic resonance spectroscopy and as contrast agents for magnetic resonance imaging (MRI).[5–7] Meanwhile, low oxidation state complexes were found to be active as chemo- and stereoselective catalysts.[8–10] On top of that, the remarkable photophysical properties of trivalent lanthanide complexes make them appealing for various luminescent applications, being the focus of the current research work.

I.1.1 Lanthanide-based luminescence

The luminescence in Ln-based compounds occurs due to the parity-forbidden* 4f-4f transitions in Ln(III) ions, which makes the lifetimes of the excited state extremely long (*e.g.* the millisecond range for europium- and terbium-based compounds). Besides, lanthanide luminescence is easily recognizable and the emission bands are sharp with the full width at half maximum (FWHM) ~ 10 nm.[11] However, all lanthanide complexes are rather weak emitters, when the ions are excited directly through parity-forbidden f-f transitions which possess a very low molar extinction coefficient (typically $< 3\ \mathrm{M^{-1}s^{-1}}$) and hence, absorb energy very weakly. To overcome this problem, the energy is often transferred from the surroundings (antenna) to the metal ion (either an inorganic matrix or an organic ligand) in a multi-step mechanism (Figure I.1-1a). First, an antenna is excited from its ground state to its singlet excited state. This is followed by intersystem crossing (ISC) to the antenna's triplet state and intramolecular energy transfer (ET) from the antenna to the lanthanide excited state. The latter then emits light, for instance, with the maximum at 545 nm for Tb and 611 – 618 nm for Eu. Similarly, an indirect excitation of erbium, neodymium and ytterbium *via* the triplet or low-lying singlet state of a nearby antenna results in luminescence in the near infrared (NIR) region. The discovery of this effect opened up new horizons in lanthanide chemistry, allowing to obtain Ln-based compounds with a luminescence brightness way higher than their inorganic analogs.

* Also known as LAPORTE-forbidden

Though, most Ln(III) ions are luminescent, some are more emissive than others. The emissive properties of a lanthanide ion are governed by the population of the excited state(s) (maximizing η_{sens}) and by the minimization of non-radiative deactivation pathways (hence, maximizing Q_{Ln}^{Ln}), following the equation:

$$Q_{Ln}^{L} = \eta_{sens} \cdot Q_{Ln}^{Ln} \qquad \text{(eq. I.1)}$$

where Q_{Ln}^{L} is the overall photoluminescence quantum yield, η_{sens} is the energy transfer efficiency, which involves the efficiency of several processes: energy absorption, ISC and ET and Q_{Ln}^{Ln} is the f-f radiative transition efficiency. The intrinsic quantum yield Q_{Ln}^{Ln} essentially depends on the value of the energy gap ΔE between the lowest lying excited (emissive) state of the metal ion (Figure I.1-1b, drawn in red) and the highest sublevel of its ground state multiplet (Figure I.1-1b, drawn in blue). The smaller this gap, the easier its closing by non-radiative deactivation may occur, for instance, through vibrations of bound ligands, or, particularly, through high energy vibrations, such as O-H vibrations. With respect to the energy gap requirement, it is obvious that Eu(III), Gd(III) and Tb(III) are the most suitable ions, with ΔE = 12300 cm^{-1} ($^5D_0 \rightarrow {}^7F_6$), 32200 cm^{-1} ($^6P_{7/2} \rightarrow {}^8S_{1/2}$) and 14800 cm^{-1} ($^5D_4 \rightarrow {}^7F_0$), respectively.

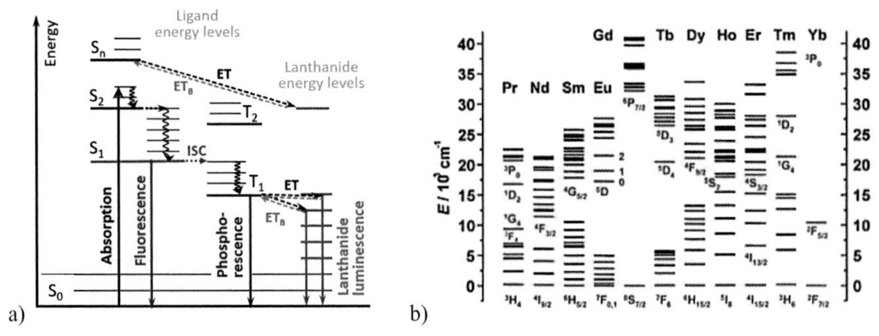

Figure I.1-1 a) Jablonski diagram, showing the main energy transfer pathways in a lanthanide complex. ET – energy transfer, ET_B – back energy transfer, ISC – intersystem crossing, S_0 – ground state, S_n – ligand singlet state, T_n – ligand triplet state; b) partial energy diagrams for the lanthanide aqua-ions. The main luminescent levels are drawn in red, the fundamental levels are indicated in blue. Adapted with permission from J.-C. BÜNZLI *et al.*[12] *Chem. Soc. Rev.* 2005, *34*, 1048–1077. Copyright© 2018 Royal Society of Chemistry.

In this respect, NIR-emitting lanthanides (*e.g.* Nd^{3+}, Er^{3+} and Yb^{3+}), possessing the smallest gap between the emissive state and the ground state multiplets, exhibit very low luminescence efficiency, practically two orders of magnitude lower than those for europium and terbium complexes. Nevertheless, low Q_{Ln}^{L} values of NIR-emitting lanthanide complexes have not reduced the interest in these compounds. Over the past two decades, a significant progress has been made in the design of the ligands, aiming at increasing the quantum yields and together with it a deeper understanding of the processes, affecting the luminescence quenching.

To maximize the energy transfer efficiency η_{sens}, the efficiency of each involved process should be considered, namely the ligand absorption, ISC and ET (Figure I.1-1a). The overall quantum-mechanical description of the energy transfer processes in lanthanide complexes was formulated by B.R. JUDD[13] and G.S. OFELT[14] in 1962. Their theory is based on the assumption that the f-f transitions are becoming less forbidden, when the wave functions of the opposite parity are mixing with the lanthanide ion derived wave functions. Thus, the wave functions, corresponding to the ground and excited states, are not of the same parity, which eliminates the symmetry prohibition on this transition.

Unfortunately, the JUDD-OFELT theory doesn't allow developing a set of rules on how to design Ln-based complexes with the optimal photophysical parameters (luminescence efficiency, lifetime of the excited state) because of the sophisticated mathematical apparatus and the difficulties in calculating the polarizability of atoms by reliable methods. For this reason, the most reliable data, allowing the prediction of the photophysical properties of the lanthanide complex, were obtained by accumulating and generalizing experimental data for a large number of compounds. M. LATVA *et al.*[15] was able to successfully generalize these data into the rule of thumb. This rule concerns the optimal energy gap between the ligand triplet level and the lanthanide resonance level (2500 – 3500 cm^{-1} for Eu^{3+} and 2500 – 4000 cm^{-1} for Tb^{3+}), which provides the maximum quantum yield of photoluminescence for lanthanide complexes (Figure I.1-1a).

The optimal energy gap between the ligand triplet level and the corresponding lanthanide resonance level (maximizing Q_{Ln}^{Ln}), together with the elimination of the most obvious ways of luminescence quenching (maximizing η_{sens}), however, still does not always guarantee a high photoluminescence quantum yield Q_{Ln}^{L}, according to equation I.1. For some lanthanides (Eu^{3+}, Sm^{3+} and Yb^{3+}) it is crucial to take into account the appearance of a band with charge transfer which can be located close to the ligand triplet level and, as a result, compete with the lanthanide 4f level to receive the energy.[16,17] However, the presence of such a state may be turned into an advantage. Indeed, if the luminescence of a Eu(III) complex is quenched by a ligand-to-metal charge transfer (LMCT) state, an Yb(III) complex with the same ligand may be highly luminescent.

I.1.2 Quenching mechanisms

In the preceding chapter, two main mechanisms for luminescence quenching in lanthanide coordination compounds are mentioned: (i) quenching through the bond vibrations and

(ii) quenching by means of charge-transfer state. However, while the latter mechanism is manifested only in certain lanthanide ions, the first mechanism is universal and should be taken into account, when choosing the ligand. The most effective "quenchers" of luminescence are O-H bonds due to their high vibrational energy (around 3000 – 3200 cm^{-1}) and as a consequence, high probability of non-radiative deactivation of the excited state of the lanthanide ion. To cancel the radiative transition with an energy *E*, it is necessary to have *n* phonons, where $n = E/E_v$ (E_v is the energy of the bond vibrations). The smaller the value of *n*, the higher the probability that some number of phonons ($\geq n$) collide with an excited atom and hence, the higher the binding energy will be. It is for this reason that in order to obtain compounds with high photoluminescence quantum yield, it is necessary to exclude the presence of water molecules, as the most common source of O-H oscillations.

The exclusion of the high-energy bond vibrations (such as O-H) is necessary for lanthanide ions, emitting in the near infrared range. For instance, for C-H bonds, the energy of the radiative transitions is too low for comparable values of E_v, decreasing the value of *n* and, therefore, the probability of luminescence quenching increases. One of the most common methods for solving this problem is to obtain fluorinated analogs of ligands because of the lower energy of C-F bond vibrations.[18] At the same time, the quenching efficiency is inversely proportional to the distance to the radiating center, so that we can confine ourselves to replace the ligand atoms, located in the immediate vicinity of the lanthanide ion.

Indeed, fluorination allows a considerable reduction of quenching *via* high-frequency C-H vibrations, resulting in an increase of the luminescence efficiency. For instance, ROH *et al.*[19] observed a 1.5-fold increase of NIR emission of erbium pentafluorobenzoate over erbium benzoate. YE *et al.*[20] proved that fluorination of the organic ligand leads to an increase of the luminescence efficiency.

However, additional studies conducted by M. SEITZ[21–23] demonstrated that substitution of hydrogen by heavier atoms, such as deuterium, may have the opposite effect, namely, the decrease in the luminescence quantum yield. It is shown that, in addition to the number of overtones, capable of non-radiatively deactivating the excited state, a very important role is played by the condition of observing the resonance proximity of the overtone energy to the transition energy. As an example, europium complexes with deuterated ligands are given. A sharp drop in the luminescence intensity in comparison to the non-deuterated analogs is explained by the closeness of the energy of the $^5D_0 \rightarrow {}^7F_5$ (13300 cm^{-1}) transition and the sixth overtone of C-D bond vibrations in the ligand (Figure I.1-2b).

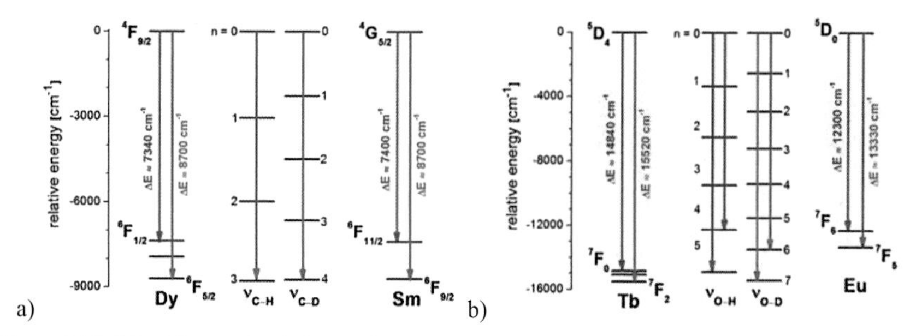

Figure I.1-2 a) Partial energy level diagram for dysprosium(III) and samarium(III), showing the energy gaps *ΔE*, according to the energy gap law (red) and the ones proposed here to be relevant for aromatic C-(H/D) oscillators (green); b) partial energy level diagram for terbium(III) and europium(III), showing the energy gaps *ΔE* relevant for O-H (red) and for O-D (green) oscillators in methanol solutions. Reprinted with permission from C. DOFFEK *et al.*[24] *Inorg. Chem.* 2014, *53* (7), 3263–3265. Copyright© 2018 American Chemical Society.

Quenching the luminescence does not always have a negative effect which must be eliminated. It can be useful, when the detection of water molecules in the lanthanide coordination environment is necessary. For instance, if one of the coordinated ligands is protonated by lowering the pH of the analyzed water system, the lanthanide ion coordinates the water molecule to preserve the saturation of the coordination sphere. This, consequently, leads to a decrease in the luminescence intensity of the complex in solution, which is a basis of pH sensors.[25]

An effective "quencher" can be any energy state close to either the ligand triplet level or lanthanide resonance level. For example, such a state can be the aforementioned one with charge transfer. MALTA *et al.*[26] constructed a curve, demonstrating the dependence of the luminescence quantum yield on the position of the state with charge transfer, using quantum mechanical modeling. As a result, he came to the conclusion that the quantum yield decreases sharply as the LMCT energy approaches the triplet level of the ligand. Additionally, effective ions of non-radiative deactivation can be transition metal ions and other lanthanide ions, especially in heterometallic complexes.

Although concentration damping of the lanthanide luminescence in the complexes is not usually observed (due to the typically large Ln…Ln distance), a concentration quenching of the luminescence of the ligands themselves in the crystalline form may well occur. Providing this fact conductive polymer matrices are used to "dilute" luminescent coatings based on lanthanide complexes (for example, for active layers of organic light-emitting diodes).[27]

I.1.3 Energy transfer mechanisms

As of now, two basic mechanisms of energy transfer have been proposed: FÖRSTER[28] (Figure I.1-3a) and DEXTER[29] (Figure I.1-3b), whose preferred implementation in each particular case is determined by the distance between the energy donor and the energy acceptor. If in DEXTER mechanism, a direct electron transfer is envisaged due to the close arrangement between the donor and the acceptor, the FÖRSTER mechanism implies a noncontact transfer of the energy due to the dipole-dipole interaction. In the case of lanthanide complexes, it is assumed that the energy transfer within the Ln-based complex is carried out by the DEXTER mechanism due to (i) the short distance between the Ln ion and the donor atoms, participating in the complexation and (ii) the absence of overlapping of f-orbitals.[30,31]

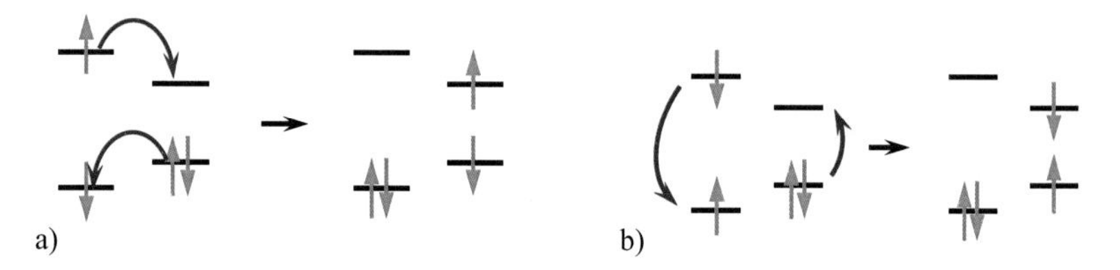

Figure I.1-3 Schematic representation of energy transfer mechanisms proposed by a) FÖRSTER and b) DEXTER.

A successful energy transfer between the donor and the acceptor requires a nonzero overlap of the emission spectrum of the donor (the ligand) and the absorption spectrum of the acceptor (lanthanide ion). Since there are no STOKES' shifts in lanthanide ions, this rule is transformed into a rule for overlapping the emission spectra of the ligand and the lanthanide.

Determining the exact path of energy migration is a non-trivial task. For example, in the case of europium complexes, three states $^5D_0$, $^5D_1$ and $^5D_2$ can act as resonance levels of the lanthanide, making only time-resolved luminescence spectroscopy capable of giving the final answer. On top of that, it was suggested that the specific mode of energy transfer is affected by the class of the ligand used.[32]

Sometimes when low luminescence efficiency of the lanthanide complex is caused by an inefficient energy transfer because of too large energy gap between ligand and metal energy levels, an additional energy transfer step (or energy state) might be introduced and its energy must be positioned in-between the ligand triplet level and the lanthanide emissive level. For this purpose either an additional metal[33] or an additional ligand can be used. The second approach is way more common in literature, because the additional ligand can protect Eu(III) ion from non-radiative deactivation,[34,35] increase the light absorption and decrease the symmetry of Ln ion environment.[34,36–39]

Thus, ligand-sensitized luminescent lanthanide(III) complexes are of considerable importance because of their unique photophysical properties (microsecond to millisecond lifetimes, characteristic narrow emission bands and large STOKES' shifts). This chapter scratches the surface of the potential ways to improve the properties of Ln-based luminescent probes by minimizing all the parameters which may lead to luminescence quenching. The best way to minimize vibration-induced deactivation processes is to design a rigid metal-ion environment, free of high-energy vibrations and protecting the Ln(III) ion from a solvent interaction. If ligand-to-metal energy transfer efficiency should be increased, an additional sensitization step may be introduced.

I.2 Methods for elucidating the structures of lanthanide complexes in powders, crystals and solutions

I.2.1 X-ray diffraction methods and the analysis of structural databases

Single-crystal X-ray diffraction and the analysis of structural data

The development of instrumental capabilities and mathematical apparatus made X-ray diffraction methods widely used in chemistry and biology. Besides, the analysis of the structural data, accumulated in the databases over the decades, is becoming more powerful, when the prediction of structure-activity relationship (SAR) or structure-property relationship (SPR) is needed.[40–43]

Analyzing Ln-based compounds, the particular interest is often drawn towards the determination of the preferential ligand-to-metal coordination mode, which is usually related to the metal complex stability and particularly crucial for Ln-based luminescent probes. According to the Cambridge Structural Database (CSD) analysis of lanthanide carboxylates, there are four common coordination modes which are shown in Figure I.2-1. The ligands may possess

(i) a monodentate chelating function κ^1 (Figure I.2-1a), if one oxygen atom coordinates Ln(III) ion;
(ii) a bidentate chelating function κ^2 (Figure I.2-1b), if one Ln(III) coordinates both oxygen atoms from one carboxylic group;
(iii) a bridging function μ_2-κ^1 (Figure I.2-1c), if the carboxylic group bridges two Ln(III) ions, each of them coordinating one oxygen atom;
(iv) a semi-bridging function μ_2-κ^1:κ^2 (Figure I.2-1d), if a κ^2-coordinated ligand possesses also a bridging function relatively to another Ln(III) ion.

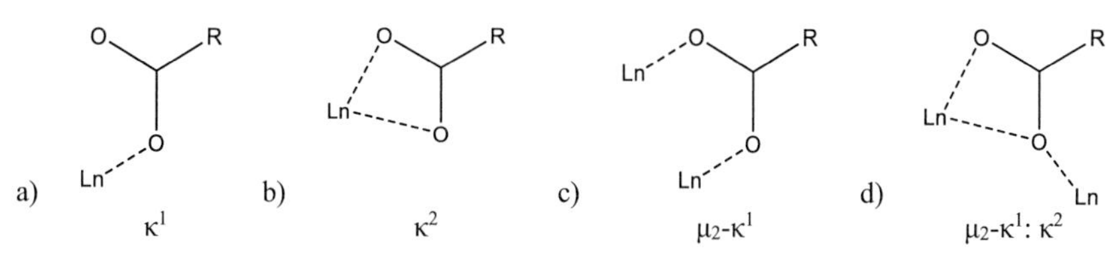

Figure I.2-1 The most common ligand coordination modes for lanthanide carboxylates.

The preference for a particular ligand coordination is often affected by the type of the metal. For instance, 3d-metals often prefer κ^1-coordinated ligands, except copper(II) complexes, where the bridging μ_2-κ^1 function is usually observed.[44] In the case of lanthanide complexes, the ligands most often possess κ^2, μ_2-κ^1 or μ_2-κ^1:κ^2 functions.[45] Because of large coordination numbers, specific for lanthanides, and the low denticity of the carboxylic groups, these complexes most often exhibit a polymeric structure and the same metal can form different polymorphs with different number of coordinating solvent molecules (solvatomorphs). Statistical CSD analysis showed that less than 9% of lanthanide carboxylates are monomers (for example[46,47]), while for others polymerization tends to occur. Most of the monomeric carboxylates are ternary complexes or homogeneous ligand complexes with very bulky ligands, where polymerization is unlikely because of steric reasons.

Thus, lanthanide carboxylates tend to form polymer chains, which is mainly due to the bridging functions of the ligands. The length of the Ln–O bond strongly depends on the type of the ligand coordination mode: the average Ln–O distance for κ^1-coordinated ligands is 2.40 Å, for κ^2-coordinated ligands 2.47 Å, for μ_2-κ^1-coordinated ligands 2.38 Å and for μ_2-κ^1:κ^2-coordinated ligands (Figure I.2-1d, from left to right) it is 2.47 Å, 2.58 Å and 2.40 Å, correspondingly.

The preference for the particular ligand coordination should depend on the lanthanide contraction, since a decrease in the Ln–O bond length increases the repulsion between the ligands, so that the amount of κ^2-coordinated ligands should decrease. This phenomenon is also discussed in Chapter III.1.1. This statistical CSD analysis of the structures might also be useful for interpreting the data for the research methods that indirectly provide structural information (such as EXAFS, IR, etc.).

Powder X-ray diffraction

The main difference between powder X-ray diffraction (PXRD) single-crystal X-ray diffraction (SC-XRD) is that each powder pattern (diffractogram) is a one-dimensional snapshot of a three-dimensional reciprocal lattice of the crystal, while the single-crystal data analysis gives the 3D structure. Since many compounds can be obtained only in the form of polycrystals, PXRD becomes the only reliable method for extracting the information regarding their crystal structures. In this case, the intensity of the scattered radiation is measured as a function of one independent variable, namely the BRAGG angle. Different structural features of the material can be revealed as additional effects on the powder diffraction pattern.

Although quite a bit of information can be obtained from PXRD data, in some cases the presence of a qualitative powder pattern and the correct structural model allows getting important structural parameters or even to refine the crystal structure (Table I.2-1). The most important factor complicating this task is the overlapping of several BRAGG peaks. Besides, with an increase of the unit cell volume, the difficulty in interpreting powder data also increases.

Table I.2-1 Information, withdrawn from the powder pattern components.

powder pattern component	instrumental parameter that influences the powder pattern component	information, withdrawn from the powder pattern components	
		crystal structure	sample properties
peak position	radiation (wavelength), instrument / sample calibration, axial beam divergence	the parameters of the crystal lattice (a,b,c,α,β,γ)	absorption, texture
peak intensity	geometry and configuration of the measurement, radiation (LORENTZ, polarization)	atomic parameters (x, y, z ...)	optical axis, absorption, texture
peak from	radiation (spectral purity), geometry, processing of the primary beam	crystallinity, distortion, defects	the size of crystallites, stress, deformation

Despite the rapid development of the PXRD method, refining the structures by powder data remains a very difficult task and usually requires the preliminary built model as a starting point[48] that will approach the actual structure during the refinement. For this purpose, the following model can be used: (i) the data of a real single-crystal experiment; (ii) the structure of isostructural compounds or (iii) the structural model from quantum chemical calculations. However, each of these models has its drawbacks. The single-crystal structure of a compound often does not coincide with its powder structure because of the different kinetics of the product formation. For instance, in the work of S. KWON *et al.*,[49] the unit cell volume, obtained from PXRD data for *trans-(R,R)*-2-aminocyclopentane carboxylic acid turned out to be more than four folds larger than the cell volume of a single crystal. An assumption regarding the isostructural nature of two compounds may be incorrect, though sometimes works very well. For instance, in the work of S. M. F. VILELA *et al.*,[50] where the single-crystal structure of a praseodymium complex was used as a model for refining the powder structure of a europium complex with the same ligand. Quantum chemical calculations may give quite accurate structures of relatively small organic molecules, however, are often not applicable for proteins or heavy metal complexes. Generally, a set of methods (PXRD in combination with DFT calculations) is used to obtain a powder structure with the maximum degree of reliability.[51]

Despite the tremendous progress already achieved in the field of PXRD structure refinement, the CDS database contains only a few dozens of PXRD structures of lanthanide complexes, whilst the atomic coordinates are mainly determined with a significant error (up to 0.5 Å for Ln–O distances).

XAFS spectroscopy

The X-ray absorption fine structure (XAFS) has been used since the mid-1970s to solve structural problems that are beyond the capacity of traditional X-ray analysis. It is applied primarily to the structures with defects or partially ordered systems (for example, intercalated graphite, in which foreign atoms are arranged disorderly between the layers of carbon atoms[52,53]). When a photoelectron is removed from an absorbing ion (*i.e.* Ln^{3+}), it can interact with its surroundings in two ways. First, it can experience strong multiple scattering from surrounding ligands, which will lead to a change in the frequencies of the absorbed waves located in the spectrum near the absorption edge. These modulations are due to the interference of the primary scattered wave and the waves arising during the multiple reflection of the photoelectron. The part of the absorption spectrum, where these modulations appear, is called XANES (X-ray Absorption Near-Edge Structure). Secondly, the photoelectron can undergo a single reflection from the ligand atom, after which it leaves the atomic system, causing a slight perturbation of the shorter wavelength part of the spectrum (EXAFS – Extended X-ray Absorption Fine Structure).

Based on the X-ray absorption spectra, it is possible to determine the composition and parameters of the first coordination sphere of the absorbing atom (atom types and their distances to an absorbing Ln^{3+} ion), which is very important for objects with no long-range order such as solutions or amorphous substances.

XANES spectroscopy is very sensitive towards the oxidation state and coordination number changes. For instance, in the work of I.J. PICKERING *et al.*,[54] the difference between K-edge energies for sulfur in different oxidation states is demonstrated (Figure I.2-2a). Unlike XANES, EXAFS allows determining the nearest environment of the absorbing atom (Figure I.2-2b). Generally the heavier and closer the atoms in this environment, the more accurate the simulation *via* this method may be performed[55]. This is why EXAFS is quite often used to determine metal-metal distances but sometimes does not allow to distinguish between light elements (C, N, O).

In the case of lanthanide carboxylates, the formation of single crystals is often difficult due to their tendency to polymerize and hence, low solubility. Besides, the elucidation of the structures of lanthanide complexes using PXRD is extremely difficult due to the presence of the heavy atom. Therefore, EXAFS spectroscopy is often used to determine the local structures of lanthanide-based compounds. Since usually in complexes lanthanides have a constant oxidation state of +3, XANES is not considered in the structural study, because the difference in oxidation states is the main focus of this method. Interestingly, EXAFS spectroscopy is often related to the structural study of solutions.[56] This interest is related to the prospects for practical applications (including biological ones) of the solutions of lanthanide-based compounds. Since solutions do not have a long-range order, the geometry of the compound in solution is called "local structure". Though EXAFS spectroscopy is unable to directly distinguish between light atoms, they can still be divided by the characteristic distances from the metal ion. For example, C. TERRIER *et al.*[57] showed that for the aqueous environment of the lanthanum ion, the distance Ln–O is approximately 2.5 Å in the first coordination sphere and more than 4 Å in the second one (Figure I.2-3a). The modeling of such a structure is shown in Figure I.2-3b.

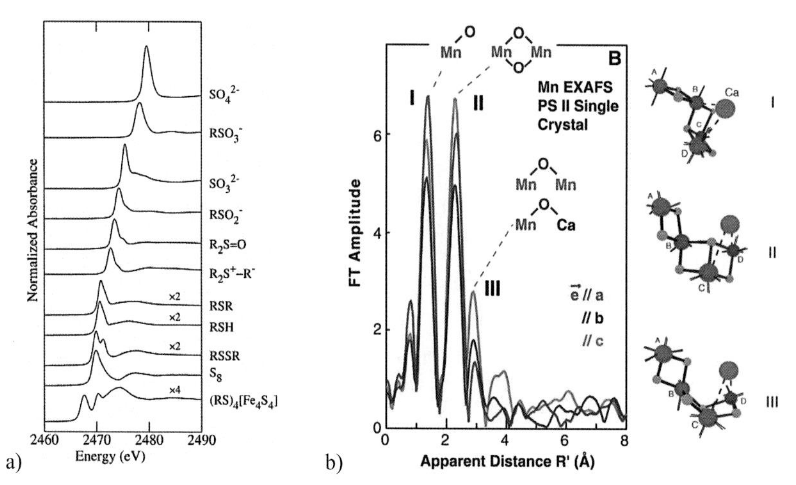

Figure I.2-2 (a) Sulfur K-edge XANES: changes in the absorption spectra with a change in the oxidation state. A series of biologically relevant compounds was discovered: (from top to bottom): sulfate pH 8.2, cysteic acid, sulfite, cycteine sulfinic acid, methionine sulfoxide, dimethylsulfoniopropionate pH 4.0, methionine, cysteine, oxidized glutathione, solid rhombic sulfur and *Clostridium pasteuranium* ferredoxin. Adapted with permission from I.J. PICKERING *et al.*,[54] *FEBS Letters* 1998, *441*, 11-14, Copyright© 2018, JOHN WILEY and Sons; b) elucidation of the Mn coordination environment in the photosystem II which consists of heteronuclear Mn-Ca clusters *via* EXAFS spectroscopy. Adapted with permission from J. YANO *et al.*,[55] *Science* 2006, *314*, 821–825, Copyright© 2018 American Association for the Advancement of Science.

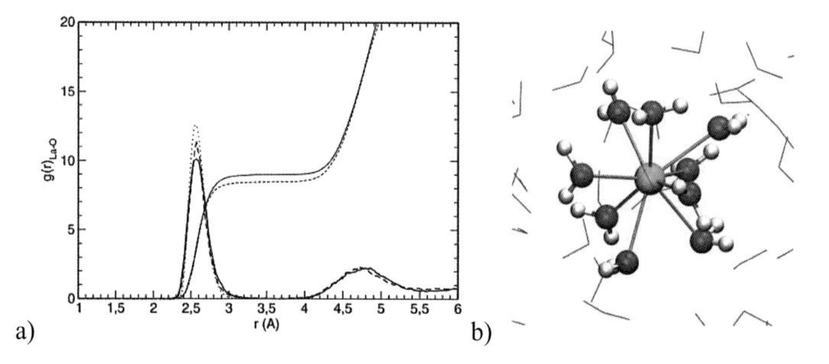

Figure I.2-3 a) Ln–O distance distribution functions obtained using different simulation methods; b) snapshot from an MD simulation of La^{3+} in bulk water. Water molecules of the first hydration shell are shown in ball-and-stick and outer-sphere molecules in lines. Adapted with permission from C. TERRIER *et al.*,[57] *J. Chem. Phys.* 2010, *133*, 044509, Copyright© 2018, AIP Publishing.

Based on these data, the scattering pattern in the EXAFS spectrum for Ln(III) ion with solely aqueous coordination environment has the form of two peaks, as shown in Figure I.2-3a. Therefore, any signals in the distance range 2.8 – 4.2 Å in the EXAFS spectra of lanthanide complexes in aqueous solutions are not associated with the water oxygen coordination. Such distances are characteristic for nitrogen or carbon atoms, bound to the lanthanide through the oxygen atom, namely NO_3^-, CO^-, COO^- groups.[58]

I.2.2 NMR spectroscopy in the presence of paramagnetic nuclei and BLEANEY's theory

The presence of paramagnetic nuclei in a molecule often significantly complicates the interpretation of NMR spectra, since this leads to broadening and shift of signals. However, in some cases it is exactly the presence of paramagnetic nuclei, which provides an additional information about the structure of the compounds in solution (Figure I.2-4). The use of so-called shifting agents, which were discovered by HINCKLEY *et al.*,[59] makes it possible to simplify the analysis of NMR spectra of macromolecules by dividing overlapping signals and also, for example, to determine the optical purity of the compounds.[60] Moreover, the paramagnetic effects manifested in NMR spectra provide powerful restraints for the determination of the three-dimensional structure of proteins,[61] open new possibilities for the analysis of protein-protein and protein-ligand interactions and offer widely applicable tools for sensitivity enhancement of NMR experiments, resonance assignments and studies of conformational heterogeneity and exchange.

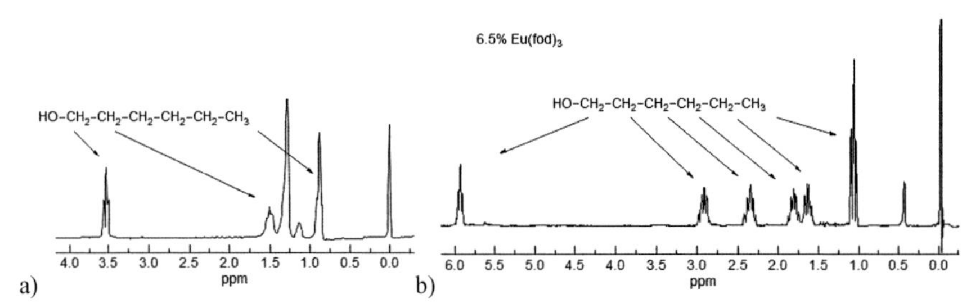

Figure I.2-4 a) ^1H NMR spectrum of 1-hexanol; b) ^1H NMR spectrum of 1-hexanol in the presence of a shifting agent. Adapted from https://www.chem.wisc.edu/areas/reich/nmr/08-tech-07-lis.htm.

The net lanthanide-induced shift for a nucleus of a ligandб coordinate, to a Ln^{3+} ion, may be represented as the sum of three contributions, namely the diamagnetic shift (δ_{dia}), the contact shift (δ_{con}) and the pseudocontact shift (δ_{PCS}).[62]

$$\delta = \delta_{dia} + \delta_{con} + \delta_{PCS} \qquad \text{(eq. I.2)}$$

The diamagnetic shifts are usually small and often neglected, except for atoms, directly coordinated to the Ln^{3+} ion. They originate from conformational changes of inductive effects and direct field effects and can be determined directly or by interpolation from shifts, induced by diamagnetic La^{3+} and Lu^{3+} ions. Contact (or FERMI) shifts involve a through-bond transmission of the unpaired electron density at the Ln^{3+} ion to the nucleus of interest. The contact shift δ_{con} (ppm) is given by the equation:[63]

$$\delta_{con} = A\langle S_z\rangle = Ag_L(g_L - 1)J(J+1)\left(\frac{\beta H_0}{3kT}\right) \qquad \text{(eq. I.3)}$$

where $\langle S_z\rangle$ is the projection of the total electron spin magnetization of the lanthanide on the direction of the external magnetic field, g_L is the LANDÉ g-factor, J is the resultant electron spin angular momentum in ħ units of the ground level, β is the BOHR magneton, H_0 is the intensity of the external magnetic field, k – BOLTZMANN constant, T – temperature.

The pseudocontact contribution (or dipolar shift) is a result of a through-space interaction between the magnetic moments of the unpaired electrons of Ln^{3+} and the nucleus under study. It is expressed by the equation:[63]

$$\delta_{PCS} = \frac{\vartheta_0\beta^2}{60kT^2}\frac{3cos^2\theta - 1}{r^3}2A_2^0\langle r^2\rangle g_L^2J(J+1)(2J-1)(2J+3)\langle J\|\alpha\|J\rangle \qquad \text{(eq. I.4)}$$

where ϑ_0 is the resonance frequency, r is the distance between the central metal and the ligand, θ is the angle between the vector r and the principal axis of symmetry of the complex, A_2^0 are ligand field coefficients which are constant for a given complex geometry and independent of the lanthanide. The last term $g_L^2J(J+1)(2J-1)(2J+3)\langle J\|\alpha\|J\rangle$ was calculated by BLEANEY.[64]

However, in the present case the contact shift has a very low influence on the overall paramagnetic NMR shift because of the large distance between the detected nuclei and the lanthanide. At such distances only the pseudocontact shift significantly influences the overall NMR shift value.

Given that the molecular geometry is independent of the lanthanide ion, in 1972 BLEANEY[64] deduced the equation which allows to check structural similarity of lanthanide complexes in solution by the following NMR paramagnetic shift dependencies:

$$\delta = \delta_c + \delta_d \qquad \text{(eq. I.5)}$$

where δ_c is a contact chemical shift, being a constant and δ_d is a dipolar chemical shift, defined as

$$\delta_d = D \cdot G_i \qquad \text{(eq. I.6)}$$

A simple substitution of equation I.6 into equation I.5 leads to

$$\delta = D \cdot G_i + C \qquad \text{(eq. I.7)}$$

where δ is the overall chemical shift, D is the characteristic of the lanthanide, listed in[63,65] (Figure I.2-5a), G_i is the geometrical function of the ith complex which is a constant for isomorphous compounds, C is a constant. Therefore, linear dependence of δ on D-factor (Figure I.2-5a) for lanthanide complexes within the lanthanide row would mean G_i, being a constant, indicating the same geometry of given lanthanide complexes in solution. Thus, BLEANEY's theory may be used for identification of the local structure similarities for the complexes with lanthanides within the lanthanide row.

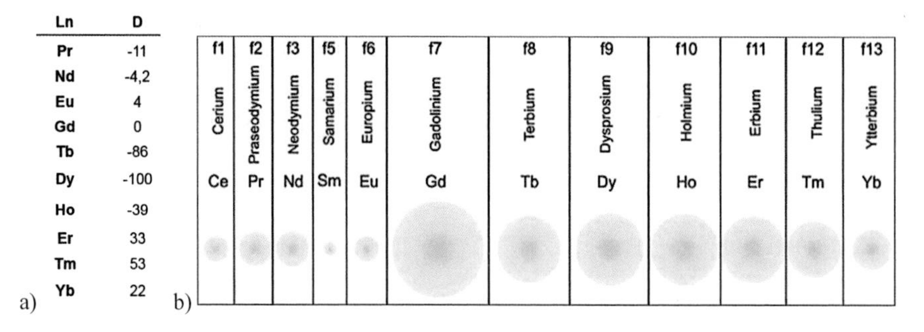

Figure I.2-5 a) D-value of each Ln^{3+} ion; b) values of ^1H NMR signal broadening depending on the Ln ion. The radii of the yellow spheres indicate the distances from the metal ion at which ^1H NMR signals of macromolecules with a rotational correlation time of 15 ns broadened by a frequency of 80 Hz at a magnetic field strength of 18.8 T. Reprinted with permission from G. OTTING *et al.*,[65] *J. Biomol. NMR* 2008, *42*, 1–9. Copyright© 2018, Springer Nature.

Despite the fact that this theory was introduced by BLEANEY back in 1972,[64] it has been very rarely used so far. This seems to be related to the following difficulties: (i) the limitations of the applicability of this method are not established (the range of allowable concentrations; the range of distances from the paramagnetic center to the detectable nucleus, on which this effect acts; the sensitivity to the influence of paramagnetic ions of heavier nuclei, such as carbon and fluorine) and (ii) strong broadening of the NMR signals which arises from the interaction with the paramagnetic nucleus (Figure I.2-5b).

I.2.3 Luminescence spectroscopy

Luminescence spectroscopy provides valuable data on the structure and the properties of lanthanide complexes, namely: the data on the stoichiometry of chemical compounds,[66–70] the stability constants of the complexes,[71,72] the distance between the metal ion and the donor ligand atoms[73,74] and the ligand exchange rate.[75] For europium compounds all these data can be obtained from the luminescence spectra, normalizing the integral intensities of all transitions to the intensity of the pure magnetic dipole (MD) transition $^5D_0 \rightarrow {}^7F_1$ which is independent on the Eu^{3+} ion environment (Figure I.2-6).

Evaluation of the coordination environment symmetry by calculation of the radiative lifetime of the excited state from the luminescence spectrum

Figure I.2-6a demonstrates the luminescence spectra of solutions, containing Eu^{3+} ions and dipicolinate anions.[76,77] The magnetic dipole (MD) transition ${}^5D_0 \rightarrow {}^7F_1$ and the electric dipole (ED) transition ${}^5D_0 \rightarrow {}^7F_4$ are usually the most intense bands in the emission spectrum in the absence of ligand luminescence (Figure I.2-6a, Figure I.2-6b). An increase of the amount of ligands substantially raises the relative intensity of the "supersensitive" transition ${}^5D_0 \rightarrow {}^7F_2$, leaving the relative intensity of the transitions ${}^5D_0 \rightarrow {}^7F_1$ and ${}^5D_0 \rightarrow {}^7F_4$ almost unchanged. Since for Eu(III) compounds the transitions with a pure magnetic and a pure electric dipole character can be distinguished, the radiative lifetime τ_R can be calculated directly from the emission spectrum, using the integral intensity ratio between the MD transition peak and the overall integral intensity of all transitions:

$$\tau_R = \frac{1}{A_{MD,0}n^3\left(\frac{I_{tot}}{I_{MD}}\right)} \quad \text{(eq. I.8)}$$

where n is the refractive index of a medium, $A_{MD,0}$ is the probability of a spontaneous emission for the transition ${}^5D_0 \rightarrow {}^7F_1$ in vacuum (calculated value 14.65 sec^{-1}), $\frac{I_{tot}}{I_{MD}}$ is the integral intensity ratio between the overall integral intensity of all transitions and the integral intensity of the MD transition peak (Figure I.2-6c).

In the general case for all other lanthanides, these calculations may be done applying the JUDD-OFELT theory, however, quite comprehensive quantum chemistry calculations are required:

$$\tau_R = \frac{A(\psi J, \psi' J')}{B(\psi J, \psi' J')} \quad \text{(eq. I.9)}$$

Thus, the higher the relative intensity of the MD transition ${}^5D_0 \rightarrow {}^7F_1$ (the higher the symmetry of the Ln ion environment), the higher the radiative lifetime τ_R will be. In the given example the radiative lifetime of the excited state decreased monotonically with an increasing amount of the ligands, together with an increase in the transition intensity ${}^5D_0 \rightarrow {}^7F_2$.

Equation I.8 is derived exclusively from theoretical considerations. Experimentally, τ_R is determined by measuring the lifetime of the excited state of the solution of the complex in deuterated solvent at low temperature (most non-radiative processes are "frozen out" under such conditions) (Figure I.2-6b). In the case of other lanthanide ions, all transitions contain both, ED and MD components, so there is no such "internal standard" for them.

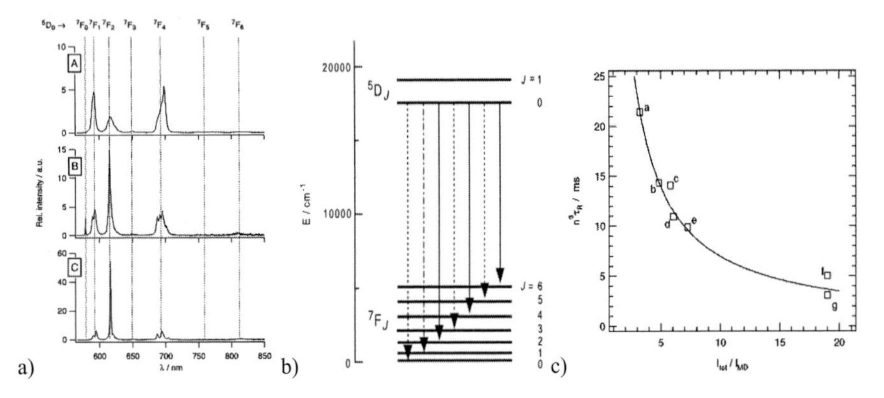

Figure I.2-6 a) Luminescence spectra of aqueous Eu^{3+} in different ligand environments (A) Eu^{3+}; (B) $Eu(DPA)^{+}$ (HDPA); (C) $Eu(DPA)_3$. The spectra are normalized to the intensity of the $^5D_0 \rightarrow {}^7F_1$ band (vertical axes have different scales); b) scheme of the levels of Eu^{3+} and the corresponding transitions between the excited state $^5D_0$ and the ground state ^7F_J. The straight lines correspond to electric dipole transitions (ED), the dash-dot line is the magnetic dipole transition (MD), the dashed line is the transitions where ED and MD components are present; c) calculated dependence $n^3\tau_R\left(\frac{I_{tot}}{I_{MD}}\right)$ (solid line) and experimental values of the radiative lifetime of the excited state for some Eu-containing compounds obtained in[78]. Figures are reprinted with permission from M.H.V WERTS *et al.*,[78] *Phys. Chem. Chem. Phys.* 2002, *4*, 1542–1548. Copyright© 2018, Royal Society of Chemistry.

However, if the absorption spectrum, corresponding to the radiating transition, is known, it becomes possible to determine the strength of the oscillator and consequently, the radiative lifetime of the excited state from the spectrum, using the formula:

$$\frac{1}{\tau_R} = 2303 \times \frac{8\pi c n^2 \acute{v}_{ul}^2}{N_A} \frac{g_l}{g_u} \int \varepsilon(\nu) d\nu \qquad \text{(eq. I.10)}$$

where c is the speed of light in vacuum, ν is the transition frequency, n is the refractive index of a medium, N_A is AVOGADRO's number, $\varepsilon(\nu)$ is the absorption spectrum of the transition, g_l and g_u are the degeneracy of the ground and excited states, respectively, which are equal to *2J+1* for lanthanide ions. It is worth noting that in equation I.10 it is assumed that the *(2J+1)* levels are degenerated or at least have the same population, which in turn can be violated due to the STARK splitting. The STARK sublevels are differently populated, leading to deviations up to 20%.[79]

Estimation of the number of inner-sphere water molecules

The luminescence of the lanthanide ion is very effectively quenched by high-frequency oscillations, for example, O-H vibrations of water molecules.[80–84] It was shown by Y. HAAS *et al.*[85] and J.L. KROPP *et al.*[86] that water molecules act independently of each other, making the degree of quenching directly proportional to the number of O-H oscillations. At the same time, O-D oscillations in fact almost do not participate in the quenching processes. Therefore, by comparing the luminescent properties of a lanthanide complex in H_2O with those of the similar complex in D_2O one can estimate the contribution of water to the quenching processes:[87]

$$n_{H_2O} = 1.05\left(\frac{1}{\tau_{H_2O}} - \frac{1}{\tau_{D_2O}}\right) \qquad \text{(eq. I.11)}$$

Estimation of lanthanide ion coordination number in complexes with organic ligands

The transition $^5D_0 \rightarrow {}^7F_0$ (0-0) is very sensitive to the coordination environment of the europium ion. The nefloauxetic effect, manifested in the luminescence spectra – the change in the position of the 0-0 transition with the change in the environment of the central europium ion – was the basis of the work performed by CHOPPIN and WANG,[88] where on the basis of a large array of experimental data a relationship between the magnitude of this effect and the europium(III) ion was found:

$$CN = 0.237\Delta\nu + 0.628 \qquad \text{(eq. I.12)}$$

where $\Delta\nu = \nu_{0-0} - 17276$ cm^{-1}. This linear dependence very well describes the experimentally observed values (the error in CN determination is ±1.5 was revealed for 98% of the complexes analyzed, ± 1.0 for 88% of the complexes and ± 0.5 for 58% of the complexes).

I.3 Applications of lanthanide complexes

Lanthanide ions exhibit fascinating properties and are therefore, widely used for catalysis and high-technology applications, such as energy-saving lighting devices, displays, optical fibers, amplifiers, lasers, responsive luminescent stains for biomedical analyses or *in cellulo* sensing and imaging, MRI contrast agents and others.[89] Herein only some of them are discussed, paying a particular attention on lighting (OLED) and biological (cellular imaging) applications.

I.3.1 Lanthanide-based emitting layers for OLEDs

The remarkable photophysical properties of Ln(III)-based compounds make them very attractive for being used as OLED emitting layers. A typical OLED consists of a number of layers deposited on either solid (glass) or flexible (polymer) substrates (Figure I.3-1a). Thin films can be obtained by different techniques, however, the most commonly used are vacuum deposition and solution-processable methods such as spin-/deep-coating and ink-jet printing. When a voltage bias is applied to the diode, holes are injected from an anode and electrons from a cathode. After a series of transport steps (Figure I.3-1b) they recombine in the emission layer with the formation of excitons which are then deactivated with a release of light through the transmissive anode and the substrate.[18]

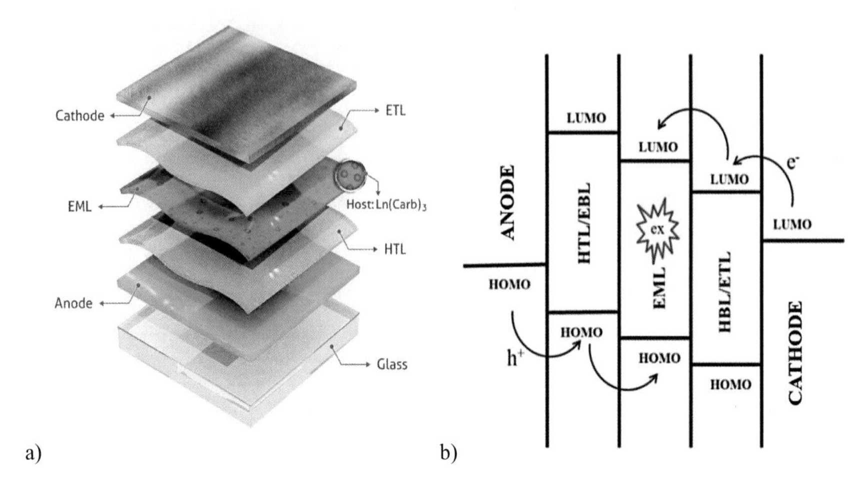

Figure I.3-1 a) Typical structure of an OLED device with lanthanide carboxylate as an emitting layer. Adapted with permission from A. Kalyakina *et al.*,[90] *Organic Electronics* 2016, *28*, 319-329 Copyright© 2018. American Chemical Society; b) scheme of injection, transport and recombination of charge carriers in OLEDs.

In the OLED structure, layers with electron and hole transporting/blocking properties are additionally introduced (HTL – hole transport layer, ETL – electron transport layer; HBL – hole blocking layer). Their purpose is, firstly, to create an intermediate stage in the OLED power circuit between the electrode and the active layer, which facilitates the injection of charge carriers in the case of a large barrier and secondly, to prevent the flow of carriers through the structure, which improves the balance between electron and hole currents (Figure I.3-1b). As these layers organic compounds, containing conjugated aromatic and heterocyclic groups, are used.[91]

Unlike photoexcitation in the case of electroexcitation both, singlet and triplet states, are excited, wherein singlet-induced fluorescence represents only a small fraction (about 25%) of the total excited-state population (the remainder are triplet states).[92] Further OLED efficiency improvements require that both, singlet and triplet excited states, contribute to luminescence. It has been proposed that ISC in lanthanide complexes may achieve this with an intramolecular energy transfer from the triplet state of the organic ligand to the 4f energy state of the Ln(III) ion.[93]

Three main classes of anionic organic ligands are common for OLED applications to visible-emitting OLEDs: β-diketonates and their derivatives, pyrazolates and carboxylates.[18] As a result, OLEDs with a very high brightness can be fabricated (up to 3000 cd/m^2 for Eu-based OLEDs and 12000 cd/m^2 for Tb-based OLEDs).

Near infrared luminescence has attracted particular interest in view of exciting applications in telecommunications, NIR-emitting LED/OLED devices, as well as in biosciences. For instance, a good portion (ca. 650 – 1700 nm) of the electromagnetic spectrum, mainly belonging to the NIR range, has only few interferences with biomaterials, so that photons can penetrate deeply into biological samples, cells and tissues, which makes NIR-light an interesting asset for both, bioanalyses and bioimaging.[94] Recently, there have been attempts to use NIR tomography to examine deep tissues, with the idea of developing highly sensitive methods for an early detection of cancer.[95] Another possible application is the use in telecommunications,[96] where the "transparency window" of silica optical fibers lies in the range of 1000 – 1500 nm. Therefore, a particular interest is drawn towards lanthanide ions with NIR-emission in this range (Nd^{3+}, Yb^{3+}, Er^{3+}). A wide variety of organic ligands have been used to sensitize NIR-emitting lanthanide ions and the highest quantum yields have been reached for ytterbium compounds.[96–99] However, there still have been very few NIR-emitting lanthanide-based OLEDs, reported up to date,[97,100,101] because of the difficulties to protect NIR-emission from quenching. Therefore, the researhcers are working on the design of new lanthanide-based NIR-emitting complexes with higher electroluminescence performance as well as on new strategies to fabricate OLED devices. Summarizing these attempts, one can identify the main strategies, being explored to sensitize the luminescence of NIR-emitting Ln(III) ions:[94]

(i) transfer from an organic ligand with singlet, triplet or intra-ligand charge-transfer states (ILCT), being the main donors;
(ii) transfer from a d-state of a transition metal cation such as Cr(III), Ru(II), Ir(III);
(iii) sensitization by another Ln(III) ion, for instance Yb(III)-to-Er(III) transfer;
(iv) for Yb(III) only: excitation of the $^2F_{5/2}$ state by electron transfer since $E_0(Yb^{3+}/Yb^{2+}) = -1.05$ V.

I.3.2 Lanthanide-based catalysts

Another important application of lanthanide complexes is devoted to their catalytic activity in various organic transformations. The whole set of Ln-based catalysis may be divided into two main groups:

(i) organolanthanide compounds as catalysts for the transformation of olefins, dienes and alkynes. In this case, lanthanide metallocenes have been proven to be very efficient for olefin transformations, such as hydrogenation, polymerization, hydroamination, hydrosilylation, hydroboration and hydrophosphination.
(ii) lanthanide alkoxides, triflates and halogenides which are used for Lewis acid-catalyzed organic reactions.[9] For instance, lanthanide triflates [$Ln(OTf)_3$] are very active catalysts for aldol, MICHAEL, allylation, DIELS–ALDER and glycosylation reactions, as well as for FRIEDELS–CRAFTS acylations. Lanthanide alkoxides [$Ln(OR)_3$] have proven to be useful

catalysts for the MEERWEIN–PONNDORF–VERLEY reduction and the hydrocyanation, whereas lanthanide shift-agents, such as [Eu(fod)$_3$] (Eu(III)-tris(1,1,1,2,2,3,3-heptafluoro-7,7-dimethyl-4,6-octanedionate) can be used as a catalyst for DIELS–ALDER and hetero-DIELS–ALDER reactions.[9]

I.3.3 Lanthanide complexes as NMR shift agents and probes for magnetic resonance imaging

Ln(III)-based compounds became very popular, when their use as shift agents was established. Being used as shift reagents in NMR, they were initially developed for spectral simplification and chiral analysis in both, organic and aqueous media. For this purpose hexacoordinated Ln(III) species with LEWIS bases (Ln = Eu, Yb, Pr) were discovered, where the anisotropic spatial distribution of unpaired f-electrons gives rise to a dipolar lanthanide-induced shift for the bound LEWIS base in solution (pseudocontact shift).

The Gd(III) ion lies at the midpoint of the lanthanide series. It has the maximum number of unpaired electrons (seven) among all lanthanide ions and this, together with long electronic relaxation time (7.9 BM), makes Gd^{3+} the best candidate as a relaxation agent. However, free Gd^{3+} is extremely toxic both, *in vivo* and *in vitro,* because its ionic radius (1.078 Å) is nearly equal to that of Ca^{2+} (1.140 Å). Therefore, in living organisms, it competes with Ca^{2+}, resulting in a strong binding with biological systems, where Ca^{2+} is needed. In order to ensure the safe clinical use of Gd^{3+}, it is complexed with organic ligands.[6] When a ligand binds the paramagnetic Gd(III) center, the rates of longitudinal (R_1) and transverse (R_2) relaxation are considerably enhanced, giving rise to an extensive line-broadening in the NMR spectra. This discovery stimulated a great deal of activity on the application of the aqua lanthanide ions as shift and relaxation probes for NMR, culminating in the development of both, gadolinium contrast agents for clinical magnetic resonance imaging (MRI) and the exploration of various shift agents for magnetic resonance spectroscopy *in vivo.* An example of the latter application is a real-time ^{31}P- or ^{23}Na-NMR analysis of perfused cells or intact animals, using wide-bore magnets or surface-coil NMR probes. The MRI application of Gd^{3+} complexes still do not have any reasonably competing analogs, though NMR shift agents are much less frequently used because of the advent of high-field pulsed multidimensional NMR techniques.

The choice of the organic ligand for Gd-based MRI contrast agents is based on its geometry and denticity. Since the Gd^{3+} coordination number in aqueous media is nine, the ligand should be octadentate to create one unoccupied coordination site for a water molecule. As such ligands, DTPA (diethylenetriaminepentaacetic acid), DOTA (1,4,7,10-tetracarboxymethyl-1,4,7,10-

tetraazacyclododecane) and some of their derivatives have already been approved for clinical use (Figure I.3-2).[6]

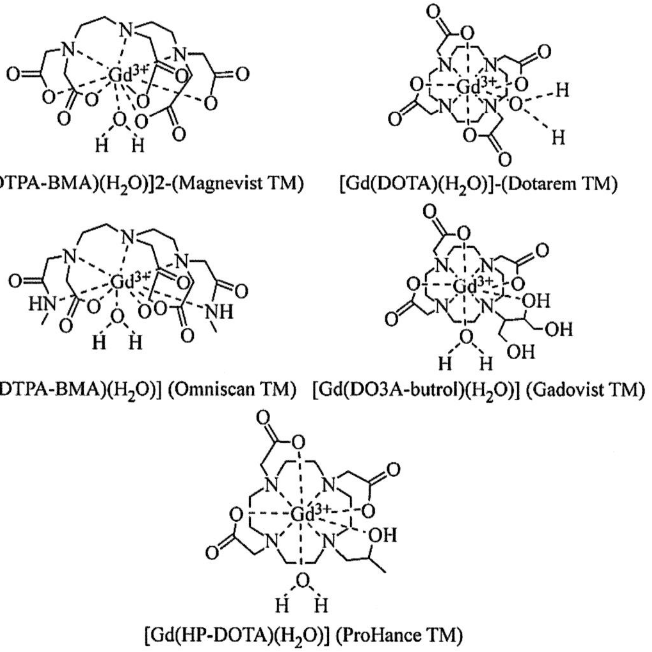

[Gd(DTPA-BMA)(H_2O)]2-(Magnevist TM) [Gd(DOTA)(H_2O)]-(Dotarem TM)

[Gd(DTPA-BMA)(H_2O)] (Omniscan TM) [Gd(DO3A-butrol)(H_2O)] (Gadovist TM)

[Gd(HP-DOTA)(H_2O)] (ProHance TM)

Figure I.3-2 Structures of some MRI contrast agents used for clinical purposes.[6] Reprinted with permission from H. U. Rashid, K. Yu and J.Zhou., *J.Struct. Chem.*, 2013, *54*, 223-249 Copyright© 2018. John Wiley & Sons, Ltd.

I.3.4 Lanthanide complexes for biological assays

A huge variety of Ln-based compounds were discovered for optical biological applications. Given that many lanthanide complexes are cell-permeable and the techniques of time-resolved detection in microscopy are very well mastered, scientists have used the unique spectroscopic properties of Ln(III) ions to get the images of cells, for instance, in the context of the follow-up of cancer therapy.[18] As these complexes are used in biological medium, which is a luminescence quencher, the design of very stable luminescent lanthanide complexes is crucial, where the lanthanide coordination sphere is saturated and no solvent molecules coordinate lanthanide ion.

Time-resolved fluorometry (TRF) with lanthanide chelate labels is a well-established technology in diagnostics. Recently TRF has attracted a great interest as a tool for applications in a range of assay formats in drug screening.[102] Lanthanide complexes are widely used as luminescent labels in immunology, particularly in a fluoroimmunoassay, which becomes more and more competitive to the commonly used radioimmunoassay. One of the first assay techniques developed was the dissociation enhanced lanthanide fluoroimmunoassay (DELFIA). This assay, however, is not ideal, because it is heterogeneous and the signal could not be traced until the end of the assay.

After more than 15 years of the development, homogeneous time-resolved fluorescence (HTRF) assays have become widespread. One of the most frequently used generic assay is fluorescence resonance energy transfer (FRET), of which the general format is shown in Figure I.3-3. When only the Ln(III)-labeled antigen is present, it binds the antibody and an efficient FRET between the lanthanide and acceptor takes place, resulting in an enhanced emission of the organic dye. When the target antigen is present, the labeled antigen is displaced and the ratio of the organic dye to the Ln(III) emission is altered.[103] With the long-lived luminescence of a lanthanide ion, the sensitivity of the assay can be vastly improved by using time-gated excitation and detection techniques. This minimizes false signals due to direct excitation of the organic dye and increases sensitivity by removing background autofluorescence from other components, present in biological media.

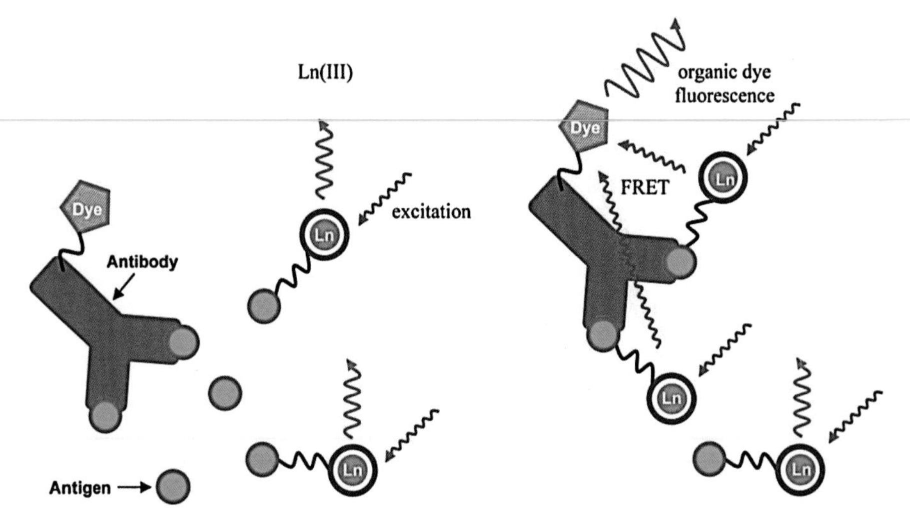

Figure I.3-3 General principle of homogenous time-resolved fluoroimmunoassay (HTRF) in the presence (left) and absence (right) of the target antigen. Reprinted with permission from E.G. MOORE *et al.*[103], *Acc. Chem. Res.* 2009, *42*, 542–552. Copyright© 2018. American Chemical Society.

When a lanthanide-based label is developed for a homogeneous assay, the chromophoric chelate strategy should be employed. Such a ligand utilizes one or more aromatic groups which absorb incident light and, besides, contain coordinating atoms. As such, these chromophores are directly bound to the Ln(III) cation. Early examples are the trisbipyridyl cryptands, developed by J.- M. LEHN *et al.*[104,105] (Figure I.3-4, compound **1**), which have been commercially developed by CISBIO (Figure I.3-4, compound **2**). Nowadays, the constant search for new Ln(III)-based compounds leads to the development of new complexes with better parameters. By now, there are two families of lanthanide-based complexes commercially available, being far beyond the market competition for FRET assays, namely (i) Lumi4Tb® family, developed by K.N. RAYMOND *et al.,*[106] based on octadentate "cages" of Tb(III) 2-hydroxyisophthalamides and (ii) Eurotracker® family, developed by D. PARKER *et al.,*[107] based on a *N*-substituted 1,4,7-triazonane moiety.

Development of the Lumi4Tb® probe was carried out in several "generations". The 1st generation of this ligand includes the development of Lumi3Tb®, bearing three 2-hydroxyisophthalamide (IAM) moieties per one ligand. This study revealed the "cage" size in this case being too small to hold Tb(III) ion inside (Figure I.3-4, compound **3**). The 2nd generation corrected this drawback and the larger "cage", though not the "closed" one, was developed (Figure I.3-4, compound **4**). The 3rd generation of Lumi3Tb® includes the "closed" analog of the compound **4** (Figure I.3-4, compounds **5-6**), as well as its modified version with a linker (Figure I.3-4, compounds **7-9**) which was named Lumi4Tb®.

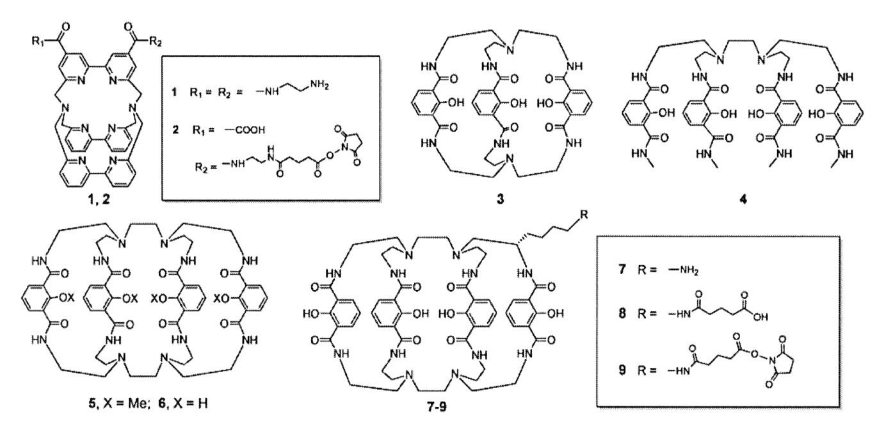

Figure I.3-4 Development of chromophoric chelating ligands. Compound **1** was originally developed by J.M. LEHN *et al.,*[104,105] compound **2** is a commercial product of CISBIO, compounds **3** and **4** were developed by RAYMOND *et al.,*[108] compounds **5** and **6** represent the Lumi4Tb® probe without a linker, compounds **7-9** are the Lumi4Tb® probe with a linker.[106] Reprinted with the permission from J. XU *et al.*[106] Copyright© 2018. American Chemical Society.

Eurotracker® is a europium complex based on aromatic sulfonate derivatives of phosphinate ligands appended to a 9-N_3 macrocyclic core, employing strongly absorbing arylalkynyl chromophores. In the recent study of D. PARKER *et al.,*[109,110] a quantum yield of 32% in water is reported, which is the highest ever reported value for a europium complex in aqueous solution so far. Its commercially available analog, depicted in Figure I.3-5, includes three important scaffolds, each bearing its functionality, namely a 9-N_3 macrocyclic core with a preliminary attached linker (Figure I.3-5, compound **1**), a sensitizing core (Figure I.3-5, compound **2**), a spacer attached to the linker, as well as the SNAP-tag vector (Figure I.3-4, compound **3**) which is responsible for transporting properties.

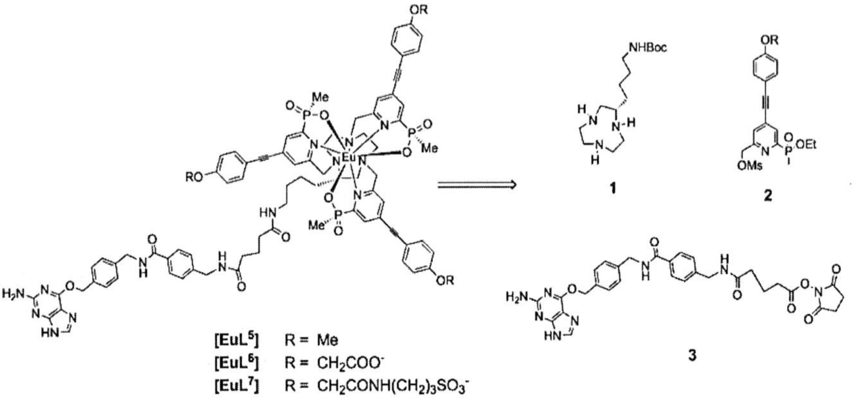

Figure I.3-5 Luminescent probes from Eurotracker® family, modified with a linker, ready for conjugation. **1**,**2**,**3**-motifs were used to construct this probe. Reprinted with permission from M. DELBIANCO *et al.*,[109] *Angew. Chem. Int. Ed.* 2014, *53*, 10718-10722. Copyright© 2018. JOHN WILEY and Sons.

I.4 Lanthanide-peptoid conjugates

Another way to ensure the transporting properties of the lanthanide complexes is the use of specially designed targeting peptides which may not only provide cell-penetrating properties but also may have a specific localization within a cell. Such transporters are able to transport covalently bound cargos of a diverse chemical nature into living cells. As lanthanide complexes may be both, the cargo and the label, the design of lanthanide-based conjugates is a meaningful step towards the development of the new probes for advanced bioimaging.

So far, several different lanthanide-peptide conjugates have been reported (for instance, Pr,[111] Sm,[112] Tb[113,114]), which may be used without further labeling with common fluorophores.[114] At the same time, *N*-substituted glycines (or peptoids) represent a promising class of biocompatible compounds recapitulating the functions and capabilities of natural polypeptides while retaining *in vivo* metabolic stability and remarkable structural diversity.[115] They are mostly used as peptidomimetics, however, their transporting properties were recently discovered, also in the group of Prof. S. BRÄSE.[116,117] The term "peptoids" is related to a specific family of oligomers of *N*-substituted glycines. They are based on an unnatural repetitive unit that swaps the amide hydrogen atom for the α-carbon side chain, making the peptoid backbone achiral (Figure I.4-1).

Figure I.4-1 Comparison of peptide and peptoid structural motifs.

To synthesize the peptoids, the solid-phase synthetic approach is usually applied. It was introduced by ZUCKERMANN[118] as a completely automated solid-phase synthetic method and relies on the two-step assembly of a monomer unit directly on solid phase out of two submonomers: bromoacetic acid and the desired primary amine which determines the side chain of the peptoid. Alternating the cycles of acylation/amination, the peptoid chain is grown on solid support and detached when the desired length is reached.

The ability of peptoids as peptidomimetics to coordinate different metal ions was demonstrated in several research works,[83,119,120] however, they are mostly devoted to the coordination with alkali[83,120,121] or d-metals (Figure I.4-2a).[122,123] Only a few examples of Ln-peptoid conjugates are known from literature (Figure I.4-2b,c),[119,124,125] although, these few reports illustrate their remarkable properties.

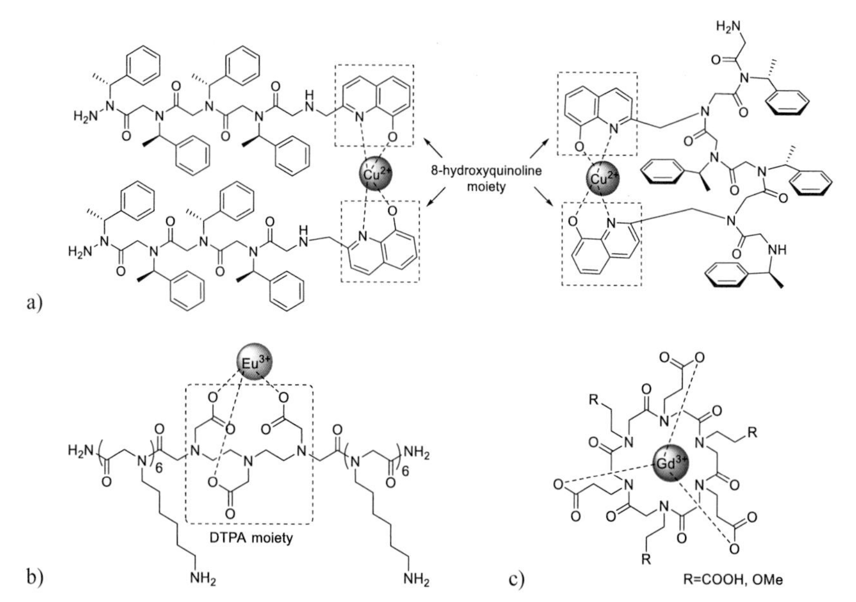

Figure I.4-2 a) Example of a peptoid-metal conjugate with the chromophore chelator attached as a side chain;[122] b) example of the incorporation of the DTPA chelator into the peptoid backbone, further conjugated with Eu^{3+};[124] c) example of a cyclic peptoid, coordinating a Gd^{3+} ion.[119]

Three main strategies for the synthesis of lanthanide-peptoid conjugates are known so far. The first strategy implies the use of the chromophoric chelator as one of the peptoid side chains which will be only responsible for coordinating a metal ion and will not affect the cellular permeability and transporting properties of the peptoid (Figure I.4-2a).[122] The second strategy is the incorporation of the chelating ligand into the peptoid backbone, which is basically the modification of the chelating ligand with the peptoids moieties. The strategy was developed by Dr. D. KÖLMEL from the group of Prof. S. BRÄSE,[124] where diethylenetriaminepentaacetic acid (DTPA) was used as a chelator (Figure I.4-2b), as it forms very stable complexes with trivalent lanthanide ions (stability constants $\log K \approx 20$[126]). The third strategy is the use of a cyclic peptoid with side chains, bearing coordinating carboxylic moiety.[119] It was found that the size of the cavity inside of the cycle of a hexameric cyclic peptoid is favorable to coordinate a Ln(III) ion (for instance, Gd^{3+} ion,[119] Figure I.4-2c).

II. Aim of the Project

The potential of lanthanide coordination compounds in biological and lighting applications can hardly be overestimated due to their fascinating and unique optical properties, such as long luminescence lifetimes, large STOKES' shift and distinctive sharp luminescence bands.

This research work targets the synthesis of the lanthanide-based coordination compounds, the investigation of their structure and properties with the further use in (i) cellular imaging; (ii) ink-jet printing; (iii) OLEDs, encompassing NIR emission and (iv) FRET-assays. In the following chapters, each subchapter reveals detailed implementations and tricks to shape the design of these probes in accordance with each field. In order to achieve this goal, seven specific aims where determined.

To improve the functional properties of lanthanide aromatic carboxylates (benzoates), a novel approach to ligand fluorination can be applied. On the one hand, fluorination is assumed to increase the luminescence efficiency and solubility of lanthanide benzoates and to reduce luminescence quenching. On the other hand, it decreases the thermodynamic stability of carboxylates. This makes stable Eu(III) benzoates showing little luminescence, while pentafluorobenzoates are highly luminescent but unstable. The present research is aimed to find a middle ground, *i.e.* the optimum ligand fluorination degree to obtain photostable, non-toxic and luminescent compounds suitable as bioprobes.

Aim 1: to investigate the influence of the ligand fluorination degree on the structure, photophysical properties and the solubility of lanthanide fluorobenzoates, revealing the corresponding structure-properties relationships.

Several different strategies to improve the efficiency of Eu(III) complexes are explored. One of them is the synthesis of the heterometallic Tb/Eu complexes that serve as an additional energy transfer step between the ligand (L) and the Eu ion. Another strategy is the incorporation of an auxiliary ligand which participates in energy transfer process, as well as prevents polymerization of the complex. Herein 1,10-phenanthroline and bathophenanthroline served as auxiliary ligands.

Aim 2: to increase the luminescence efficiency of europium fluorobenzoates by applying two strategies: the introduction of a two-step sensitization path $L \rightarrow Tb^{3+} \rightarrow Eu^{3+}$ and the introduction of a sensitizing auxiliary ligand and thereby determine the conditions of their applicability.

It is a great challenge to obtain NIR-emitting lanthanide complex. Since NIR emission has lower energy than visible one, it can be easily quenched through non-radiative pathways, requiring a completely different design of a luminophore. This work investigates a bulky ligand anthracene-9-carboxylate with highly π-conjugated ring system for NIR luminescence sensitization, the complexes with which are assumed to form anhydrous lanthanide complexes.

Aim 3: to expand the research topic on the design of NIR-emitting lanthanide complexes, optimizing the necessary parameters to obtain the efficient probes, as well as their testing as OLED emitting layers.

In addition to the development of the new luminescent probes, the optimization of the synthesis of the existing ones is also a crucial task. As the synthesis of a common FRET-donor Lumi4Tb® probe is reported to be very challenging, its optimization is required. On top of that, this research contributes an attempt to broaden the scope of the linkers for this compound. This part of the thesis was planned as a research stay at the Harvard Medical School - Massachusetts General Hospital, Boston, USA under supervision of Prof. Dr. RALPH MAZITSCHEK.

Aim 4: to optimize the synthetic route towards the Lumi4Tb® probe and to propose alternative strategies for a linker incorporation.

As lanthanide fluorobenzoates are promising compounds for bioapplications, they potentially can be coupled with biomolecules or specific targets. Their molecular geometric properties define that, being the farthest to the Ln^{3+} ion, the *p*-position of the benzoate is assumed to exhibit the lowest impact towards the lanthanide ion emission. However, the way how such a modification influences the functional properties should be further studied.

Aim 5: to synthesize *p*-substituted lanthanide fluorobenzoates, investigate their photophysical properties and thereby determine the structure-property relationship.

Expanding the scope of this work, the most promising discovered lanthanide-based labels can be further conjugated with poly-*N*-substituted glycines (peptoids) to create novel perspective probes for bioimaging, using different synthetic approaches and exploring the optimization of synthetic routes.

Aim 6: to synthesize lanthanide-peptoid conjugates, exhibiting intense cellular luminescence.

Finally, in order to increase the thermodynamic stability of lanthanide-peptoid conjugates, cyclic peptoids might be used as chelating ligands to tightly coordinate the lanthanide ion. To obtain highly luminescent lanthanide-peptoid conjugates, a heterocyclic moiety might be additionally introduced into the peptoid backbone.

Aim 7: to synthesize a library of cyclic peptoids, bearing a heterocyclic functionality in the backbone and at least three carboxylic functionalities on the side chains and hence, use them as ligands for the conjugation with Eu(III).

III. Results and Discussion

Numbering system

In the present research work the following numbering system is used:

1) The number of the compound relates to the chapter number, where it appears. For instance, if the compound number is **2**.1, its features are discussed in Chapter III.**2**.
2) For the compounds, bearing COOH-functionality, the number also contains "H" as a symbol, for instance, **1.9-H** means that this compound is a carboxylic acid from Chapter III.**1**.
3) If the carboxylic group is protected, the type of protecting group is reflected in the number. For instance, for compound **2.12-Me** it becomes clear that this is carboxylic acid **2.12-H**, which was protected with methyl group, so **2.12-Me** is a methyl ester.
4) If ligands are discussed, they are usually present in the deprotonated form of the corresponding acid, *i.e.* ligand **2.14** is the deprotonated carboxylic acid **2.14-H**.
5) Metal complexes are designated as **M-X.Y**, where **M** – the type of the metal; **X.Y** – ligand. For instance, compound **Eu-1.4** is europium complex with ligand **1.4** and the properties of this compound are discussed in Chapter III.**1**. These numbers, however, do not distinguish between mono-, bi- or tricoordinated ligands and do not show the hydrate composition. Thus, a potassium salt with one metal per ligand will be denoted as **K-1.9,** while tricoordinated Eu(III) 2-fluorobenzoate dehydrate with the general formula Eu(**1.9**)$_3$(H_2O) is still denoted as **Eu-1.9**. Though it might look a little confusing, such a system is still convenient, since the research work mostly contains tricoordinated lanthanide complexes, studied in solutions and therefore, the hydrate composition of the powders is not essential. In the cases where it is indeed important, it is indicated in the main text.
6) If the single-crystal structures are discussed, the label contains **–sc,** for instance, the label **Eu-1.4-sc** tells that it is a europium complex with ligand **1.4**, grown as a single crystal.
7) If for one complex several different phases are observed, the phases are distinguished by a Greek letter. For instance, for a single crystals of samarium complex with ligand **1.1**, two phases **α-Sm-1.1-sc** and **β-Sm-1.1-sc** are obtained. If the phase is determined for a powder, the same strategy is applied, for instance, **δ-Eu-1.1** is powder of europium complex with ligand **1.1** and its exact powder structure is determined.
8) If two different and sophisticated ligands are used, the complex gets a new number to avoid confusion, however, the designation still indicates that this is a Eu complex. For instance, Eu complex **Eu-2.8** with peptoid **2.53** is denoted as **Eu-2.54**. When an auxiliary ligand is "simple", the complex gets its name after the hyphen, for instance, **Eu-1.9-Phen** is complex **Eu-1.9**, where 1,10-phenanthroline (Phen) is additionally coordinated.

9) Compounds which are used on solid phase are designated, using the pattern **number-SP**, where **SP** stands for "solid phase". For instance, **2.56-SP** is peptoid **2.56** on solid support.
10) If the amine functionality is protected, *e.g.* with Boc-group (*tert*-butyloxycarbonyl group), it is usually indicated as superscript and designates that all the amines are Boc-protected. For instance, **2.46^{Boc}-SP** is a compound on solid support, where **all** the side chains, bearing amine functionality, are Boc-protected.
11) Where a compound, bearing two amine functionalities, is mono-Boc-protected, it gets a new number to avoid confusion. For instance, compound **2.41** is mono-Boc protected analog of diamine **2.38**.
12) For cyclic peptoids, the letter "c" is added prior to the number. For instance, peptoid **2.90-*t*Bu** (where carboxylic functionalities are *t*Bu-protected) is cyclized to obtain cyclic peptoid **c-2.90-*t*Bu**. Deprotection of this compound affords compound **c-2.90-H** (according to rules 2 and 3).
13) The complexes with cyclic peptoids are denoted in a same way as the complexes with benzoates. For instance, **Eu-c-2.90** is europium complex with deprotonated cyclic peptoid **c-2.90-H**.
14) In exotic cases, the label may contain some additional information. For instance, the byproduct **2.98-*t*Bu-dimer** is denoted as a dimer. Another example is **δ-Eu-1.1$_d$**, where the subscript "d" indicates that the compound was synthesized using deuterated solvents.

Nomenclature of the peptoids

As for now there is still no consistent nomenclature for peptoids. Once the amines in the side chains become more complex, the nomenclatures may differ, depending on the author and usually have no systematic reference of the name to the submonomer. Due to the greater diversity of submonomers used in this work, a systematic naming of peptoids has been introduced, based on the system, previously developed in the group of Prof. S. BRÄSE (Figure III.1a). Besides, several new names have been introduced (Figure III.1b). In some cases, however, the adopted nomenclature system has to be improved. For instance, the benzoic acid derivatives with different substituents were numbered as *N*0b1, *N*0b2, *N*0b3 in order to keep the names simple (Figure III.1b). Besides, the existing nomenclature does not indicate the geometry of the triazole group. Therefore, for the triazoles, generated from 3-azidopropan-1-amine, the superscript "N" is used (Figure III.2a), while the superscript "C" is used for the triazoles, synthesized from prop-2-yn-1-amine (Figure III.2b). To name the peptoids, the monomer abbreviations, separated by a hyphen, are strung together from the *N* to the *C* terminus. An "H" at the beginning of the sequence indicates a free amine at the *N*-terminus. If the terminal amine is functionalized, it is indicated by the corresponding abbreviation, for example with "Ac" for an acetyl moiety. Accordingly, "OH" or "NH_2" at the end of the sequence represents a free carboxylic acid or a carboxylic acid amide at the *C*-terminus.

Cyclic peptoids are marked by a prefixed "c". If some moieties are coupled with the side chains of the peptoids, *italic font* is used. If the moiety is introduced to the backbone, the standart straight font is used. In such a way the name ***H-Phen1*-(*N*3tz^{N})-(*N*1ph)-(*N*1ay)-(*N*1Pc)-*NH*$_2$** indicates that Phen1 moiety (Figure III.2b) is coupled with a peptoid *via* the CuAAC reaction following by the formation of a triazole ring on a side chain (Figure III.3a), while the name ***H*-(*N*3cx^{tBu})-Phen-(*N*3cx^{tBu})-(*N*1ph)-(*N*3cx^{tBu})-*OH*** indicates that 1,10-phenanthroline is introduced to the backbone as shown in Figure III.3b. The abbreviature *Cyo* were used for cyclooctine moiety. For pyridine, bipyridine and terpyridine moieties the abbreviations **Pyr, BPyr** and **Terpy** were used.

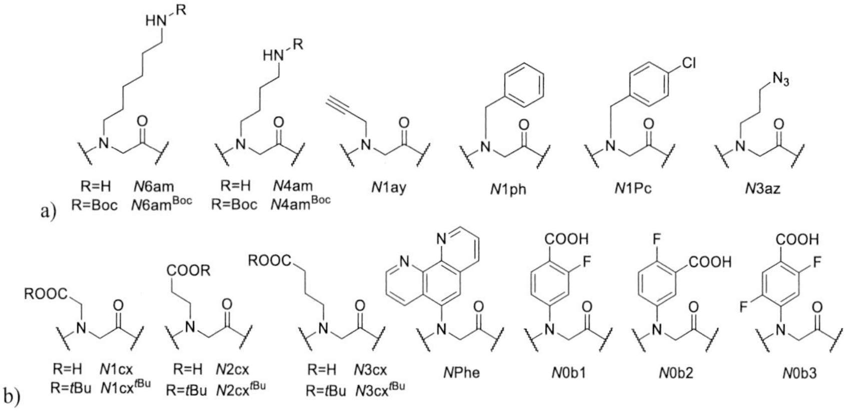

Figure III.1 Example of the nomenclature of the peptoid monomers synthesized in this work.

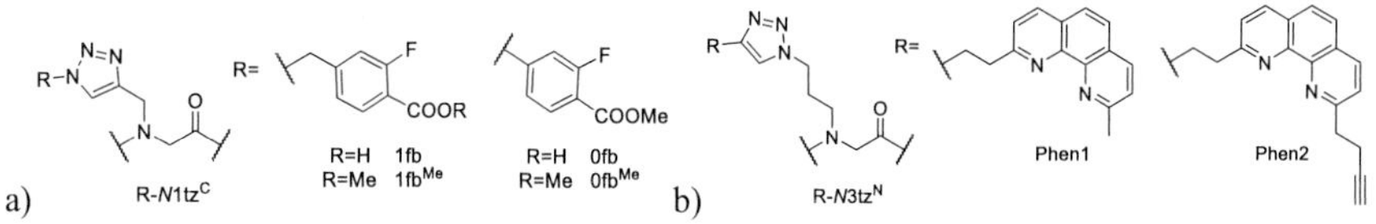

Figure III.2 New monomers, bearing triazole moiety, used in this work: a) triazoles, synthesized from 3-azidopropan-1-amine; b) triazoles, synthesized from prop-2-yn-1-amine.

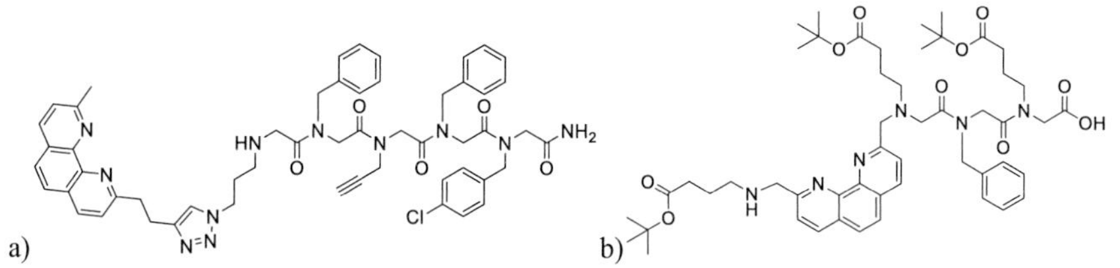

Figure III.3 Representation of peptoids with an 1,10-phenanthroline moiety: a) coupled with the side-chain; b) incorporated to the backbone.

III.1 Lanthanide complexes

A growing interest for lanthanide luminescence is accelerated by the increased demand for luminescent materials for a broad range of applications that include telecommunications,[89,94,127] lighting, electroluminescent devices,[89,128] biosensors[18,129–132] and bioimaging probes.[18,133,134] Their rebirth and increasing importance in biology and material research has been highlighted by numerous reviews.[1,89,132,135,136] While most commercially available compounds for bioimaging are of purely of organic origin,[137] some lanthanide coordination compounds (LCCs) have already found applications as highly sensitive and selective probes in immuno- and DNA-assays.[1,135,136] However, there is still plenty of room for the further improvement of existing Ln-based assays, as well as for the discovery of new ones, *e.g.* for new diagnostic techniques.[38,138] This research work is devoted to the study of lanthanide-based probes, exhibiting different functional properties and their use for biological and lighting applications. Depending on the application, Ln(III) complexes with different ligands, bearing different functional groups (FGs), should be designed. Each subchapter covers a set of tricks for the development of Ln(III)-based probes, suitable for the particular application.

Chapter III.1.1 implies a novel approach of ligand fluorination to improve the functional properties of Ln(III)-based benzoates. The introduction of fluorine-substituents in different positions may alter not only the luminescent characteristics of the benzoates, but also their thermodynamic stability and solubility. The connections between these functional properties are thoroughly studied.

Chapter III.1.2 contains general statements regarding the efficiency of Ln-based luminescent probes and demonstrates how the efficiency can be improved by choosing a favorable geometry of the complex together with minimizing the energy losses by introduction of fluorine substituents. This chapter also shows, how the alteration of auxiliary ligands may affect the suitability of the probe for particular applications, *i.e.* bathophenanthroline derivatives are solely suitable for OLEDs, while 1,10-phenanthroline derivatives are solely suitable for bioimaging. Such a "compound-to-material" transformation is very crucial for any practical application.

Chapter III.1.3 reveals, how the selection of an appropriate ligand for NIR-emitting lanthanides (NIR – near infrared) toghether with the optimization of an OLED electronic structure may lead to the record characteristics of NIR-emitting OLEDs. Lanthanide 9-anthracenates were successfully tested as emitting layers in host-free OLEDs, exhibiting pure NIR-luminescence.

In addition, Chapter III.1.4 demonstrates the optimiazation of the synthetic route towards Tb(III)-based luminescent probes for FRET-assays. Such probes are irreplaceable for many biological studies, due to the features of Ln(III)-based luminescence, such as large STOKES' shift and long lifetimes of the excites states (in the ms range). However, such probes are still inaccessible due to the sophisticated synthesis and hence, relatively high price.

III.1.1 Lanthanide fluorobenzoates with different fluorination degrees

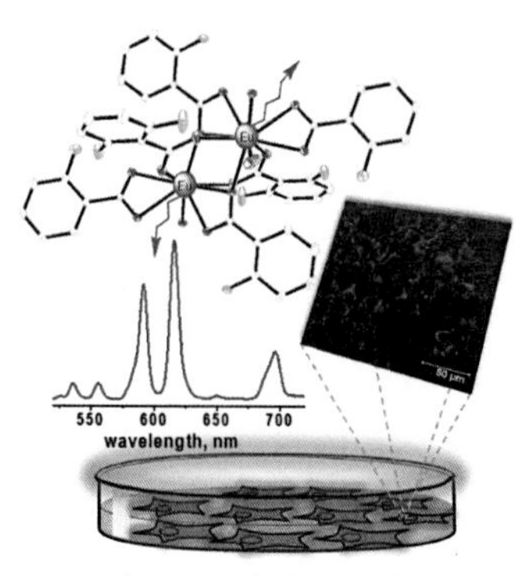

The present chapter represents a concise overview of the chemical properties of lanthanide fluorinated benzoates. It was shown that the ligand fluorination increases the solubility of lanthanide benzoates and reduces luminescent quenching,[139,140] *but decreases the thermodynamic stability of the carboxylates.*[12,18,139–142] *Thus, stable Eu benzoates show little luminescence, while perfluorobenzoates are luminescent but unstable.*[139] *The aim of the current research was to find a middle ground, i.e. the optimum fluorination degree to obtain photostable, non-toxic and luminescent compounds, suitable for the use as bioprobes. Herein, the thorough study of lanthanide fluorinated benzoates (Eu, Tb, Nd, Er, Yb, Gd, La, Lu) has been conducted revealing a structure-properties relationship and the potential of their use as bioprobes. A close attention was paid to alterations in the crystal structure (5 phases) which arise in complexes going along the lanthanide row from praseodymium to lutetium (excluding the radioactive promethium). As additional and complementary tools,* ^{19}*F NMR and EXAFS spectroscopy were employed to obtain information about the local structures of the complexes in solutions. The photophysical properties of complexes emitting in visible (Ln = Tb, Eu) and NIR regions (Ln = Nd, Er, Yb) were also thoroughly studied. To evaluate the potential of such complexes as luminescent materials, their solubility and film forming properties were estimated.*

This section was previously published by the author ALENA S. KALYAKINA *et al.* in *Chemistry – A European Journal* **2015**, *21*, 17921-17932[139] and in *Chemistry – A European Journal* **2017**, *23*, 14944-14953,[33]

The current research study presents the synthesis of lanthanide complexes with nine fluorinated benzoate ligands, shown in Figure III.1-1. The structures and variability of functional properties by alternating the number and position of fluorine atoms in the benzene ring were studied. Since the vibration energy of C-F bonds is lower, than those of C-H bonds, the introduction of fluorine substituents is assumed to increase the luminescence efficiency of Ln complexes by decreasing non-radiative relaxation. The selected ligands have at least one *ortho*-fluorine atom, since this position is the closest one to the Ln(III) coordination-sphere and has higher influence on quenching through bonds vibrations,[20,143] leading to the luminescence enhancement. There are two acids, **1.3-H** and **1.8-H** (Figure III.1-1) that were considered to evaluate the effect of other substituents in the *ortho*-position to the carboxylic group.

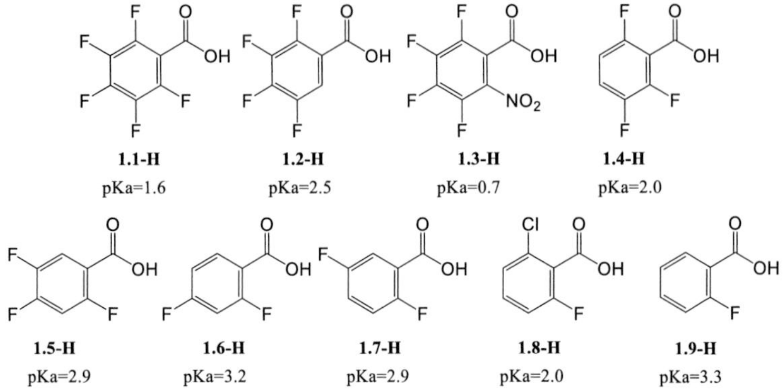

1.1-H	1.2-H	1.3-H	1.4-H
pKa=1.6	pKa=2.5	pKa=0.7	pKa=2.0

1.5-H	1.6-H	1.7-H	1.8-H	1.9-H
pKa=2.9	pKa=3.2	pKa=2.9	pKa=2.0	pKa=3.3

Figure III.1-1 List of acids examined in this work. All the acids were purchased from SIGMA ALDRICH. The pKa values were estimated using Advanced Chemistry Development (ACD/Labs) Software V11.02 (© 1994-2016 ACD/Labs).

Luminescent properties were studied in detail in powders for the complexes of lanthanides, emitting in NIR (Nd, Er, Yb) and visible (Eu, Tb) regions, as well as for non-emitting ones (Gd, La, Lu). Luminescence in aqueous solutions have been studied for Tb and Eu complexes.

The lanthanide complexes having the composition of $Ln(L)_3(H_2O)_x$ are further abbreviated as **Ln-L**, where Ln stands for the selected lanthanide and L = **1.1** – **1.9** represents the ID of the ligand according to Figure III.1-1, while the single crystals of the corresponding LCCs are denoted as **Ln-L-sc**.

Synthesis

Lanthanide carboxylates can be synthesized following two routes.[12,139–142] The first route is an exchange reaction between aqueous solutions of an alkali metal carboxylate and a lanthanide inorganic salt.[140,142] The second route involves the interaction of a freshly prepared lanthanide hydroxide $Ln(OH)_3$ with a solution of a carboxylic acid in organic medium.[139] Given that all chosen ligands show high enough acidity (Figure III.1-1, pKa = 0.7 – 3.3), both routes are proved to be feasible. The choice of the synthetic approach has been performed depending on the water solubility of the formed lanthanide complexes. The exchange reaction served for the synthesis of poorly soluble **Ln-1.3**, **Ln-1.5**, **Ln-1.6**, **Ln-1.7** and **Ln-1.9**, while water-soluble **Ln-1.1**, **Ln-1.2**, **Ln-1.4**, **Ln-1.8** complexes were synthesized, using the corresponding hydroxides and acids in organic medium, followed by recrystallization from water.[139]

It is not surprising that in a latter synthetic route the hydrate composition of products strongly depends on (i) the solvent used for the synthesis and (ii) the isolation method. Therefore, for the syntheses of lanthanide complexes, the existing synthetic procedure was modified and unified on the example of lanthanide pentafluorobenzoates **Ln-1.1**. Using the general procedure **GP1b**, described in the experimental part, the microcrystalline powders of the general formula $Ln(pfb)_3(H_2O)_x$ were obtained. The hydrate composition of powders was determined by thermogravimetric analysis (TGA) and seems to decrease while going within the Ln(III) row: $x = 2$ for Ln = Pr, Nd; $x = 1$ for Ln = Sm, Eu, Gd, $Tb_{0.5}Eu_{0.5}$, Tb, Dy, Ho, Er; $x = 0$ for Ln = Yb, Tm, Lu (Figure III.1-2). However, the *slow* evaporation of the aqueous solution of the corresponding powders at room temperature gave prismatic and needle-shaped single crystals (**α-Ln-1.1-sc,** where Ln = Nd, Sm, Eu, Gd, $Tb_{0.5}Eu_{0.5}$; **β-Ln-1.1-sc,** where Ln = Sm, Gd, $Tb_{0.5}Eu_{0.5}$; **β'-Ln-1.1-sc,** where Ln = Lu and **γ-Ln-1.1-sc,** where Ln = Tb, Dy, Ho, Er, Tm). The single-crystal phases have a significantly higher water content ($x = 5 – 6$) than powder phases ($x = 0 – 2$) and tend to degrade in air. The presence of these phases with higher water content has never been observed in the powder samples.

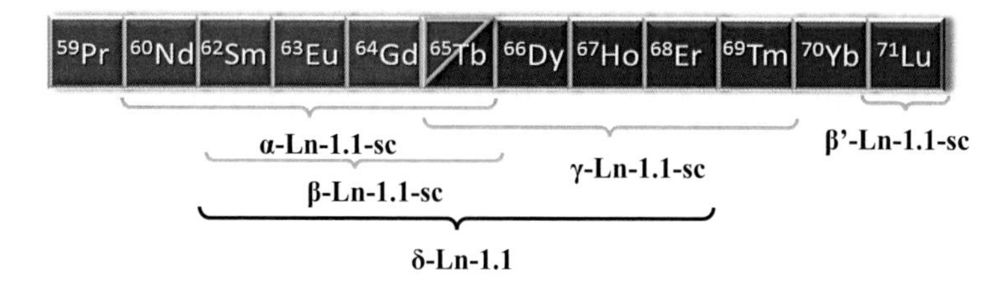

Figure III.1-2 Overview of the phases, obtained for Ln(III) pentafluorobenzoates.

X-ray diffraction data

The structures of the studied complexes were determined from single-crystal and powder X-ray diffraction patterns. In order to reveal the influence of the Ln(III) ion type on the structure of the Ln(III) complex, 14 single-crystal and 13 powder diffraction experiments* in total were carried out for lanthanide pentafluorobenzoates (different Ln(III) ions, the same ligand). For other lanthanide fluorobenzoates, 10 single-crystal and 8 powder diffraction experiments were performed (the same Ln(III) ion, different ligands) to reveal how the ligand composition may affect the structure of the Ln(III) complex. Having such a massive data set in hands, it was possible to independently determine the influence of the type of Ln(III) ion and the ligand fluorination degree on the structure of Ln fluorobenzoates.

* Single-crystal and powder X-ray diffraction was carried out together with Dr. I.S. Bushmarinov, A.N. Nesmeyanov Institute of Organoelement Compounds, Moscow, Russia

The structural study of Ln pentafluorobenzoates **Ln-1.1** was meant to determine, how the type of the Ln(III) ion affects the crystal structure. The structure of Eu pentafluorobenzoate **δ-Eu-1.1** was determined usin powder X-ray diffraction data (PXRD), followed by periodic DFT calculations to obtain more reliable geometric parameters. Afterwards, the powder pattern of the determined structure was compared to the powder patterns of all studied complexes **Ln-1.1**, (Ln = Sm, Eu, Gd, Tb, Dy, Ho and Er). This phase was observed in all of them, though, in different amounts. The structure of **δ-Eu-1.1** is shown in Figure III.1-3a.

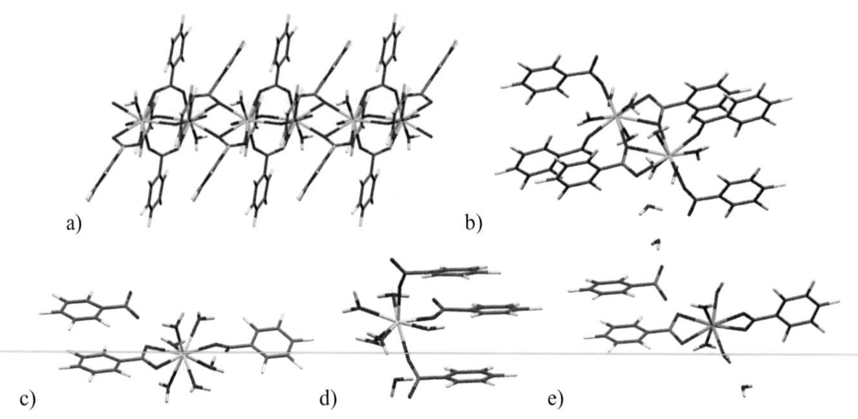

Figure III.1-3 Single-crystal structures for different phases of lanthanide pentafluorobenzoates: a) **δ-Eu-1.1** b) **α-Eu-1.1-sc**; c) **β-Gd-1.1-sc**; d) **γ-Er-1.1-sc**; e) **β′-Lu-1.1-sc**. Atom assignment: C-grey, O-red, F-lime green, H-white, Eu-cyan, Gd-light green, Er-light blue, Lu-green.

Table III.1-1 Summary of Ln crystal structures. All polyhedra are approximated and/or distorted.

	α-Ln-1.1-sc	**β-Ln-1.1-sc**	**γ-Ln-1.1-sc**	**β′-Ln-1.1-sc**	**δ-Ln-1.1**
found in	single crystal	single crystal	single crystal	single crystal	powder[a]
composition	$[Ln_2(pfb)_6(H_2O)_8](H_2O)_2$	$[Ln(pfb)_2(H_2O)_6]^+(pfb)^-$	$[Ln(pfb)_3(H_2O)_5](H_2O)$	$[Ln(pfb)_2(H_2O)_4]^+(pfb)^-(H_2O)_2$	$[Ln(pfb)_3(H_2O)]_n$
CN	9	9	8	9	9
polyhedra	monocapped tetragonal antiprism	monocapped tetragonal antiprism	tetragonal antiprism	tetragonal antiprism	monocapped tetragonal antiprism
nuclearity	dimer	monomer	monomer	monomer	polymer
formula unit	$C_{42}H_{20}F_{30}O_{22}Ln_2$	$C_{21}H_{12}F_{15}O_{12}Ln$	$C_{21}H_{12}F_{15}O_{12}Ln$	$C_{21}H_{12}F_{15}O_{12}Ln$	$C_{21}H_2F_{15}O_7Ln$
ligand coordination	$Ln_2(\mu_2\text{-}\kappa^1\text{:}\kappa^2\text{-pfb})_2(\kappa^1\text{-pfb})_4(H_2O)_8$	$Ln(\kappa^2\text{-pfb})(\kappa^1\text{-pfb})(H_2O)_6$	$Ln(\kappa^1\text{-pfb})_3(H_2O)_5$	$Ln(\kappa^2\text{-pfb})_2(H_2O)_6$	$[Ln(\mu_2\text{-}\kappa^1\text{:}\kappa^2\text{-pfb})_2(\mu_2\text{-}\kappa^1\text{-pfb})_4(H_2O)]_n$
observed for	Nd, Sm, Eu, Gd, (TbEu)	Sm, Gd, (TbEu)	Tb, Dy, Ho, Er, Tm	Lu	Sm, Eu, Gd, Tb, Dy, Ho, Er

[a] powder patterns for Pr, Nd, Sm, Eu, Dy, Ho, Tm, Yb, Lu also contained unidentified phases

Analyzing the structural data for pentafluorobenzoates **Ln-1.1** (both, single-crystal and powder XRD), five structural types with x = 6, 5 or 1 were obtained (Figure III.1-3, Table III.1-1). Interestingly, the same lanthanide can form different hydrates, depending on the crystallization conditions, indicating that the preferred water content is not significantly affected by the cation (*e.g.* **α-Ln-1.1-sc**, **β-Ln-1.1-sc** and **δ-Ln-1.1** structural types were observed for Sm). It turned out that several structural parameters are not dependent on the Ln(III) ion used, but on crystallization conditions, such as the nuclearity of the complex (since different nuclearities are observed for the same Ln(III) cations) and the hydrogen bonding patterns (due to similar reasons). In the present study among the synthesized pentafluorobenzoates monomeric phases were also detected, though ususaly they are rarely observed. According to Cambridge Crystallography Database (CSD),[144] less than 9% of Ln(III) carboxylates in CSD are monomeric, while other 91% of these structures tend to be polynuclear*. Besides, monomers might be observed mostly in the cases, where another ligand with strong donating properties is present,[47] which is not the case for Ln(III) pentafluorobenzoates.

The ligand coordination mode seems to be significantly metal-specific. The $(pfb)^-$ ligand can coordinate the metal by one of its oxygen atoms (further denoted κ^1) or by both (further κ^2), as demonstrated in Figure III.1-4a. The preference for the specific coordination mode should be affected by the lanthanide contraction, since shortening of the Ln–O bond increases steric repulsion between the ligands, leading to the decrease of the probability of κ^2 coordination. The changes in the preference for the coordination obey the following trend: ligands with κ^2 coordination in **α-Ln-1.1-sc** and **β-Ln-1.1-sc** phases are observed for the Nd – Gd range, while exclusively κ^1 ligand coordination in the **γ-Ln-1.1-sc** phase is specific for the Dy – Tm range. For terbium which is in-between these two ranges, both structural types, **α-** and **γ-Tb-1.1-sc**, are possible. The structure **α-Tb-1.1-sc** is known from literature[145] and was obtained in this work as well as **γ-Tb-1.1-sc**. In different crystallization conditions, the κ^2 coordination mode in the **δ-Ln-1.1** phase is found in powder patterns in the Sm – Er range (Figure III.1-3a), but not for Pr, Nd, Tm and Yb. Lutetium seems to break this trend, as **β′-Lu-1.1-sc** exhibits two κ^2 ligands (Figure III.1-3). However, the analysis of CSD for lanthanide aromatic carboxylates (Figure III.1-4b) revealed that this is a feature of Lu. Though the preference for κ^2 coordination exhibits a downward trend in the Eu – Yb range with statistically insignificant variations, Lu complexes show a higher rate of κ^2 coordination than any other lanthanides, which is statistically significant, even despite the small amount of Lu structures in the CSD.

Thus, the preference for either κ^1 or κ^2 coordination of the $(pfb)^-$ ligands is determined by the cation. The preferred coordination number in solid state is 8 or 9 and monomeric, dimeric or polymeric complexes can be obtained by varying the crystallization conditions. The monomeric

* CSD analysis was carried out together with Dr. I.S. BUSHMARINOV, A. N. Nesmeyanov Institute of Organoelement Compounds, Moscow, Russia.

form is dominant when crystallizing at low temperatures, while the polymeric one is dominant at higher temperatures.

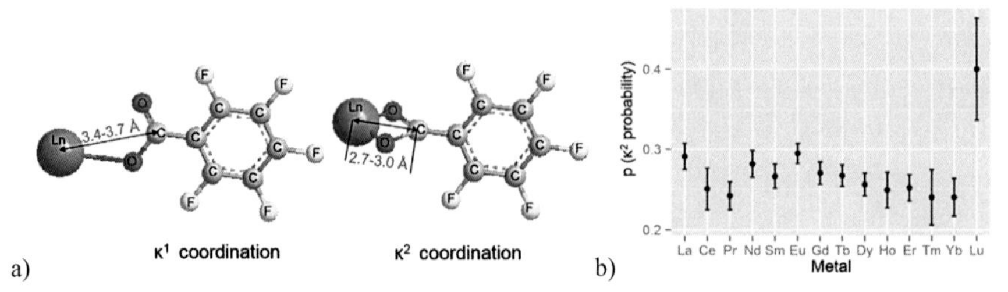

Figure III.1-4 a) Two possible coordination modes for a Ln-(pfb) fragment; b) the fraction of κ^2 coordinated ligands among Ln complexes with aromatic carboxylates; the error bars correspond to relative e.s.d. of underlying binomial distributions, calculated as $\sqrt{p(1-p)/N}$, where p is the κ^2 probability and N is the number of observations.

To analyse the influence of the fluorination degree on the structural features of lanthanide fluorobenzoates, ten single-crystal structures of the other fluorobenzoates **Ln-1.2 – 1.9** (Figure III.1-1) were determined. It seems that the nature of the ligand has no significant effect on the immediate Ln environment in the crystal (Figure III.1-5): regardless of the ligand variation, most of the studied complexes fall into two similar groups. One of them contained **Er-1.2-sc, Er-1.4-sc, Er-1.5-sc** and **Eu-1.8-sc**, while the other one contained **Eu-1.5-sc, Eu-1.6-sc** and **Eu-1.7-sc** which were virtually isomorphous and different in number and position of fluorine atoms and crystal system (Figure III.1-5). In contrast to **Ln-1.1**, these complexes contain no more than two inner-sphere water molecules (as shown above, Figure III.1-3b-d). Interestingly, the single-crystal structure **Eu-1.4-sc,** crystallized at 4 °C, turned out to be similar to the powder structure of **δ-Eu-1.1**, though the latter has never been obtained at low temperatures. With exception of **Eu-1.9-sc** (Figure III.1-6), all studied complexes are polymeric even in single crystals obtained at low temperature, unlike **Ln-1.1-sc** which is monomeric with higher water count (5 to 6 water molecules per lanthanide ion) (Figure III.1-3). The likely reason for this difference is the high LEWIS acidity of the lanthanide pentafluorobenzoates **Ln-1.1**.[146–149] The donor abilities of **1.1** are limited and the positive charge of the Ln in **Ln-1.1** is compensated by coordination of additional water molecules.

The most important factor affecting the crystal structures of partially fluorinated benzoates turned out to be the number of non-hydrogen *ortho*-substituents of the benzene ring (n_{ortho}). In **Eu-1.4-sc** ($n_{ortho} = 2$) the aromatic ring tends to form a 30° angle relative to the plane of the carboxylic group (Figure III.1-6b), while in **Eu-1.5-sc** ($n_{ortho} = 1$, Figure III.1-6c) it is oriented almost in parallel to COO^-. This trend is also observed for other complexes with ligands with $n_{ortho} = 2$ (ligands **1.4**, **1.8**) and $n_{ortho} = 1$ (ligands **1.2, 1.5, 1.6, 1.7, 1.9**) (Figure III.1-7). A likely reason for that is the different steric repulsion,[150] meaning that the lower steric hindrance within the ligands with $n_{ortho} = 1$ allows the complexes to be more flexible and might further stabilize them in solution.

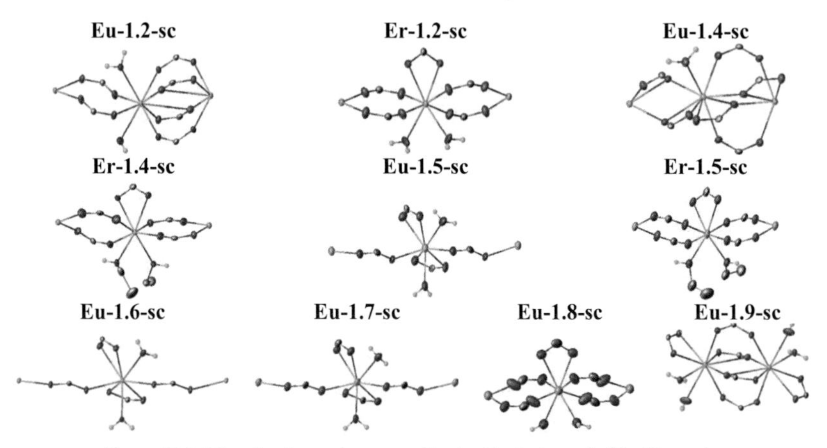

Figure III.1-5 Coordination environment of lanthanides in the studied **Ln-N** complexes.

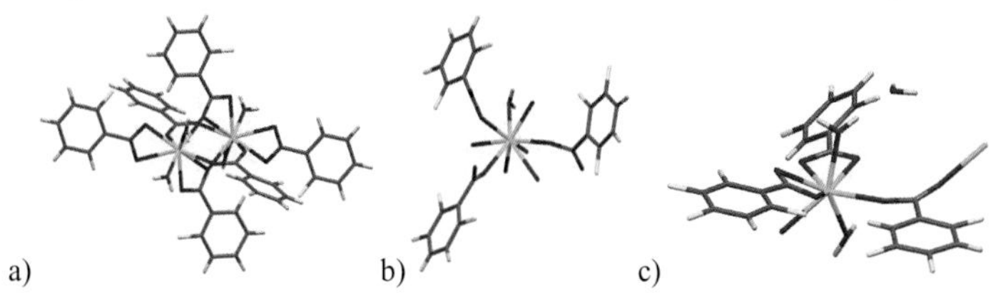

Figure III.1-6 Molecular structures of a) **Eu-1.9-sc**; b) **Eu-1.4-sc**; c) **Eu-1.5-sc**.
Atom assignment: C-grey, O-red, F-lime green, H-white, Eu-cyan.

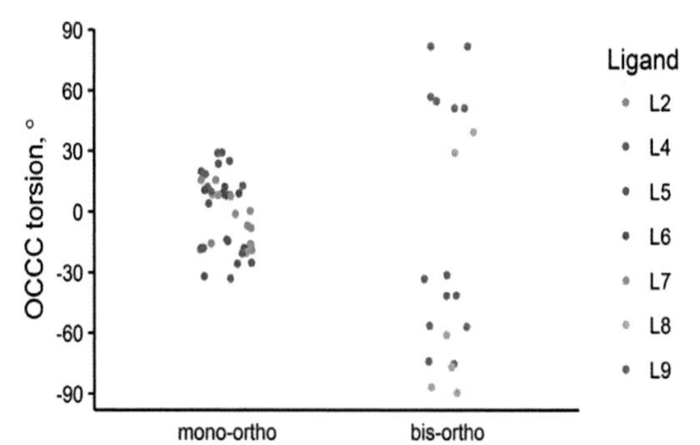

Figure III.1-7 OCCC torsion angles (< 90°) in structures of all studied complexes.
Small jitter was added along the x-axis to reduce overplotting*.

The analysis of PXRD data revealed the importance of the n_{ortho} values. During the study of phase composition of the obtained powders, it turned out that the phase pure compounds were rarely

* CSD analysis was performed together with Dr. I.S. BUSHMARINOV, A. N. Nesmeyanov Institute of Organoelement Compounds, Moscow, Russia.

formed, which was likely due to high acidity of fluorobenzoic acids, leading to high lability of the formed complexes. Despite this, the phase composition of each lanthanide fluorobenzoate is reproducible, leading to a possibility to conduct a detailed phase analysis of powders in order to reveal the general trends.

Given that the phase-pure powder is formed, the single-crystal structure is not necessarily the prevalent one in corresponding powder structure. For instance, the **Gd-1.2** structure which is virtually isomorphous (up to a fluorine atom) to **Eu-1.4-sc** (Figure III.1-6b) was found to contribute to powder patterns of most complexes and is considered to be one of the most stable phases for lanthanide fluorobenzoates. This phase was also found in some **Ln-1.1**[139] complexes, however, its major contribution took place in the complexes with $n_{ortho} = 2$. Virtually isomorphous **Eu-1.5-sc**, **Eu-1.6-sc** and **Eu-1.7-sc** demonstrate another major contribution to the powder structures. These compounds were different, not only in terms of chemical composition, but also in terms of symmetry. Furthermore, the belonging to a particular crystal system allowed to distinguish the found phases between **Eu-1.5-sc**-like and **Eu-1.6-sc**-like. However, this difference is negligible, allowing us to consider **Eu-1.5-sc** as the second most stable phase (Figure III.1-6c). It is worth to note that this phase is usually more stable in the complexes with $n_{ortho} = 1$.

Thus, the analysis of the crystal structures revealed the most stable and reproducible structural motifs, depending on the number of *ortho*-fluorine substituents, which has a predictive ability and can be extended to other fluorobenzoates.

NMR spectroscopy and BLEANEY's theory implementation

Despite the difficulties in the use of BLEANEY's theory for the characterization of Ln-based compounds in solutions, described in Chapter I.2.2, this theory seems to be very useful in the present work. The BLEANEY's theory indicates that the chemical shift for the whole set of lanthanide complexes with a selected ligand can serve as a reliable indicator of the similarity in their coordination environment. This similarity is validated by a linear dependence $\delta(D)$ of the chemical shift δ vs. the dipolar constant (or D-factor) which is not dependent on the local environment G_i and is an intrinsic feature of lanthanide ions (Chapter I.2.2).

Since lanthanide pentafluorobenzoates **Ln-1.1** do not have protons to investigate their ^{1}H NMR spectra, the BLEANEY's theory was applied to ^{19}F NMR spectra, which has never been done before. For that 0.01 M solutions of **Ln-1.1** in deuterium oxide-d_2 and methanol-d_4 were analyzed in comparison with those of Hpfb. As expected, a signal ratio of 2:1:2 was observed for all the complexes in D_2O.

All signals belong to fluorine atoms in the aromatic system and no additional signals were detected. However, we could not observe any paramagnetic ^{19}F NMR shift: ^{19}F NMR signals in all the complexes were completely unaffected by changes of the cation (Figure III.1-8b, Table III.1-2). The ligand exchange in lanthanide complexes is often extremely fast,[151] but the the averaged chemical shifts of fluorine atoms in coordinated ligands should still be affected by the paramagnetic Ln ion. Thus, the amount of $(pfb)^-$ bound to Ln is negligible in relatively diluted 0.01 M D_2O solutions. As the use of a non-aqueous solvent should increase the thermodynamic stability of the complexes,[152,153] the same experiment was run, using methanol-d_4 solutions with the same concentration. Indeed, in methanol-d_4 the dependence of the chemical shift on D-factor was observed due to the influence of the paramagnetic Ln ion (Figure III.1-8c, Table III.1-2).

The δ(D) dependencies were examined for all three fluorine signals (Figure III.1-8c) and appeared to be linear, witnessing the same compositions of pentafluorobenzoates for the whole lanthanide row in methanol-d_4 solution. Since the dipolar shift depends strongly on the distance between the nucleus of interest and the paramagnetic nucleus ($\delta_d \sim r^{-3}$),[62,65] the *o*-substituent signal is mostly affected by the presence of a paramagnetic ion. This corresponds to the highest slope of the δ(D) dependence for *o*-fluorine signals (Figure III.1-8c, Figure III.1-9) in comparison with *m*- and *p*-fluorine signals. Therefore, one can conclude that:

(i) the BLEANEY's theory can be applied for the analysis of ^{19}F NMR spectra of Ln complexes;
(ii) **Ln-1.1** local structures in methanol solutions are almost independent on the lanthanide. This result was also confirmed by X-ray absorption spectroscopy in the EXAFS-region, which will be discussed later.

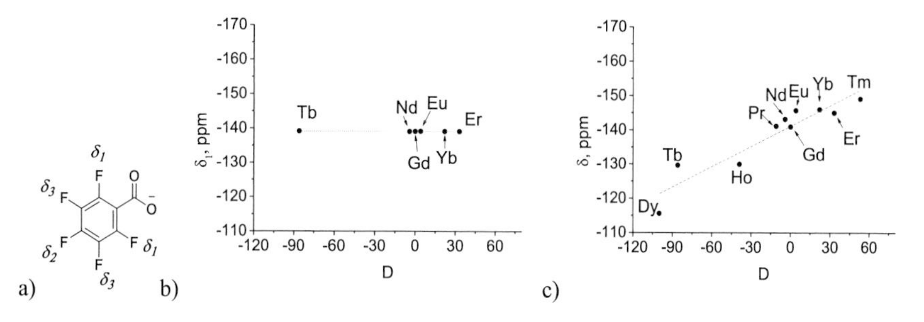

Figure III.1-8 a) Structure of pentafluorobenzoate anion with the assigned ^{19}F NMR chemical shifts and δ(D) dependencies for: b) *ortho*-fluorine ^{19}F NMR signals (δ_1) of 0.01 M D_2O solution of **Ln-1.1**; c) *ortho*-fluorine ^{19}F NMR signals (δ_1) of a 0.01 M solution of **Ln-1.1** in methanol-d_4.

Table III.1-2 ^{19}F NMR chemical shifts and dipolar constant D for **Ln-1.1** in CD_3OD and **Ln-1.1** in D_2O.

Compound	**D**[¥]	**^{19}F NMR shifts for Ln-1.1 in CD_3OD**			**^{19}F NMR shifts for Ln-1.1 in D_2O**		
Ln=		δ_1	δ_2	δ_3	δ_1	δ_2	δ_3
Pr	-11	141.14	157.14	163.82			
Nd	-4.2	143.16	157.66	164.59	139.03	154.69	162.35
Sm	n/a	142.90	157.01	164.93			
Eu	4	145.68	159.37	167.71	139.11	154.74	162.45
Gd	0	140.93	157.53	166.68	139.07	154.67	162.40
Tb	-86	129.62	153.62	161.80	139.14	154.77	162.48
(Tb/Eu)	-82	135.74	155.50	163.44			
Dy	-100	115.56	149.58	156.48			
Ho	-39	129.93	153.26	161.10			
Er	33	145.00	158.25	166.57	139.07	154.63	162.34
Tm	53	149.09	159.77	168.20			
Yb	22	145.99	158.21	166.30	139.06	154.70	162.40
Lu	n/a	143.17	157.14	165.11	139.05	154.68	162.39
"free" ligand							
Hpfb	n/a	141.77	153.60	164.28	140.54	150.94	161.43

[¥] Values of the dipolar constant D are taken from[64].

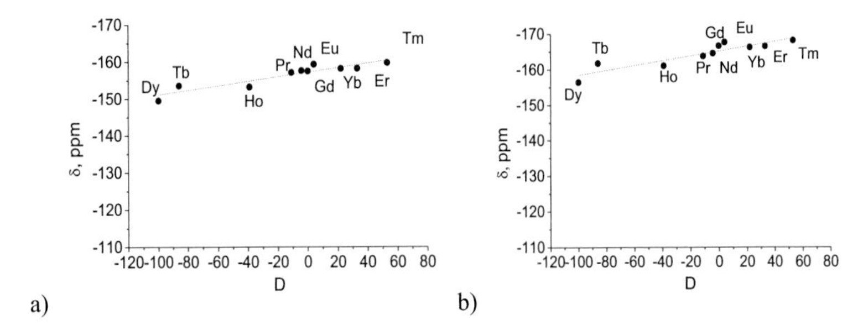

Figure III.1-9 Dependence of chemical shift on D-factor for: a) *para*-fluorine signals (δ_2, Figure III.1-8a) of **Ln-1.1** in methanol-d_4; b) *meta*-fluorine signals (δ_3, Figure III.1-8a) of **Ln-1.1** in methanol-d_4.

The BLEANEY's theory was additionally applied to other fluorobenzoates, however, in this case ^{1}H NMR was applicable, although often due to the strong signal broadening the evaluation of NMR spectra was difficult, especially for Gd, Tb, Er, Nd complexes*. ^{1}H and ^{19}F NMR spectra were recorded for all synthesized lanthanide complexes, DMSO-d_6 was used as a solvent, to ensure the solubility of all the compounds. In order to evaluate the similarities in the Ln ion coordination environments for DMSO-d_6 solutions, the BLEANEY's theory was applied for both, ^{1}H and ^{19}F NMR spectra.

---

* This effect is discussed in the current study (Chapter I.2.2).

However, in the case of compounds **Ln-1.3**, the same δ(D) dependency as with aqueous pentafluorobenzoates **Ln-1.1** was observed, demonstrated a very little slope which indicated the dissociation of these compounds. On one hand, the zero-slope of the linear dependences δ vs. D obtained from ^{19}F NMR spectra for **Ln-1.3** reveals the insensitivity of the chemical shift towards the lanthanide used and hence, indicates that ligands are not coordinated by lanthanide ions. Besides, relatively low acidity constants of other ligands did not promote their dissociation in DMSO, as seen by plotting δ vs. D graphs for all the other complexes. The slope of the curves is dependent on the position of fluorine atoms, which was monitored due to a variation in the distances between different fluorine atoms and paramagnetic lanthanide ion. Therefore, BLEANEY'S theory also proved to be a useful tool for the signal assignment in ^{19}F NMR spectra, where otherwise an unambiguous ascription is hindered.

^{1}H NMR data for complexes **Ln-1.9** could not be sufficiently analysed due to the large broadening of the signals of three protons in the ^{1}H NMR spectra. However, the signal of the last proton, showing up at 6.74 – 8.72 ppm, was detectable, for which δ(D) dependence turned out to be linear. These data were further augmented by ^{19}F NMR spectra, which yielded a linear dependence with a relatively high slope, though additional phase correction for **Gd-1.9**, **Yb-1.9** and **Er-1.9** was needed. Moreover, for complex **Eu-1.9**, the dependence of the chemical shift on the concentration was followed in order to determine if the dissociation occurs in complexes at a certain concentration (Figure III.1-10).

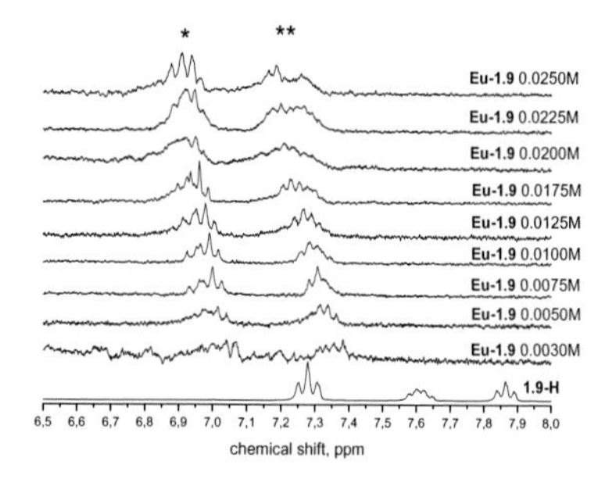

a)

Sample conc, M	Peak **, ppm	Peak *, ppm
"free ligand"	7.86, 7.60	7.28
0.0030	7.33	7.04
0.0050	7.32	7.01
0.0075	7.30	6.99
0.0100	7.28	6.99
0.0125	7.27	6.98
0.0175	7.23	6.96
0.0200	7.21	6.95
0.0225	7.20	6.95
0.0250	7.18	6.93

b)

Figure III.1-10 a) ^{1}H NMR spectra for solutions of **Eu-1.9** in D_2O at different concentrations; b) values of the chemical shifts for **Eu-1.9** solution in D_2O at different concentrations in comparison with **1.9-H**.

Photophysical properties

The phase composition of each particular lanthanide fluorobenzoate was constant after each synthesis, which made a meaningful determination of their luminescent properties possible, even though most of the powders were not phase-pure. Notably, **Eu-1.3** demonstrated no luminescence, therefore, ligand **1.3** was excluded from further consideration. We assign the absence of **Ln-1.3** luminescence to a non-radiative luminescence quenching by NO_2-group.[154,155] This might be due to the appearance of charge-transfer state which should be of intraligand origin, because the quenching is observed for all complexes **Ln-1.3** where Ln = La, Nd, Eu, Gd, Tb, Er, Yb, Lu.

The luminescence properties of lanthanide fluorobenzoates are barely found in literature: only the emission of **Eu-1.1**, **Tb-1.1**[145] and **Er-1.1**[19] was demonstrated, though no detailed research was performed. To start the systematical examination of luminescent properties of lanthanide fluorobenzoates, firstly, the triplet state energy level was estimated for gadolinium complex **δ-Gd-1.1**. In the low temperature spectrum (77 K), a short-wavelength band with a maximum at 410 nm was observed additionally to the band with a maximum at 490 nm (Figure III.1-11a). It was assumed that the band at 490 nm (left edge at 455 nm), observed at both temperatures, corresponds to phosphorescence, while the fluorescence at 410 nm only occurs at low temperature. Such a high efficiency of inter-system crossing (ISC) in gadolinium complexes at room temperature is unusual but seems plausible since the STOKES' shift between the emission band at 490 nm and the absorption band at 260 nm is too high (Figure III.1-11b). To prove that the luminescence lifetime was recorded at room temperature and appeared to be 10.2(4) μsec, proving the aforementioned statement (Figure III.1-11b, insertion). The luminescence spectrum, measured with a 1.0 μsec delay (Figure III.1-11a), also showed the band at 490 nm. The triplet state value calculated from the deconvoluted phosphorescence spectrum is 22000 cm^{-1} (Figure III.1-11a, insertion) which falls within the gap suitable for Eu^{3+} sensitization and is slightly higher than Tb^{3+} resonance level.

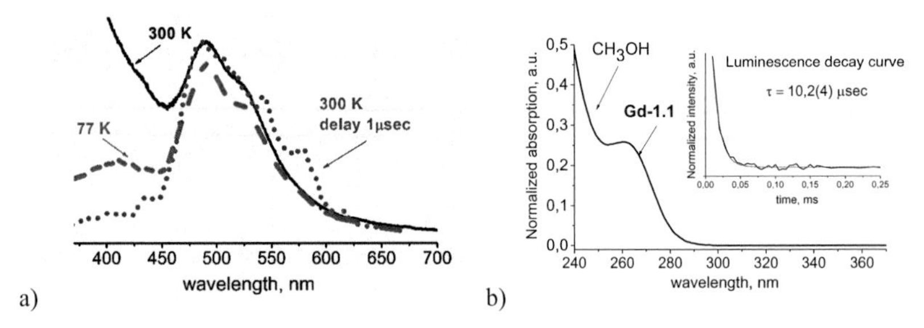

Figure III.1-11 a) Steady state luminescence spectra (at 77 K and 300 K) of **δ-Gd-1.1** with λ_{ex} = 337 nm; insertion: the deconvoluted luminescence spectra of **δ-Gd-1.1**; b) absorption spectrum of **δ-Gd-1.1** in methanol solution, insertion: luminescence lifetime of **δ-Gd-1.1**.

The triplet levels of other ligands **1.2, 1.4 – 1.9** turned out to be roughly the same (ca. 20500 cm^{-1}), independent on the nature, number and position of substituents in the benzene ring. However, the luminescent properties of complexes with emitting lanthanides are quite different. Quantum yield values (Q^{L}_{Ln}) vary from 10% (**Tb-1.7**) to 70% (**Tb-1.4**) for terbium complexes and from 5%

(**Eu-1.6**) to 40% (**Eu-1.2**) for Eu complexes (Table III.1-3). The Eu(III) luminescence spectroscopy[156] was applied to determine, which of the following processes leads to rather low Q_{Ln}^{L} values in the case of Eu fluorobenzoates:

(i) low ligand-to-metal transfer efficiency (η_{sens}), or

(ii) low efficiency of f-f transitions within the metal itself (Q_{Ln}^{Ln}), according to equation I.1 $Q_{Ln}^{L} = \eta_{sens} \cdot Q_{Ln}^{Ln}$, where Q_{Ln}^{L} is the total PLQY, η_{sens} is the efficiency of the energy transfer from the ligand to the lanthanide and Q_{Ln}^{Ln} is the through-metal quantum yield which determined the efficiency of the f-f energy transfer.

Table III.1-3 Quantum yields* ($Q_{Ln}^{L} \pm 3$, %) and lifetimes ($\tau \pm 0.1$, ms) of visible emitters.[a]

L	Tb		Eu				$Tb_{0.5}Eu_{0.5}$				
	τ, ms	Q_{Ln}^{L}, %	τ, ms	η_{sens}	Q_{Ln}^{L}, %	Q_{Eu}^{Eu}	τ_{545}, ms	τ_{612}, ms	η_{sens}	Q_{Ln}^{L}, %	Q_{Eu}^{Tb}
1	1.40	40	0.70	0.5	15	30		0.90	1.0	30	29
2	2.10	35	1.00	0.9	40	45		0.70	0.6	20	30
4	1.30	70	0.70	0.6	20	30		0.52	0.3	10	31
5	0.74	55	0.29	0.5	10	25	0.47	0.34	n/a	15	n/a
6	0.62	40	0.29	0.2	5	25	0.51	0.41	n/a	25	n/a
7	0.66	10	1.25	0.2	10	75	0.53	0.44	n/a	10	n/a
8	1.24	30	0.42	0.4	10	15		0.75	0.3	15	30
9	1.03	50	0.52	0.6	15	25	0.75	0. 52	0.6	20	30

[a] Given accuracy, corresponding to the reproducibility, was estimated for a multiple-assay experiment. However, according to H. Ishida *et al.*[157], the errors might be on the order of 10% for Q_{Ln}^{L} and of 2-3% for τ.

The impact of these two factors was evaluated on the example of **Ln-1.1** series. For instance, low intrinsic quantum yield Q_{Ln}^{Ln} may be caused by quenching *via* coordinated water molecules (one molecule per Eu(III) ion in **δ-Eu-1.1**). To eliminate this reason, the deuterated complex **δ-Eu-1.1d** with the composition $Eu(pfb)_3(D_2O)$ was synthesized and its photophysical characteristics were determined. Indeed, deuteration led to the rise of radiative lifetime ($\tau_{obs} = \tau_R$), resulting in an increase of through-metal quantum efficiency up to Q_{Ln}^{Ln} = 100% which was determined by the ratio between observed and radiative lifetimes (see equation III.2 later in this chapter). However, the overall through-ligand quantum efficiency Q_{Ln}^{L} did not change, witnessing that the energy transfer efficiency is still rather low (Table III.1-4). Moreover, a decrease in the sensitization efficiency upon deuteration was observed, which may be attributed to the saturation of the ligand to lanthanide energy transfer pathway.

To increase the energy transfer efficiency, the sensitization path $(pfb)^{-} \rightarrow Tb^{3+} \rightarrow Eu^{3+}$ was introduced. For this purpose the bimetallic complex **δ-(TbEu)-1.1** with the ratio Tb:Eu = 1:1 was synthesized. The efficiency of such an approach has already been successfully demonstrated for

* PLQY and τ values were measured together with Dr. D. Volz, Cynora GmbH, Bruchsal, Germany.

both, inorganic systems and complexes with organic ligands,[158] but has not been widely used so far. The spectrum of **δ-(TbEu)-1.1** almost exclusively shows typical europium luminescence bands with negligible traces of terbium luminescence (Figure III.1-12a). The lifetime of europium $^5D_0$ excited state coincided for both, **δ-Eu-1.1** and **δ-(TbEu)-1.1**, while the lifetime of the excited state, measured at the maximum of Tb-based luminescence (λ_{em} = 545 nm) significantly decreased (2.00 ± 0.05 ms for **δ-Tb-1.1**, 0.12 ± 0.05 ms for **δ-(TbEu)-1.1**), which proves almost complete $Tb^{3+} \rightarrow Eu^{3+}$ energy transfer. The integral intensity ratio and therefore, radiative lifetime as well as through-metal quantum efficiency also remained the same because of the identity of coordination environments in these compounds (Table III.1-4). Nevertheless, almost twofold increase of Q_{Ln}^{L} up to 28.5% was observed due to the improved efficiency of the energy transfer. Single crystals of **γ-Tb-1.1-sc**, **α-Eu-1.1-sc** and **α-(TbEu)-1.1-sc** differ from the powders of the corresponding **δ-Ln-1.1** by hydrate composition (one water molecule per one metal ion in **δ-Ln-1.1** powders, four water molecules per one metal ion in **α-Ln-1.1-sc** crystals and five water molecules per one metal ion in **γ-Tb-1.1-sc**) and consequently, demonstrate very low luminescence intensity due to the efficient quenching of Eu-based luminescence by coordinated water molecules.

High sensitivity of luminescence to coordinated water molecules is in line with the dependence of luminescence on deuteration in powders and was confirmed by luminescence quenching in aqueous solutions. Indeed, the substitution of H_2O with D_2O led to a tenfold increase of ionic luminescence intensity of 0.07 M D_2O solution of **δ-Eu-1.1**$_d$ comparing to 0.07 M H_2O solution of **δ-Eu-1.1** (Figure III.1-12b). Another reason of luminescence intensity decrease in aqueous solution comparing to luminescence in powders is the dissociation of the complex, leading to a decrease of antenna effect, which is witnessed by the appearance of the ligand luminescence at 410 nm. Additionally for aqueous solution of europium complexes, the fluorescence band at 410 nm is observable at room temperature, which is in line with the assignement of luminescence bands for **δ-Gd-1.1** complex (410 nm – fluorescence, 490 nm – phosphorescence) (Figure III.1-11a).

Nevertheless, ionic luminescence of aqueous solutions of **Eu-1.1** and **Tb-1.1** is still 3 – 4 times more intensive than that of the corresponding chlorides (Figure III.1-12b, insertion, Figure III.1-12c), witnessing the presence of through-ligand sensitization and thus, incomplete dissociation, which is in line with EXAFS data. Different asymmetry ratios (r_{asymm}) of 0.07 M aqueous solution of $EuCl_3$ and 0.07 M aqueous **Eu-1.1**, showing the ratio between the integrated intensities of $^5D_0 \rightarrow {}^7F_1$ and $^5D_0 \rightarrow {}^7F_2$ bands, is also an evidence of the difference in coordination environments of aqueous solutions of Eu chlorides and Eu pentafluorobenzoates (Table III.1-4, Figure III.1-12b, insertion).

Table III.1-4 Observed (τ_{obs}) and calculated (τ_{rad}) lifetimes of europium $^5D_0$ excited state and the values of quantum yield upon excitation through-ligand (Q^L_{Ln}) and through-metal (Q^{Ln}_{Ln}) of europium-containing samples in comparison to aqueous $EuCl_3$. The concentration of aqueous solutions is 0.07 M.

State	Compound	T, K	λ_{ex}, nm	$r_{asymm} \pm 0.2$	$\tau_{obs} \pm 0.05$, ms	$\tau_{rad} \pm 0.07$, ms	$Q^L_{Ln} \pm 1, \%$	$Q^{Ln}_{Ln} \pm 10, \%$
Powder	**δ-Eu-1.1**	300	280	5.8	0.65	1.00	15.0	65
	δ-Eu-1.1$_d$	300	280	5.9	1.20	1.00	15.0	≈100
	δ-Eu-1.1	77	280	5.3	0.77	0.80	15.2	88
	δ-Eu-1.1$_d$	77	280	5.6	1.25	0.90	15.2	≈100
	δ-(TbEu)-1.1	300	280	5.6	0.67	0.90	28.5	75
Single crystal	**α-Eu-1.1-sc**	300	337	3.2	<0.15	0.52	**	<30
	α-(TbEu)-1.1-sc	300	337	2.5	<0.15	0.41	**	<40
Aqueous solution	**Eu-1.1** in H_2O	300	290	1.0	<0.15	0.34	**	<40
	Eu-1.1$_d$ in D_2O	300	290	1.1	<0.15	0.36	**	<40
	(TbEu)-1.1 in H_2O	300	290	0.9	<0.15	0.26	**	n/a
	$EuCl_3$ in H_2O	300	290	0.8	0.11	0.35	**	31

** not measured

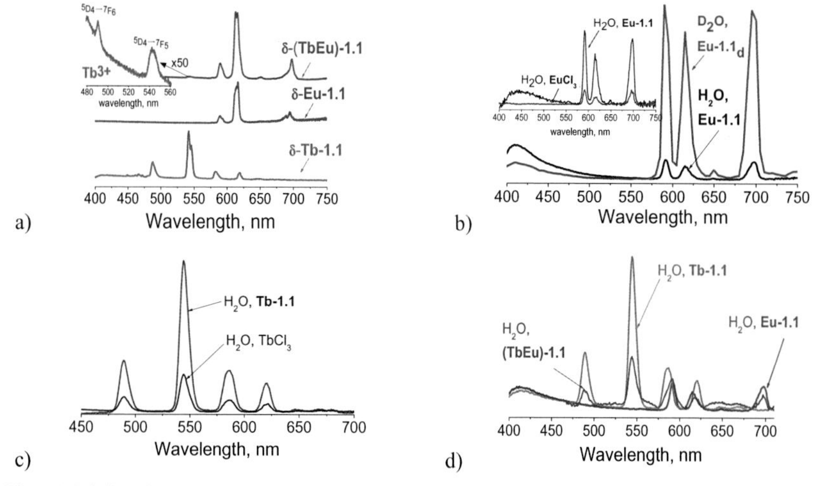

Figure III.1-12 a) Corrected emission spectra of **δ-Eu-1.1**, **δ-Tb-1.1** and **δ-(TbEu)-1.1** with $\lambda_{ex} = 280$ nm at 300 K; b) the comparison of luminescence spectra of aqueous **Eu-1.1**, D_2O solution of **Eu-1.1$_d$**, instertion: the spectrum of aqueous **Eu-1.1** compared to aqueous **$EuCl_3$**, $\lambda_{ex} = 280$ nm; c) luminescence spectrum of aqueous **Tb-1.1** compared to aqueous **$TbCl_3$**, $\lambda_{ex} = 280$ nm; d) comparison of luminescence spectra of aqueous **Eu-1.1**, aqueous **Tb-1.1** and aqueous **(TbEu)-1.1**. Rather concentrated aqueous solutions (0.07 M) were used for the measurements in b)-d).

In a same way, the overall quantum yield (Q^L_{Ln}) for other fluorobenzoates may be increased by clarifying which process is responsible for its decrease. Table III.1-3 demonstrates that **Eu-1.7** possess the highest Q^{Ln}_{Ln} value, although, according to the TGA data, its powder structure contains two inner-sphere H_2O molecules in contrast to the complexes with one inner-sphere H_2O molecule (**Eu-1.1 – Eu-1.4** and **Eu-1.8**). Meanwhile, in the case of all the compounds from **Ln-1.7** series, the powder structure is unknown and does not contain the corresponding single-crystal structure. This leads to the fact that we cannot relate the high luminescence performance of **Eu-1.7** compound with structural data. The limiting factor for the total luminescence process is ligand-to-metal charge transfer η_{sens}, with a notable exception of **Eu-1.7**, while almost all complexes possess low η_{sens} (except of **Eu-1.2**, Table III.1-3). Employing a proven above effective approach on the example of **Ln-1.1**,[139] heterometallic complexes were synthesized to study the possibility of increasing the luminescence efficiency of Eu(III) complexes by a two-step sensitization path $L \rightarrow Tb^{3+} \rightarrow Eu^{3+}$,[31,158–160] where Ln = $Tb_{0.5}Eu_{0.5}$. The powders of heterometallic complexes have the same phase composition as the corresponding Tb and Eu complexes, which was proved by PXRD data. The isotope distribution in such complexes was studied by means of high resolution mass spectra for dimeric fragments and was shown to be within the statistical error as expected Eu_2:TbEu:Tb_2 equals to 1:2:1 (Figure III.1-13).

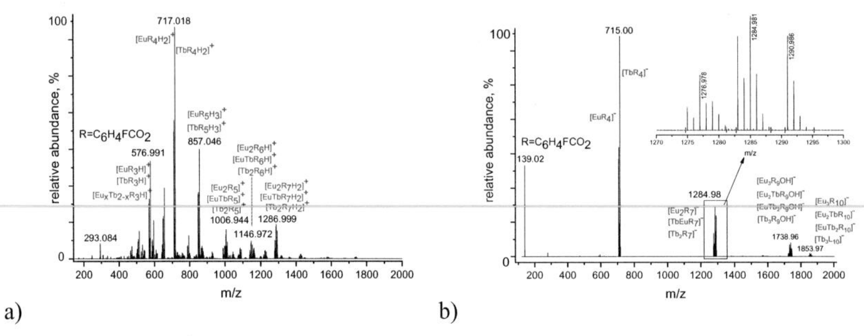

a) b)

Figure III.1-13 ESI-MS* data for heterometallic complexes: a) positive mode; b) negative mode, insertion: integration of the isotope distributions of $[Eu_2R_7]^-$, $[EuTbR_7]^-$ and $[Tb_2R_7]^-$ units for the range 1270 to 1300.

A high efficiency of $Tb^{3+} \rightarrow Eu^{3+}$ energy transfer could be observed in almost all heterometallic complexes, where predominantly Eu(III)-based luminescence was observed. This can be explained by the multiplicity of this transition: in contrast to the competitive process of Tb^{3+} radiative relaxation ($^5D_4 \rightarrow {}^7F_J$), the transition Tb^{3+} ($^5D_4$) $\rightarrow$ Eu^{3+} ($^5D_0$) do not require the change of spin. Therefore, it is supposed to be faster and would more likely occur, though the probability would highly depend on the distance between Tb^{3+} and Eu^{3+} ions. Only the compound **(TbEu)-1.6** was an exception and exhibited both, Tb(III) and Eu(III)-based orange luminescence, where both ions radiated equally (Figure III.1-14a). Generally, Tb^{3+} possesses a relatively short lifetime and quite high Q_{Ln}^{L} and therefore, it may emit quickly and effectively, indicating that the $Tb^{3+} \rightarrow Eu^{3+}$ energy transfer hardly occurs in this case. The 1:1 ratio of both metals in the heterometallic complex explains why the Q_{Ln}^{L} value of 25% in **(TbEu)-1.6** is roughly the half of **Eu-1.6** Q_{Ln}^{L} plus the half of **Tb-1.6** Q_{Ln}^{L}.

* ESI-MS was performed by Dr. P. WEIS, Institute of Physical Chemistry, Karlsruhe Institute of Technology, Karlsruhe, Germany.

The substitution of the half of the europium ions with terbium ions led to roughly 50% decrease of the quantum yield of Eu^{3+} luminescence in heterometallic complexes with ligands **1.2**, **1.4** and **1.7**. The sensitization strategy L→Tb^{3+}→Eu^{3+} does not apply to these complexes due to the fact that the partial substitution of Eu^{3+} ions by Tb^{3+} only decreases the amount of the emitting centers (Eu^{3+}) and does not affect the energy transfer efficiency. The increase of the efficiency η_{sens}, in heterometallic complexes should be more than twice that of the monometallic one to cover the difference in the number of europium ions.

Heterometallic complexes with other ligands **1.1**, **1.5**, **1.8**, **1.9** showed an increase of the Q_{Eu}^{L} in comparison with those for homometallic Eu(III) complexes (Table III.1-5). A key to the explanation of this different behaviour is the ligand-to-metal energy transfer efficiency. Indeed, introducing terbium ions and forming the heterometallic complex in some cases led to an increase in the overall ligand-to-europium energy transfer (complexes with **1.1**, **1.8** and **1.9**; Figure III.1-14b), while in other cases it decreased (complexes with **1.2** and **1.4**).

Due to the overlap of terbium and europium emission bands, it was impossible to calculate the energy transfer efficiency for complexes with **1.5**, **1.6** and **1.7**. However, for some heterometallic lanthanide complexes, the metal-to-metal (η_{ET}) energy transfer efficiency was calculated.[18,161,162] Tb^{3+} is considered as energy donor while Eu^{3+} as energy acceptor, assuming that the energy transfer in heterometallic complexes goes *via* the path L→Tb^{3+}→Eu^{3+}. Given this assumption we calculated the efficiency of the Tb^{3+}→Eu^{3+} energy transfer by measuring the lifetimes of the donor in the presence of the acceptor (τ_{obs} of Tb^{3+} in heterometallic complexes) $\tau_{het}(Tb^{3+})$ and in its absence (τ_{obs} of Tb^{3+} in Tb complex) $\tau_{hom}(Tb^{3+})$[18]:

$$\eta_{ET} = 1 - \frac{\tau_{het}(Tb^{3+})}{\tau_{hom}(Tb^{3+})} \quad \text{(eq. III.1)}$$

Due to the fact that most heterometallic complexes possess only one-metal-centred luminescence, this calculation was possible only in the case of complexes with ligands **1.5**, **1.6**, **1.7** and **1.9**, where the emission spectra contain both, terbium and europium luminescence bands. In all the heterometallic complexes, the Tb^{3+} lifetime decreased compared to the respective homometallic terbium complexes, which points to the presence of an additional sensitization pathway *via* Eu^{3+} channel.[163,164] The growth of Eu^{3+} luminescence quantum efficiency in **(TbEu)-1.5** and **(TbEu)-1.9** compared to **Eu-1.5** and **Eu-1.9**, respectively, might occur due to the high Tb^{3+}→Eu^{3+} energy transfer efficiency (Table III.1-5, Figure III.1-14b).[163] On top of that, an additional sensitization of Eu^{3+} is probably due to the influence of highly paramagnetic Tb^{3+}, whose presence makes the energy transfer S_1→T_1 less forbidden.[139]

Table III.1-5 Values of τ_{het}, τ_{hom} and η_{ET} for heterometallic Tb/Eu complexes with ligands **1.5**, **1.6**, **1.7** and **1.9**. The arrows indicate an increase/decrease of the Q_{Ln}^{L} compared to Eu^{3+} homometallic complexes.

Ligand	τ_{het}, ms	τ_{hom}, ms	η_{ET}, ms	Q_{Ln}^{L}
1.5	0.47	0.74	0.36	↑
1.6	0.51	0.62	0.18	↓
1.7	0.53	0.66	0.20	↓
1.9	0.75	1.03	0.27	↑

Thus, the introduction of Tb^{3+} ion into europium complex may lead to either a decrease of the overall quantum yield (complexes with ligands **1.2**, **1.4** and **1.7**, Table III.1-3) or its increase, wherein it may be both, due to an additional Tb-centered luminescence (**(TbEu)-1.6**, Figure III.1-14a) or due to the prescence of Tb^{3+}→Eu^{3+} energy transfer pathway (complexes with ligands **1.1**, **1.5**, **1.8** and **1.9**, Figure III.1-14b).

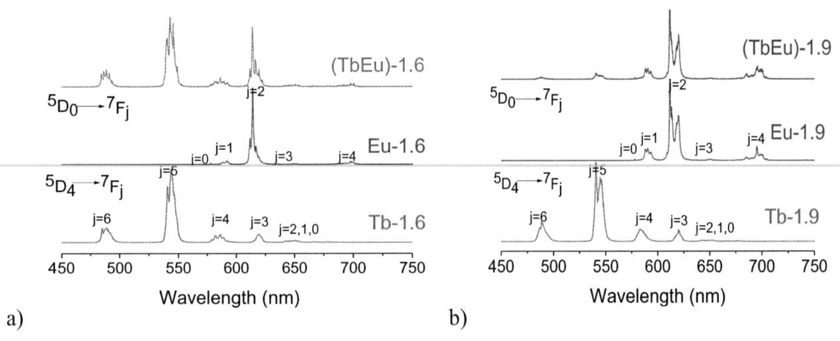

Figure III.1-14 Comparison of the luminescence spectra for terbium and europium complexes with heterometallic Tb/Eu complex: a) Tb^{3+} and Eu^{3+}-centered luminescence in complex (**TbEu)-1.6**; b) mostly solely Eu^{3+}-centered luminescence due to the Tb^{3+}→Eu^{3+} energy transfer in complex (**TbEu)-1.9**.

The luminescence intensity is different for all complexes with NIR-emitting lanthanides, except of those within **Ln-1.3** series. The characterization of the luminescence intensity was performed, paying attention on two factors, due to the difficulties with the measurement of quantum yields in the IR region. Firstly, the higher the lifetime of the excited state τ_{obs}, the higher the internal quantum efficiency, *i.e.* the Q_{Ln}^{Ln} upon through-metal excitation.[76,165]

$$Q_{Ln}^{Ln} = \tau_{obs}/\tau_{rad} \qquad \text{(eq. III.2)}$$

where τ_{rad} – radiative lifetime which is generally 0.8 – 1.0 ms for ytterbium ionic luminescence in coordination compounds. Secondly, the greater the ratio of the intensities of through-ligand excitation (250 – 320 nm) to through-metal excitation (set of narrow bands), the higher the efficiency of ligand-to-metal energy transfer η_{sens}. The evaluation of η_{sens} was carried out for neodymium complexes: for complexes **Nd-1.3** – **Nd-1.9** η_{sens} does not exceed 0.5%, whereas for **Nd-1.2** it reaches 8%.

Luminescence intensity studies

An essential characteristic of the luminescence intensity in photoluminescent compounds is brightness that is proportional to the number of emitted photons or its analog *luminosity* which can be defined usig the equation:[108,140,166,167]

$$L = \varepsilon \times Q_{Ln}^{Ln} \qquad \text{(eq. III.3)}$$

where ε is the molar extinction coefficient. Therefore, the design of new phosphorescent materials must be aimed not only at increasing their quantum yield, but also at enriching ligand absorption. Ligand molar extinction coefficients were determined from absorption spectra of diluted aqueous solutions of their potassium salts, assuming that they were equal to the absorption spectra of the free ligand ion (Figure III.1-15a) due to complete dissociation.

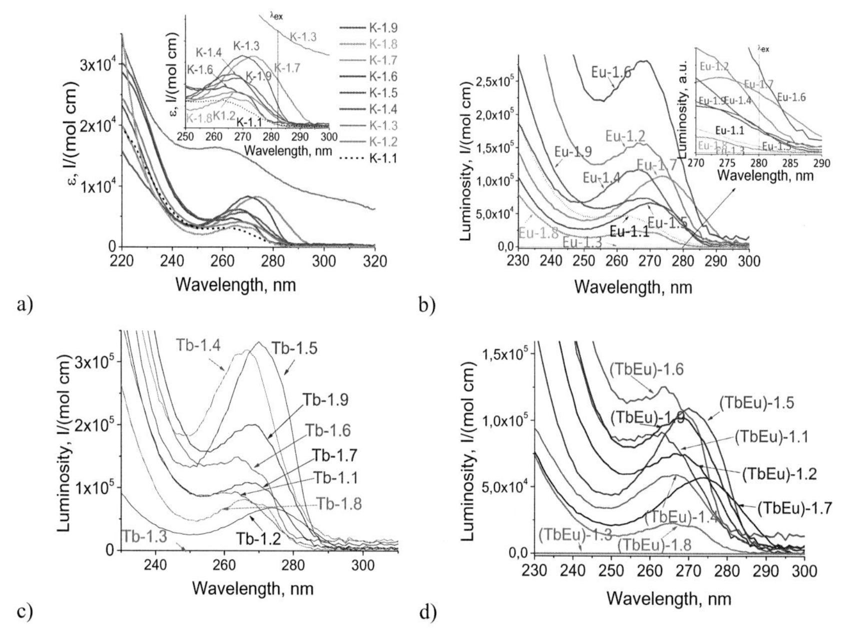

Figure III.1-15 a) Absorption spectra of **K-L**; b) luminosity of **Eu-L** (N = **1.1** – **1.9**); c) luminosity of **Tb-N** (N = **1.1** – **1.9**); d) luminosity of **(TbEu)-L** (L = **1.1** – **1.9**).

Ligand **1.3** has the highest absorption associated to the presence of the conjugated nitro group but its lanthanide complexes possess no luminescence and thus, the Q_{Ln}^{L} of **Eu-1.3** is equal to zero, leading to the luminosity value being also zero (Figure III.1-15a). At the same time, due to lower absorption, the luminosity of **Eu-1.1** (λ_{ex} = 280 nm) is lower than that of **Eu-1.9**, **Eu-1.5**, **Eu-1.7** and **Eu-1.6**, despite the Q_{Ln}^{L} of **Eu-1.1** being higher (Table III.1-5). Thus, the highest luminosity value under excitation of 280 nm was obtained for complexes with one *ortho*-substituent (**Eu-1.6**, **Eu-1.7** and **Eu-1.2**), while the lowest values were obtained for the complexes with two *ortho*-substituents, such as **Eu-1.3**, **Eu-1.1** and **Eu-1.8**. Given the same conditions, complexes **Eu-1.4**, **Eu-1.5** and **Eu-1.9** exhibited almost the same value of luminosity. The same calculations were performed for Tb(III) and heterometallic complexes (Figure III.1-15c,d). This result indicates that the complexes with n_{ortho} = 1 are indeed more promising for biological applications.

Solubility, ink-jet printing and thin film deposition

Given all factors mentioned above it might be expected that the thermodynamic stability of fluorobenzoate complexes decreases with an increase of the number of fluorine atoms.[139,168,169] The latter may also affect solubility, which generally follows the opposite trend (the more fluorinated the compound, the less it is soluble in water).[140] During the evaluation of the aqueous solubility of the lanthanide fluorobenzoates, it was discovered that their solubility indeed depends on the number and positions of fluorine atoms in the benzene ring, as well as the presence of other substituents (Figure III.1-16a).

Evaluation of the aqueous solubility was carried out for europium(III) complexes. Interestingly, carboxylates with the highest solubility turned out to be less thermodynamically stable, so that their high solubility was caused by dissociation of the complex, according to the NMR data. This dissociation occurred mostly in benzoates with two non-hydrogen *ortho*-substituents (**Eu-1.1**, **Eu-1.4** and **Eu-1.8**), while **Eu-1.3** turned out to be completely insoluble. The most promising candidates for applications in solution were complexes **Eu-1.6**, **Eu-1.7** and **Eu-1.9**, possessing solubility values of 15, 30 and 25 mmol/l at 25 °C, respectively.

Complexes **Tb-1.9** and **Eu-1.9** were tested as luminescent dyes for ink-jet printing due to their high solubility in organic solvents. Solutions of the aforementioned compounds in ethanol were placed inside the printer instead of its inks and the pictures were printed on a usual paper and then placed under UV lamp with the λ_{ex}=254 nm. Bright green and red luminescence was observed for both pictures, printed using Tb- and Eu-based dyes, correspondingly (Figure III.1-16b).

Pentafluorobenzoates **Ln-1.1**, however, tends to diccociate in solution and therefore, are not promising for biological applications (Figure III.1-16a). However, their high solubility not only in water (68.0 ± 2.1 mM), but also in common organic solvents (up to 0.2 M in ethanol/benzene mixture) makes them attractive for solid-state applications, such as ink-jet printing or those, where thin films are required.

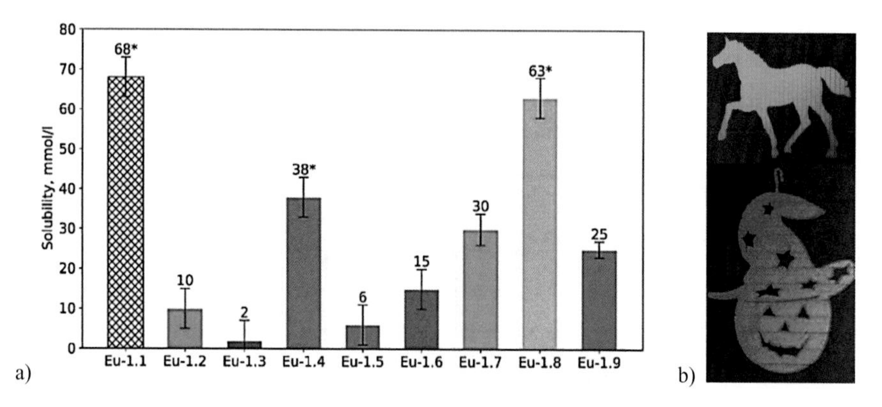

Figure III.1-16 a) Aqueous solubility of Eu(III) complexes with selected ligands. The star indicates the complexes, where high solubility value is caused by high dissociation degree; b) pictures, printed using green emitting complex **Tb-1.9** (top picture) and red-emitting complex **Eu-1.9** (bottom picture) as luminescent dyes.

Thin films of **δ-Tb-1.1**, **δ-Eu-1.1** and **δ-(TbEu)-1.1** were deposited from the corresponding solutions in water, methanol, ethanol, chloroform, acetone, benzene and ethanol:benzene (1:1) mixture. The uniform films with the roughness of 3 nm (Figure III.1-17) were only obtained when deposited from ethanol:benzene mixture. The luminescence spectra of the obtained thin films are shown in Figure III.1-18 and turned out to be identical with the spectra of the **δ-Tb-1.1**, **δ-Eu-1.1**, **δ-(TbEu)-1.1** powders (Figure III.1-12a).

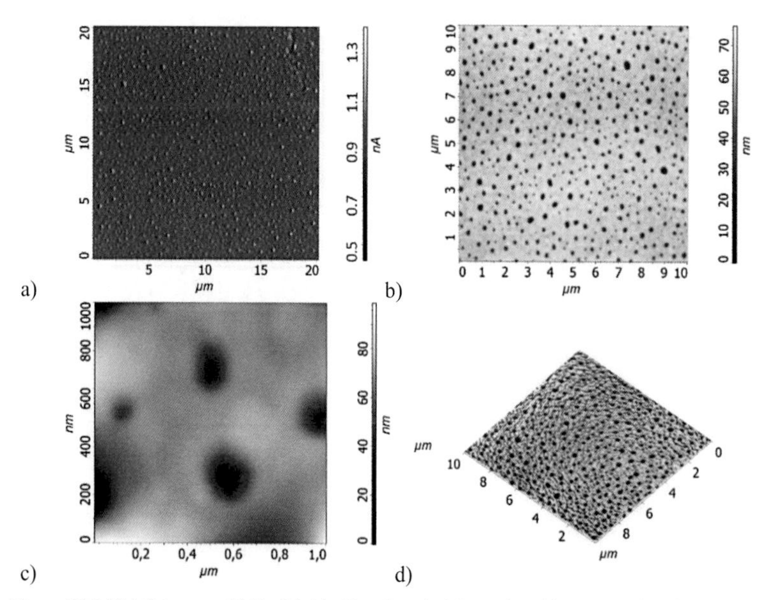

Figure III.1-17 AFM scans of **δ-Eu-1.1** thin films deposited from ethanol:benzene (1:1) solution: a) 20x20 μm, Sa = 3 nm; b) 10x10 μm; c) 1x1 μm; d) 3D view of the 10x10 μm scan.

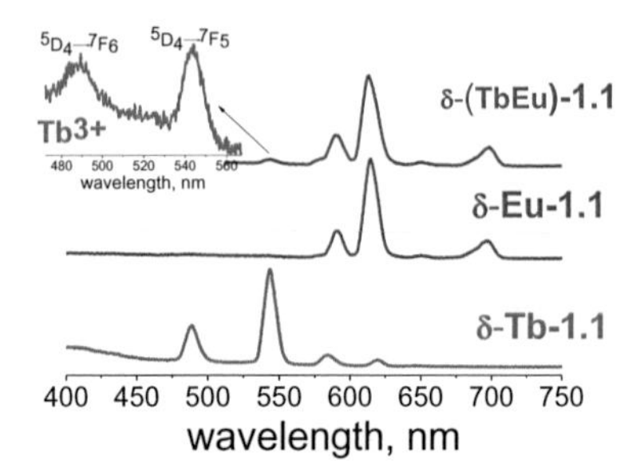

Figure III.1-18 Luminescence spectra of **δ-Tb-1.1**, **δ-Eu-1.1**, **δ-(TbEu)-1.1** in the thin film state, deposited from ethanol:benzene solution (1:1), λ_{ex} = 280 nm, room temperature.

Luminescence of visible emitters in aqueous solutions and cellular experiments

The behaviour of the studied complexes Eu^{3+}, Tb^{3+} and heterometallic Tb/Eu fluorobenzoates *in cellulo,* is proxied by the luminescence of the studied complexes in aqueous solution. By considering their luminescence at concentration of 3 mM, their applicability for ink-jet printing and bioimaging was estimated. Figure III.1-19 summarizes the results for Eu complexes.

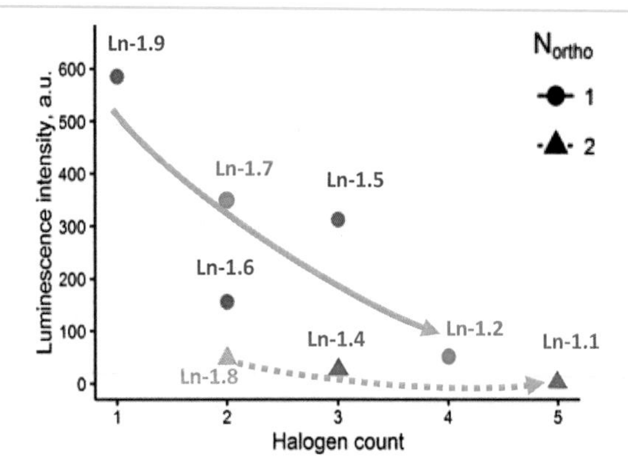

Figure III.1-19 Integral intensity of luminescence (integrated region 500–750 nm) for 3 mM aqueous solutions of Eu(III) complexes, depending on the ligand nature.

The same as the data for the dissociation in solution, the maximum luminescence intensity decreases with an increase in the number of halogen substituents. The ligands with two *ortho*-substituents clearly receive an additional penalty to their stability. It can be assumed that no complex except **Eu-1.1** fully degrades in water to $[Eu(H_2O)_6]^{3+}$, the luminescence of which would be negligible (Figure III.1-20). However, it is likely that complexes with a higher amount of halogen atoms tend to coordinate more water molecules due to the following two factors:

(i) an increased stability of "free" anions;

(ii) an increased LEWIS acidity of the complex.

Indeed, the predicted pK_a of the ligand explains 56% of the observed variation in the luminescence intensity ($I \approx 264pK_a - 482$, $R^2 = 0.56$, p-value = 0.02). The difference in the Eu environment between complexes is visible from the varied $I_{0\text{-}2}/I_{0\text{-}1}$ ratio across the ligands (Figure III.1-20).

Complex **Eu-1.9** possess the highest intensity among Eu(III) fluorobenzoates (Figure III.1-20). Complex **Eu-1.3** was poorly soluble in water (Figure III.1-16) and was therefore excluded from the study. Notably, the integral intensity of the band $^5D_0 \rightarrow {}^7F_2$ in the spectra of several complexes turned out to be less than the intensity of $^5D_0 \rightarrow {}^7F_1$, which is unusual for lanthanide complexes.[76,77] The intensity of the $^5D_0 \rightarrow {}^7F_1$ transition ($I_{0\text{-}1}$) is believed to be insensitive to the local environment around Eu^{3+}. At the same time, the transition $^5D_0 \rightarrow {}^7F_2$ ($I_{0\text{-}2}$) is hypersensitive and is low, when Eu^{3+} is situated at a centrosymmetric site (*i.e.* $[Eu(H_2O)_6]^{3+}$). When the symmetry of the Eu^{3+} coordination environment decreases, the intensity of this hypersensitive path increases.[34,170–172]

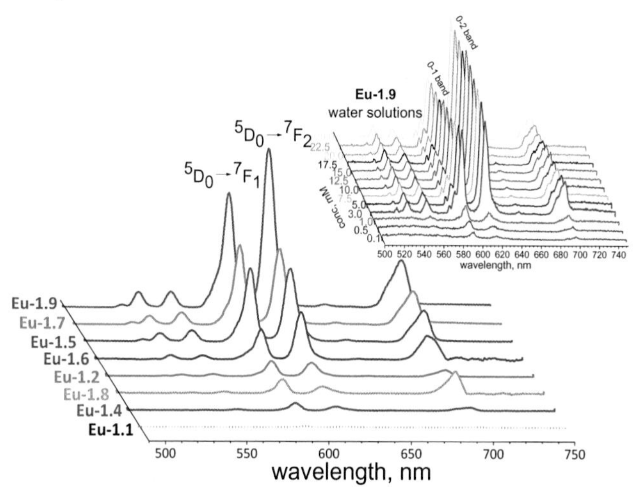

Figure III.1-20 Emission spectra of aqueous **Eu-L** (3 mM); insertion: concentration dependence of the luminescence spectra of **Eu-1.9** complex in aqueous solution.

In such a manner, different $I_{0\text{-}1}/I_{0\text{-}2}$ ratios for aqueous solutions of Eu(III) complexes are due to the different symmetry of the Eu^{3+} local environment. This ratio indicates the different dissociation degree: it reaches a maximum in the case of **Eu-1.1 ($I_{0\text{-}1}/I_{0\text{-}2}$ = 1.60)**, where only $[Eu(H_2O)_6]^{3+}$ is present, decreases for **Eu-1.2, Eu-1.4** and **Eu-1.8** ($I_{0\text{-}1}/I_{0\text{-}2}$ = 1.10 to 1.50) and has the lowest value in the case of **Eu-1.5**, **Eu-1.6**, **Eu-1.7** and **Eu-1.9** ($I_{0\text{-}1}/I_{0\text{-}2}$ = 1.00 to 0.70).

To prove this assumption, we recorded the luminescence spectra of **Eu-1.9** solutions of different concentrations (Figure III.1-20, insertion). As expected, at higher concentrations when there is no dissociation, the luminescence intensity is high due to the efficient sensitization by the ligand and hence, $I_{0-1} < I_{0-2}$. At lower concentrations (below 1.0 mM), the intensities of these two peaks are inversed, *i.e.* $I_{0-1} > I_{0-2}$, which indicates a high symmetry of the environment and probably the loss of coordinated ligand(s). According to Figure III.1-19, the value of 3.0 mM is a threshold value, at which the intensity of both bands rises sharply and then increases smoothly as the concentration increases. These data support the decrease of the Eu^{3+} environment symmetry with increasing the ligand coordination. Simultaneously, the overall luminescence integral intensity rises stepwise, which confirms that high luminescence intensity is due to the presence of an undissociated complex and not due to the increased amount of europium ions (radiative centers).

We selected **Eu-1.7** and **Eu-1.9** solutions (demonstrating the highest luminescence intensity at the concentration of 3.0 mM) for further testing *in cellulo.* They combine high aqueous solubility (Figure III.1-16) and non-toxicity with the lowest concentration threshold at which undissociated fragments of the complex still exist. Human cervix carcinoma (HeLa) cells were incubated with 1.0 mM solutions of the selected probes. In agreement with the previous results, compound **Eu-1.7** revealed detectable but weak signals, whereas compound **Eu-1.9** exhibits strong luminescence (Figure III.1-21).

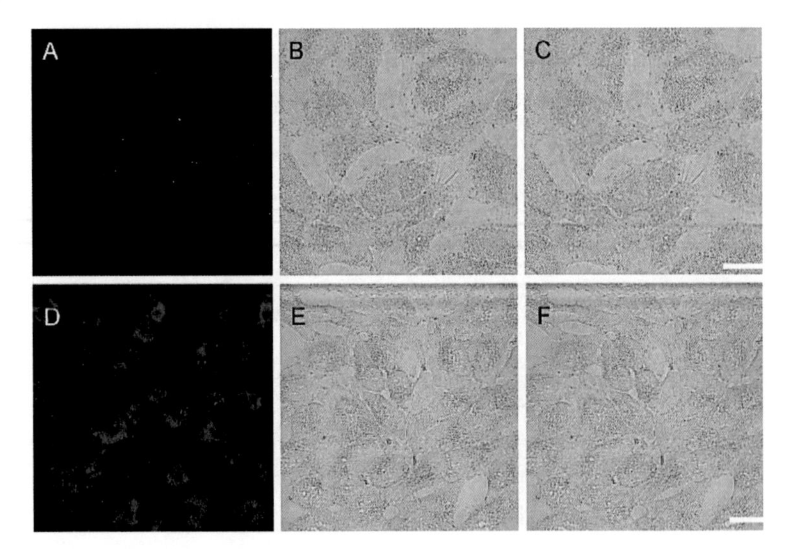

Figure III.1-21 Cellular uptake of **Eu-1.7** and **Eu-1.9** in HeLa cells: $1 \cdot 10^4$ cells were treated with 1 mM **Eu-1.7** (top row) and **Eu-1.9** (bottom row) for 24 h at 37 °C; the scale bar is 25 μm; A,D – fluorescence channels, B,E – brightfield channels, C,F – overlay.

To determine a suitable working concentration in cell culture, we performed an MTT viability assay. The LD_{50} value turned out to be 4.0 mM for **Eu-1.9** and 2.0 mM for **Eu-1.7** in HeLa cells. In the case of **Eu-1.7**, the luminescence intensity was hardly detectable (Figure III.1-21a-c). In contrast, **Eu-1.9** possesses high luminescence intensity even in diluted 0.25 mM solutions (Figure III.1-21d-f), which is probably due to the effect of the solution concentrating inside the cells. This complex was used for the cellular experiments at concentrations of 0.25 – 3.00 mM and the optimal concentration range was found to be 0.50 – 2.00 mM.

Conclusions

Thus, the thorough study of fluorinated lanthanide benzoates revealed the strong structure-property correlation. The structural study was performed based on 24 single-crystal structures and 38 powder patterns. In some cases, crystallization conditions (temperature and rate) affected the nuclearity and water content of crystal structures, for instance, unusual for lanthanide carboxylates mononuclear complexes were obtained. It was also founded that the major factor affecting the crystal structures of fluorobenzoates is the number of non-hydrogen *ortho*-substituents of the benzene ring (n_{ortho}): the complexes with $n_{ortho} = 1$ possess significantly higher luminescence intensity and thermodynamic stability than the complexes with $n_{ortho} = 2$. This is due to the influence of the substituents on the orientation of the benzene rings in the ligand (different OCCC torsion angle). Hence, the complexes with $n_{ortho} = 1$ are more flexible and possess higher stability in solution, which was observed for the case of water solutions, where more favorable geometry is achieved for the complexes with $n_{ortho} = 1$.

Relatively low quantum yield of europium complexes is mostly due to the low efficiency of ligand-to-metal energy transfer and is not related to the quenching *via* coordinated water molecules. Therefore, an additional Tb^{3+} energy step was introduced, utilizing sensibilization strategy $L \rightarrow Tb^{3+} \rightarrow Eu^{3+}$, which led to the twofold increase of the efficiency of Eu^{3+} emission.

III.1.2 Lanthanide ternary complexes

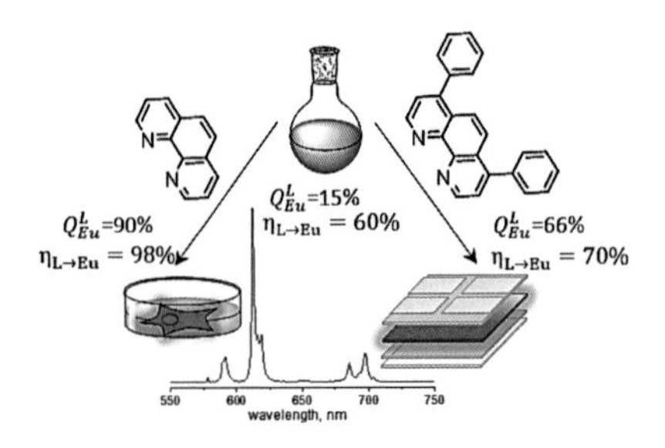

The following chapter describes the synthesis and characterization of Eu(III) ternary complexes, possessing record photoluminescence yields up to 90%. This high luminescent performance traces to the absence of quenching moieties in the Eu coordination environment and to an efficient energy transfer between ligands, combined with a particular symmetry of the coordination environment.

This section was previously published by the author ALENA S. KALYAKINA *et al.* in *Chemistry Communications* **2018,** *54* (41), 5221-5224[173] Copyright© 2018, with permission from Royal Society of Chemistry.

The design of new highly luminescent materials crucially depends on the understanding of the aspects of their photophysical behaviour. The broad scope of possible practical applications of Eu(III)-based luminophores makes them the most intriguing and mainly used among other lanthanides.[12,174] The brightness and the efficiency of Eu(III)-based luminescence can be increased drastically by coordinating the Eu(III) ion with a chromophore ligand.[18,139,140,174–176] The highest reported photoluminescence quantum efficiency (Q^L_{Ln}) for Eu(III)-based emitters has been achieved for an inorganic-organic hybrid material (Q^L_{Ln} = 96%).[177] As for lanthanide coordination compounds (LCC), Eu(III) ternary β-diketonates, known for being highly luminescent, possess a Q^L_{Ln} as high as 80 – 85% in the solid state.[35,178] Apart from them, benzimidazole-substituted pyridine-2-carboxylic acids can also serve as efficient sensitizers of Eu(III)-based luminescence with a Q^L_{Ln} up to 71%.[174] However, since red Eu(III) emission can be easily quenched by high energy X-H vibrations (X=O, N, C), such high Q^L_{Ln} values for the LCC are barely achievable.

The current chapter reports on new highly emissive LCCs, exhibiting record Q^L_{Ln} values up to 90%, which, to the best of our knowledge, is unprecedented for Eu(III) coordination compounds. These complexes were successfully used as OLED emitting layers and as luminescent probes for bioimaging. As shown in Chapter III.1.1 Eu fluorobenzoates with one *ortho*-fluorine substituent (n_{ortho} = 1) are promising luminescent materials due to their UV-stability and solubility.[33] This work was aimed to increase Q^L_{Ln} of Eu fluorobenzoates, using 1,10-phenanthroline (Phen)[180,181] and bathophenanthroline (BPhen) as auxiliary ligands (Figure III.1-22). These ligands can protect the Eu(III) ion from non-radiative deactivation,[34,35] increase the light absorption and decrease the lanthanide ion symmetry.[34,36–39] These compounds were synthesized, using the standart procedure, described in **GP2**. The lanthanide carboxylate and the auxiliary ligand were dissolved in ethanol, mixed together and reluxed for 2 h. After cooling, the precipitate was collected and dried in vacuo.

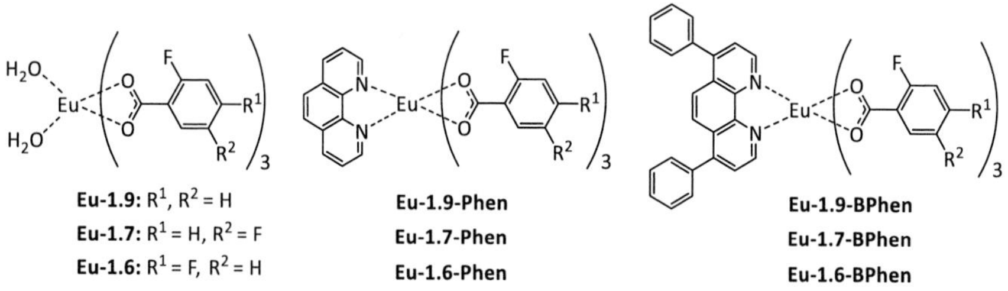

Eu-1.9: R^1, R^2 = H
Eu-1.7: R^1 = H, R^2 = F
Eu-1.6: R^1 = F, R^2 = H

Eu-1.9-Phen
Eu-1.7-Phen
Eu-1.6-Phen

Eu-1.9-BPhen
Eu-1.7-BPhen
Eu-1.6-BPhen

Figure III.1-22 Structural formulae of **Eu-1.9**, **Eu-1.7**, **Eu-1.6** and their ternary complexes with 1,10-phenanthroline **Eu-1.9-Phen**, **Eu-1.7-Phen**, **Eu-1.6-Phen** and with bathophenanthroline **Eu-1.9-BPhen**, **Eu-1.7-BPhen**, **Eu-1.6-BPhen.**

X-ray diffraction data

The structures of the synthesized ternary complexes (TCs) were determined by means of single-crystal X-ray diffraction. It has been shown previously,[33] that the most important factor affecting the crystal structures of homo-ligand partially fluorinated benzoates is the number of non-hydrogen *ortho*-substituents of the benzoic acid derivative (n_{ortho}). Thus, the complexes with the same n_{ortho} values tend to be isostructural up to the number of F-substituents. The same trend is shared by the TCs, synthesized in this work (Figure III.1-23). The molecular structures were determined for complexes **Eu-1.9-Phen**, **Eu-1.7-Phen** and **Eu-1.9-BPhen**.

Single crystals of **Eu-1.9-Phen**, **Eu-1.7-Phen** were grown from ethanol, **Eu-1.9-BPhen** was grown from toluene. The crystal structure of **Eu-1.9-Phen** turned out to be similar as reported previously,[182] however, revealing only one type of Eu complex without coordinated solvent molecules (the reported crystal structure contains two dimeric units, one of which contains two coordinated ethanol molecules)*. Refined structures of **Eu-1.9-BPhen** and **Eu-1.7-Phen** are centrosymmetric with the Eu...Eu distance of 4.042 Å for **Eu-1.7-Phen** and 4.146 Å for **Eu-1.9-BPhen**. Though the structures look similar (Figure III.1-23, Figure III.1-24), their geometries are slightly different (Table III.1-6), which in turn leads to different ligand functions. The carboxylic ligand with the chelating-bridging function in the case of **Eu-1.7-Phen** is not

* Structural data: chemical formula $C_{66}H_{40}Eu_2F_6N_4O_{12}$, M_r = 1498.94, crystal system monoclinic, space group $P2_1/n$, a = 14.3944 (7) Å, b = 13.0642 (7) Å, c = 15.4137 (9) Å, β = 103.999 (2)°, V = 2812.5 (3) $Å^3$, Z = 2, F(000) = 1480, D_x = 1.770 Mg m^{-3}, MoKα radiation, β= 0.71073 Å, μ = 2.30 mm^{-1}, T = 123 K, 0.06 × 0.04 × 0.01 mm.

chelating in the case of **Eu-1.9-BPhen** (denoted with * in Table III.1-6) which has a bidentate bridging function.

Table III.1-6 Selected bond lengths and ligand coordination modes for the structures of complexes **Eu-1.9-BPhen** and **Eu-1.7-Phen**.

Bond	Coordination mode	Eu-1.9-BPhen	Eu-1.7-Phen
Eu(1)-O(1)	μ^2-κ^1:κ^2	2.356(3)	2.380(2)
Eu(1)-O(2)	μ^2-κ^1	2.355(3)	2.380(2)
Eu(1)-O(3)	κ^2	2.418(3)	2.470(3)
Eu(1)-O(4)	μ^2-κ^1:κ^2	2.406(3)	2.468(2)
Eu(1)-O(5)	κ^2	2.472(3)	2.457(2)
Eu(1)-O(6)	μ^2-κ^1	2.382(3)	2.361(2)
Eu(1)-O(7)	μ^2-κ^1:κ^2	3.158(4)*	2.708(2)
Eu(1)-N(1)	κ^1	2.557(4)	2.598(3)
Eu(1)-N(2)	κ^1	2.608(4)	2.601(3)

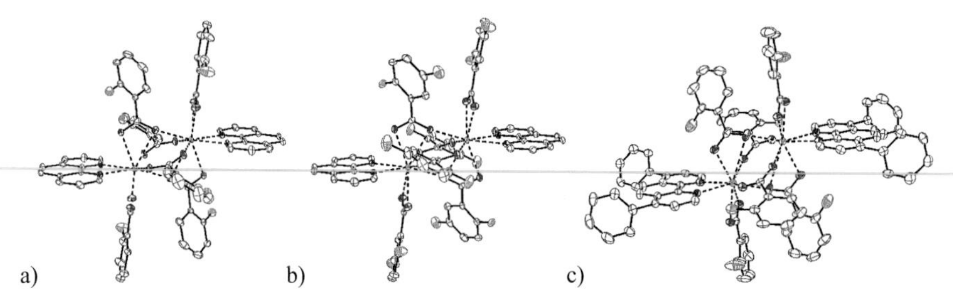

Figure III.1-23 ORTEP drawings of the structures of the studied complexes with 50% thermal ellipsoids. Atom assignments: Eu-green, C-grey, O-red, F-lime green, N-purple. Hydrogen atoms are excluded for clarity. a) **Eu-1.9-Phen**; b) **Eu-1.7-Phen**; c) **Eu-1.9-BPhen**.

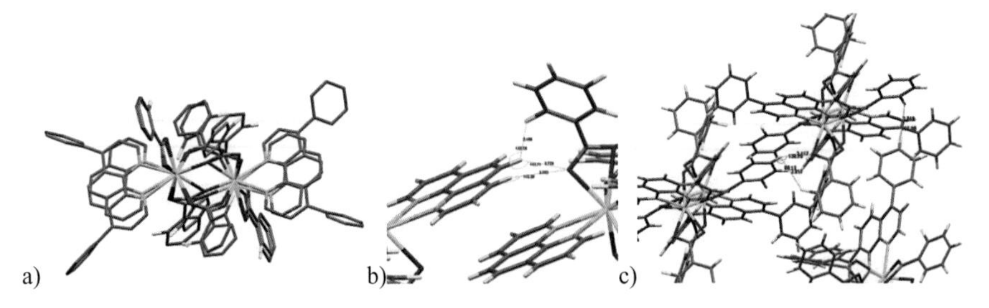

Figure III.1-24 a) Overlay of the structures **Eu-1.7-Phen** and **Eu-1.9-BPhen**; b) H-bonding in **Eu-1.7-Phen**; c) H-bonding in **Eu-1.9-BPhen**. Atom assignments: Eu-cyan, C-grey, O-red, F-lime green, N-blue, H-white.

The compounds are isostructural within the Phen-series (**Eu-1.9-Phen** and **Eu-1.7-Phen,** up to R^2-substituent). The structure of **Eu-1.9-Phen** is the same as reported for Eu pentafluorobenzoate complex[182] and its Tb analog.[183] The structure of **Eu-1.9-BPhen** (CCDC number 1818246)* is

* Single-crystal X-ray diffraction experiment and structure refinement was carried out by A.S. GOLOVESHKIN, A.N. Nesmeyanov Institute of Organoelement Compounds, Moscow, Russia.

also dimeric, exhibiting similar coordination modes of the ligands (Figure III.1-23, Table III.1-6), however, possessing a slightly different geometry compared to **Eu-1.9-Phen*** and **Eu-1.7-Phen**† (CCDC number 1818247). This observation traces to the differences in short contacts and hydrogen bonds (Figure III.1-24).

None of these structures **Eu-1.9-Phen**, **Eu-1.9-BPhen** and **Eu-1.7-Phen** are similar to the previously reported structure of lanthanide tetrafluorobenzoate ($n_{ortho} = 2$) with Phen as a ligand, reported by V. UTOCHNIKOVA *et al.*[184] This complex crystallizes in monomeric units with two Phen moieties per Ln ion due to the increase of n_{ortho}, resulting in the decrease of the tetrafluorobenzoate anion coordination ability. However, the fluorobenzoates studied in this chapter, possess relatively high thermodynamic stabilities and hence, have higher tendencies to polymerize.[33,185,186] The coordinated Phen or BPhen molecules (one per Ln(III) ion) prevent polymerization, forcing the dimer formation.[25,26]

Geometry studies

Coordination of Phen or BPhen molecule led to the change of Eu(III) coordination number and polyhedron shape. The analysis of the CSD‡ showed that, contrary to an assumption encountered in literature that CN = 9 is the most common coordination number for Eu(III) complexes, CN = 8 is as much as popular (Figure III.1-25).[144] However, some groups reported Ln complexes with an odd CN to be strong luminescent materials due to lower symmetry around the lanthanide ion.[187] Besides, when cumulative coordination lowers the symmetry around the Ln ion, the intensity of LAPORTE-forbidden f–f transitions is increasing by mixing of ligand and lanthanide orbitals.[34] This can influence both, luminescence brightness and color, as well as the rate of radiative f–f transitions.[34,188] This work aims to remove quenching H_2O molecules from the 8-coordinated complexes **Eu-1.9**, **Eu-1.7** and **Eu-1.6** by means of *N,N*-donor auxiliary ligands and achieve 9-coordinated polyhedrons. The coordination polyhedron of complexes **Eu-1.9**, **Eu-1.7** and **Eu-1.6** was determined to be a distorted snub disphenoid (J-SD[189], Figure III.1-26a), while the structure of **Eu-1.7-Phen** (as well as the ones of **Eu-1.9-Phen** and **Eu-1.9-BPhen**) is a distorted tricapped trigonal prism (J-TCTPR[190], Figure III.1-26b). These results indicate that unlike **Eu-1.7** complex **Eu-1.7-Phen** has no inverted centre in the crystal field, resulting in an increase in electron transitions in the 4f-orbitals due to odd parity[191] and hence, in an increase of the Q_{Ln}^{L}.

* SC-XRD experiment and structure refinement were carried out by Dr. M. NIEGER, University of Helsinki, Helsinki, Finland.

† SC-XRD experiment and structure refinement were carried out together with Dr. I.S. BUSHMARINOV, A.N. Nesmeyanov Institute of Organoelement Compounds, Moscow, Russia.

‡ CSD analysis was performed together with Dr. V.V. KORZINOV, Karlsruhe Institute of Technology, Karlsruhe, Germany.

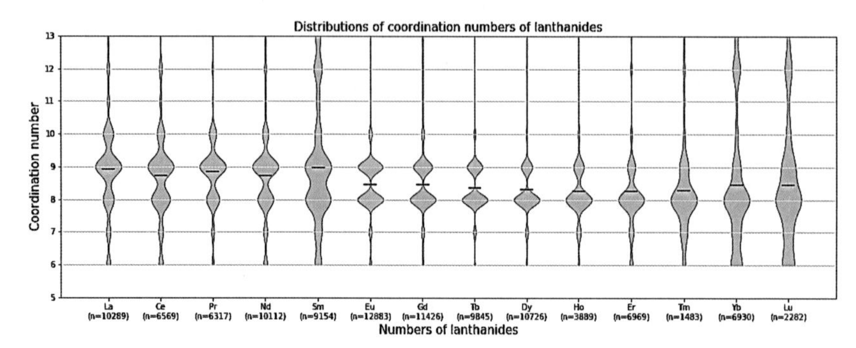

Figure III.1-25 Distribution of Ln coordination numbers according to the CCDC analysis using a violin plot. The horizontal line indicates the mean value.

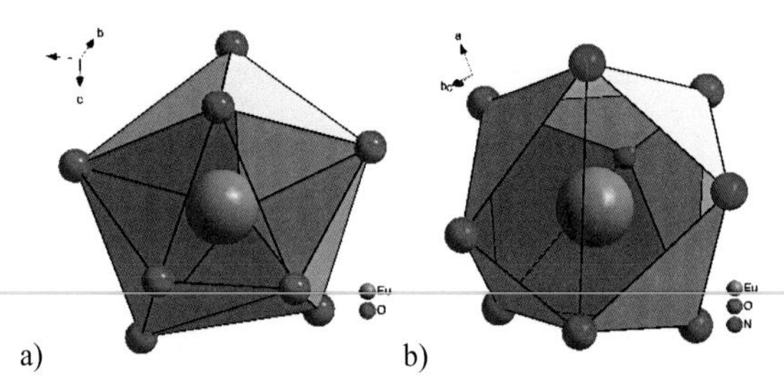

Figure III.1-26 Coordination polyhedron of a) **Eu-1.7** and b) **Eu-1.7-Phen**. Eu(III) ion is in the middle, coordinating oxygen from carboxylic ligands (red balls) and nitrogen from the 1,10-phenanthroline ligand (blue balls).

Photophysical properties

The Q_{Ln}^{L} value for the ternary complexes (TCs) within Phen- and BPhen-series increased from 25% up to 100% compared to complexes **Eu-1.9**, **Eu-1.7** and **Eu-1.6** discussed in Chapter III.1.1[33] due to the increase of the intrinsic quantum yield Q_{Eu}^{Eu}. The Q_{Eu}^{Eu} value was calculated as a ratio of the lanthanide luminescence lifetime τ_{obs} and its pure radiative lifetime τ_{rad}, according to equation III.2. The value of τ_{obs} may be obtained by the direct luminescence decay measurements, while τ_{rad} may be determined from the corrected emission spectrum (Figure III.1-27), according to the method, described by WERTS *et al.*[78] (Chapter I.2.3, equation I.8). Therefore, Q_{Eu}^{Eu} is higher than 90% for all ternary complexes (Table III.1-7). The explanation for such high intrinsic quantum yields lies in the absence of non-radiative pathways, causing by the absence of high energy bond vibrations in Eu(III) coordination environment. Hence, assuming the dependency $Q_{Ln}^{L} = \eta_{sens} \cdot Q_{Ln}^{Ln}$ (Chapter I.1.1, equation I.1), the only factor which can decrease the overall Q_{Ln}^{L} is the energy transfer efficiency η_{sens}. In the case of **Eu-1.9-Phen** η_{sens} is as high as 98%, which leads to a record Q_{Ln}^{L} value of 90% for this complex, though it is found to be within the range of 65-85% for other TCs (Table III.1-7).

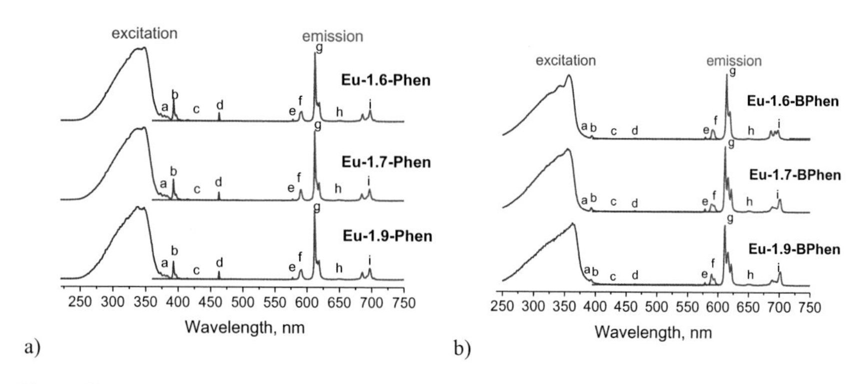

Figure III.1-27 Luminescence spectroscopy for Eu(III) complexes with a) 1,10-phenanthroline (Phen) and b) bathophenanthroline (BPhen): left, blue color – excitation monitored at 612 nm (uncorrected spectrum); right, red color – emission excited at 347 nm and corrected for detector sensitivity. Peak assignment for the excitation: a = $^5G_6$, $^5G_5$←$^7F_1$, b = $^5L_6$←$^7F_0$, c = $^5D_3$←$^7F_1$, d = $^5D_2$←$^7F_0$, for the emission: e = $^5D_0$→$^7F_0$, f = $^5D_0$→$^7F_1$, g = $^5D_0$→$^7F_2$, h = $^5D_0$→$^7F_3$, I = $^5D_0$→$^7F_4$.

Table III.1-7 Photophysical properties of **Eu-1.9-Phen**, **Eu-1.7-Phen**, **Eu-1.6-Phen** and **Eu-1.9-BPhen**, **Eu-1.7-BPhen**, **Eu-1.6-BPhen**. Estimated errors from Q_{Eu}^{L} and Q_{Eu}^{Eu} values are 10% and 5%, respectively*. Calculated as described in literature.[78]

	Eu-1.9-Phen	Eu-1.7-Phen	Eu-1.6-Phen	Eu-1.9-BPhen	Eu-1.7-BPhen	Eu-1.6-BPhen
τ_{obs}, ms	1.45	1.50	1.46	1.30	1.40	1.45
τ_{rad}, ms	1.58	1.51	1.61	1.41	1.40	1.45
Q_{Eu}^{Eu}, %	91	99	91	92	>99	>99
Q_{Eu}^{L}, %	90	82	77	66	70	67
η_{sens}, %	98	84	85	70	70	65

It is known that Q_{Ln}^{L} of a previously reported TC of Ln(III) 8-coordinated tetrafluorobenzoate, containing two Phen moieties per Ln(III) ion, only reaches 45%,[184] which confirms that Phen ligand is responsible for a high quantum yield not only due to an increase of the η_{sens} value but also due to the formation of a proper coordination environment which affects Q_{Eu}^{Eu}.

The excitation processes in TCs was analysed by investigating ternary complex **Eu-1.9-Phen** by means of time-resolved step-scan FTIR (trFTIR)[192–193] between 1900 and 800 cm^{-1} (shown is only the relevant part from 1750 to 1200 cm^{-1})†. The measurement gives a strong evidence of the population of a long-lived electronically excited state after excitation with a 355 nm laser pulse (Figure III.1-28). Negative (due to the ground state depopulation) and positive (due to the population of an electronically excited state) absorption bands are observed in the step-scan difference spectrum directly after excitation. The experimental difference spectra show the

* τ_{obs} and Q_{Eu}^{L} were measured by Dr. A.M. Kaczmarek and Prof. Dr. R. Van Deun, Ghent University, Ghent, Belgium.
† trFTIR was performed by Dr. M. Zimmer and Prof. Dr. M. Gerhards, TU Kaiserslautern, Kaiserslautern, Germany.

strongest change in absorbance directly after excitation of the sample followed by a slow decay of the signal intensity without shifts in the band positions. This slow decay is interpreted to be the repopulation of the electronic ground state from a long-lived excited state. Therefore, it was assumed that only one excited electronic state is populated on the time-scale from 100 ns to 500 μs. It is worth mentioning that the absolute Δ absorbance value is less than 1 mOD, which is rather weak compared to similar complexes measured with this apparatus.[192–193] Nonetheless, a band assignment is feasible due to the low noise level.

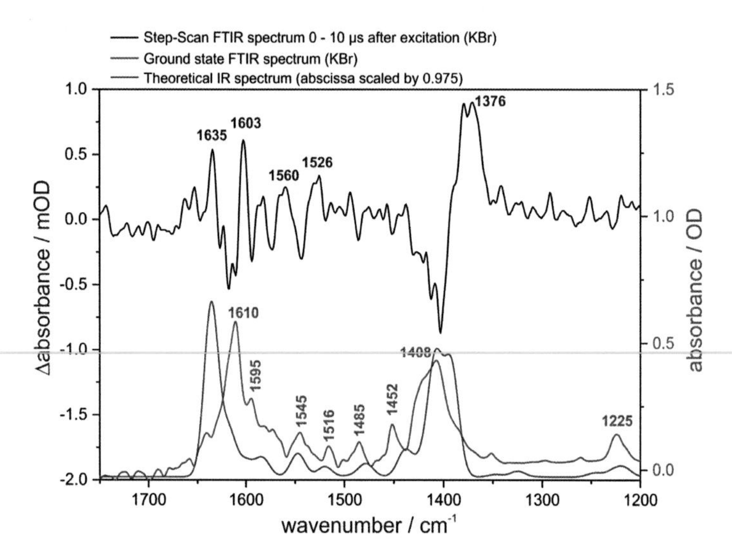

Figure III.1-28 Experimental FTIR spectrum of **Eu-1.9-Phen** in the electronic ground state (red) and the calculated IR spectrum in the electronic ground state (blue); the step-scan FTIR difference spectrum averaged over the first 10 ms after laser excitation (black).

The simulated IR spectrum of **Eu-1.9-Phen**, using DFT calculations* (Figure III.1-29) coincides with the experimental one. Also, the simulation of the absorption spectra *via* TD-DFT showed that the first excitation (313 nm) represents a combination of a $\pi - \pi^*$ excitation in the Phen ligand and an intraligand charge-transfer (ILCT) from the bridging fluorobenzoate to the Phen moiety. The most intense transition (295 nm) results mainly from an ILCT from the non-bridging fluorobenzoate to the Phen, indicating the strong participation of the auxiliary ligand Phen in the energy transfer processes, serving as an additional step of energy transfer between anionic ligand L and emissive Eu(III) ion. Such a phenomenon eventually leads to an increase of the $\eta_{L \to Eu}$ value up to 98% (Table III.1-7).

* DFT calculations were performed by Dr. F. DIETRICH, TU Kaiserslautern, Kaiserslautern, Germany.

Combination of such factors as similar sensitization efficiency, similar Eu coordination polyhedron and even similar absorption intensity[33,184] results in similar photophysical properties of complexes with Phen and BPhen. However, these two ligands dramatically differ in terms of practical application, which was demonstrated by testing of the corresponding Eu(III) complexes as bioimaging probes and as emitting layers in OLEDs.

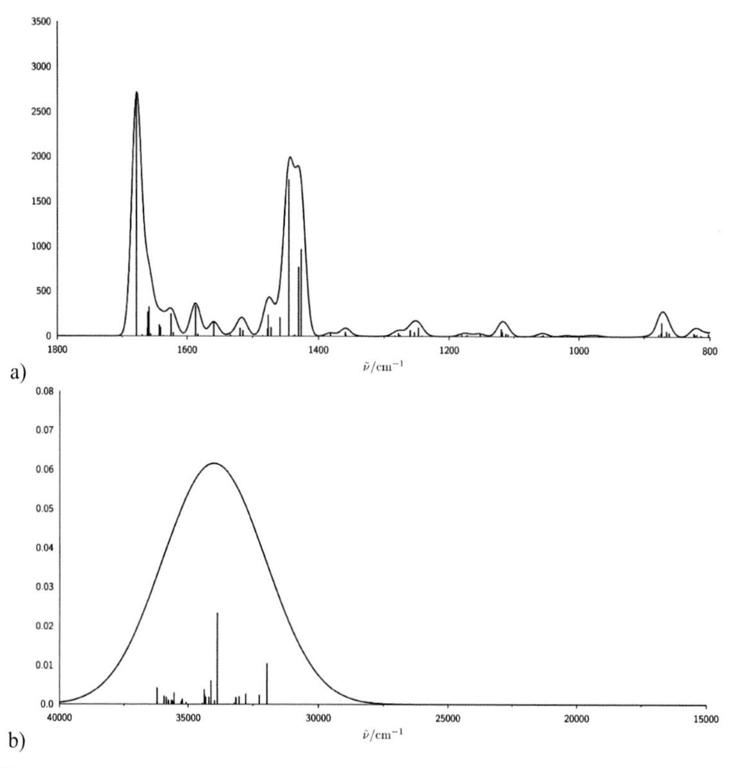

Figure III.1-29 a) Calculated IR spectrum of **Eu-1.9-Phen**; b) calculated absorption spectrum of **Eu-1.9-Phen**. Both results are consistent with the experimental data.

OLED fabrication and cellular experiments

For an application in OLEDs, the most efficient complexes **Eu-1.9-Phen**, **Eu-1.7-Phen**, **Eu-1.9-BPhen** and **Eu-1.7-BPhen** were tested in the heterostructure PEDOT:PSS/PVK/EML/TPBi/Al (EML = emission layer)*. In the spectra of the TCs **Eu-1.9-Phen** and **Eu-1.7-Phen** containing Phen, exhibiting the highest Q_{Ln}^{L} values, the $^5D_0 \rightarrow {}^7F_2$ band of europium luminescence was present, however, both spectra were dominated

* OLEDs were fabricated by Dr. V.V. UTOCHNIKOVA and Dr. A.A. VASHCHENKO, Lebedev Physical Institute, Moscow, Russia.

by the broadband luminescence of transport layers (Figure III.1-30a, left). The explanation of this behaviour, shown recently,[184] is insufficient electron mobility of Phen in combination with its too high LUMO level (1.2 eV), leading to a moderate performance of **Eu-1.9-Phen** and **Eu-1.7-Phen** as emitting materials in OLEDs despite the record value of Q_{Ln}^{L}. In contrast to this, the use of the BPhen-containing complexes **Eu-1.9-BPhen** and **Eu-1.7-BPhen** resulted in pure ionic Eu(III)-based luminescence (Figure III.1-30a, right). The only possible explanation is the improved charge transport properties of BPhen in comparison to Phen, given the similar structures of the complexes and almost the same I-V curves of OLEDs with these compounds as EML (Figure III.1-30b).

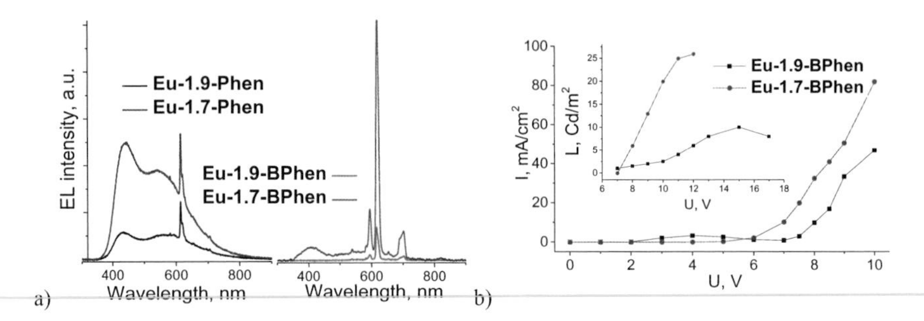

Figure III.1-30 Characteristics of OLEDs PEDOT:PSS/PVK/EML/TPBi/Al: a) EL spectra of **Eu-1.9-Phen**, **Eu-1.9-BPhen** and **Eu-1.7-Phen**, **Eu-1.7-BPhen**; b) I-V curves and L-V curves (insert) of OLEDs with **Eu-1.9-BPhen** and **Eu-1.9-BPhen** as emitting layers.

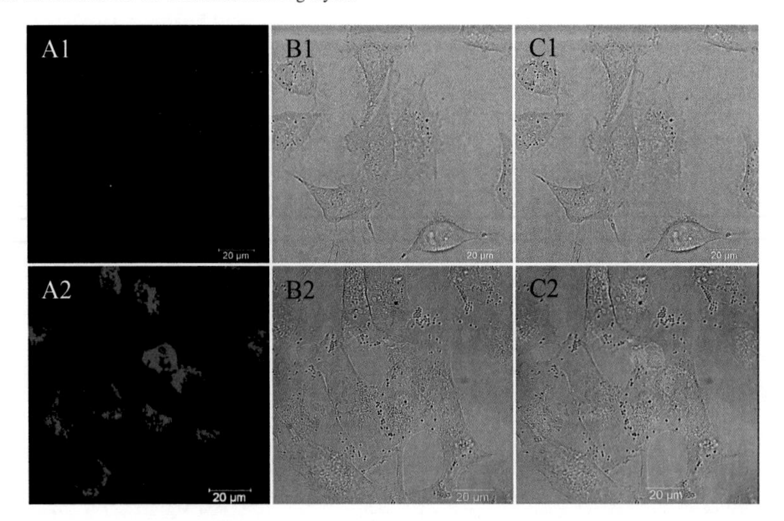

Figure III.1-31 Confocal microscopy images of HeLa cells. Control (top row): A1 – fluorescence channel, λ_{ex} = 405 nm, laser power = 30%, detection range 580 – 750 nm, B1-brightfield channel, C1-overlay; **Eu-1.9-Phen**, concentration 0.1 mM (bottom row): A2 – fluorescence channel, λ_{ex} = 405 nm, laser power = 30%, detection range 580 – 750 nm, B2 – brightfield channel, C2 – overlay.

However, cellular tests of complexes within Phen- and BPhen-series showed the opposite behaviour. DMSO-solutions of TCs with BPhen, tested at safe concentrations (0.1 mM), showed very low cellular permeability after incubation of human cervix carcinoma (HeLa) cells. This cell line was then incubated with the Phen-containing **Eu-1.9-Phen**, **Eu-1.7-Phen** and **Eu-1.6-Phen** probes, whereas compound **Eu-1.9-Phen** showed the strongest luminescence due to its protruding Q_{Ln}^{L} (Figure III.1-31). Thus, ternary complexes with Phen turned out to be very promising for cellular imaging, caused by their high solubility (5 – 7 mM in EtOH/H_2O 2/1, v/v) and very low toxicity (safe concentration* is 0.5 mM).

Conclusions

In summary, this work demonstrated exceptionally high values of the total photoluminescence quantum yield (Q_{Ln}^{L}) of Eu(III) ternary complexes, being the highest reported values by its time as well as showed principal applicability of these complexes for both, optoelectronic and biological applications.

* Concentration, at which more than 80% cells are alive.

III.1.3 Lanthanide 9-anthracenates

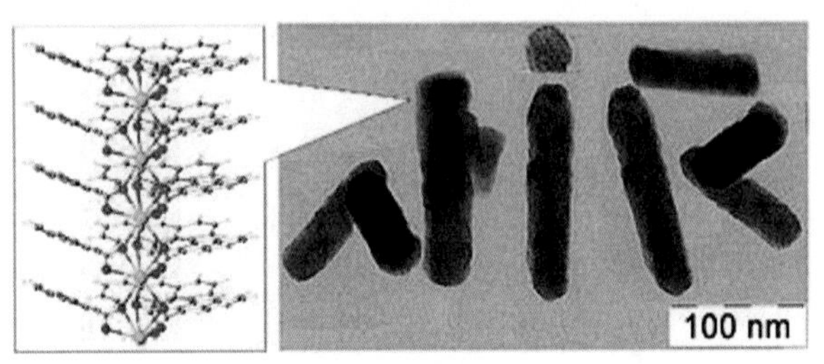

While searching for new NIR-emitting materials, lanthanide 9-anthracenates Ln(ant)$_3$ were synthesized and thoroughly characterized. Among NIR-emitting 9-anthracenates, ytterbium complex Yb(ant)$_3$ demonstrated the highest NIR luminescence efficiency. It was successfully tested as an emission layer for a host-free OLED, possessing solely an f-f emission band at 1000 nm with an electroluminescence (EL) quantum yield reaching 0.21%. Such a relatively high performance could be achieved due to (i) the high electron mobility caused by the extended stacking interaction in the crystal structure of the emitting material and (ii) the high photoluminescence (PL) quantum yield of Yb(ant)$_3$ of 1.5% which was increased up to 2.5% by partial substitution of Yb^{3+} by non-emitting Lu^{3+} ions (dilution). Besides, the first blue gadolinium-based PHOLED was prepared based on Gd(ant)$_3$ where purely ligand triplet luminescence is observed.

This work was carried out as a joint project with Dr. VALENTINA V. UTOCHNIKOVA, Lomonosov MSU (Moscow, Russia). The section was previously published by VALENTINA V. UTOCHNIKOVA and the author ALENA S. KALYAKINA *et al.* in *Journal of Materials Chemistry* **2016**, *4*, 9848-9855,[128] Copyright© 2017, with permission from Royal Society of Chemistry.

Numerous applications of NIR organic light-emitting diodes (OLEDs) draw a growing interest to their development especially in the fields of optical communication and biomedical imaging.[89,94,100,101,194,195] However, in sharp contrast to their importance, the number of efficient NIR electroluminescent OLED devices with emission spectra above 1000 nm is negligible. Different classes of materials have served as emitting layers in OLEDs, among them NIR-emitting lanthanide ions are of great interest, because their emission lies in the range 1000 – 1500 nm which matches the "transparency window" of silica optical fibres.[96] A wide variety of organic ligands has been used to sensitize luminescence of NIR-emitting lanthanide ions, whereas ytterbium complexes usually exhibit the highest quantum efficiency,[96–99] which is, unfortunately, still far below useful values. Very few NIR-emitting lanthanide-based OLEDs that has been reported up to date[97,100,101] cover most of the examples available in literature. Usually EL quantum yield for NIR-emitting OLEDs is below 1%.[196–200] Such a poor performance of NIR-emitting lanthanide complexes is caused by two reasons: (i) poor PL or (ii) poor charge carrier mobility. The first problem may be solved by the selection of an appropriate anionic ligand which ensures efficient ligand-to-metal energy transfer. The second problem is usually solved by the selection of an appropriate host material, possessing high charge carrier mobility. In this case, the emitting material should be doped into the host matrix. However, this approach often decreases the efficiency of Ln-based luminescence due to an inefficient host-to-emitter energy transfer, making the first problem prevalent. In order to avoid these losses, a host-free emission layer is proposed in this work. Therefore, the emitter has to exhibit a high charge carrier mobility additionally to high luminescence efficiency. This approach has been successfully tested on the example of lanthanide 9-anthracenates (further denoted as $Ln(ant)_3$ or **Ln-1.10**).

Anhydrous complexes of $Ln(ant)_3$ (Ln = lanthanides, except Ce, Pm) precipitated quantitatively after the mixing of aqueous solutions of the potassium salt of anthracene-9-carboxylic acid **1.10-H** (Figure III.1-32) and lanthanide chloride $LnCl_3 \cdot 6H_2O$.

X-ray diffraction data

Single crystal of the Yb(III) complex was grown from DMSO and its crystal structure was determined by means of SC-XRD*. The complex co-crystallized with solvent molecules and the composition of the single crystals corresponded to the formula $Yb(ant)_3(DMSO)_3 \cdot 0.16\ H_2O$, where DMSO is coordinated directly to the Ln ion, while water molecules are outer-sphere and not present in every unit cell.

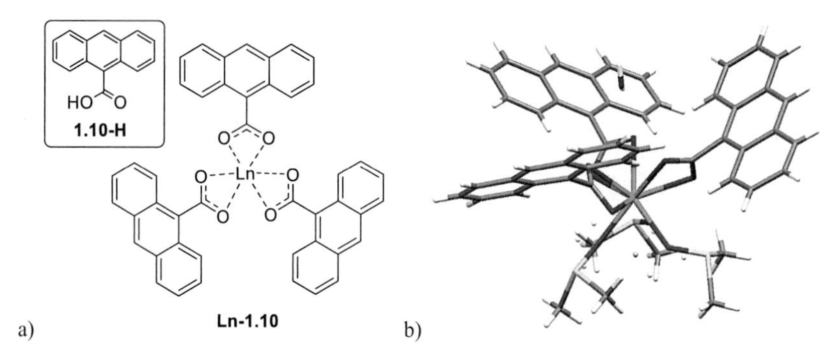

Figure III.1-32 a) Structural formula of compex **Ln-1.10** and ligand **1.10-H**; b) single-crystal structure of **Yb-1.10-sc** with the formula $Yb(ant)_3(DMSO)_3 \cdot 0.16H_2O$.

The structure of $Yb(ant)_3(DMSO)_3 \cdot 0.16\ H_2O$ consists of different monomeric units, where the Yb^{3+} ion is coordinated by six oxygen atoms of three κ^2 ant^- ligands and three oxygen atoms of DMSO molecules, giving a total coordination number of nine (Figure III.1-32). The Yb–O bonds

*SC-XRD was performed by Dr. I S. Bushmarinov, A.N. Nesmeyanov Institute of Organoelement compounds, Moscow, Russia

with ant⁻ ligands have a length of 2.405(2) and 2.415(2) Å and O–Yb–O angles are equal to 54.05(2)°. The Yb–O bond length with DMSO is equal to 2.322(1) Å. The outer-sphere water molecule participates in H-bonding network and connects monomeric units in chains.

The solids of synthesized lanthanide 9-anthracenates were studied by means of PXRD*. The reflections of the *hk*0 zone of a trigonal or hexagonal space group (a ≈ 27.5 Å) were observed in PXRD patterns for most of the **Ln-1.10**. These complexes might be called "solid liquid crystals", since they did not crystallize in 3-dimensional crystal structures. Fortunately, europium 9-anthracenate **Eu-1.10** made an exception and its powder pattern, corresponding to the true 3D structure, was indexable with trigonal unit cell parameters $a = b = 28.0947(13)$, $c = 3.71428(16)$, space group R3. The structure was solved from powder data, the RIETVELD refinement of this solution was restraint-consistent[201] with a half uncertainty window (HUW) of 0.17(13) Å, indicating a reliable determination of the bonding system (Figure III.1-33).

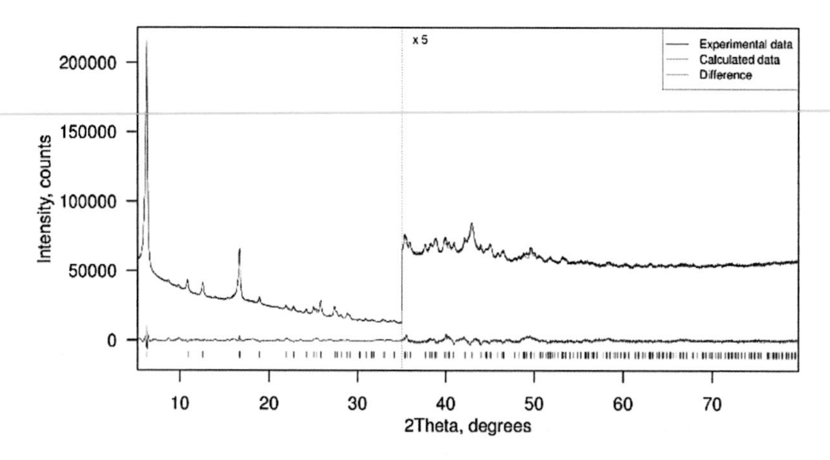

Figure III.1-33 RIETVELD fit of Eu(ant)$_3$.

The conventional R values were acceptably low ($R_{wp}/R'_{wp}/R_p/R'_p/R_{Bragg} = 1.77/2.88/1.34/2.26/0.63\,\%$ at rms $\Delta d = 0.01$ Å) and the difference curve was featureless. With the adjustment for the unit cell parameter a and exclusion of reflections with $l \neq 0$, the same structural model successfully described the **Yb-1.10** pattern, indicating that the structures are essentially isomorphous, with **Yb-1.10** differing by lack of the apparent order in the c direction. The amorphous halo at $d \approx 3.50$ Å in the Yb(ant)$_3$ and Lu(ant)$_3$ powder patterns most likely

* Powder diffraction patterns were measured and analyzed together with Dr. IVAN S. BUSHMARINOV, Nesmeyanov Institute of Organoelement compounds, Moscow, Russia.

appeared due to non-BRAGG X-ray scattering often observed in partially ordered phases. This possible lack of order can be explained by the crystal packing in **Eu-1.10**, discussed below.

The structure of **Eu-1.10** consists of $\{Eu(ant)_3\}_n$ polymeric chains, only loosely packed by weak H...H interactions (Figure III.1-34a). Within each chain carboxylate anions adopt a half-bridge (μ_2-κ^1:κ^2) coordination mode and each Eu atom within the tricapped trigonal prism coordination polyhedron has a coordination number of (6+3) (Figure III.1-34b). The distance between neighbouring Eu atoms is equal to 3.71 Å. This type of lanthanide coordination polymer has been reported previously exactly once for lanthanum phenylacrylate, according to the Cambridge Structural Database (CSD).[202] However, in this compound the La...La distance was 3.99 Å, which leads to an assumption that in the case of **Eu-1.10** Eu atoms of are forced into a closer contact by stacking interactions between **1.10** ligands. Indeed, the ligand planes are parallel, with an anthracene separation of 3.71 Å, indicating a rather strong interaction (Figure III.1-34b) (compared to 3.77 Å in fluoroanthracenic acid and 3.90 Å in pure **1.10-H**).[203]

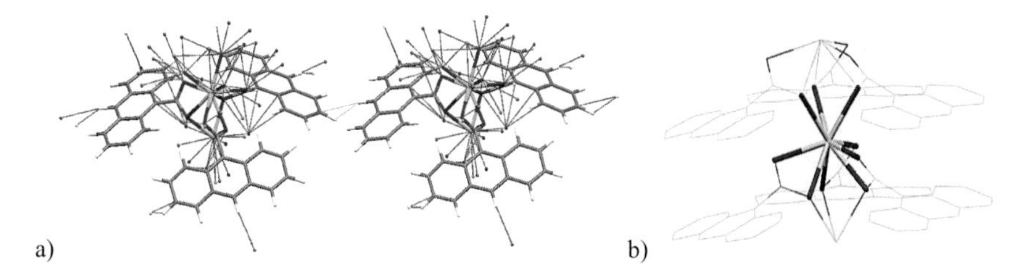

Figure III.1-34 a) C-H...H angle in **Eu-1.10**; b) polymeric chains and packing fragment in **Eu-1.10**.

Thus, the crystal of **Eu-1.10** consists of strongly bound $\{Eu(ant)_3\}_n$ rods (≈ 1.5 nm in diameter), linked to each other only by weak H...H interactions. This fully explains, why most of the complexes form “solid liquid crystals”. The formation of an ordered structure for **Eu-1.10** was possible due to the strongest H...H interaction that was observed in this very structure. Indeed, the strength of H...H interactions depends on geometry and, at a certain C-H...H angle this strength reaches its maximum. In turn, the C-H...H angle depends on the lanthanide ion radius and, though slightly, must change for different complexes. Thus, in **Eu-1.10** this angle turns out to be optimal (119.44°), resulting in a 3D ordering, while in other **Ln-1.10** complexes the rods are freely sliding against each other.

Given that the 00*l* reflections are weak and overlapped, even for **Eu-1.10** and the crystallite size for **Ln-1.10** could be only estimated from *hk*0 reflections. According to the DOUBLE-VOIGT approach,[204] the mean domain size in this direction was 35.6(2) nm. In practice, the transmission electron microscopy (TEM) data have shown that the material consists of nanorods of a nearly uniform diameter of 30(3) nm and varying length (Figure III.1-35a), thus, making the material with this shape ready for the fabrication of nanodevices as synthesized*.

The EXAFS spectra, measured at the lanthanide L_{III}-edge, were used to assess the nearest coordination environment of the lanthanides in **Ln-1.10** in solid state and to provide reliable Ln–O distances for **Ln-1.10** compounds lacking a 3D order†. Observations demonstrated by the powder patterns of all aforementioned complexes seem to be similar. Thus, the first coordination environments of Ln^{3+} ions are the same and consist of nine oxygen atoms, each belonging to COO^--groups. The differences in EXAFS data of **Ln-1.10** (Ln = Eu, Gd, Tb, Dy, Er, Yb, Lu) appeared mostly due to the different quality of the spectra: high-quality spectra were obtained for **Lu-1.10**, **Er-1.10** and **Yb-1.10**, whereas, for the highest quality spectrum of **Lu-1.10**, the k-range of 2.0 – 14.0 was registered. For these complexes, it was possible to distinguish oxygen atoms by distances: the first shell contains six nearest oxygen atoms at the distances R = 2.23 – 2.28 Å and three furthest oxygen atoms at the distances R = 2.38 – 2.44 Å, which is in line with PXRD data.

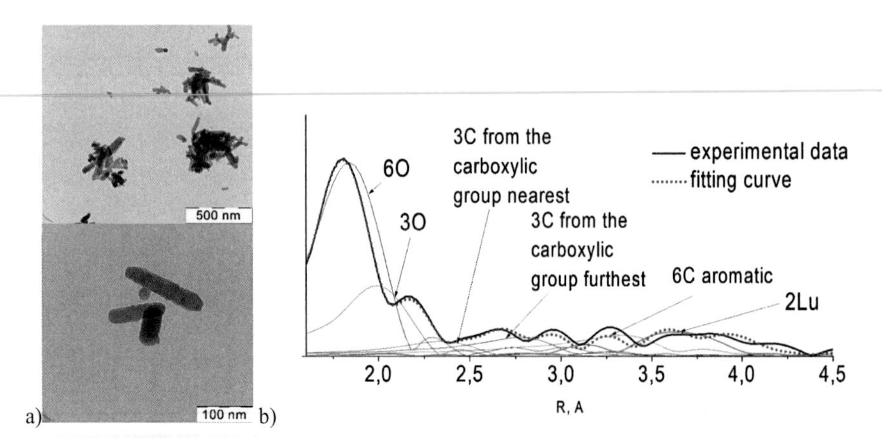

Figure III.1-35 a) TEM images of **Tb-1.10**; b) EXAFS spectra of **Lu-1.10**: experimental curve, fitting curve and assignment of the “heaviest” paths.

For the lighter lanthanides (Eu, Gd, Tb, Dy) only the average Ln…O distances were determined which turned out to be R = 2.36 – 2.38 Å. The average distances Ln…O for the entire Eu – Lu row are in line with lanthanide contraction but seem to be almost similar within the group of the lightest lanthanides from Eu to Dy (R = 2.36 – 2.38 Å) and the heaviest ones from Er to Lu (R = 2.28 – 2.31 Å). The Ln…Ln distances are almost the same for all the complexes and turned out to be 3.7 – 3.8 Å which is unusually short for lanthanide complexes.

* TEM analysis was performed by Dr. V.V. Utochnikova, Lomonosov MSU, Moscow, Russia
† EXAFS spectra were measured by the author in the National research center “Kurchatov Institute”, Moscow, Russia

Photophysical properties

The study of photophysical properties* was carried out for complexes with NIR-emitting lanthanides **Nd-1.10**, **Er-1.10** and **Yb-1.10**, as well as for those with **Gd-1.10**, **Lu-1.10** and **Tb-1.10**, emitting in the visible range. The comparison of **Lu-1.10** and **Gd-1.10** spectra shows a large band shift, pointing out that the fluorescence occurs in the case of **Lu-1.10** (Figure III.1-36b, fluorescence) and phosphorescence in the case of highly paramagnetic **Gd-1.10** (Figure III.1-36a, Figure III.1-36b, phosphorescence). This fact together with the long lifetime of the excited state of **Gd-1.10** at room temperature (7.8 μsec) proved that **Gd-1.10** exhibits a pure room temperature phosphorescence, with a PLQY of 3.8% (Table III.1-8). Such phosphorescent complexes and their derivatives may find application for OLED production and substitute expensive iridium compounds.

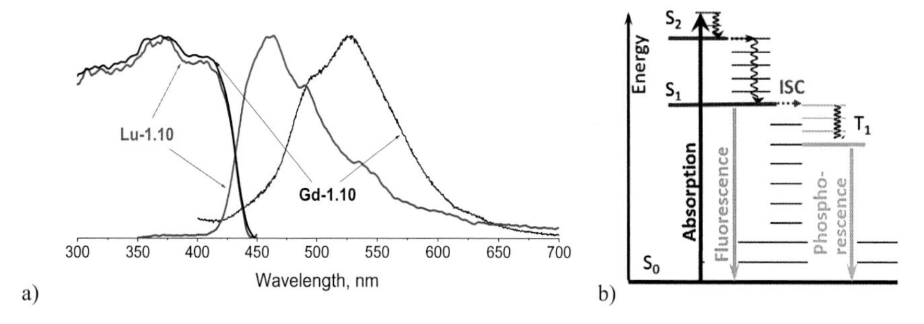

Figure III.1-36 a) Steady-state excitation and photoluminescence spectra of **Lu-1.10** and **Gd-1.10** at room temperature; b) simplified energy transfer scheme for Gd(III) complex luminescence: singlet luminescence (fluorescence), intersystem crossing (ISC) and triplet luminescence (phosphorescence).

The singlet and triplet excited state values calculated by measuring luminescence of **Lu-1.10** and **Gd-1.10** complexes turned out to be 22400 cm^{-1} and 20200 cm^{-1}, respectively. Therefore, no Tb(III)-based ionic luminescence was observed for **Tb-1.10**. The excited $^5D_4$ state of Tb(III) ion is 20400 cm^{-1} which cannot be sensitized by the ligand triplet state with the lower energy 20200 cm^{-1}. However, the absence of the ionic europium luminescence of **Eu-1.10** (the energy of the excited $^5D_0$ state of Eu^{3+} is 17200 cm^{-1}) is surprising and may be caused by the presence of ligand-to-metal charge transfer state (LMCT).

Nd-1.10, **Er-1.10** and **Yb-1.10** complexes of NIR-emitting ions exhibited bright ionic luminescence (Figure III.1-37). There is an intensive through-ligand excitation indicated by

* Photophysical properties were investigated by Dr. V.V. UTOCHNIKOVA, Lomonosov MSU, Moscow, Russia, Dr. A.M. KACZMAREK, Prof. Dr. R.VAN DEUN, Ghent University, Ghent, Belgium and Dr. L. MARCINIAK, Institute of Low Temperature and Structure Research, Wrocław, Poland

steady-state excitation spectra of these compounds that coincided with each other and with the excitation spectra of **Ln-1.10** (Ln = Gd, Lu). Table III.1-8 represents quantum yields and lifetimes of the excited state of **Ln-1.10** (Ln = Gd, Nd, Er, Yb), measured on the luminescence maximum at the through-ligand excitation (365 nm).

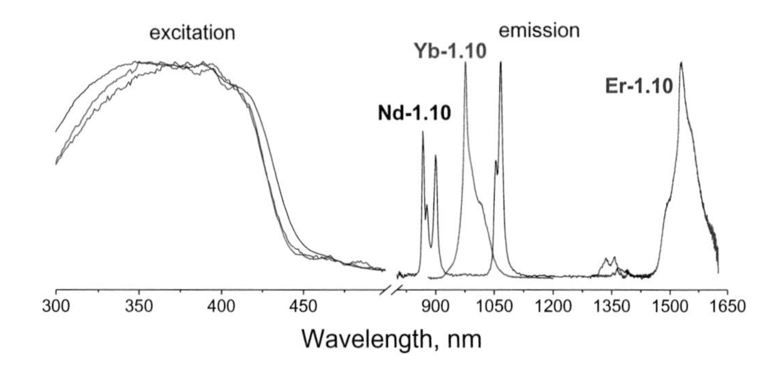

Figure III.1-37 Normalized steady-state excitation and photoluminescence spectra of **Nd-1.10** (black), **Yb-1.10** (red) and **Er-1.10** (blue) at room temperature.

Table III.1-8 Luminescent characteristics of **Ln-1.10** (Ln = Gd, Nd, Er, Yb).

	Gd(ant)3	Nd-1.10	Er-1.10	Yb-1.10
τ, μsec	7.8	17	13	52
PLQY, %	3.8	n/a	n/a	1.5

The pure radiative lifetime (τ_{rad}) was calculated from the absorption spectrum *via* EINSTEIN's equation[205] and turned out to be equal to 0.68 ms (in DMSO solution). Using this value the internal quantum yield of **Yb-1.10** was calculated according to equation III.2 as $Q_{Yb}^{Yb} = \frac{\tau_{obs}}{\tau_{rad}}$ equals to 7.9%. The external quantum yield Q_{Yb}^{L} was measured to be 1.5%, which is rather high compared to the typical one for ytterbium complexes (ca. 1%).[206] The sensitization efficiency was measured, using equation I.1 as $\eta_{sens} = \frac{Q_{Yb}^{L}}{Q_{Yb}^{Yb}} = 20\%$. This indicates that L→$Yb^{3+}$ energy transfer is not very efficient, which may be explained by the huge difference in the energies of the ligand excited state (20200 cm^{-1}) and the ytterbium resonance level (10200 cm^{-1}), which leads to the necessity of searching for another sensitization mechanism. Given all these facts, sensitization through LMCT state seems plausible. The presence of this state was proved by the comparison of a diffuse reflectance spectrum (DRS) of **Gd-1.10** (which do not have LMCT state) with a DRS of **Yb-1.10** and **Eu-1.10**. Therefore, it is reasonable to search the efficient NIR-emitters among the ytterbium complexes with ligands already known to quench europium luminescence by LMCT state.

Another quenching mechanism, which should be considered, is the concentration quenching.[207] Therefore, a series of solid solutions of the formula $Yb_xLu_{1-x}(ant)_3$ were synthesized*, which was possible due to the isostructurality to monometallic complexes. Thus, the introduction of a Lu(III) ion is meant to dilute the number of emitting Yb(III) centres and hence, decrease the concentration quenching. The lifetimes of the excited state increase linearly upon dilution†, proving the presence of concentration quenching in all the examined samples (Figure III.1-38, black line). With a decrease of the concentration quenching, the quantity of emitting species (Yb^{3+} ions) was decreasing, which indicates the necessity of a direct measurement of the PLQY. In the range of x = 0 – 0.1, the PLQY showed an almost linear increase due to an increase of the amount of emitting species, *i.e.* ytterbium ions (Figure III.1-38, red line, growing). In the range x = 0.1 – 0.2 this increase is slowing down due to the concentration quenching and at x = 0.2 the maximum PLQY of 2.5% is reached which is then decreased down to 1.5% due to the concentration quenching (Figure III.1-38, red line, decline).

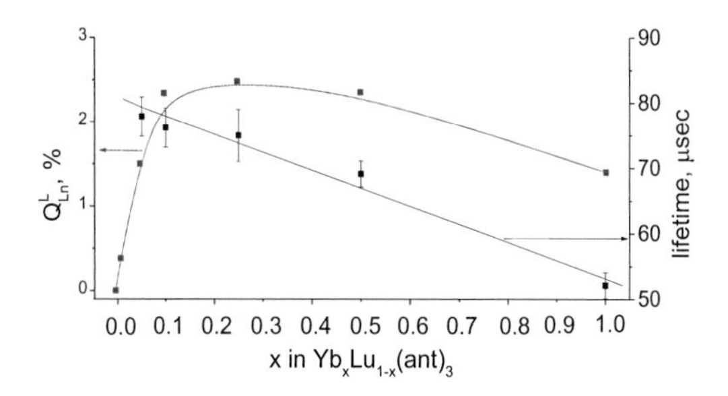

Figure III.1-38 Steady-state lifetimes of the excited state and luminescence quantum yields of $Yb_xLu_{1-x}(ant)_3$.

The resulting work not only proves that concentration quenching is crucial in the coordination of NIR-emitting compounds, but also allowed the increase of PLQY up to 2.5% by partial substitution of Yb^{3+} with non-emitting Lu^{3+}.

OLED fabrication‡

The extended stacking observed in the structures of **Ln-1.10** should result in a high electron conductivity of these compounds, which together with their excellent photoluminescent properties in the NIR range makes it interesting to test these compounds as emissive layers in OLEDs. It

* Synthesis was performed by Dr. V.V. UTOCHNIKOVA, Lomonosov MSU, Moscow, Russia.

† Lifetimes of the excited state and PLQYs of $Yb_xLu_{1-x}(ant)_3$ were measured by Dr. L. MARCINIAK, Institute of Low Temperature and Structure Research, Wroclaw, Poland.

‡ OLEDs were fabricated by Dr. V.V. UTOCHNIKOVA and Dr. A.A. VASHCHENKO, P.N. Lebedev Physical Institute, Moscow, Russia.

should be also noted that according to the TGA and DSC data **Ln-1.10** are extremely stable, since they undergo neither weight loss nor phase transition in the temperature range from 77 up to 620 K.

Based on the on single-component emission layers from **Nd-1.10, Er-1.10** and **Yb-1.10**, a series of NIR OLEDs with the structure ITO/PEDOT:PSS/**Ln-1.10**(~ 30 nm)/TPBi(10 nm)/Al was fabricated. Here PEDOT:PSS was chosen as a standard hole-injecting layer and TPBi as hole-blocking material.

The device based on **Nd-1.10** with the same structure demonstrated neodymium ionic luminescence with the presence of a ligand phosphorescence band due to lower efficiency (Figure III.1-39a). Because of the instrumental limitations we were unable to detect the 1.5 μm band of erbium-based luminescence from the device based on **Er-1.10**, but the presence of another luminescence band at 1.0 μm indicates that erbium ionic luminescence is also present.

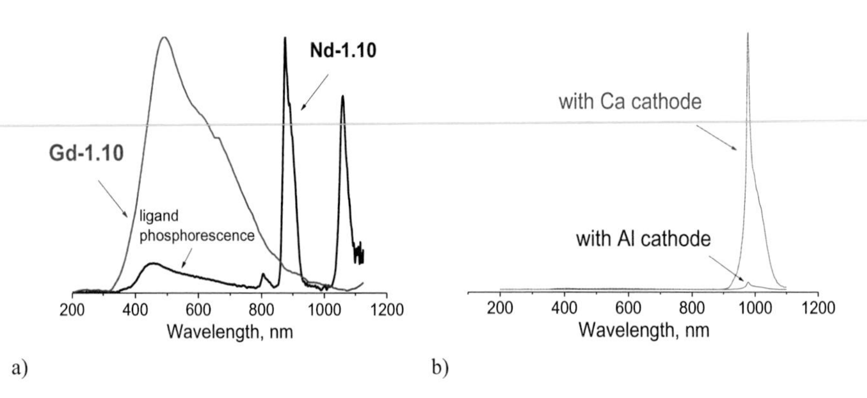

Figure III.1-39 Electroluminescence spectra of the devices: a) ITO/PEDOT:PSS/**Ln-1.10**/TAZ/Al (Ln = Nd, Gd) and b) ITO/PEDOT:PSS/**Yb-1.10**/TAZ/M (M = Al, Ca/Al) at 18 V.

The OLED device with **Er-1.10** showed a better performance. It exhibited pure NIR luminescence without any additional bands of ligand emission. The switch-on voltage of this device corresponded to 4 V, with its EL spectrum consisting of the ytterbium ionic emission band (with a maximum at 978 nm) with low intensity band in the visible range at 530 nm which originated from the ligand phosphorescence (Figure III.1-40a). The variation of the operating bias voltages from 5 to 12 V did not significantly change the EL spectrum (Figure III.1-40).

Another candidate for an active layer material is **Gd-1.10**, due to the presence of pure room temperature phosphorescence. To ensure its EL properties, a device with the structure ITO/PEDOT:PSS/**Gd-1.10**/TPBi/Al was fabricated, demonstrating an intense luminescence

(Figure III.1-39a). This observation makes gadolinium complexes another promising candidate to substitute iridium compounds as OLED emitters.

Figure III.1-40b displays the current-voltage characteristics for the device ITO/PEDOT:PSS/**Yb-1.10**/TPBi/Al. The diode demonstrated low resistance, while no shortening occurred up to the high current density of *approx.* 500 mA/cm^2. It witnesses high stability of **Yb-1.10**, even under electric current, which might be a common property of lanthanide aromatic carboxylates.[90]

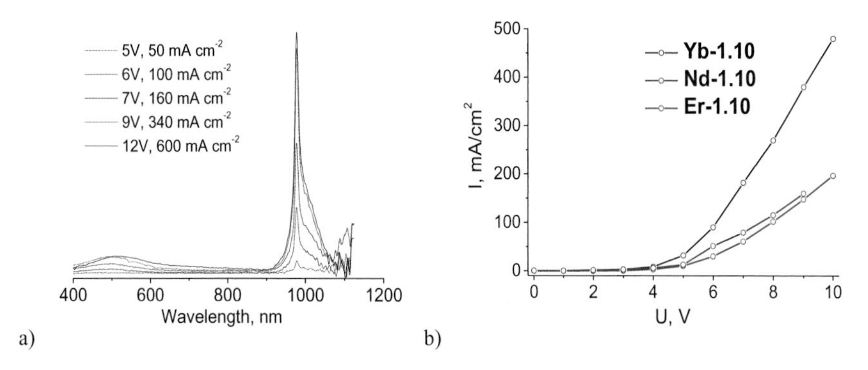

Figure III.1-40 a) Electroluminescence spectra of the device ITO/PEDOT:PSS/**Yb-1.10**/TPBi/Al at different voltages and b) current-voltage curves of the devices ITO/PEDOT:PSS/**Ln-1.10**/TPBi/Al (Ln = Nd, Er, Yb).

The substitution of TPBi as an electron-blocking layer with TAZ allowed to reach the efficiency of EQE = 0.14% at 14 V. Replacing the Al cathode by Ca/Al let to a further increase of OLED performance due to the lower work function of Ca (Figure III.1-39b). The maximum efficiency of 0.21% was achieved at 12 V. The highest published value for NIR-emitting diodes is 1%, though less than 20% of the emission spectrum of the published device with an EQE of 1.0% is in the NIR range. In our case, the EQE is lower, but the luminescence is fully in the NIR range, meaning that an **Yb-1.10** device does not only exhibit at least comparable efficiency in the NIR range, but also avoids the undesirable visible emission. Given the results and the high performance of the obtained device, one can expect that these lanthanide complexes as emissive materials would have a significant impact on the development of NIR-emitting OLEDs.

Conclusions

This work demonstrates a manufacturing process of the first fully phosphorescent OLED based on the gadolinium(III) 9-anthracenate complex $Gd(ant)_3$ which possesses room temperature phosphorescence. Besides,NIR-emitting OLEDs, based on erbium, neodymium and ytterbium 9-anthracenates were successfully fabricated. In this study the best luminescence performance was achieved in the OLED device based on ytterbium(III) 9-anthracenate complex $Yb(ant)_3$. The photophysical study revealed that the quantum yield in the NIR range for this complex reaches 1.5% for and can be potentially increased up to 2.5% by the formation of a solid solution of $Yb_{0.3}Lu_{0.7}(ant)_3$ and hence, reducing the concentration quenching.

Such a high luminescence performance of NIR-emitting lanthanide 9-anthracenates may be explained by two factors: (i) the absence of quenching coordinated water molecules and (ii) the efficient energy transfer through LMCT state.

The high electron conductivity of lanthanide 9-anthracenates may be explained by their structural features, studied by means of powder X-ray diffraction. The complexes consist of the long chains, stabilized by stacking interactions, which further defines their morphology as nanorods. This makes lanthanide 9-anthracenates ready for the fabrication of nanodevices as synthesized. On top of that, they possess high thermal stability (up to 350 °C with no phase transition), as well as high stability under electric current (up to 500 mA/cm^2).

The combination of these properties leads to a high electroluminescence efficiency, also in the NIR range, peaking at 0.21% at 14 V for $Yb(ant)_3$-based OLEDs.

III.1.4 Lanthanide complexes for FRET-assay

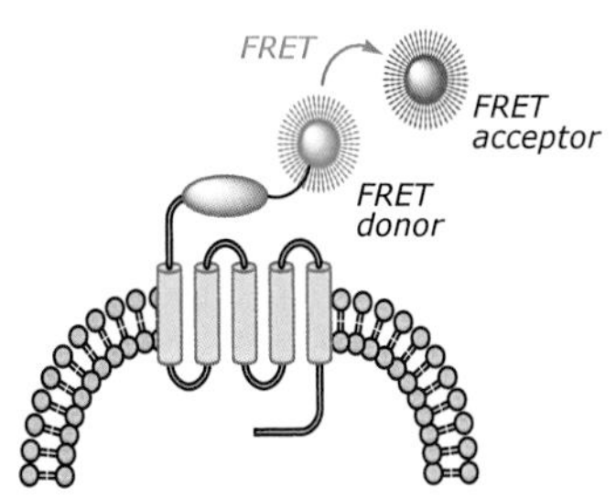

This chapter demonstrates the optimization of the synthetic route for Lumi4Tb®, the most popular FRET-donor for biological assays. Lumi4Tb® is a Tb(III) complex with an octadentate, macrotricyclic ligand that features a bicapped topology and 2-hydroxyisophthalamide chelating units. The optimized synthetic route to obtain this compound includes 13 steps to obtain Lumi4Tb® without a linker and 17 steps to obtain functionalized Lumi4Tb®, suitable for derivatization and protein bioconjugation.

This work was executed during the research stay from 11th of February 2018 through 05th of May 2018 at the Harvard Medical School - Massachusetts General Hospital (Boston, USA) under supervision of Prof. Dr. RALPH MAZITSCHEK.

The favorable luminescent properties of Ln-based emitters (long lifetimes, narrow luminescent bands, large effective STOKES' shifts) make them unique and irreplaceable as FRET-donors for a variety of biomedical applications. Nowadays, the FRET-assay is becoming a powerful technique for studying molecular interactions inside living cells with improved spatial (angstrom) and temporal (nanosecond) resolution, distance range, sensitivity and a broader range of biological applications. It can be useful, for instance, for the analysis of protein–protein interactions with high spatial and temporal specificity (*e.g.* clustering), in the study of conformational changes, in the analysis of binding sequences and in applications such as high-throughput screening. FRET is the radiationless energy transfer from an excited fluorescent donor to a nearby (< 10 nm) acceptor, whose absorption spectrum overlaps with the donor emission spectrum.[208] The changes in a FRET-signal are highly sensitive to distance changes over the length scale of proteins, because the energy-transfer efficiency inversely varies with the sixth power of the separation between the donor (D) and the acceptor (A). FRET-based biosensors for live-cell imaging often incorporate two differently colored fluorescent proteins, as D and A*, to read out changes in the protein conformation or interaction (Figure III.1-41a). The imaging of the intermolecular interactions between two labeled proteins may be limited by (i) crosstalk (direct acceptor excitation by light used to excite the donor); (ii) bleed through (partial overlap of donor and acceptor emission wavelengths) and (iii) non-unitary ratios of donor- and acceptor-labeled proteins (Figure III.1-41b).

* usually CFP and YFP.

The use of lanthanides, especially Tb(III), as FRET donors along with time-gated detection (TGD) (often referred to a lanthanide-based FRET, or LRET) offers distinct advantages over conventional FRET: (i) bleed-through is minimized because the narrow emission bands of Tb(III) can be spectrally isolated from sensitized acceptor emission signals (Figure III.1-41c); (ii) crosstalk and autofluorescence background signals are fully eliminated by using TGD with a microsecond delay (Figure III.1-41c) due to the long luminescence lifetime (millisecond range); (iii) Tb(III)-based FRET imaging affords the possibility of multiplexed FRET imaging, where Tb(III) can sensitize the emission of two or more differently colored acceptors. To fully implement time-gated FRET microscopy for protein-protein interaction study three components are required: (i) suitably bright and stable lanthanide complexes that can be selectively targeted to proteins in living cells; (ii) methods to deliver these complexes into specific cellular compartments without unduly perturbing cellular physiology and (iii) a time-gated luminescence microscope.

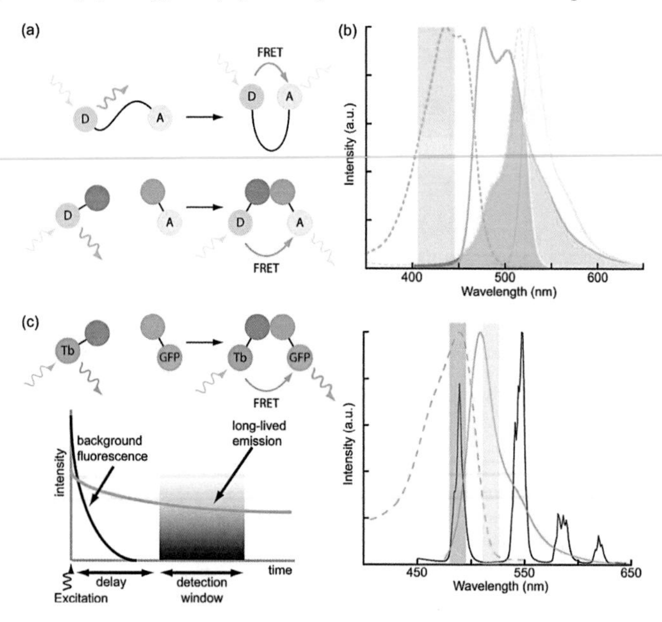

Figure III.1-41 FRET-based biosensors and effects of donor and acceptor photophysics on detection. (a) Conformational changes of a single-chain biosensor (top) or interaction of dual-chain biosensor (bottom) components that bring donor (D) and acceptor (A) fluorophores within the FRET distance (<10 nm). (b) Excitation (dotted) and emission (solid) spectra of CFP (cyan) and YFP (yellow), a common donor–acceptor pair for live-cell FRET imaging. Overlap of CFP emission and YFP excitation spectra (green) allows sensitized YFP emission (yellow band) to be detected upon the excitation of CFP (cyan band). Crosstalk, or direct excitation of YFP in the CFP band (blue) and bleed-through of CFP emission into the YFP band (orange), obscures the true FRET signals, necessitating multiple measurements with different filter sets. (c) With a long-lifetime Tb(III) donor and short-lifetime GFP acceptor, Tb(III) and Tb(III)-sensitized GFP emission can be separated, using narrow-pass emission filters, eliminating bleed-through. Crosstalk is eliminated by time-gated detection (TGD) of long-lifetime Tb(III)-to-GFP FRET. Adapted with permission from L.W. MILLER *et al.*[208], *Inorg. Chem.* 2014, *53*, 1839–1853. Copyright© 2018 American Chemical Society.

There are several requirements to the lanthanide complexes to be practically used in such a FRET-based bioassay:

(i) kinetic inertness with respect to metal binding;

(ii) high extinction coefficient (> 10000 M^{-1} cm^{-1}) and emission quantum yield (> 0.1);

(iii) long-wavelength (> 350 nm) ligand absorption maximum;

(iv) resistance to photobleaching;

(v) one or more functional groups to conjugate with biomolecules or targeting moieties.[209]

While hundreds of luminescent Tb(III) and Eu(III) complexes have been reported, only relatively few meet all of the above-mentioned requirements.[208]

Lumi4Tb® is the most popular FRET-donor which incorporates all aforementioned properties and is, therefore, widely used in many biomedical assays. However, these unique properties as well as the sophisticated synthesis of a Lumi4Tb® probe makes it very expensive and inaccessible for many research purposes. This work demonstrates the optimization of the synthesis of Lumi4Tb® and proposes a novel approach to the linker incorporation, so that Lumi4Tb® can be easily attached to biomolecules or targeting moieties.

Synthesis of the macrocyclic ligand

Lumi4Tb® is a Tb(III) complex with macrocyclic bicapped H(2,2) ligand, bearing 2-hydroxyisophthalamide (IAM) moiety, acting as both, an antenna chromophore and a chelating group (Figure III.1-42). This predisposes the metal-binding units into an environment ideal for f-element cation coordination. Such bicapped macrocyclic complexes retain their exceptional stability and extremely bright photophysical properties under a large variety of solution conditions at nanomolar concentrations and even in the presence of strongly competitive anions and cations, which makes these complexes applicable in biologically relevant homogeneous assay conditions.

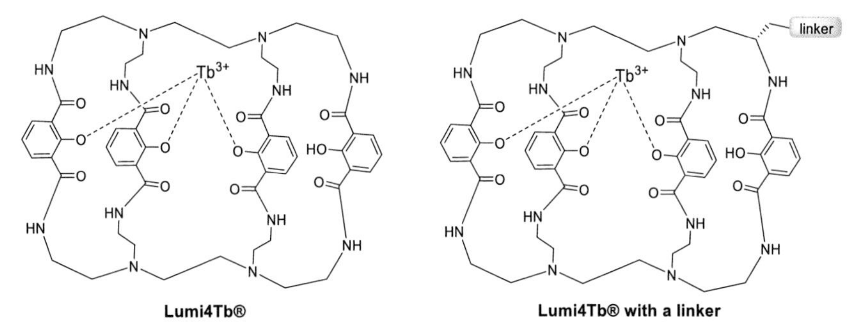

Figure III.1-42 Structural formulae of compound Lumi4Tb® and its analog with the linker.

The synthetic route proposed for the synthesis of a Tb(III) complex with the cage-type bicapped macrocyclic ligand contains several important steps:

(i) synthesis of the “caps” based on a pentaethylene tetraamine (PT) scaffold (Figure III.1-43a);

(ii) synthesis of the monoactivated “cages” based on methyl 2-methoxyisophthalate (IAC-OMe) scaffold (Figure III.1-43b);

(iii) conjugation of the PT with IAC-OMe, resulting in the formation of a monocapped “cage”, followed by saponification of methyl ester protecting groups (Figure III.1-43c);

(iv) conjugation of the monocapped “cage” with the second “cap”, following by the deprotection of the methoxy groups (Figure III.1-43d);

(v) synthesis of Tb(III) complexes (Figure III.1-43e).

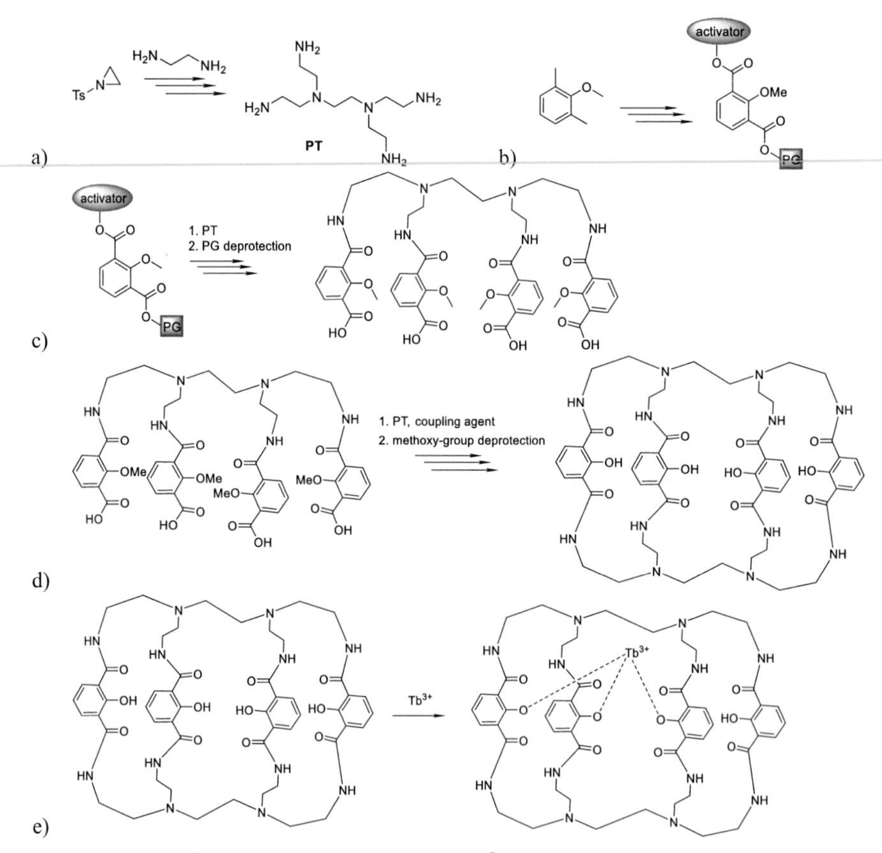

Figure III.1-43 General scheme for the synthesis of Lumi4Tb® probe without a linker: a) synthesis of the “cap”; b) synthesis of the monoactivated “cage”; c) synthesis of the monocapped “cage”, where PG=protecting group; d) synthesis of the bicapped “cage”; e) synthesis of Tb(III) complex.

The "cap" **1.17** was synthesized, using an optimized literature procedure[210] based on the original preparation.[211] The synthetic route starts with the protection of 2-chloroethylamine hydrochloride **1.11*** with tosyl group (Figure III.1-44a), using tosyl chloride **1.12** in water and K_2CO_3 as a base.[212,213] The reported conditions for this reaction gave no full consumption of tosyl chloride **1.12**, caused by its poor solubility in water and it was isolated together with product **1.13** (Table III.1-9, entry 1). An increase of the reaction time did not result in any further improvement of the reaction yield or the product purity. Besides, HCl molecules, generated in the reaction, may prevent the reaction from going further by forming amine hydrochloride, therefore, basic conditions are required. Alterating the amount of the reactants, the pure product was finally obtained and no further purification was needed (Table III.1-9, entry 7).

Table III.1-9: Summary of tested tosylation reactions with 2-chloroethylamine hydrochloride **1.11**, the amount of tosyl chloride **1.12** is taken as 1.0 equiv.

Entry	Eq. amine	Eq. base	Purity [%]
1	1.0	1.0 (K_2CO_3)	60
2	1.0	1.2 (K_2CO_3)	66
3	1.0	2.0 (K_2CO_3)	72
4	1.0	2.5 (K_2CO_3)	74
5	1.0	3.0 (K_2CO_3)	74
3	1.0	1.5 (KOH)	– [a]
6	1.4	2.0 (K_2CO_3)	91
7	1.4	2.5 (K_2CO_3)	100

[a] product was not isolated.

An increase of the amount of amine **1.11**, ensuring the full consumption of TsCl **1.12**, led to the formation of the desired product of decent purity (Table III.1-9, entry 6). The optimal reaction conditions, namely 1.4 equiv. of the amine and 2.5 equiv. of K_2CO_3, afforded desired product **1.13** of high purity.

The obtained sulfonamide **1.13** was then converted to tosyl-aziridine **1.14** [213] (Figure III.1-44, b). The subsequent ring-opening during the reaction with ethylene diamine **1.15** (EDA) resulted in the formation of tosyl-protected PT **1.16** (Figure III.1-44, c). The latter step was performed, using benzene as a solvent, as described in the original papers,[210,211] though the patented[214] protocol

* ~ 12 € for 5 g, offered by SIGMA ALDRICH GmbH, 20.05.2018

of the Lumi4Tb® synthesis suggests to use a toluene/acetonitrile mixture with heating. The slightly higher solubility of tosyl-aziridine **1.14** in benzene in comparison with a toluene/acetonitrile mixture allowed using lower temperatures for this reaction, which decreased the rate of tosyl-aziridine degradation and eventually, increased the reaction yield (22% using toluene/acetonitrile at 65 °C, as described in the patent[214] vs. 66% using benzene, 0 – 30 °C, as described in[211,215]). The last step involved deprotection of tosyl protecting groups by refluxing **1.16** in 33% HBr solution in acetic acid for 2 days, to afford compound **1.17**.[216,217] The use of aqueous HBr with acetic acid, as described in the patent,[214] afforded the product in a low yield (46% vs. 90% for non-aqueous HBr, used in this work). The overall yield for the "cap" synthesis was 55% over 4 steps (Figure III.1-44).

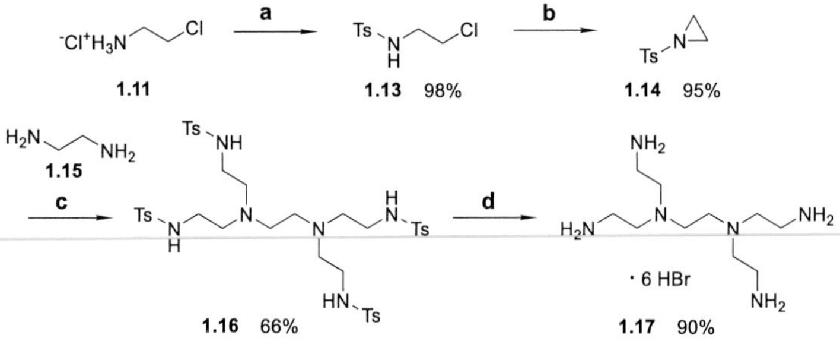

Figure III.1-44 Synthesis of the "cap" based on the pentaethylene tetraamine scaffold: (a) TsCl (**1.12**), K_2CO_3, H_2O, ovrn; (b) NaOH, H_2O, 0 – 10 °C, ovrn; (c) EDA (**1.15**), benzene, 0°C – rt, ovrn; (d) HBr in AcOH, 48 h, reflux.

The synthesis of the "cages" is shown in Figure III.1-45, using the modified procedure from literature.[218] In order to obtain a bicapped ligand, the carboxylic groups of the "cages" should be activated stepwise to prevent the polymerization during the first "cap+cage" coupling (Figure III.1-43c).

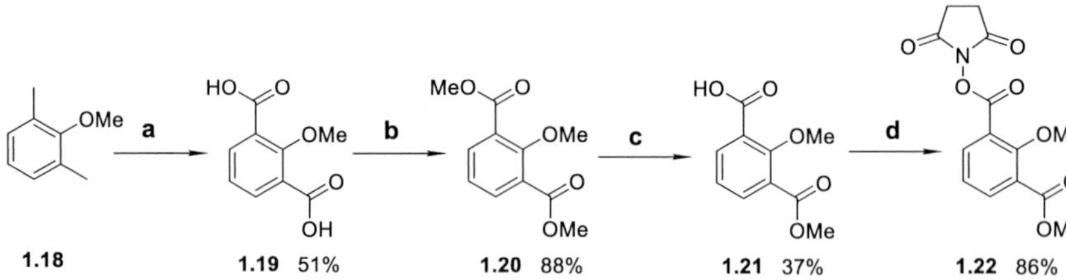

Figure III.1-45 Synthesis of the "cages" based on the IAC-OMe scaffold: (a) 1. KOH, $KMnO_4$, 80 °C, ovrn, 2. HCl, H_2O; (b) H_2SO_4, MeOH, 4 h, reflux; (c) 1. KOH (1.0 equiv.), MeOH, 50 °C, 4 h, 2. HCl, H_2O; (d) NHS, EDC, acetonitrile.

The synthetic route started with an oxidation of 2,6-dimethylanisole **1.18**, a cheap and easily available reactant,* with $KMnO_4$ in water at 80 °C, which afforded dicarboxylic acid **1.19** in 51% yield (Figure III.1-45a). Monoprotection of the dicarboxylic acid was performed in two steps:

(i) protection of both carboxylic groups by the formation of compound **1.20** with two methyl ester groups, using FISCHER esterification with H_2SO_4 in MeOH (Figure III.1-45b);

(ii) monodeprotection of one carboxylic group to afford compound **1.21** (Figure III.1-45c) as a result of a mild saponification with 1.0 equiv. of KOH in MeOH at room temperature.[219]

Since this reaction is not selective, the crude product contained all three possible products: double-, mono- and non-protected 2-methoxyisophthalates (**1.20**, **1.21** and **1.19**, respectively).

Limiting the amount of byproduct **1.19** by adding only 1.0 equiv. of the base, hydroxides of different alkali metals were tested (Table III.1-10). Lithium hydroxide LiOH which is known to be very effective in saponification reactions indeed turned out to be more reactive than NaOH and KOH, showing a conversion even at room temperature (Table III.1-10, entries 1 – 2). However, in this case such an efficiency is a drawback, since together with desired product **1.21**, di-acid **1.19** was formed (Figure III.1-46). Moreover, when the reaction mixture was refluxed with LiOH, the unfavorable deprotection of the ether group was observed and the reaction resulted in 2-hydroxyisophthalate **1.23** as the major product (Table III.1-10, entry 3, Figure III.1-46) (calculated according to the integration of diode array signals in LC-MS). This effect of an unusual deprotection of aromatic ethers in the presence of esters in *ortho*-position is poorly described in literature,[220] particularly, the deprotection with LiOH has never been reported before.

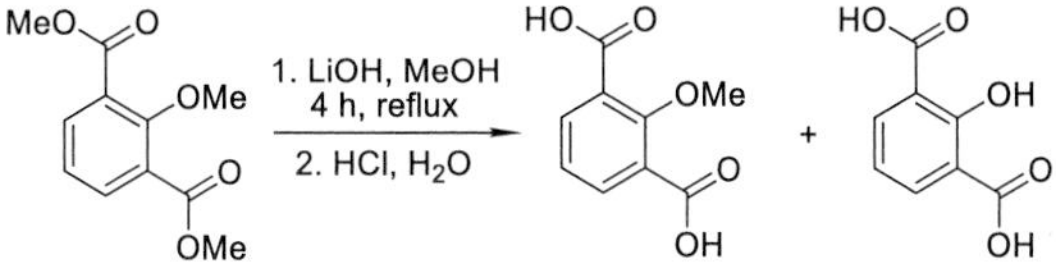

MeO O OMe O OMe — 1. LiOH, MeOH 4 h, reflux; 2. HCl, H_2O → HO O OMe O OH + HO O OH O OH

1.20 — **1.19** 30% — **1.23** 70%

Figure III.1-46 Deprotection of the methyl ether together with methyl esters in *ortho*-position, refluxing **1.19** with LiOH in MeOH.

When NaOH was used as a base, 20% conversion was observed after stirring the reaction mixture for 2 h at room temperature (Table III.1-10, entry 4). When the reaction time was increased up to 4 h, the amount of the product (P) **1.21** increased (25%, Table III.1-10, entry 5), however, the amount of the byproduct (BP) **1.19** increased as well, making these reaction conditions less

* ~ 32 € for 25 g, offered by SIGMA ALDRICH GmbH, 20.05.2018

efficient. Shortening of the reaction time to 1.5 h and the reaction temperature increase (Table III.1-10, entry 6) led to a sole formation of BP **1.19**.

Potassium hydroxide was found to be more favorable for this reaction. The first 8 h stirring at room temperature resulted in little conversion (10%) (Table III.1-10, entries 7 – 9) and no byproduct **1.19** was formed. However, after 24 h, the amount of product **1.21** did not significantly change, while BP **1.19** started to appear (Table III.1-10, entry 10). It was found that heating of the reaction mixture for at least 2 h at 50 °C led to an improvement of the conversion (Table III.1-10, entry 11). On top of that, the yield was slightly higher when water had been eliminated and the reaction was performed in MeOH (Table III.1-10, entries 12 – 14). After explorative research, the optimal reaction time was found and they are displayed in Table III.1-10, entry 13.

Table III.1-10: Summary of tested conditions for saponification reaction with dimethyl 1,2-methoxyisophthalate **1.20**. The amount of the base is 1.0 equiv., unless otherwise stated.

Entry	Base	Solvent	Time	T	P(1.21)/SM(1.20)/BP(1.19) ratio, %
1	LiOH	MeOH/H_2O	2 h	rt	20/60/20
2	LiOH	MeOH	2 h	rt	30/45/25
3	LiOH[a]	MeOH	2 h	reflux	-/-/30[b]
4	NaOH	MeOH/H_2O	2 h	rt	20/70/10
5	NaOH	MeOH/H_2O	4 h	rt	25/50/25
6	NaOH	MeOH/H_2O	1.5 h	reflux	-/60/40
7	KOH	MeOH/H_2O	1 h	rt	-/100/-
8	KOH	MeOH/H_2O	2 h	rt	-/100/-
9	KOH	MeOH/H_2O	8 h	rt	10/90/-
10	KOH	MeOH/H_2O	24 h	rt	10/70/20
11	KOH	MeOH/H_2O	4 h	50 °C	40/40/20
12	KOH	MeOH	2 h	50 °C	45/40/15
13	KOH	MeOH	4 h	50 °C	55/25/20
14	KOH	MeOH	5 h	50 °C	45/20/35
15	KOH	CH_3CN	10 min	rt	20/60/20
16	KOH	THF	15 min	rt	30/40/30

[a] 4.0 equiv. [b] 2-hydroxyisophthalate **1.23** was obtained.

Due to the limited solubility of potassium benzoates in common organic solvents, the choice of an appropriate solvent might make the product to precipitate and hence, increase the overall reaction yield and simplify the purification, *e.g.* as reported for malonic acid.[221] Therefore, saponification was performed in acetonitrile and THF at room temperature. (Table III.1-10, entries 15 – 16). In both cases product **1.21**, indeed, was precipitating as the respective potassium salt. However, after reaching the maximum, the amount of precipitate significantly decreased within 2 minutes and only a little amount of the product could be isolated (less than 10% in both cases), resulting in a clear solution still containing all three compounds (SM **1.20**, P **1.21**, BP **1.19**).

Experimentally found optimized reaction conditions are listed in Table III.1-10, entry 13, namely stirring of compound **1.20** with 1.0 equiv. of potassium hydroxide at 50 °C for 4 h, giving 55% of desired product **1.21**, 20% of di-acid BP **1.19** and 25% of unreacted SM **1.20**, according to LC-MS data. The separation of these products was performed by multiple extractions: firstly, the extraction of methyl esters with EtOAc, so that mono- and bis-potassium salts remained in the aqueous layer. Secondly, after the latter was acidified, monoprotected acid **1.21** was extracted with DCM, while non-protected dicarboxylic acid **1.19** stayed in the aqueous layer. Thus, monomethyl ester **1.21** was isolated in 37% yield.

Compound **1.21** was directly used in the coupling reaction, using PyBOP as a coupling agent (Figure III.1-47). The reaction was completed within 5 minutes, however, the "monocapped" product **1.24** had the same retention time as the PyBOP-related byproduct (phosphine oxide) in hexanes/EtOAc and DCM/MeOH solvent mixtures. Therefore, the attempts to purify the compound *via* flash column chromatography were not successful.

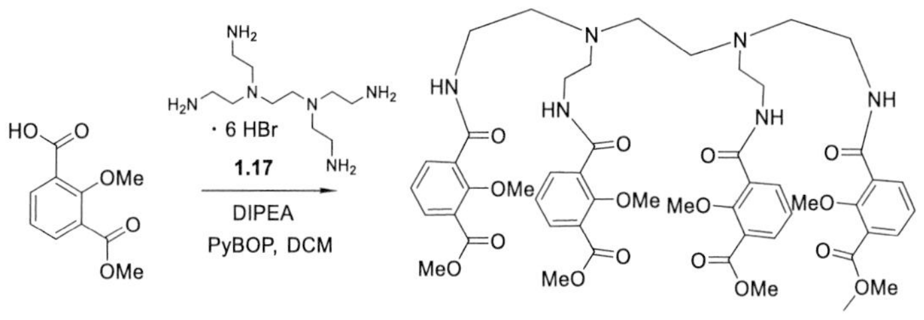

Figure III.1-47 Coupling of the monoprotected acid "cage" **1.21** with PT (**1.17**) using PyBOP as a coupling agent.

To conduct the coupling reaction without any additional coupling agents, monomethyl ester **1.21** had to be preactivated. The activation process consisted of the formation of an *N*-hydroxysuccinimidyl ester (NHS-ester) by consecutively reacting carboxylic acid with carbodiimide and NHS. The pure NHS-ester **1.22** was recrystallized from methanol in a very good yield (Figure III.1-45d). The subsequent reaction of the NHS-ester with PT (**1.17**, Figure III.1-48a) afforded monocapped compound **1.24** in 77% yield. The saponification of **1.24** afforded monocapped compound **1.25**, bearing carboxylic acid moieties, ready for the next coupling step.

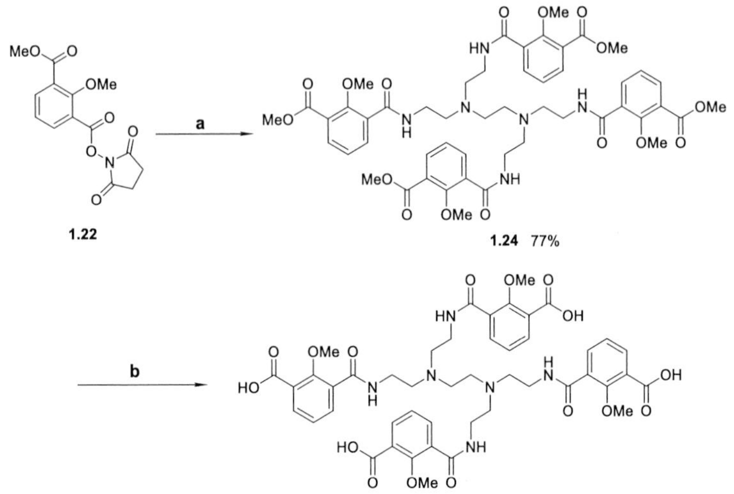

Figure III.1-48 Coupling of the NHS-preactivated **1.22** with PT (**1.17**), followed by the deprotection of the methyl esters: a) PT (**1.17**), DIPEA, DCM; b) 1. KOH, MeOH, 4 h, reflux, 2. HCl.

The coupling of monocapped compound **1.25** with PT is a very delicate reaction, as it should be performed for two compounds with four reactive sites each, leading to numerous byproducts. In order to minimize the amount of byproducts, such reactions are usually performed in high dilution and a very low rate of supply of the starting materials (for instance, dilution 0.25 mmol/L, reaction time 7 days, yield 52%, reported for compound **1.26**[106] or dilution 1.9 mmol/L, reaction time 24 h, yield 46%, reported for an analogous compound[222]). Herein, a new protocol was established, allowing to perform the reaction in 2 h with a lower dilution (5.0 mmol/L).

Equimolar solutions of starting materials **1.25** and **1.17** with DIPEA in DMF were added portionwise to a stirred solution of PyBOP, adding a 500 μL portion of each solution simultaneously every 5 min (ten portions in total). After ten portions had been added, the reaction mixture was left stirring for 1 h and then the solvent was evaporated under reduced pressure. The simultaneous addition of two starting materials to a moderately diluted solution of PyBOP ensured the high probability to obtain desired product **1.26**, because of the high rate of amide bond formation, using PyBOP activation (Figure III.1-49).

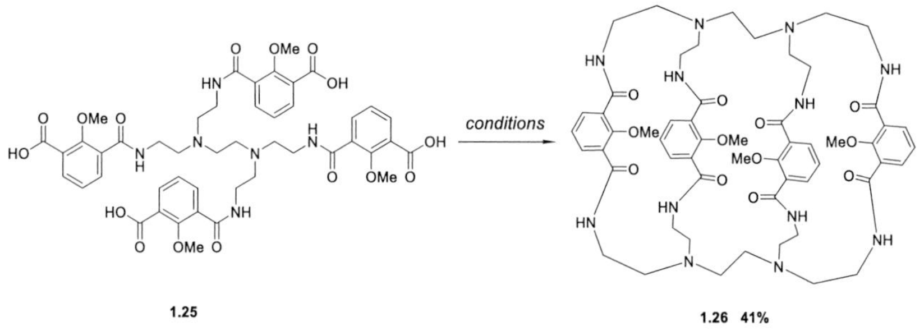

1.25 → conditions → 1.26 41%

Figure III.1-49 Coupling of the monocapped acid **1.25** with PT (**1.17**), using PyBOP as a coupling agent to afford the bicapped ligand **1.26**; conditions: PT (**1.17**), PyBOP, DMF, portionwise, 2 h, rt.

The next step involved the demethylation of the bicapped "cage" **1.26**, which has been usually done, using BBr_3 in anhydrous DCM.[106,218] Since this reaction has been reported to be extremely slow (2 – 4 days), other conditions were also tested and monitored by LC-MS. The conversion degree was calculated based on the integration of the diode array signals.

The experimental evidence demonstrated the reaction with BBr_3 being slow. After 2 days only 18% conversion was detected (Table III.1-11, entry 1), whereas an increase of temperature and additional stirring for one day afforded only 26% of the product. ZHU *et al.*[223] reported that sodium azide promoted the hydrolysis of the anisole, thus NaN_3 in DMF in the presence of 1% water with heating (80 °C), was able to deprotect methyl ethers. Indeed, after 2 h a little conversion was observed (Table III.1-11, entry 3). However, after 4 h, neither starting material **1.26**, nor partially deprotected intermediates, nor product **1.27** were detected on LC-MS. The reaction with LiOH in MeOH which was shown to deprotect methyl ethers earlier (Figure III.1-46), also did not afford the product even after 8 h of reflux and in this case SM **1.26** was fully recovered (Table III.1-11, entry 5). Full conversion was, however, achieved by refluxing of **1.26** with LiI in 2,6-lutidine for 4 h[224] (Table III.1-11, entry 6). Using the same reaction conditions, compound **1.24**, bearing ether and ester groups, was fully deprotected to afford compound **1.28** in 69% yield (Figure III.1-50). This compound was synthesized for the further complexation with Tb(III) and the investigation if this coumpound could relplace Lumi4Tb®.

Despite high symmetry, ^{1}H and ^{13}C-NMR spectra of bicapped ligand **1.27** turned out to be very sophisticated and were difficult to resolve, suggesting the coexistence of several conformers/isomers. Therefore, HPLC together with mass spectrometry were used for the characterisation of these compounds. In both cases, the HPLC chromatogram of **1.27** showed only one sharp peak, identified by mass spectroscopy as the desired macrotricyclic ligand.

Table III.1-11. Summary of tested reaction conditions of **1.26** demethylation.

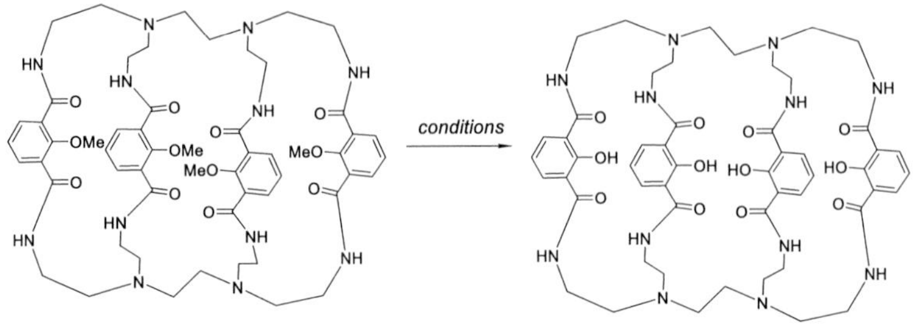

1.26 **1.27**

Entry	Conditions	Time	Conversion[a] [%]
1	BBr_3, anhydrous DCM, 0 °C – rt	2 d	18
2	BBr_3, anhydrous DCM, 0 °C – 40 °C	3 d	26
3	NaN_3, DMF, 80 °C	2 h	40
4	NaN_3, DMF, 80 °C	4 h	–[b]
5	LiOH, MeOH, reflux	8 h	–[c]
6	LiI, 2,6-lutidine, reflux	4 h	100

[a] calculated based on LC-MS results; [b] product was not isolated, SM was lost; [c] SM was recovered.

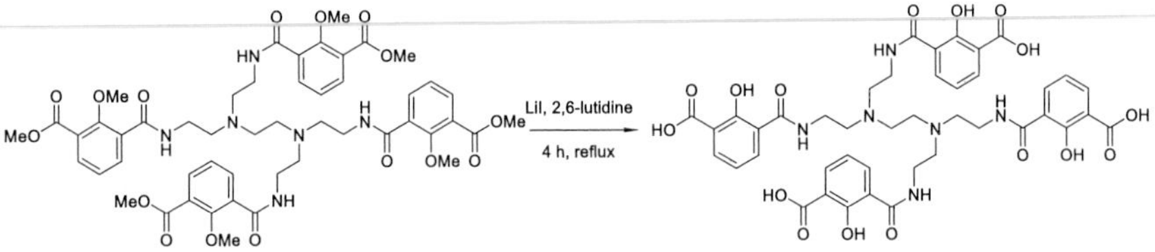

1.24 **1.28** 69%

Figure III.1-50 Deprotection of ester and ether groups using LuI in 2,6-luthidine.

The mild reflux of compound **1.27** with methanol solution of Tb(III) chloride afforded product **Tb-1.27** (Lumi4Tb® probe) in a quantitative yield (Figure III.1-51).

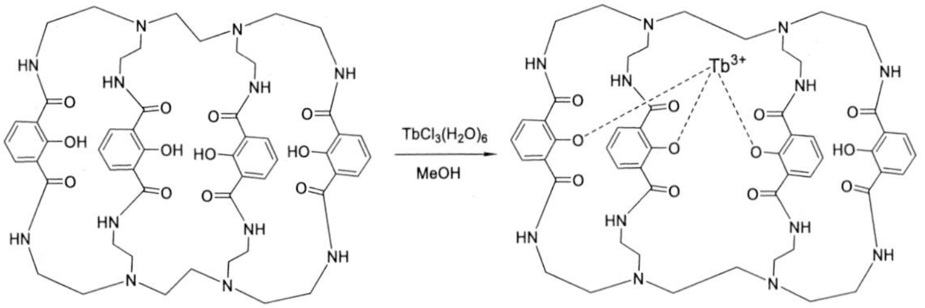

1.27 **Tb-1.27**

Figure III.1-51 Synthesis of Tb(III) complex **Tb-1.27** with bicapped ligand **1.27**.

Monocapped ligand **1.25** and its demethylated derivative **1.28** were also used for the complexation with Tb(III). Complexes **Tb-1.25** and **Tb-1.28** were therefore synthesized, using a standard two-step procedure for the synthesis of lanthanide aromatic carboxylates, namely the deprotonation of carboxylates, followed by an addition of terbium chloride. As a result, in the case of **Tb-1.28**, the slurry precipitated from MeOH solution, which is due to the lower solubility of this complex in MeOH in comparison to bicapped complex **Tb-1.27**. At the same time, complex **Tb-1.25** precipitated immediately and turned out to be very poorly soluble, not only in water, but also in common organic solvents (MeOH, EtOH, toluene, acetonitrile), which is due to the immediate formation of coordination polymers (Figure III.1-52). This result indicates that hydroxy groups indeed participate in the coordinative bonding with Tb(III) ion in the cases of **Tb-1.27** and **Tb-1.28**, making them highly luminescent and soluble (Figure III.1-53), while in the case of **Tb-1.25**, Tb(III) coordinates terminal carboxylic groups and did not coordinate donor atoms inside the "cage", probably due to the steric hindrance.

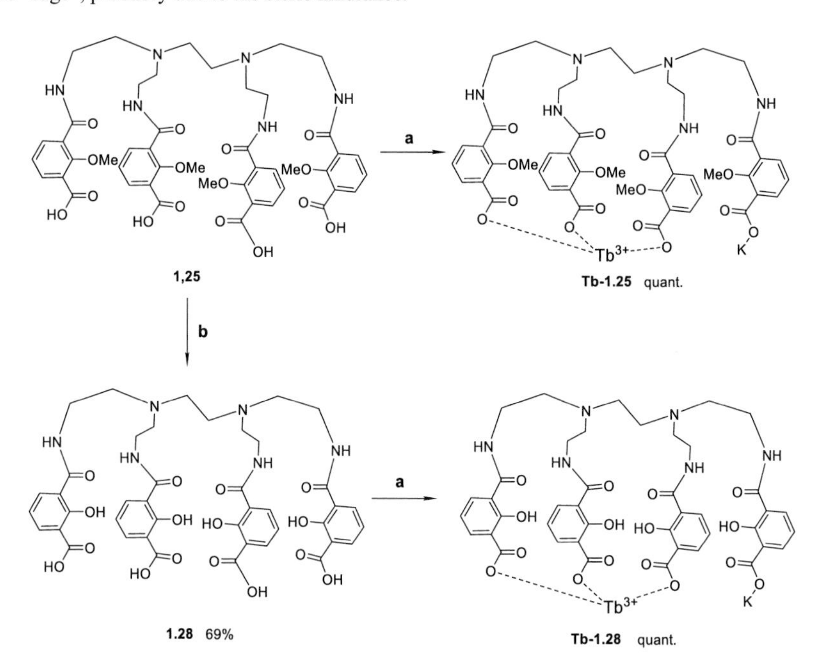

Figure III.1-52 Synthesis of Tb(III) complexes with monocapped carboxylic ligands: a) 1. KOH in MeOH (1.00 equiv.), 2. $TbCl_3(H_2O)$ in MeOH, reflux; b) LiI, 2,6-lutidine, 4 h, reflux.

Figure III.1-53 Comparison of the luminescence intensity under UV-lamp with $\lambda_{ex} = 254$ nm for 10 nM aqueous solutions of Tb chloride, **Tb-1.27**, **Tb-1.28** (from left to right).

Linker incorporation and outlook

To make the Lumi4Tb® probe useful for biological assays, it should contain a linker that can be used for the further conjugation to biomolecules or targeting moieties. In commercially available Lumi4Tb® probes, such a linker is incorporated to one of the "caps" resulting in the formation of an asymmetric, chiral probe (Figure III.1-54a). The synthesis of Lumi4Tb® probe with such a linker is unnecessary complicated and in this work we propose the linker incorporation by an amide alkylation (Figure III.1-54b). According to the crystal structure of compound **1.26** refined by K. RAYMOND *et al.*[106] (Figure III.1-55), this approach should make no difference in the coordination ability of the ligand. The lone pairs of the four tertiary amines on the "caps" are all directed inwards, while all four methyl groups are directed outwards. The H-bonding network results in a nearly planar structure of the ligand. The four diamido IAM groups adopt an up → down → up → down arrangement with pseudo-symmetry, as shown in Figure III.1-55a.

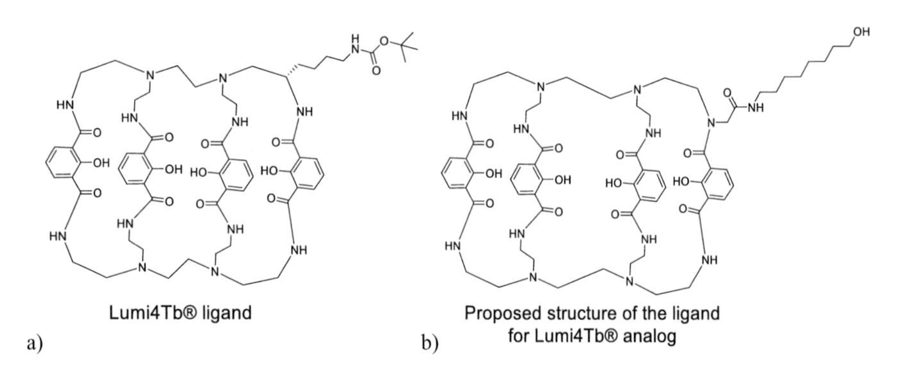

Figure III.1-54 a) Lumi4Tb® ligand structure with an attached linker; b) bicapped "cage" structure with a linker, proposed in this work.

As shown in Figure III.1-55b, four out of eight carbonyl moieties are indeed directed outward from the "cage", which makes them uncapable to coordinate the Tb(III) ion. The neighboring nitrogen atoms are positioned too far away to form a coordination bond with the lanthanide in the center of the "cage". Therefore, if one of these positions was modified, it might not affect the geometry of the "cage" and its coordinating ability.

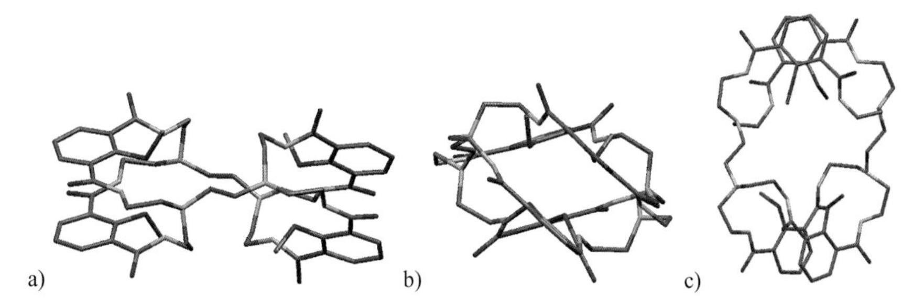

Figure III.1-55 Molecular structure of **1.26**:[106] a) front view; b) side view; c) top view.

Synthetic route, which was proposed for the synthesis of the "cap" with the attached linker, is shown in Figure III.1-56. After selective monoprotection of *N1*-(2-aminoethyl)ethane-1,2-diamine, ring opening reaction with tosylaziridine and subsequent phthalimide deprotection affords an asymmetric PT compound, bearing one amino group, ready for a selective linker incorporation (by any reaction which is more favorable for primary amines vs. sulfonamides).

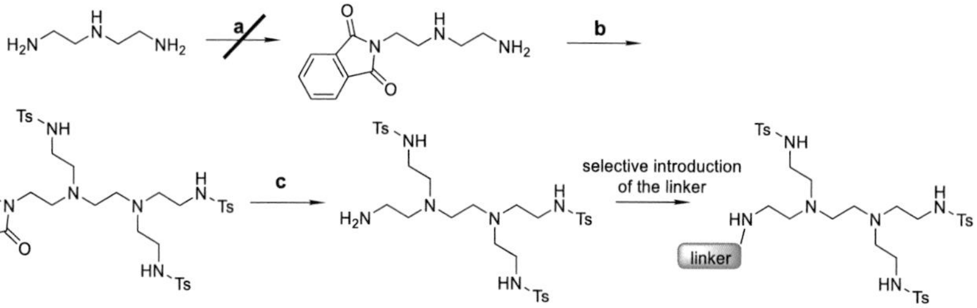

Figure III.1-56 Synthetic route for the synthesis of the "cap" with the attached linker, proposed in this work: a) phthalic anhydride, solvent, heating; b) tosylaziridine, benzene, rt; c) 1. $H_2NNH_2 \cdot H_2O$, MeOH, rt, 2. HCl.

Unfortunately, the first step of this reaction was not successful and the desired product has never been isolated, though, according to LC-MS data, it was formed in small amounts (< 10%). The optimization of the reaction conditions by varying the amount of the SM, solvents, temperature and reaction times, however, did not lead to any improvement. Therefore, this approach was not implemented in this work.

Alternatively, an alkylation of the "cap" might be performed, using Ts-protected compound **1.16**. Two synthetic strategies, MITSUNOBU- and SN_2-reactions, were tested in this work.

To investigate, if MITSUNOBU conditions can be applied to the alkylation of **1.16**, a test reaction with 1.00 equiv. of 1-butanol was performed, varying the solvent, the reagent and the temperature. The formation of the products was monitored by LC-MS (Table III.1-12).

In a deficiency of 1-butanol, the reaction only gave single- and double-substituted products (**1.29** and **1.30**, respectively) and a few if any amount of triple or quadruple substituted products were detected by LC-MS. Diethyl azodicarboxylate (DEAD) (Table III.1-12, entries 4, 5, 8, 9) and diisopropyl azodicarboxylate (DIAD) (Table III.1-12, entries 1 – 3, 6, 7) were both tested as reagents for this reaction and both showed quite similar results (Table III.1-12). The use of different solvents also made only a little difference, probably mostly due to the different solubility of SM **1.16** in these solvents (Table III.1-12, THF, entries 1 – 5; DMSO, entries 6 – 9). An increase of the temperature during the reaction indeed accelerated the reaction, however, also caused the formation of more double-substituted byproduct **1.30** (Table III.1-12, entries 3, 5).

Table III.1-12: Summary of tested MITSUNOBU reactions with **1.16**. The reaction time in all entries was 2 h.

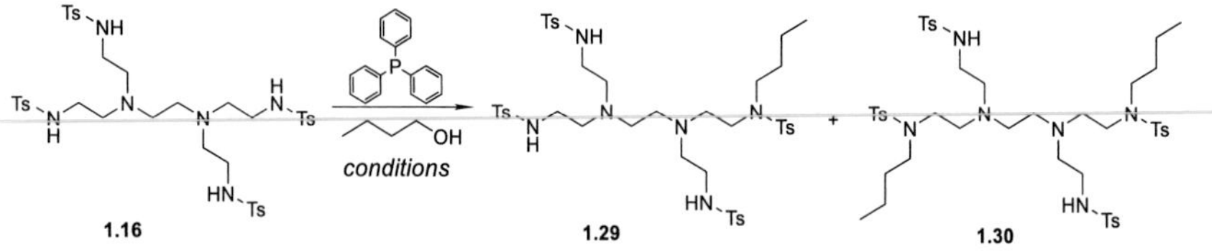

Entry	Reagents	Solvent	T, °C	P/SM/BP[a] ratio, %
1	DIAD	THF	0 °C	-/100/-
2	DIAD	THF	rt	10/40/10
3	DIAD	THF	reflux	-/25/25
4	DEAD	THF	rt	10/35/10
5	DEAD	THF	reflux	-/25/30
6	DIAD	DMSO	rt	10/50/-
7	DIAD	DMSO	reflux	15/45/10
8	DEAD	DMSO	rt	10/55/-
9	DEAD	DMSO	reflux	15/45/15

[a] Schematic representation of the double substituted product. The exact geometry of the product is not considered.

Though the possibility to use the MITSUNOBU reaction for the alkylation of compound **1.16** was demonstrated above, only little conversion was observed with *N*-Boc-aminohexanol, refluxing the latter with **1.16**, triphenylphosphine (TPP) and DEAD in DMSO for 2 h (Figure III.1-57).

Alternatively, an alkylation of the "cap" can be performed *via* an SN_2-reaction by reflux of **1.16** with 1.00 equiv. of ethyl 2-bromoacetate in acetonitrile for at least 4 h. The long reaction time has been chosen due to the poor solubility of SM **1.16** in acetonitrile. After 4 h, product **1.31** was isolated together with the doubly-alkylated BP, however, this mixture was barely separable. After several attempts to purify the desired product *via* flash column chromatography, pure product **1.31** was obtained in a very low yield (6%) (Figure III.1-58a). Therefore, subsequent reactions were only performed in a very small scale (~ 2 mg). The formation of compound **1.32** as a result of the PyBOP coupling was demonstrated, using LC-MS, a full conversion was achieved within 5 minutes.

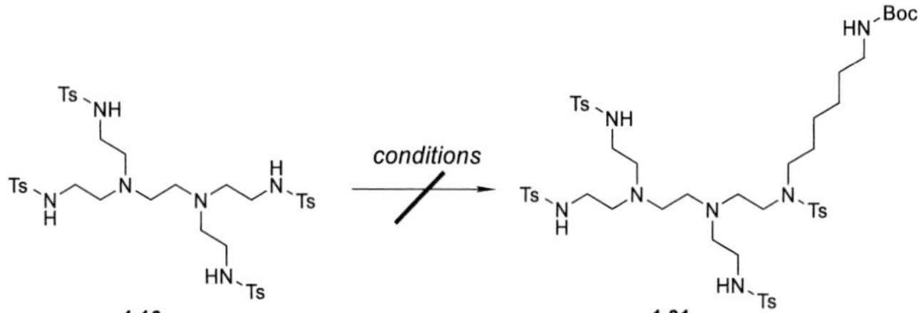

Figure III.1-57 MITSUNOBU reaction with *N*-Boc-aminohexanol; conditions: *N*-Boc-aminohexanol, DEAD, TPP, DMSO, 2 h, reflux.

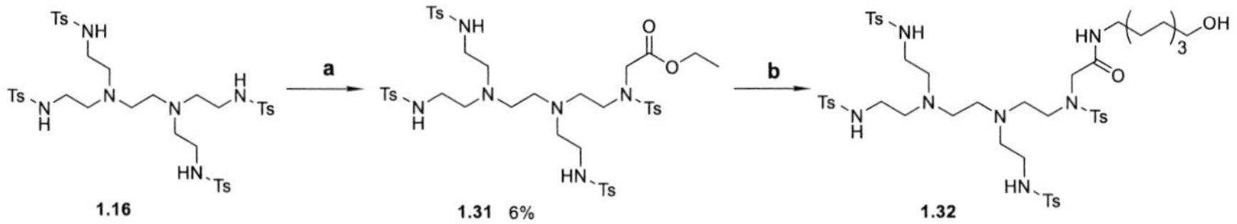

Figure III.1-58 Incorporation of the linker, using a substitution reaction with subsequent coupling to increase the length of the linker; a) ethyl 2-bromoacetate, acetonitrile, Cs_2CO_3, 4 h, reflux; b) 1. LiOH, MeOH, 1 h, rt, 2. 8-aminooctanol, PyBOP, DCM.

Though, the synthesis and purification of compound **1.31** still needs to be optimized, the possibility of the incorporation of a linker by alkylation of compound **1.16** was demonstrated.

Photoluminescent properties

The following section describes the investigation of the luminescent properties of complex **Tb-1.27**. The luminescence spectra and the lifetime of the excited state were recorded, using a custom-designed sensing circuit for photophysical measurements, described by S.A. REIS *et al.*[225] For this purpose, 100 µL of 0.7 nM solution of **Tb-1.27** in tris-buffered saline (TBS) were transferred into a semi-micro 10 mm cuvette. When recording the luminescence spectrum, a laser with the excitation of 337 nm was used and the emission was detected in the range of 400 – 700 nm. For luminescence kinetics measurements, ten measurements had been performed. The signals for all ten runs were then averaged together, filtered through a 100 kHz low-pass filter (5-tap infinite-impulse response BUTTERWORTH) to remove high frequency noise. As least-squared fit modeling a one-phase decay was used to determine the relaxation half-lives. Values are reported at 95% confidence intervals.

Complex **Tb-1.27** exhibited typical Tb(III) based luminescence, demonstrating a similar emission profile, as reported for the Lumi4Tb® probe[106] (Figure III.1-59a). The lifetime of the excited state was measured to be 0.93 ms, which is, however, different from the reported value of 2.45 ms.[106] This discrepancy probably is due to the use of different solvents for the measurements. The original paper[106] does not specify which buffer has been used, instead they report to perform the kinetics measurements in “buffered aqueous solution”. Nevertheless, the luminescence lifetime value, obtained in this work, lies in the millisecond range typical for Tb(III) complexes, which is rather high and suitable for FRET-assay.

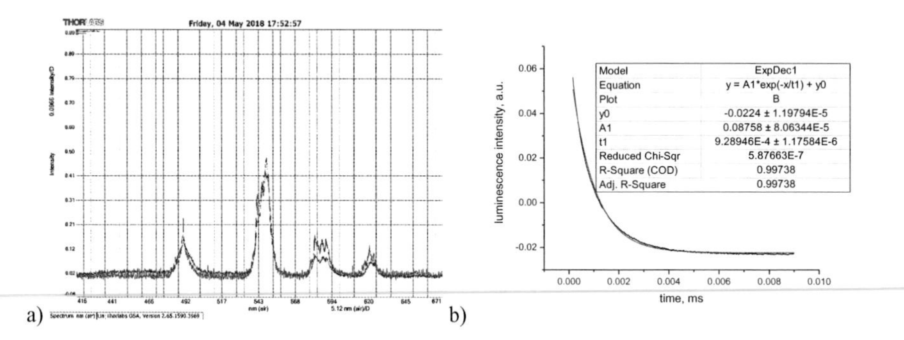

Figure III.1-59 Photophysical measurements for complex **Tb-1.27** in water (blue) and in TBS (green), concentration 0.7 nM, λ_{ex} = 337 nm: a) emission spectrum; b) luminescence decay profile, τ = 0.93 ms.

Conclusions

Thus, this work optimizes the synthetic strategy to obtain the commercially available FRET-donor Lumi4Tb®, which is a multistep process, containing various bottlenecks from chemical complexity to cost points of view. The proposed synthetic route involves cheap and easily available starting materials and reduces reaction times, maintaining high yields.

The obtained probe possesses a very high luminescence intensity, demonstrating the typical Tb(III) emission profile, similar to those reported by RAYMOND *et al.*[106] On top of that, the study of the single-crystal structure of the bicapped ligand reported previously allowed to propose a new way of the linker incorporation, namely amide alkylation, since four nitrogen atoms of the amides are positioned too far away from the center and are not able to form a coordination bond with Tb(III). The synthesis of the alkylated “cap” was performed, demonstrating the potential ability of the implementation of the proposed approach.

III.2 Lanthanide conjugates

In Chapters III.1.1 and III.1.2 of this thesis it was shown that lanthanide fluorobenzoates have high potential to serve as a basis for luminescent bioprobes. These compounds may be regarded as precursors for a coupling with biomaterials but an additional functional group (FG) needs to be introduced into the ligand structure to enable further binding of the luminescent complex to a desired cargo. The introduction of a FG to the structure of the ligand may affect the luminescent properties of Ln(III) complexes in three different ways: it may (i) change the geometry of the Ln(III) complex; (ii) change the electronic properties of the ligand and hence, change the ligand triplet level value; (iii) be a quencher and decrease the luminescence intensity due to its high-energy bond vibrations. The first two reasons can also potentially improve the emitting properties of Ln(III)-based labels, as the geometry might be changed to a more favorable one or the change in a triplet level value may increase the luminescence efficiency, while the 3rd reason can only affect negatively and should be removed or its influence should be minimized.

Therefore, the incorporation of a linking FG seems to be more reasonable to the *p*-position of the benzene ring, since this position is the farthest to the Ln(III) ion. Thus, being modified, it has a lower chance to influence the luminescent properties of the complex. Though, this influence is almost unpredictable and therefore, should be studied empirically.

As shown in Chapter III.1.2, the introduction of an auxiliary ligand may significantly increase the efficiency of lanthanide complex luminescence. Thus, Eu(III) 2-fluorobenzoates with 1,10-phenanthroline were found to be very promising for biological applications (Chapter III.1.2). For such probes, a linker can be introduced to the auxiliary ligand. The incorporation of linkers to fluorobenzoate ligands (Figure III.2-1) as well as to the 1,10-phenanthroline-based auxiliary ligand is discussed in Chapter III.2.1.

Chapter III.2.2 shows Eu(III) conjugates with peptoids and focuses on some synthetic tricks which may be applied to obtain these compounds. The optimization of the conjugation conditions has been performed and the photophysical properties and cellular luminescence were tested for the most promising compounds.

Chapter III.2.3 represents the synthesis and properties of a novel class of compounds, namely heterocycle-containing cyclic peptoids which can potentially carry the metal ion (*e.g.* Eu^{3+}) inside the cycle. In this case, the heterocyclic scaffold is responsible for the sensitization of Eu(III) emission, while the peptoid side chains, bearing the carboxylic functionality, are responsible for the metal coordination and the compensation of the charges. These complexes, combining both, luminescent and transport properties, may find various applications for targeted luminescence.

III.2.1 Functionalized ligands and lanthanide complexes with them

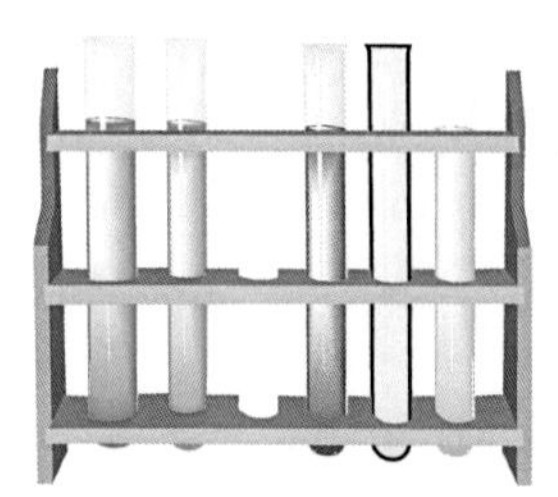

In the present chapter, the properties of lanthanide p-substituted fluorobenzoates is studied. This chapter sets its focus on the synthesis and properties of p-substituted fluorobenzoic acid derivatives, featuring the linking functionality, which is useful for further conjugation with the transporters or biomolecules. This study was also expanded on several non-linking functional groups in order to determine an empirical pattern on how different substituents of the ligands may influence the photophysical properties of lanthanide complexes with them. Moreover, 1,10-phenanthroline was modified with different linkers in order to determine if the lanthanide ternary complex can be conjugated through it.

Modification of the anionic ligand

As shown in Chapter III.1.2, among other fluorobenzoates, lanthanide 2-fluorobenzoates and 2,5-difluorobenzoates are the most promising ones for bioapplications. Therefore, these ligands have been chosen for *p*-modification (Figure III.2-1). In total, 20 ligands were tested in this part of the work. This chapter encompasses the synthesis of **2.1-H – 2.7-H, 2.15-H – 2.18-H** and **2.20-H** ligands, while *p*-substituted ligands **2.8-H – 2.13-H** and **2.19-H** were commercially available.

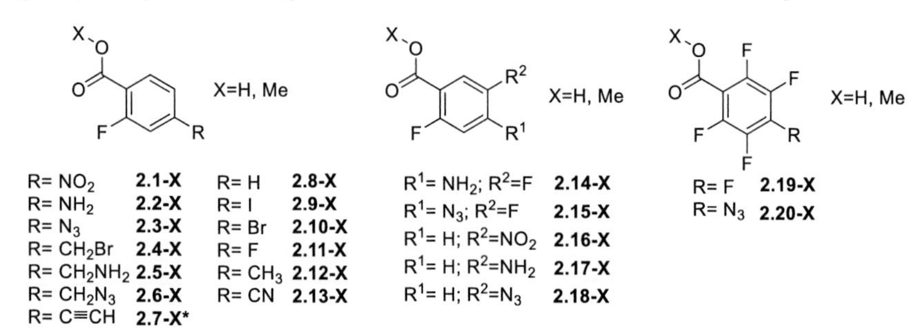

Figure III.2-1 *P*-substituted fluorobenzoates used in this work. Compound **2.7-Me** (denoted with a star) is additionally TMS-protected.

The synthetic route for the incorporation of the different linkers was proposed in this work. The synthesis started with the carboxylic acid as a precursor and involved three crucial steps (Figure III.2-2): (i) protection of the carboxylic acid with methyl group (Figure III.2-2a); (ii) modification of the functional group (FG) in *para*-position (Figure III.2-2b) and (iii) saponification of the methyl ester resulting in the formation of a carboxylic acid (Figure III.2-2c).

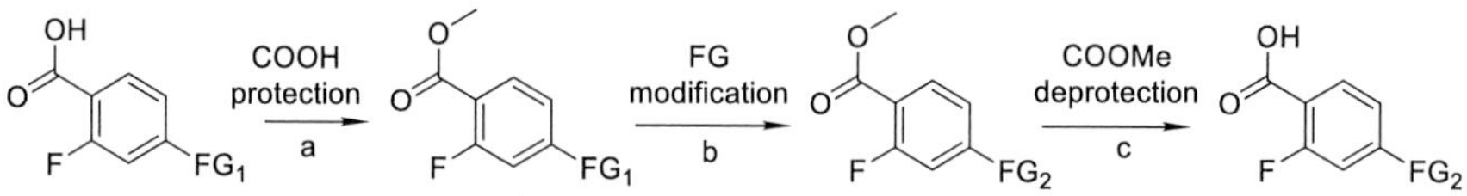

Figure III.2-2 General scheme for the *p*-substituent modification used in this work; conditions: a) MeOH, H_2SO_4, 48 h, reflux; b) modification of the *p*-substituent; c) 1. NaOH, MeOH, H_2O, 2. HCl, H_2O.

Not all the precursors were commercially available, therefore, some of them were additionally synthesized. For instance, compound **2.1-H** was synthesized in two steps:

(i) nitration of 2-fluorotoluene;

(ii) oxidation of the methyl group to the carboxylic group.

The first step, however, afforded two products **2.22** and **2.23,** which could be easily separated *via* flash column chromatography (Figure III.2-3a).

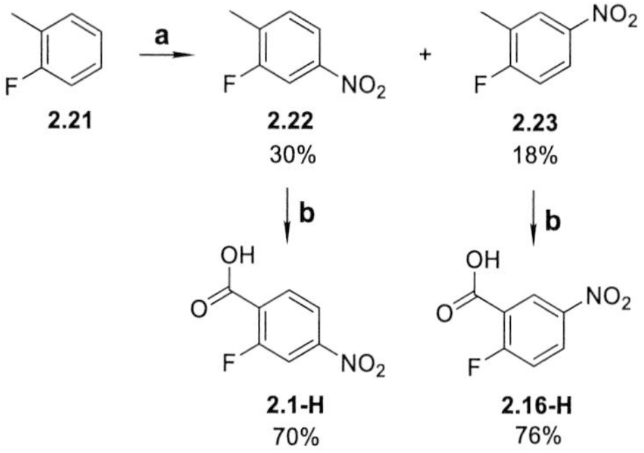

Figure III.2-3 Two-step introduction of nitro group; conditions: a) HNO_3, H_2SO_4, 60 °C; b) $K_2Cr_2O_7$, H_2SO_4, 60 °C.

After separation both compounds **2.22** and **2.23** were further oxidized (Figure III.2-3b), which afforded acids **2.1-H** and **2.16-H** in very good yields (70% and 76%, respectively).

Following the scheme described above (Figure III.2-2), different carboxylic acids were taken as a starting points, bearing -NO_2, -NH_2, -CH_3, -CH_2Br, -CN, -CH_2NH_2 and -Br functionalities in *para*-position (Figure III.2-4). Difluorobenzoic acid was also modified, following the same strategy, so that the commercially available 4-amino-2,5-difluorobenzoic acid (**2.14-H**) has been converted into 4-azido-2,5-difluorobenzoic acid (**2.15-H**) (Figure III.2-5).

The *meta*-position of byproduct **2.16-H** (Figure III.2-3) was also further modified, so that *meta*-modified fluorobenzoates, bearing amino group (**2.17-H**) and azido group (**2.18-H**), could also be tested (Figure III.2-6).

i	FG_1=				FG_2=		
i=1	NO_2	**2.1-H**	**2.1-Me**	87%	NH_2	**2.2-Me** quant.	**2.2-H** 90%
i=2	NH_2		**2.2-Me**		N_3	**2.3-Me** 87%	**2.3-H** 82%
i=3	CH_3	**2.12-H**	**2.12-Me**	83%	CH_2Br	**2.4-Me** 80%	**2.4-H** 88%
i=4	CH_2Br		**2.4-Me**		CH_2N_3	**2.6-Me** 90%	**2.6-H** 94%
i=5	CN	**2.13-H**			CH_2NH_2		**2.5-H** quant.
i=6	CH_2NH_2	**2.5-H**	**2.5-Me**	87%	CH_2N_3	**2.6-Me** 82%	**2.6-H** 90%
i=7	Br	**2.10-H**	**2.10-Me**	86%	C≡C-TMS	**2.7-Me** 67%	C≡C **2.7-H** 86%

Figure III.2-4 General scheme for the *p*-substituent modification, used in this work; conditions: a) MeOH, H_2SO_4, 48 h, reflux; b_i) modification of *p*-substituent: b_1) H_2 (balloon), Raney-Ni, MeOH, THF; b_2) HCl, $NaNO_2$, NaN_3; b_3) NBS, AIBN, $CHCl_3$; b_4) acetone/H_2O, NaN_3; b_6) HCl, $NaNO_2$, NaN_3; b_7) trimethylsilylacetylene, $PdCl_2(PPh_3)_2$, CuI, 1,4-dioxane, 60 °C, 1 h; c) 1. NaOH, MeOH, H_2O, 2. HCl, H_2O; d_5) H_2 (balloon), Pd/C, MeOH.

2.14-H	**2.14-Me**	**2.15-Me**	**2.15-H**
	86%	82%	82%

Figure III.2-5 Scheme for *p*-substituent modification of 4-amino-2,5-difluorobenzoic acid; conditions: a) MeOH, H_2SO_4, 48 h, reflux; b) HCl, $NaNO_2$, NaN_3; c) 1. NaOH, MeOH, H_2O, 2. HCl, H_2O.

i	FG_1=			FG_2=		
i=1	NO_2	**2.16-H**	**2.16-Me** 83%	NH_2	**2.17-Me** 70%	**2.17-H** 88%
i=2	NH_2		**2.17-Me**	N_3	**2.18-Me** 45%	**2.18-H** 89%

Figure III.2-6 Scheme for *m*-substituent modification of 2-fluorobenzoic acid; conditions: a) MeOH, H_2SO_4, reflux, 48 h; b_1) H_2 (balloon), Raney-Ni, MeOH, THF; b_2) HCl, $NaNO_2$, NaN_3; c) 1. NaOH, MeOH, H_2O, 2. HCl, H_2O.

In the framework of the collaboration with the group of V.V. UTOCHNIKOVA (Lomonosov MSU, Moscow, Russia) a fluorobenzoate with a higher fluorination degree was also tested, because the substitution of the benzene ring can make a huge difference on its electronic properties and hence, its suitability to sensitize the Ln(III) ion. For that, pentafluorobenzoic acid was also modified with azido group in *para*-position, following the scheme represented in Figure III.2-7.*

* This strategy was executed in the project of M.Sc. N.N. SOLODUKHIN.

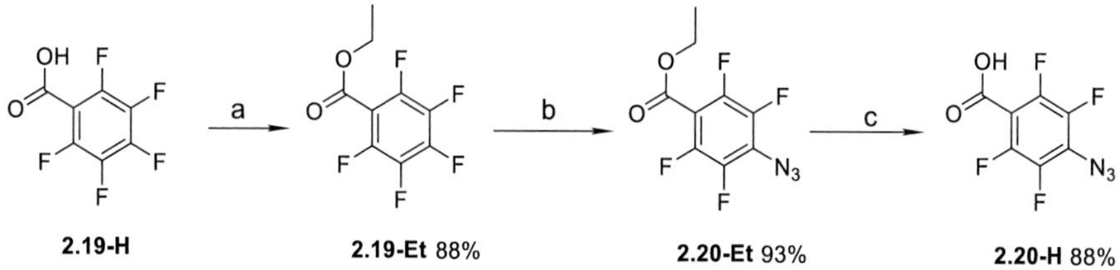

2.19-H 2.19-Et 88% 2.20-Et 93% 2.20-H 88%

Figure III.2-7 Synthetic route for the modification of the *p*-position of pentafluorobenzoic acid **2.19-H**; conditions: a) EtOH, $SOCl_2$, 0 °C – rt, 12 h; b) acetone/H_2O, NaN_3, 8 h, reflux; c) 1. NaOH, MeOH, H_2O, 2. HCl, H_2O.

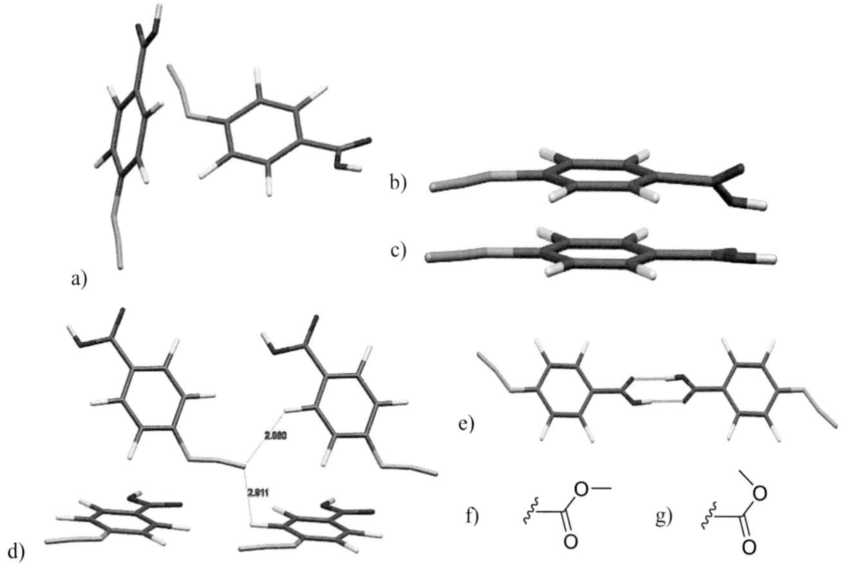

Figure III.2-8 a) Molecular structure of **2.20-H**, containing two independent fragments in the unit cell (asymmetric unit); b) side view of one of the fragment in **2.20-H**, possessing a non-planar conformation of COOH-group relatively to the benzene ring; c) side view of another fragment in **2.20-H**, possessing planar conformation of COOH-group relatively to the benzene ring; d) short contacts, stabilizing the geometry of two fragments in **2.20-H**; e) dimerization of **2.20-H** through H-bond formation between carboxylic groups; f) planar conformation of the carboxylic group; g) antiplanar conformation of carboxylic group. Structures are shown in a capped stick style, where the color indicates the atom type: grey for C, white for H. red for O, blue for N, yellow for F.

The single-crystal structure of compound **2.20-H**, refined in the work of SOLODUKHIN N.N.*, consists of two non-equivalent molecular fragments, possessing a slightly different geometry (Figure III.2-8a). These fragments are stabilized together by a short contact network, mostly between N_3-group of one fragment and F-atoms of another one (Figure III.2-8d). In one fragment the carboxylic group is, however, not co-planar to the benzene ring (Figure III.2-8b), exhibiting a torsion angle C(3A)-C(4A)-C(7A)-O(1A) equals to –26.17°. This may be due to the minimization

* SC-XRD was performed by I.V. ANANYEV, A.N. Nesmeyanov Institute of Organoelement Compounds, Moscow, Russia.

of VAN-DER-WAALS repulsion by the formation of dimers (Figure III.2-8e), which is often observed for crystal structures of carboxylic acids. In another fragment the carboxylic group turned out to be planar to the benzene ring with a torsion angle C(3)-C(4)-C(7)-O(1) equals to -7.09° (Figure III.2-8c). The conformation of the carboxylic group itself is, however, planar in both fragments (Figure III.2-8e,f), which is in line with the data from literature, showing that this conformation is energetically more favorable relatively to the anti-planar conformation (energy gain is 2 kcal/mol). The anti-planar conformation (Figure III.2-8g) may be observed only when COOH-group is involved into H-bonding network, for instance, in 1,2-disubstituted dicarboxylic acids where hydroxyl groups participate in the intermolecular bonding H…O-H.

CuAAC cycloaddition reaction

The most common method for the synthesis of 1,2,3-triazoles is the 1,3-dipolar cycloaddition of azides and alkynes (Huisgen reaction):[226–228] This reaction usually affords 1,4- and 1,5-regioisomers of triazole. The attempts to increase the control on the regioselectivity were not particularly successful until the copper(I)-catalyzed alkyne-azide cycloaddition (CuAAC) "click" reaction was discovered,[229] which allows to obtain exclusively 1,4-substituted 1,2,3-triazoles.[230,231] This reaction was implemented in this work for the conjugation of Ln-based luminescence probes with other molecules which can potentially serve as molecular transporters or being a cargo themselves.

The reaction conditions for azido-substituted fluorobenzoates were optimized*. Firstly, the standard procedure for non-fluorinated arylazides was tested (Table III.1-9, entry 1), namely CuI (20 mol%), sodium ascorbate (40 mol%) and DBU (1.0 equiv.) as a base. The reaction was monitored by ^{19}F NMR, since the ^{19}F chemical shift was clearly distinguishable from starting material **2.20-Et** and product **2.24-Et**. After 2 h, **2.20-Et** was fully consumed, however, the product was isolated in 40% yield. The use of trimethylamine as a base did not make any significant difference (Table III.1-9, entry 2). It is worth noting that the decrease of the amounts of DBU and sodium ascorbate leads to the formation of an unfavorable byproduct $[PhC{\equiv}C]_2$ as the product of side GLASER reaction.[232,233] The reaction yield was significantly increased up to 84% by using CuBr in DCM (Table III.1-9, entry 3). Besides, the workup in the latter case included only filtration and recrystallization, while the two previous attempts contained purification *via* flash column chromatography.

* The optimization was done by the intern student M.Sc. N.N. SOLODUKHIN

Table III.2-1 Summary of tested click chemistry reaction conditions with **2.20-Et** and phenylacetylene **2.25**.

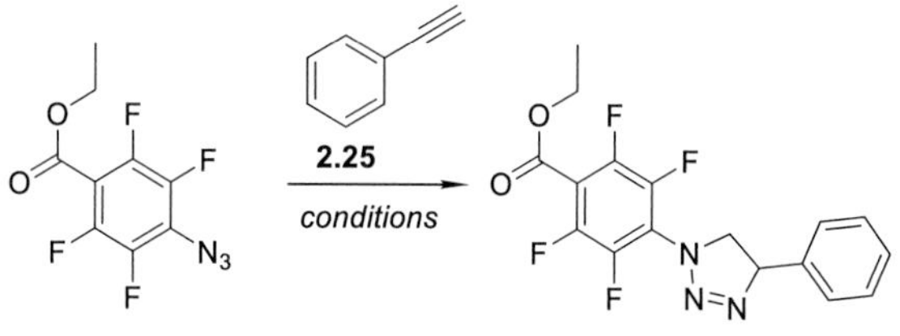

2.20-Et → (**2.25**, *conditions*) → **2.24-Et**

Entry	Conditions	Base	Conversion[a], %/ time, h	Yield, %
1	CuI, sodium ascorbate, DMF	DBU	98/2	40
2	CuI, sodium ascorbate, DMF	Et_3N	98/4	44
3	CuBr, DCM	Et_3N	98/4	84

[a] monitored by ^{19}F NMR

To expand the variety of the coupling products, two fluorinated alkynes were synthesized, following the sequence shown in Figure III.2-9*. This scheme involved palladium-catalyzed SONOGASHIRA coupling of the aryl bromide with trimethylsilylacetylene (TMS-acetylene), followed by deprotection of the TMS protecting group under basic conditions.

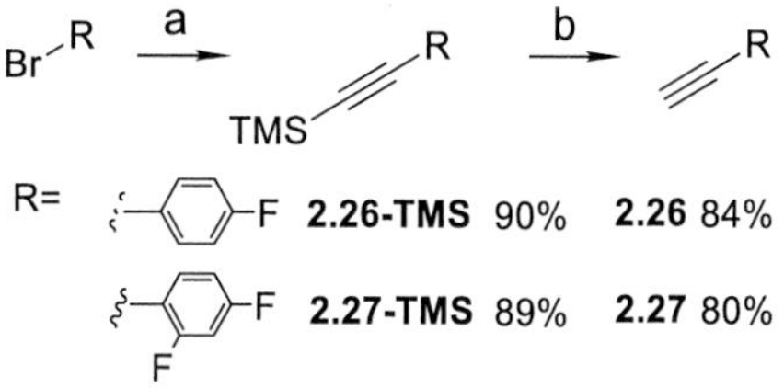

Br–R → (a) TMS–≡–R → (b) ≡–R

R = 4-fluorophenyl: **2.26-TMS** 90% **2.26** 84%

R = 2,4-difluorophenyl: **2.27-TMS** 89% **2.27** 80%

Figure III.2-9 Synthesis of substituted phenylacetylenes **2.26** and **2.27**: a) TMS-acetylene, $Pd(PPh_3)_4$, CuI, trimethylamine, dry THF, 24 h, reflux; b) NaOH, MeOH.

Following the optimized conditions (Table III.1-9), several "click"-reaction products were synthesized†. The reaction with 1-ethynyl-4-fluorobenzene (**2.26**) (Table III.2-2, entry 2) and *N*-Boc-propargylamine (Table III.2-2, entry 4) indeed gave product **2.28-Et** in a quite high yield (84% and 75%, respectively), however, the yield in the reaction with 1-ethynyl-2,4-difluorobenzene (**2.27**) to afford product **2.29-Et** was only moderate (46%) (Table III.2-2, entry 3). The implementation of the very same reaction conditions to the monofluorinated benzoate, however, afforded the coupled product (**2.31-Me**) in 68% yield (Figure III.2-10).

* This work was performed by the intern student N.N SOLODUKHIN.

† This work was performed by the intern student N.N SOLODUKHIN

Table III.2-2 Summary of tested "click" chemistry reaction conditions with **2.20-Et**.

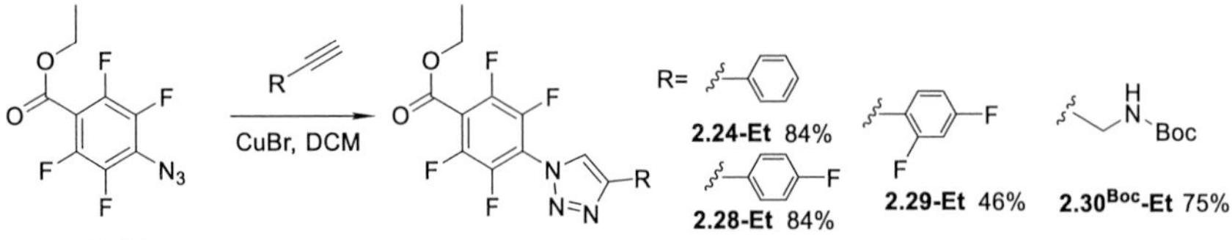

Entry	Compound	Yield, %
1	2.24-Et	84
2	2.28-Et	84
3	2.29-Et	46
4	2.30Boc-Et	75

[a] monitored by ^{19}F NMR.

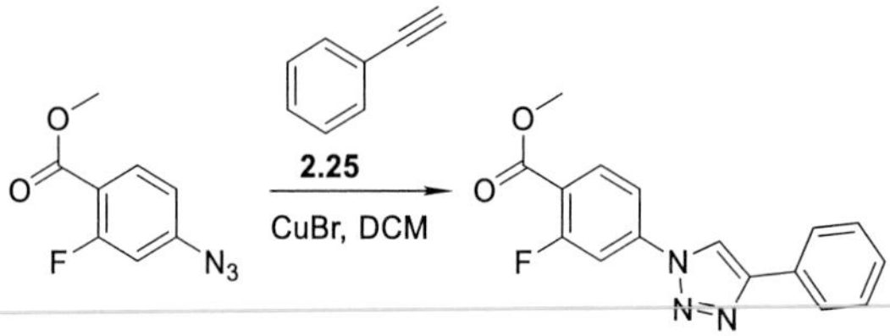

Figure III.2-10 Synthesis of **2.31-Me** as a product of "click" reaction of methyl 4-azido-2-fluorobenzoate **2.3-Me** and phenylacetylene **2.25**.

In order to use such coupled products as ligands for lanthanides, the carboxylic group was deprotected according to Figure III.2-11, so that these compounds were ready for complexation. Deprotection of **2.24-Et, 2.28-Et, 2.29-Et** and **2.31-Me** afforded carboxylic acids **2.24-H, 2.28-H, 2.29-H** and **2.31-H** in very good yields (Figure III.2-11a,c). During the saponification of **2.30^{Boc}-Et**, Boc-group was not fully removed by the treatment of the saponified product with HCl, therefore, it was removed prior to saponification by stirring with concentrated HCl for several hours (Figure III.2-11b).

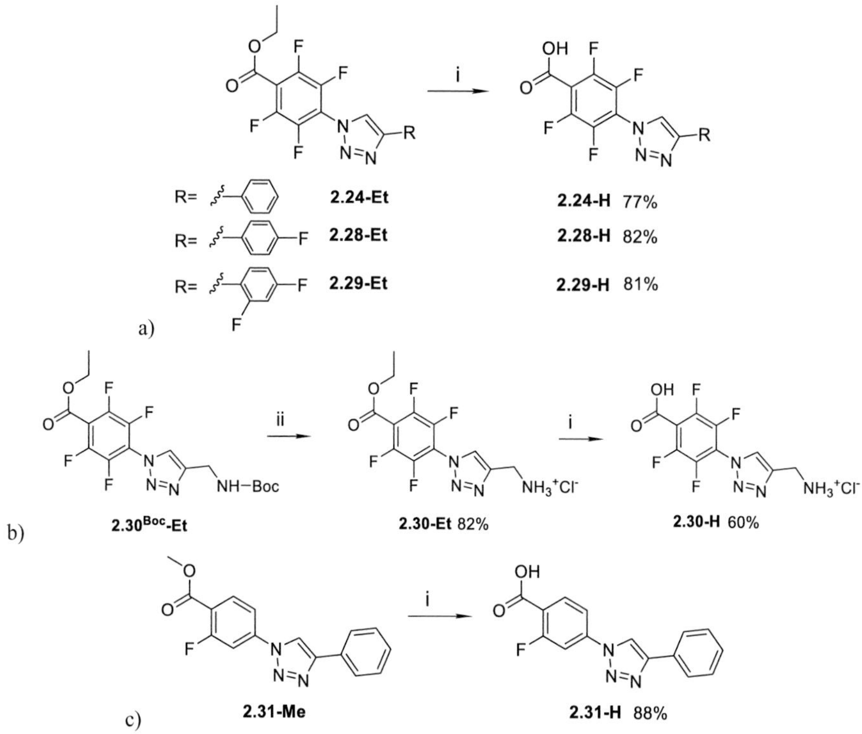

a)
R= phenyl 2.24-Et → 2.24-H 77%
R= 4-fluorophenyl 2.28-Et → 2.28-H 82%
R= 2,4-difluorophenyl 2.29-Et → 2.29-H 81%

b)
2.30^{Boc}-Et (NH−Boc) —ii→ 2.30-Et 82% ($NH_3^+Cl^-$) —i→ 2.30-H 60% ($NH_3^+Cl^-$)

c)
2.31-Me —i→ 2.31-H 88%

Figure III.2-11 Saponification of the CuAAC products: a) coupled products of the reaction of ethyl 4-azido-2,3,5,6-tetrafluorobenzoate **2.20-Et** with arylacetylene derivatives; b) Boc-deprotection and saponification of **2.30^{Boc}-Et**; c) coupled products **2.31-Me** of the reaction of methyl 4-azido-2-fluorobenzoate **2.3-Me** with phenylacetylene **2.25**; conditions: i) 1. NaOH, MeOH, ovrn, 2. HCl, H_2O; ii) HCl, 5 h, reflux.

The molecular structure was determined for compound **2.31-Me** (Figure III.2-12a)*. The structure consists of layers, packed together backwards, so that methyl ester group is able to form H-bonds with phenyl groups of two neighboring molecules: (i) the neighbor within a layer and (ii) the "flipped" molecule from another layer (Figure III.2-12b). Such a packing is additionally stabilized by the short contact of the fluorine atom with methyl group from methyl ester. In each molecule, the aromatic rings are slightly flipped to each other: the torsion angle between fluorobenzoate and triazole is equal to -155.37°, while between triazole and phenyl it is -155.19° (Figure III.2-12a).

* The measurement and refinement were performed together with I.S. BUSHMARINOV, A.N. Nesmeyanov Institute of Organoelement compounds, Moscow, Russia.

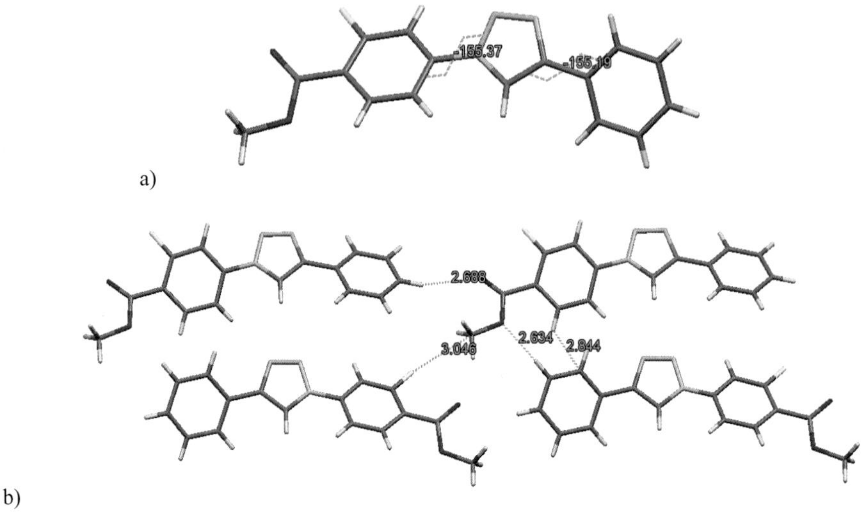

a)

b)

Figure III.2-12 a) Molecular structure of compound **2.31-H** (asymmetric unit); b) molecular package of **2.31-H** in a single crystal, demonstrating H-bonds and short contacts, stabilizing the molecular fragments in this structure. Structures are shown in a capped stick style, where the color indicates the atom type: grey for C, white for H, red for O, blue for N, yellow for F.

Modification of 1,10-phenanthroline derivatives

The derivatization of the auxiliary ligand may also be useful for the coupling of lanthanide complexes with biological relevant molecules. The great ability of 1,10-phenanthroline to coordinate Ln fluorobenzoates and form ternary complexes was already discussed in Chapter III.1.2. The modified ligand may serve as both, sensitizer of Ln(III) luminescence and linking moiety, which can help to bind the luminescent label to another molecule covalently, assuming that the complex will be still luminescent and thermodynamically stable. The modification of 1,10-phenanthroline and its derivatives had been carried out in this work. Firstly, the farthest to the binding site position was modified, following the same logic as before that lanthanide luminescence quenching through the vibration of chemical bonds decreases with an increase of the distance between the Ln(III) ion and quenching functional groups. The nitration of 1,10-phenanthroline (**2.32**) gave 5-nitro-1,10-phenanthroline (**2.33**) with a selectively functionalized 5th position which is the furthest position to the binding site (Figure III.2-13a). Reduction of the introduced nitro group afforded 5-amino-1,10-phenanthroline (**2.34**) (Figure III.2-13b).

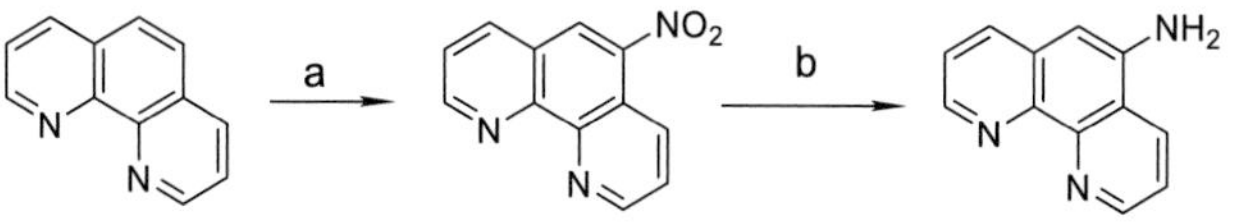

Figure III.2-13 Synthesis of 1,10-phenanthroline derivatives: a) HNO_3, H_2SO_4, 3 h, reflux; b) H_2NNH_2, H_2O, Pd/C.

Another way to functionalize 1,10-phenanthroline derivatives is the alkylation of the methyl groups of neocuproine (**2.35**) which is its 2,9-methylated analog. As only one linker should be introduced into the neocuproine, one of its methyl groups should be functionalized. For that purpose a procedure developed in the group of Prof. BRÄSE by Dr. K. PESCHKO, was implemented (Figure III.2-14). As a strong base, LDA has been shown to deprotonate methyl groups and the subsequent addition of propargyl chloride afforded mono- (**2.36**) and disubstituted (**2.37**) compounds. The amount of the base is found to be crucial for this reaction, which can control the ratio of the formed products. The maximum amount of the monoalkylated product was obtained by using 1.2 equiv. of LDA, relatively to the amount of neocuproine; though, at this point, the dialkylated product already started to form. Purification *via* flash column chromatography afforded monoalkylated **2.30** (19%) as well as dialkylated **2.31** (6%) compounds.

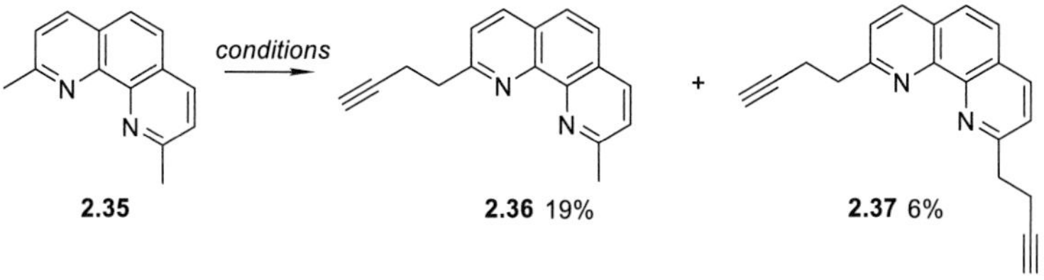

Figure III.2-14 Functionalization of neocuproine; conditions: LDA, propargyl chloride, 0 °C – rt.

Synthesis and properties of lanthanide complexes with selected ligands

Eu(III) complexes with selected ligands, discussed above, were synthesized, according to the standard procedure, namely the reaction of the potassium salt of the ligand with aqueous Eu(III) chloride. The complexes precipitate from the aqueous media, were filtered and dried in vacuo.

The influence of ligand *para*-functionalization on the luminescent properties of Eu(III) complexes was investigated. In some cases, it was also necessary to synthesize Gd(III) and Lu(III) complexes to determine the value of the ligand triplet level.

As expected, mostly all synthesized Eu(III) complexes turned out to be luminescent and the characteristic Eu(III) emission bands were detected for both, powders and 5 mM MeOH solutions.

This observation is in line with our assumption that an introduction of a FG to *para*-position has likely no influence on the luminescence properties. Even **Eu-2.1** complex, bearing quenching NO_2-group, exhibits luminescence properties in solid state, though, as it was shown before, the presence of NO_2-group in *ortho*-position absolutely quenched Eu-based emission (for compound **Eu-3.1**, Chapter III.1.1). However, no Eu(III)-centered emission was detected for **Eu-2.1** complex in MeOH solution. This is, probably, due to the flexibility of the local structures in solution, so that the distance between the Eu(III) ion and quenching NO_2-group in solution may be much lower than in solid state.

Interestingly, almost all studied lanthanide complexes with ligands, bearing azido group, turned out to be absolutely non-emissive, regardless of the position of azido group (**Eu-2.3, Eu-2.15, Eu-2.18**, **Eu-2.20**). Given that such an effect has already been observed in literature, we may consider azido group as a universal quencher of Eu(III)-based luminescence. However, in some cases Eu(III) emission can be recovered after the coupling of the azide with an alkyne, which was observed in the case of **Eu-2.30**, however, did not happen for **Eu-2.21**, **Eu-2.24**, **Eu-2.28 Eu-2.29** and **Eu-2.31**.

An MTT assay, made for compound **Eu-2.30**, using HeLa cells, revealed that the complex is not toxic at concentrations below 2 mM. Cellular imaging was performed on the example of HeLa cells, using the confocal microscopy technique after incubation of the HeLa cells with 0.5 mM **Eu-2.30** in DMEM medium with 1% DMSO for 24 h. This compound (**Eu-2.30**), demonstrated moderate luminescence intensity. Luckily, it was possible to excite the luminescence of this compound, using a greenish-blue laser with $\lambda_{ex} = 488$ nm. This allowed to carry out the experiments with nuclei staining Hoechst-33342 which is excited at another wavelength $\lambda_{ex} = 405$ nm. This experiment illustrated the endosomal concentration of compound **Eu-2.30** (Figure III.2-15).

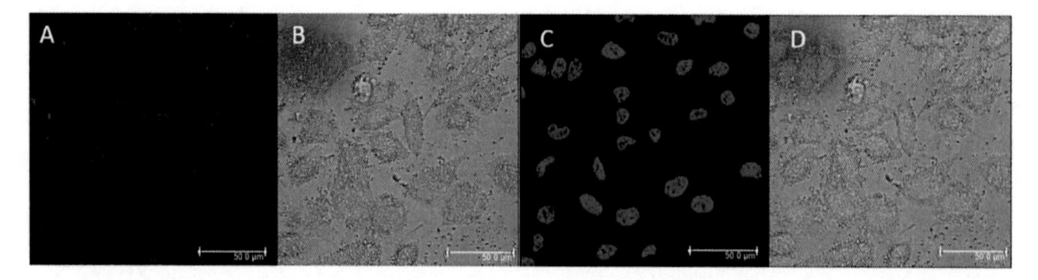

Figure III.2-15 Confocal microscopy images for 0.05 mM solution of compound **Eu-2.30**; A – fluorescence channel, $\lambda_{ex} = 488$ nm, laser power 50%, detection 580 – 750 nm (**Eu-2.30** emission); B – brightfield channel, C – fluorescence channel $\lambda_{ex} = 405$ nm, laser power 20%, detection 430 – 500 nm (Hoechst-33342 emission), D – overlay.

Despite bearing quenching azido group, compound **Eu-2.6** was luminescent. In this case, the azido group is not attached to the benzene ring directly, but through a CH_2-spacer which was assumed to make the influence of the azido group on the ligand's electronic properties almost negligible. Indeed, the luminescence intensity as well as the shape of the emission spectrum is very similar for the complexes, where the FG is attached through a CH_2-spacer, namely **Eu-2.4** (R = CH_2Br), **Eu-2.5** (R = CH_2NH_2), **Eu-2.6** (R = CH_2N_3) and **Eu-2.12** (R = CH_3) (Figure III.2-16a). Figure III.2-16 demonstrates that the emission spectra of complexes **Eu-2.12** and **Eu-2.5** coincide completely. However, their excitation bands are different, being very sensitive to changes in the ligand's electronic structure. Although the excitation spectra of these two complexes have the similar shape, the edge of **Eu-2.5** excitation is red-shifted (Figure III.2-16a). Interestingly, the change of the excitation edge does not necessaryly affect the triplet level value. The latter was measured, following the common technique, *i.e.* measuring phosphorescence of Gd complexes with these ligands at 77 K (**Gd-2.12** and **Gd-2.5**, respectively). These measurements revealed that there is only very few if any difference in the triplet level value for ligands **2.12** and **2.5**. Their triplet level value was determined to be ~ 20400 cm^{-1} (Figure III.2-16b). Assuming that the geometry of complexes **Eu-2.12** and **Eu-2.5** might be very similar, this result would fully explain the coincidence of the emission spectra of these compounds.

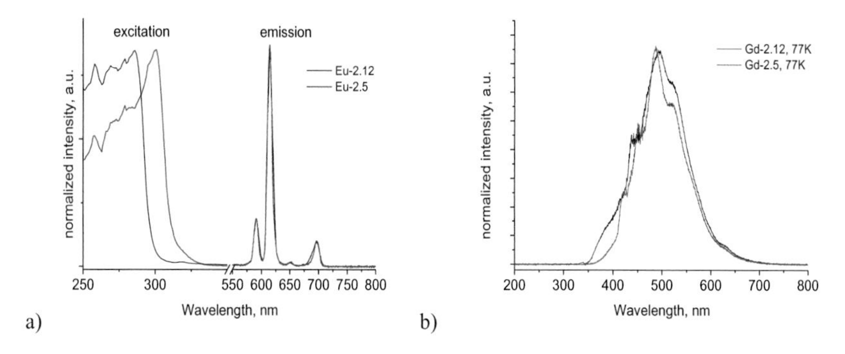

Figure III.2-16 a) Excitation and emission spectra of solids **Eu-2.12** and **Eu-2.5**. Excitation was recorded on the emission maximum, λ_{em} = 615 nm; emission spectra were recorded at the excitation maximum (λ_{ex} = 280 nm for **Eu-2.12**, λ_{ex} = 300 nm for **Eu-2.5**); b) emission spectra of solids **Gd-2.12** and **Gd-2.5**, measured at 77 K.

The structures of five *p*-modified lanthanide complexes were determined in this work to investigate how the incorporation of different substituents may affect the geometry of Ln(III) complexes. The single-crystal structures of **Eu-2.2** (*p*-amino substituted, R = NH_2) and **Eu-2.6** (*p*-azidomethyl substituted, R = CH_2N_3) turned out to be similar to the structure of non-substituted complex **Eu-2.8** (R = H). All these three structures consist of dimeric units, having an inversion center between two Eu(III) ions. The overlay of these structures (Figure III.2-17) shows that they are "almost isostructural": the coordination environment of Eu(III) for all three complexes coincides (Figure III.2-17a for the pair **Eu-2.2** and **Eu-2.6** and Figure III.2-17b for the pair **Eu-2.2** and **Eu-2.8**). Only a slight difference in the orientation of the benzene ring is observed for these structures: they are tilted because of the differences in the short contacts in which the new introduced functional groups are involved.

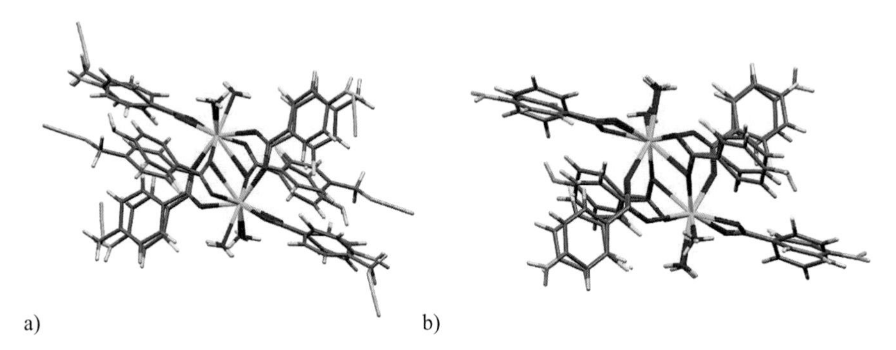

Figure III.2-17 Overlay of the molecular structures of **Eu-2.2**, **Eu-2.6** and **Eu-2.8** in pairs: a) overlay of **Eu-2.2** and **Eu-2.6**; b) overlay of **Eu-2.2** and **Eu-2.8.** Structures are shown in a capped stick style, where the color indicates the atom type: grey for C, white for H. red for O, blue for N, yellow for F, cyan for Eu.

The complexes with a similar geometry were expected to possess similar luminescence properties. Indeed, the triplet level values for ligands **2.2** and **2.8** were measured to be the same and equal to ~ 20400 cm^{-1}. Besides, both, excitation and emission spectra, coincide for compounds **Eu-2.2** and **Eu-2.8** (Figure III.2-18a). The same results were observed in the case of the polymers **Eu-2.9** (R = I) and **Eu-2.10** (R = Br). Though these complexes cannot be called isostructural, they are still very similar, so that the Eu(III) coordination environment can be overlayed for these two structures as well (Figure III.2-19a), but the benzene rings are significantly tilted. This is due to the difference in the coordination of solvent molecules (ethanol and water molecules in both structures), which leads to a slightly different short contact network, generated by them. However, since **Eu-2.9** (R = I) and **Eu-2.10** (R = Br) have the same coordination sphere, their excitation and emission spectra are very similar (Figure III.2-18b).

The molecular structure of compound **Eu-2.13**, however, also consists of polymeric chains but it is not similar to polymers **Eu-2.9** and **Eu-2.10**. All three carboxylic ligands possess a quite unusual κ^1 coordination mode, two of them adopt the bridging function μ^2:κ^1-κ^1 (Figure III.2-19b). This is due to the strong tendency of CN-group to participate in H-bonding, which is, eventually, the main force, stabilizing the ligands in such a crystal structure (Figure III.2-19b). The single-crystal diffraction data for ligands **2.20-H** and **2.31-Me** are shown in Table V.3-6, for complexes **Eu-2.2**, **Eu-2.6, Eu-2.9, Eu-2.10** and **Eu-2.13** they are represented in Table V.3-7.

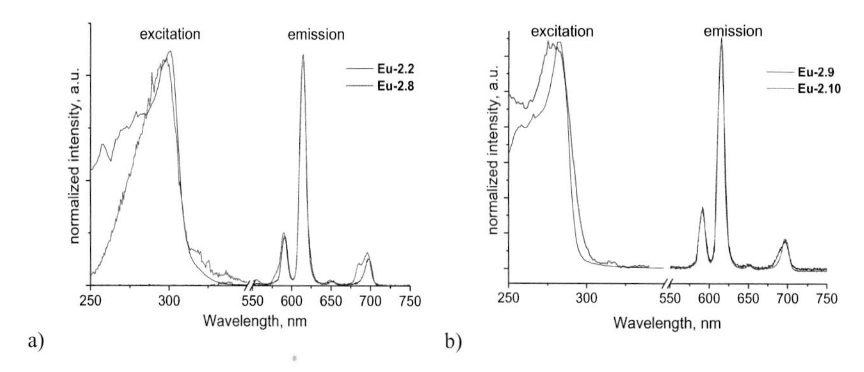

Figure III.2-18 a) Excitation and emission spectra of solids **Eu-2.2** and **Eu-2.8.** Excitation was recorded on the emission maximum, $\lambda_{em} = 615$ nm; emission spectra were recorded at the excitation maximum $\lambda_{ex} = 300$ nm; b) excitation and emission spectra of solids **Eu-2.9** and **Eu-2.10.** Excitation spectra was recorded on the emission maximum, $\lambda_{em} = 615$ nm; emission spectra were recorded at the excitation maximum ($\lambda_{ex} = 280$ nm).

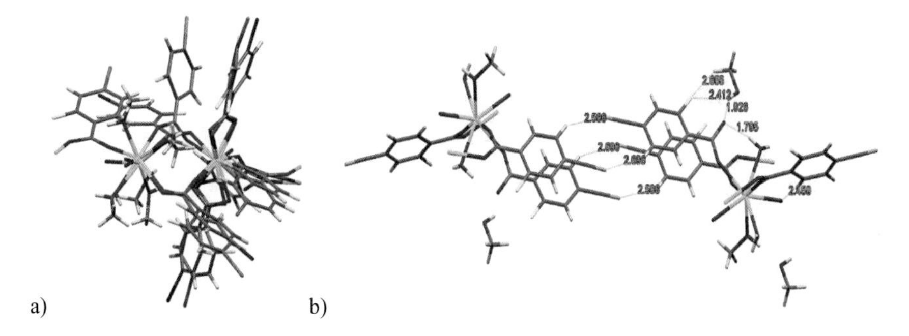

Figure III.2-19 a) Overlay of the molecular structures of **Eu-2.9** and **Eu-2.10**; b) molecular structure of **Eu-2.13.** Structures are shown in a capped stick style, where the color indicates the atom type: grey for C, white for H. red for O, blue for N, yellow for F, cyan for Eu, purple for I, brown for Br.

Thus, in this case, *p*-modification of the ligand does not affect its electronic properties (similar triplet level values) and the geometry of Ln(III) complexes (similar crystal structures), which leads to similar luminescence properties, which supports the idea to use *p*-modified Eu(III) complexes for further conjugation.

Conclusions

In this work, *p*-substituted fluorobenzoic acid derivatives were synthesized as well as Ln(III) complexes with them. The study of crystal structures of the selected compounds showed that Eu(III) 2-fluorobenzoates with -NH_2 and -CH_2NH_2 groups in *para*-position are isomorphous and reveal the same geometry as unmodified Eu(III) 2-fluorobenzoates. Besides, I- and Br-substituted benzoates also possess a similar geometry. The investigation of the luminescent properties of these Eu(III) complexes demonstrated that the complexes within these structural groups indeed possess similar emission profile, because of the same coordination environment of Eu(III) ion. Therefore, luminescence spectroscopy may serve as an indicator of the geometry changes and may be used if an additional structural study is difficult.

Azido group was found to be a quencher, when attached directly to the benzene ring. However, the luminescence properties of N_3-modified Eu(III) complexes might be recovered after the conjugation with the corresponding alkyne. Interestingly, if the azido group was introduced through a CH_2-spacer, it did not quench Eu(III)-based luminescence. These results indicate that *p*-modification of Eu(III) complexes is indeed the most reliable way to introduce the linker into the structure of the complex. The complexes with these linkers may be further used for a coupling with transporters or biologically relevant molecules.

III.2.2 Synthesis and properties of europium conjugates with linear peptoids

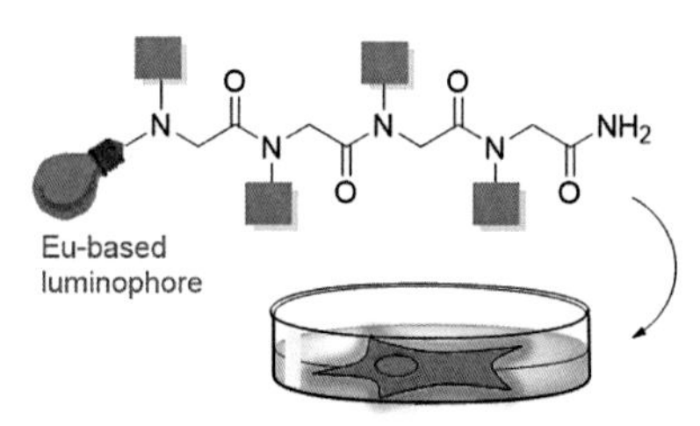

This chapter demonstrates the synthesis and properties of linear peptoids and Eu(III) conjugates with them. The optimization of the conjugation method is performed by the implementation of various conjugation conditions, namely the conjugation of a Ln(III) complex with a peptoid on solid support or with an already cleaved and purified peptoid, as well as the conjugation through different linkers, also using CuAAC and SPAAC reactions. For the luminescent Eu(III) conjugates, the photophysical properties as well as the cellular luminescence were studied.

Poly-*N*-substituted glycines, known as peptoids, are very promising candidates for the use as molecular transporters. Different types of peptoids with specific properties were synthesized, fully characterized and labeled further with Eu-based luminescent probes.

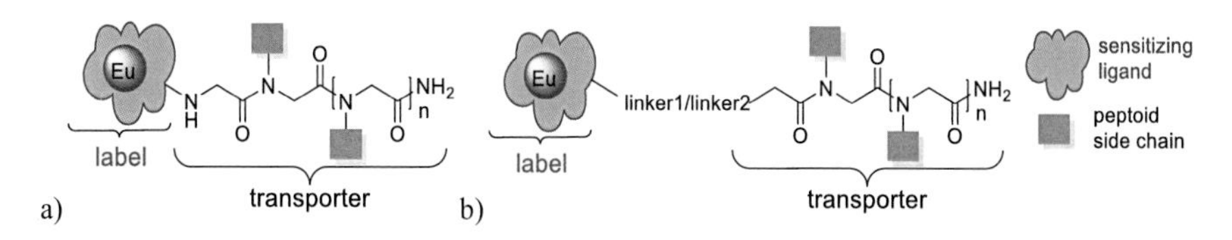

Figure III.2-20 General scheme of the label-peptoid platform, synthesized in this work using: a) label as a building block during the peptoid synthesis and b) cross-coupling reaction between the label and the peptoid.

In order to generate a "label-transporter" platform with linear peptoids, two different approaches were implemented. The first one implies the attachment of a luminescent label as a building block during the peptoid synthesis (Figure III.2-20a). The second one includes the cross-coupling CuAAC reaction between azide and alkyne linkers, each should be preliminarily attached to the label and to the peptoid, respectively (Figure III.2-20b). Though the second approach requires the incorporation of azido group which is shown to be a luminescence quencher, Eu(III) emission can be recovered after the cross-coupling reaction, as demonstrated in Chapter III.2.1.

The attachment of Eu-based luminescent label can be performed either through an anionic or through an auxiliary ligand. As illustrated in Chapter III.2.1 *p*-substituted Eu(III) fluorobenzoates are promising candidates for the conjugation following the first pathway. Substituted 1,10-phenanthroline moieties were used as an example of the second pathway.

Synthesis of *N*-substituted glycines (peptoids)

The linear peptoids were synthesized, using the submonomer method developed by ZUCKERMANN.[118] This solid-phase method is based on the reaction between halogenated acetic acids and primary amines and allows to obtain a peptoid oligomeric chain of any desired length, bearing almost all possible side chains (Figure III.2-20).

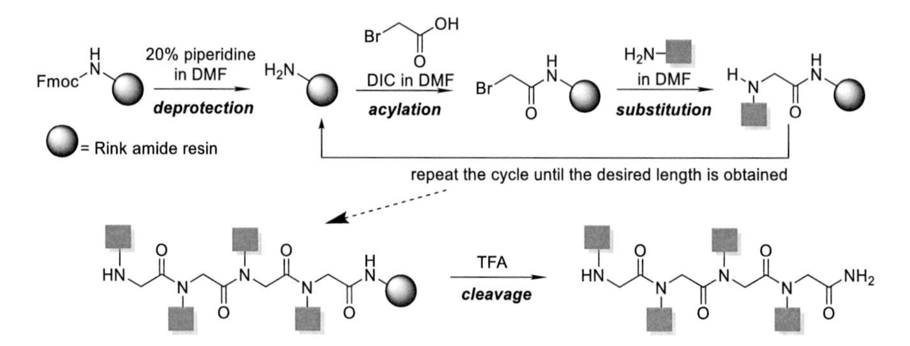

Figure III.2-21 Scheme for the construction of peptoid oligomers on the example of a tetramer, using solid-phase synthesis. Grey squares represent any side chain of the peptoid.

If the used amines bear an additional nucleophilic functionality they must be protected to avoid side reactions such as polymerization. Therefore, the side chains having a second nucleophilic amino function were protected with Boc-group (Figure III.2-22). For that purpose an excess of 1,4-diaminobutane **2.38** or 1,4-diaminohexane **2.39** was reacted with di-*tert*-butyl dicarbonate **2.43** overnight under an inert atmosphere to afford mono-Boc-substituted amines. The use of THF as a solvent instead of 1,4-dioxane used in literature[234,235] was proposed in the doctoral thesis of STEPHAN MÜNCH from the group of Prof. S. BRÄSE and allowed to obtain products **2.41** and **2.42** in high yields (68% and 62%, respectively, instead of reported 54%[235]).

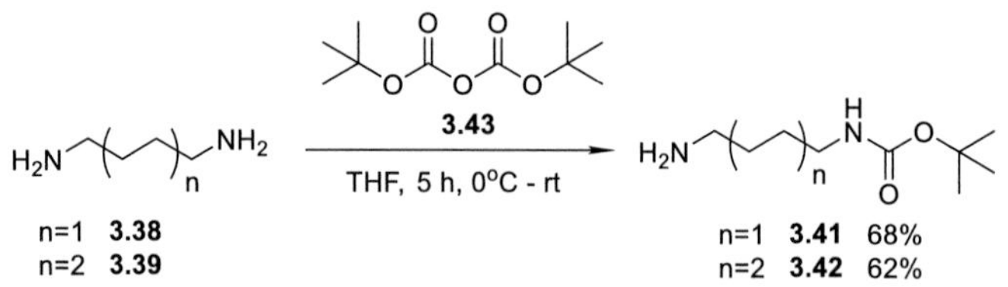

3.43
THF, 5 h, 0°C - rt

n=1 **3.38**
n=2 **3.39**

n=1 **3.41** 68%
n=2 **3.42** 62%

Figure III.2-22 Synthesis of the mono-Boc-protected 1,4-diamine **2.41** and 1,6-diamine **2.42**.

For further CuAAC coupling a submonomer, bearing N_3-functional group, was synthesized. The reaction of 3-chloropropylamine hydrochloride **2.44** with sodium azide and the subsequent conversion of HCl salt into a free amine under basic conditions afforded product **2.45** in 68% yield (Figure III.2-23).

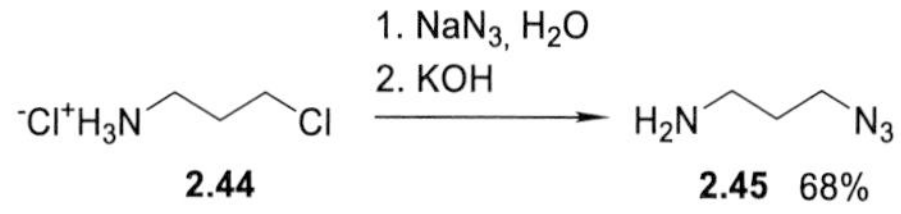

1. NaN_3, H_2O
2. KOH

$^{-}Cl^{+}H_3N$~~~Cl (**2.44**) → H_2N~~~N_3 (**2.45** 68%)

Figure III.2-23 Synthesis of 3-azidopropan-1-amine **2.45**.

Conjugation of the linear peptoids with europium complexes

To investigate the possibility to conjugate Eu(III) complexes, using solid-phase synthesis several tetrameric peptoids were synthesized. The syntheses were performed on Fmoc-protected rink amide resin. After the resin was swollen in DMF, Fmoc-group was deprotected, using 20% piperidine in pDMF (Figure III.2-21). After several washing steps, the deprotected resin was ready for the subsequent acylation and substitution reactions which were altered until the desired length of the peptoid molecule was obtained. The peptoid was then cleaved from the resin as demonstrated in Figure III.2-21.

The principal possibility to conjugate peptoids with lanthanide complexes was firstly demonstrated on the example of *H*-**R**-(*N*6am)$_3$-*NH_2* tetramers, where **R**-NH_2 are Eu(III) based complexes, bearing amino group. The synthesis of such Eu-conjugates was performed on solid phase, which is described above (Figure III.2-21). After different trimers were built, the next coupling step, using bromoacetic acid gave compound **2.46^{Boc}-SP** attached to the solid support. This compound reacted with three Eu complexes **Eu-2.2**, **Eu-2.17** and **Eu-2.14** in DMF for 24 h, to afford Ln-labeled peptoids **Eu-2.47-SP^{Boc}**, **Eu-2.48^{Boc}-SP** and **Eu-2.49^{Boc}-SP**, respectively. The reaction time in this case was significantly increased (from 30 minutes for *N6am* to 24 h), because it was found that the substitution reaction with aromatic amines during the peptoid synthesis takes longer than those for aliphatic ones. Due to the steric effect and the significant excess of the amine (Eu complex), it was expected that mostly a 1:1 substitution takes place, despite the fact that the used Eu(III) complexes bear three amino groups each (Figure III.2-24a). After the substitution, the resin was washed several times with DMF and finally DCM. It is worth noting that the resin exhibited weak red luminescence, which may serve as an evidence that the Eu-based luminescent compound indeed was attached to the peptoid on solid support. Surprisingly, cleavage of the peptoid from the resin resulted in the formation of peptoids **2.50-H**, **2.51-H** and **2.52-H**, respectively, bearing the carboxylic acid as the last submonomer instead of the Eu(III) complex (Figure III.2-24b). This result indicates that the attached Eu(III) complexes possess not enough stability and dissociate under the acidic cleavage conditions. Under these circumstances TFA cleaved the coordination bonds, so that a single fragment of the Eu(III) complex covalently bound to the peptoid was still attached (Figure III.2-24b). Indeed, the composition of the obtained peptoids was proved by MALDI-TOF-MS, showing the mass of *H*-(**R**)-(*N*6am)$_3$-fragment, where R = *N0b1* (**2.50-H**), *N0b2* (**2.51-H**), *N0b3* (**2.52-H**).

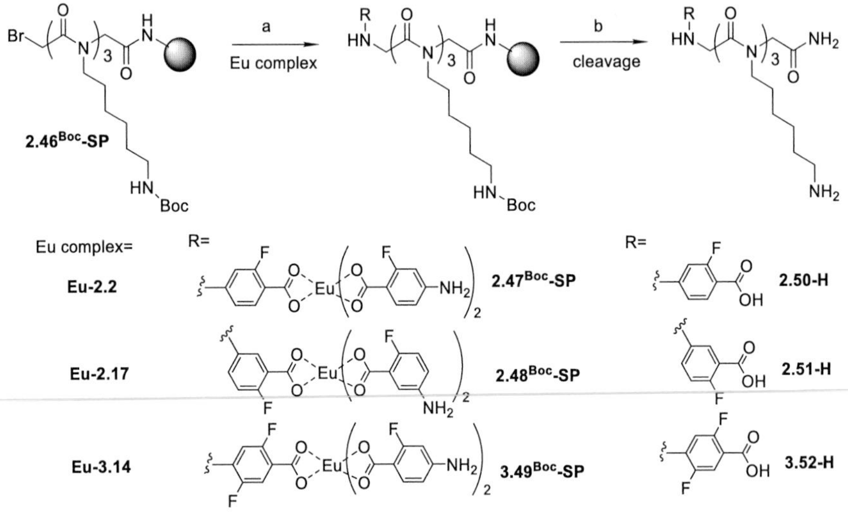

Br N 3 H N O O 2.46^{Boc}-SP HN Boc a Eu complex R HN O H N 3 O HN Boc b cleavage R HN O N 3 NH$_2$ O NH$_2$

Eu complex= Eu-2.2 Eu-2.17 Eu-3.14 R= Eu NH$_2$ 2 2.47^{Boc}-SP 2.48^{Boc}-SP 3.49^{Boc}-SP R= OH 2.50-H 2.51-H 3.52-H

Figure III.2-24 Conjugation of Eu(III)-based labels with peptoids, using Eu(III) complexes as building blocks during the peptoid synthesis: a) the substitution reaction of peptoid **2.46-SP** with Eu(III) complexes **Eu-2.2**, **Eu-2.17** and **Eu-2.14** which afforded peptoids **Eu-2.47^{Boc}-SP, Eu-2.48 ^{Boc}-SP** and **Eu-2.49 ^{Boc}-SP**; b) cleavage of the peptoid from the resin with simultaneous deprotection of Boc-protected amino groups using TFA. Dissociation of the Eu(III) complexes under acidic conditions led to the formation of peptoids **2.50-H**, **2.51-H** and **2.52-H** with respective carboxylic acids as the last submonomer.

Thus, the conjugation of the Eu(III) fluorobenzoates with peptoids, using the solid-phase synthesis was shown to be unsuccessful due to the cleavage of the coordination bonds when removing the peptoid oligomer from solid support. Therefore, the next attempts to obtain Ln-based conjugates were performed, using peptoids which were cleaved from the resin and purified by preparative HPLC prior the conjugation with the Eu(III)-based label.

Though the cleavage of the coordinating bonds was frustrating, products **2.50-H**, **2.51-H** and **2.52-H** could still be useful. After purification, they were used as ligands for an Eu(III) complex synthesis, using their carboxylic functionality (Figure III.2-25).

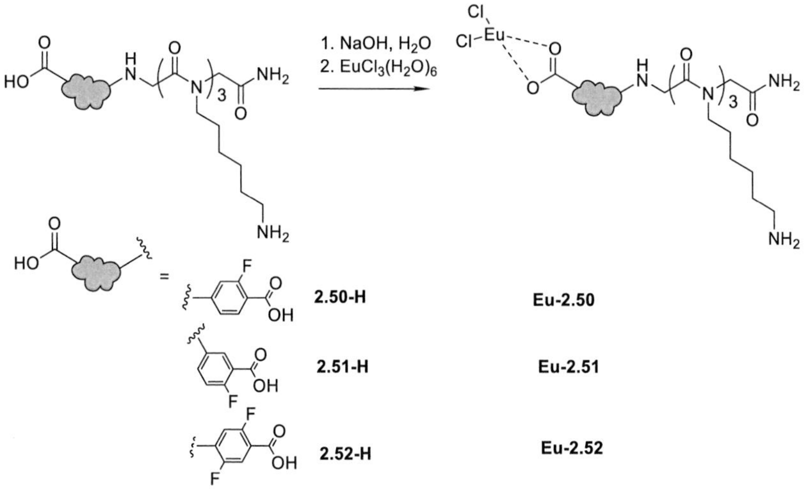

Figure III.2-25 Complexation of compounds **2.50-H, 2.51-H** and **2.52-H** with Eu.

The complexation was carried out according to the standard procedure used for Ln complexation with carboxylic acids, namely the addition of an aqueous solution of Eu(III) chloride to the preliminary deprotonated carboxylate in ethanol. The reaction mixtures were refluxed for 3 h. After that the obtained crude products were purified by preparative RP-HPLC and lyophilized. Interestingly, MALDI-TOF-MS measurement showed that in all three cases only one peptoid-based ligand was able to coordinate Eu(III) ion. The fragment $[Eu–(peptoid)–Cl]^+$ was found in all three complexes, which allows assuming that compounds **2.50-H**, **2.51-H** and **2.52-H** have the structure $[Eu(peptoid)Cl_2]$ and no fragments of "$Eu(peptoid)_3$" or "$Eu(peptoid)_2$" are present. Another evidence of the Eu-peptoid coordination is the typical Eu(III)-based red luminescence of the obtained compounds. Though the luminescence intensity of compounds **Eu-2.50**, **Eu-2.51** and **Eu-2.52** was only moderate, cellular experiments were nevertheless carried out. Surprisingly, all three compounds exhibited a very high cellular luminescence, even at low concentrations up to 1.0 μM compared to unconjugated Eu(III) fluorobenzoates (for instance, 1 mM for **Eu-1.9** complex (Figure III.1.21) and 0.1 mM for **Eu-1.9-Phen** complex (Figure III.1.31)). The best luminescence performance was observed for compound **Eu-2.50** which possessed a brighter luminescence at lower concentrations compared to the other derivatives (Figure III.2-26).

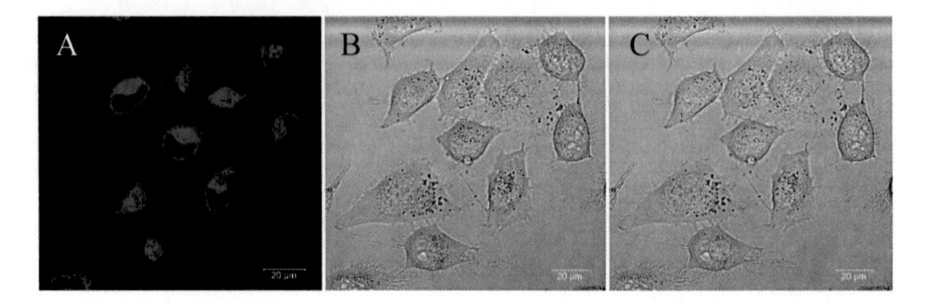

Figure III.2-26 Confocal microscopy images of HeLa cells incubated with 20 μM solution of **Eu-2.50**; A – fluorescence channel, $\lambda_{ex} = 405$ nm, laser power 30%, detection range 580 – 750 nm (**Eu-2.50** emission); B – brightfield channel, C – overlay.

This result indicates that the attachment of the peptoid to the lanthanide complex increases its cellular uptake, so that Eu-conjugates may work at lower molar concentrations than unconjugated Eu complexes.

Conjugation of Eu(III) complex with a peptoid can also be performed through an auxiliary ligand. For this purpose, 1,10-phenanthroline derivatives were synthesized, as shown in Chapter III.2.1 (Figure III.2-27).

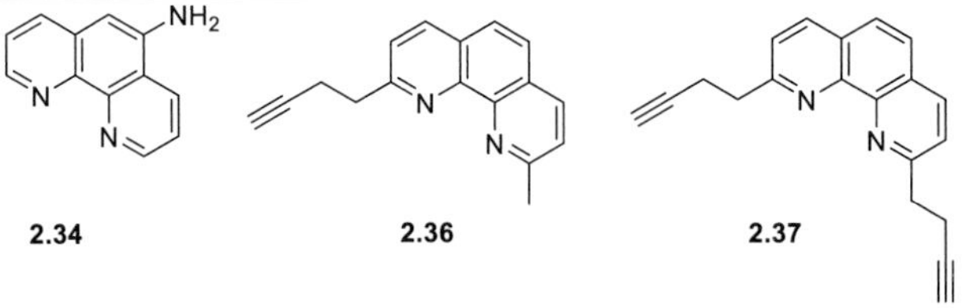

2.34 **2.36** **2.37**

Figure III.2-27 1,10-phenanthroline derivatives for the further conjugation with peptoids.

Compound **2.34,** bearing amino group, was used as a submonomer during the peptoid synthesis, namely by reacting with peptoid **2.46-SP** to afford peptoid *H*-(*N*Phe)-(*N*6am^{Boc})$_3$-*SP* (**2.53-SP**) (Figure III.2-28). This compound was cleaved from the resin to afford *H*-(*N*Phe)-(*N*6am)$_3$-*NH$_2$* (**2.53**). After it had been purified by preparative RP-HPLC, it was used as an auxiliary ligand to form a ternary complex with Eu(III) 2-fluorobenzoate (**Eu-2.8**) to afford Ln-peptoid conjugate **Eu-2.54** (Figure III.2-28). The composition of the Ln-peptoid conjugate was proved by MALDI-TOF-MS. The luminescence performance of this compound was tested on the example of 0.5 mM methanol solution. The luminescence spectrum shows only ion-luminescence of the Eu(III) ion with no additional signals observed (Figure III.2-29). Beside that, the ratio of the integral intensities of the 0-1 and 0-2 bands in the emission spectrum indicates a highly asymmetric coordination environment of Eu(III), which proves the coordination of an organic ligand to the Eu(III) ion.

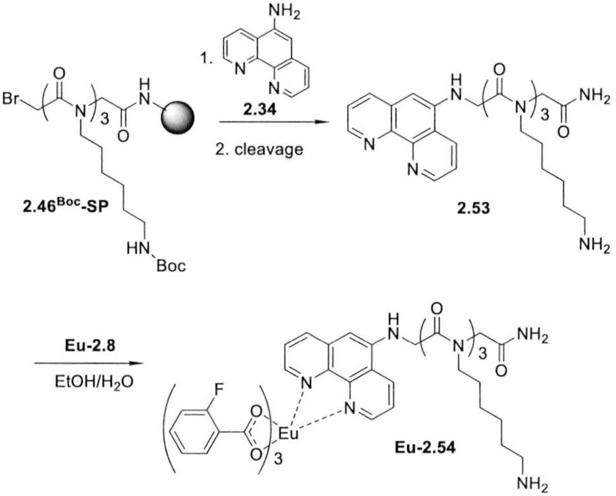

Figure III.2-28 Conjugation of 4-amino-1,10-phenanthroline with peptoid **2.46^{Boc}-SP** to afford peptoid **2.53**, which further serves as an auxiliary ligand for the complexation with Eu(III), resulting in Eu-peptiod conjugate **Eu-2.54**.

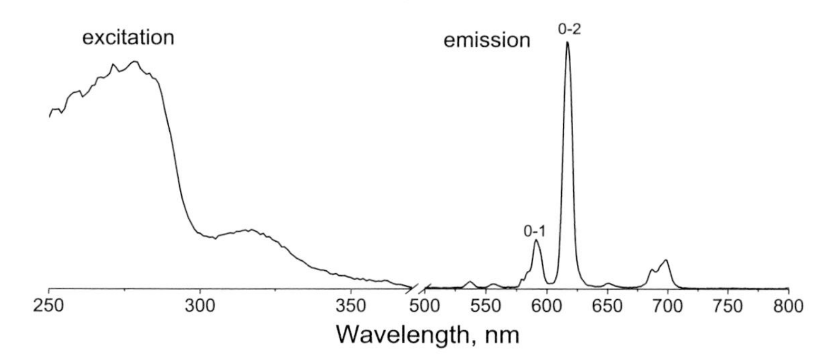

Figure III.2-29 Excitation (λ_{ex} = 280 nm) and emission (λ_{em} = 612 nm) spectra of 0.5 mM MeOH solution of the tested Ln-peptoid conjugate **Eu-2.54**.

The successful strategy was then applied to other Ln-conjugates. In the group of Prof. S. Bräse within the PhD project of Stephan Münch rhodamine-labeled peptoids, possessing specific transporting properties, were synthesized (Figure III.2-30, R = rhodamine B). Two of them exhibited an olfactory localization in zebra fish (Figure III.2-30 A and B), which was revealed in the group of Prof. Schepers. These peptoid motifs were used as a starting point for the synthesis of Ln-conjugates in this work (Figure III.2-30, R = Ln-based label).

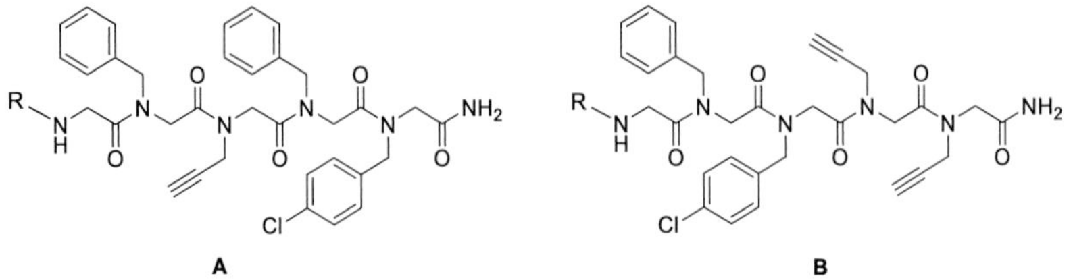

Figure III.2-30 Peptoids, possessing olfactory transporting properties in zebra fish; R = rhodamine B.

The synthesis of Eu-based conjugate with the scaffold A (Figure III.2-30) was performed in several steps by using the submonomer approach (Figure III.2-31), namely: (i) growing the peptoid chain on solid phase until compound **2.55-SP** was obtained; (ii) attachment of the 4-amino-1,10-phenanthroline moiety on solid phase; (iii) cleavage of the resin and subsequent HPLC-purification of the obtained peptoid **2.56**; (iv) formation of Eu ternary complex **Eu-2.57** (Figure III.2-31).

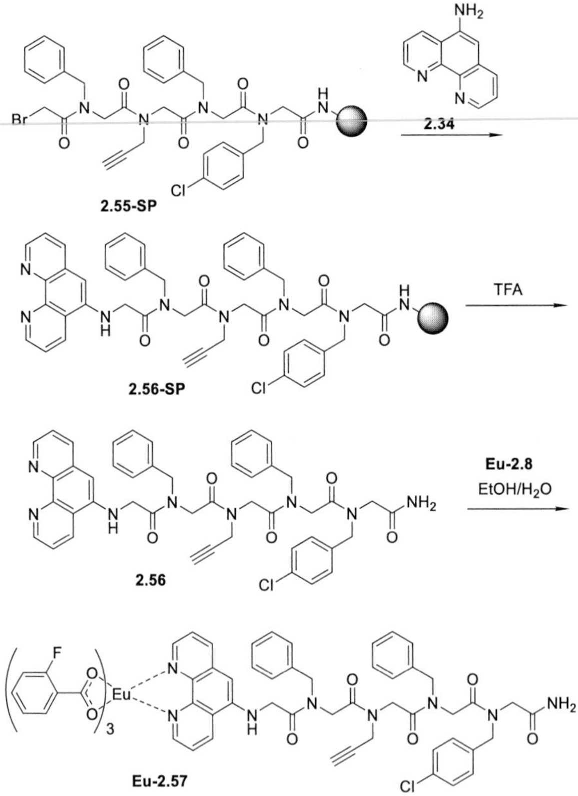

Figure III.2-31 Synthetic route to obtain the Eu(III)-labeled peptoid with the scaffold A.

The same pathway was applied to the synthesis of Eu-labeled peptoid with the scaffold B (Figure III.2-32), where 1,10-phenanthroline moiety was attached to compound **2.58-SP**, using the solid-phase synthesis to afford compound **2.59-SP** on solid support. Subsequent cleavage and purification of compound **2.59-SP** afforded peptoid **2.59**, which was used for the formation of ternary complex **Eu-2.60** (Figure III.2-32).

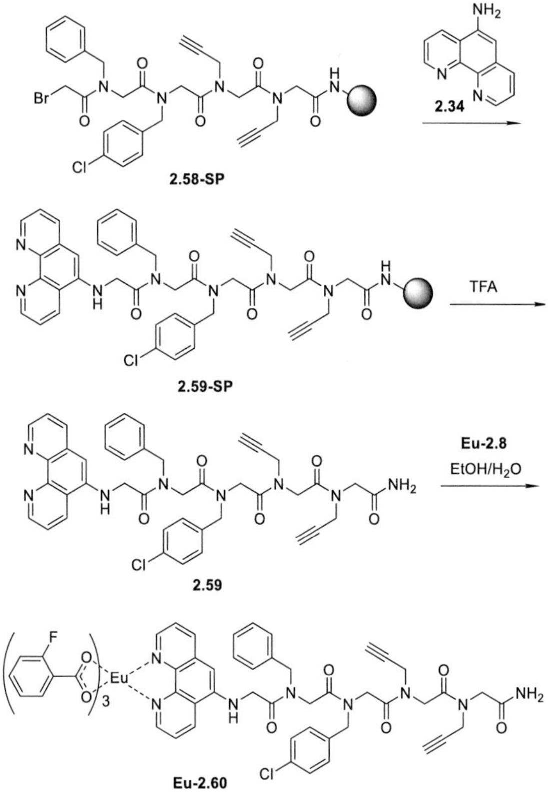

Figure III.2-32 Synthetic route to obtain the Eu(III)-labeled peptoid with the scaffold B.

Eu conjugate **Eu-2.57** exhibited a very intensive red luminescence. The excitation recorded on the different emission bands showed different excitation pathways of organic and ligand-to-metal origin. Therefore, when recording the emission spectrum at λ_{ex} = 280 nm the intensity of Eu-based luminescence was rather high, though the use of 420 nm excitation showed the broad emission band, corresponding to a pure ligand fluorescence. Indeed, peptoid **2.59** itself exhibited a rather intense fluorescence with the maximum at ~ 570 nm. Its position and shape fully coincided with the fluorescence band in the case of conjugate **Eu-2.57**. Thus, for **Eu-2.57**, the presence of two competing radiative processes was indicated, namely ligand fluorescence and ligand-to-Eu energy transfer followed by Eu-centered emission. Though Eu(III) luminescence bands were observed under excitation at lower wavelengths ($\lambda_{ex} = 280$ nm), the presence of the ligand fluorescence bands under excitation with higher wavelengths (lower energy) indicated an inefficient ligand-to-metal energy transfer.

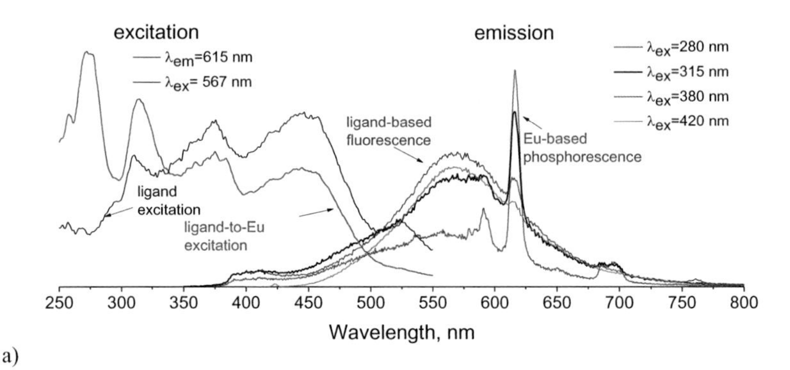

a)

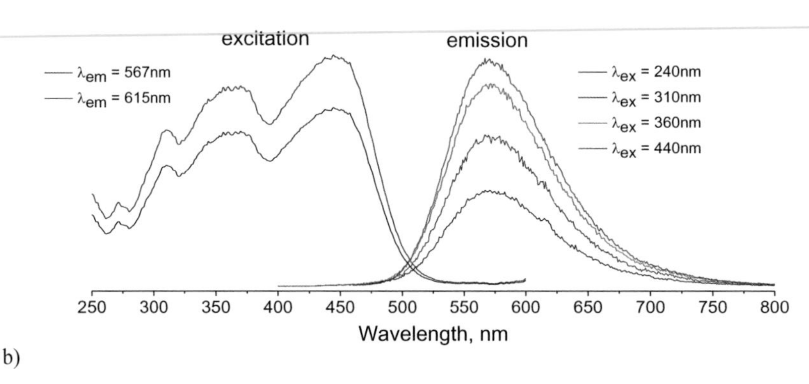

b)

Figure III.2-33 a) Emission and excitation spectra of Eu conjugate **Eu-2.57**, measured in 2 mM MeOH solution, demonstrating the presence of two competing emission processes: ligand fluorescence and Eu(III)-based phosphorescence; b) emission and excitation spectra of peptoid **2.56**, demonstrating its fluorescence.

Both conjugates, **Eu-2.57** and **Eu-2.60**, were tested in HeLa cells. For this purpose, HeLa cells were incubated with DMEM solution of each compound at the concentration of 20 μM which is far below the safe level (the safe concentration is 4 mM for **Eu-2.57** and 5 mM for **Eu-2.60**). In both cases the compounds were able to penetrate the cell membrane and were highly luminescent (Figure III.2-34). Nevertheless, according to the previously performed luminescence study, almost exclusively ligand fluorescence was observed when the compound was excited at higher wavelengths and Eu(III) luminescence peaks were barely visible (Figure III.2-33a). Therefore, this intense luminescence in cells is more likely to be of pure organic origin.

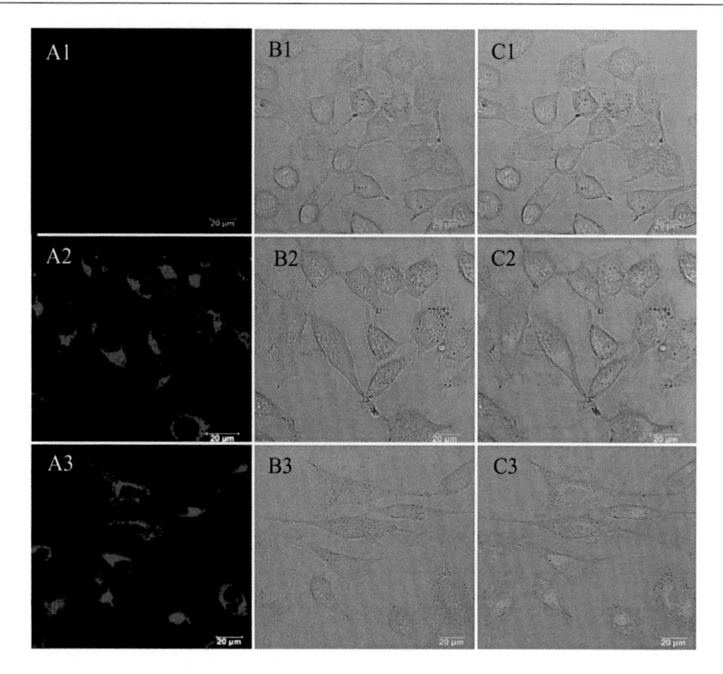

Figure III.2-34 Confocal microscopy images of HeLa cells incubated with Eu-peptoid conjugates. Control (top row): A1 – fluorescence channel, λ_{ex} = 405 nm, laser power = 30%, detection range 580 – 750 nm, B1 – brightfield channel, C1 – overlay; **Eu-2.57** (middle row): A2 – fluorescence channel, λ_{ex}= 405 nm, laser power = 30%, detection range 580 – 750 nm, B2 – brightfield channel, C2 – overlay; **Eu-2.60** (middle row): A3 – fluorescence channel, λ_{ex}= 405 nm, laser power = 30%, detection range 580 – 750 nm, B3 – brightfield channel, C3 – overlay.

In order to test the applicability of the CuAAC reaction for the conjugation of peptoids with Ln-based luminescent probes three peptoids, bearing either ethynyl or azido functional groups, were synthesized (Figure III.2-35). In the case of peptoids, based on the scaffolds A and B, only the incorporation of an azide was possible, because they already contained ethynyl functional groups in their side chains, which would make the CuAAC reaction not selective.

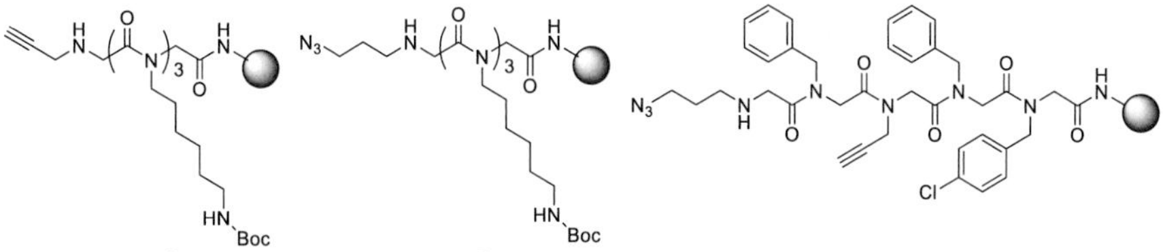

2.61^Boc-SP **2.62^Boc-SP** **2.63-SP**

Figure III.2-35 Structures of the peptoids, bearing ethynyl (**2.61^Boc-SP**) and azido (**2.62^Boc-SP** and **2.63-SP**) FGs.

The CuAAC reaction was performed on solid support in DMF, using CuI as a catalyst. The ethynyl-functionalized peptoid **2.61^Boc-SP** was further used for the coupling with 4-azido-2-

fluorobenzoic acid (Figure III.2-36, n = 0) and 4-azidomethyl-2-fluorobenzoic acid (Figure III.2-36, n=1), while peptoids **2.62^Boc-SP** and **2.63-SP** were coupled with 1,10-phenanthroline-based alkynes (Figure III.2-37 and Figure III.2-38).

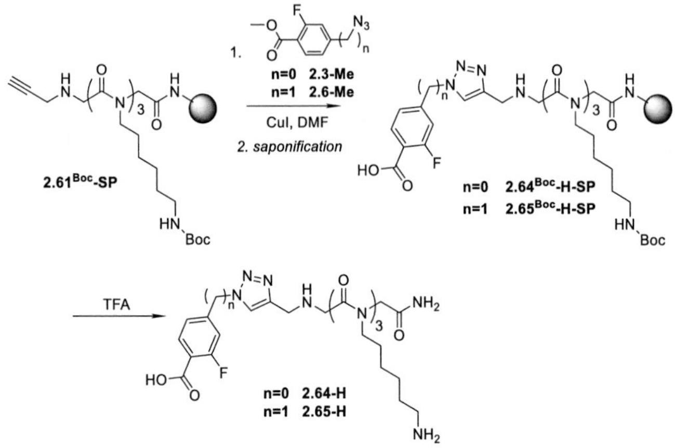

Figure III.2-36 Coupling of peptoid **2.61^Boc-H-SP** with benzoic acid derivatives **2.3-Me** and **2.6-Me**, using the CuAAC reaction on solid support to afford peptoids **2.64^Boc-H-SP** and **2.65^Boc-H-SP**, which were subsequently deprotected and cleaved from the resin to afford peptoids **2.64** and **2.65**.

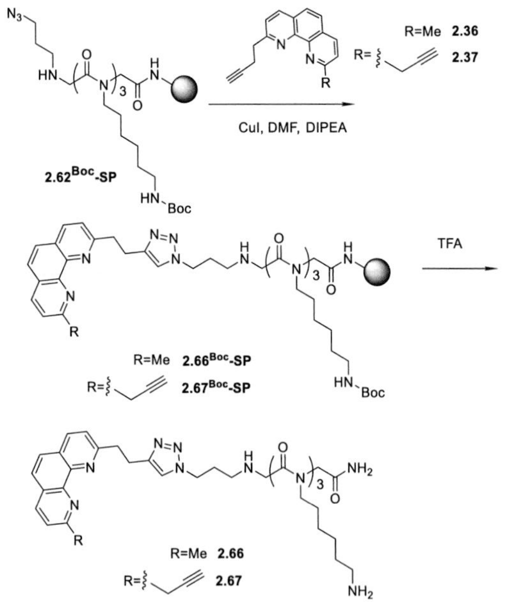

Figure III.2-37 Coupling of peptoid **2.62^Boc-SP** with 1,10-phenanthroline derivatives **2.36** and **2.37**, using the CuAAC reaction on solid support to afford peptoids **2.66^Boc-SP** and **2.67^Boc-SP**, which were subsequently deprotected and cleaved from the resin to afford peptoids **2.66** and **2.67**.

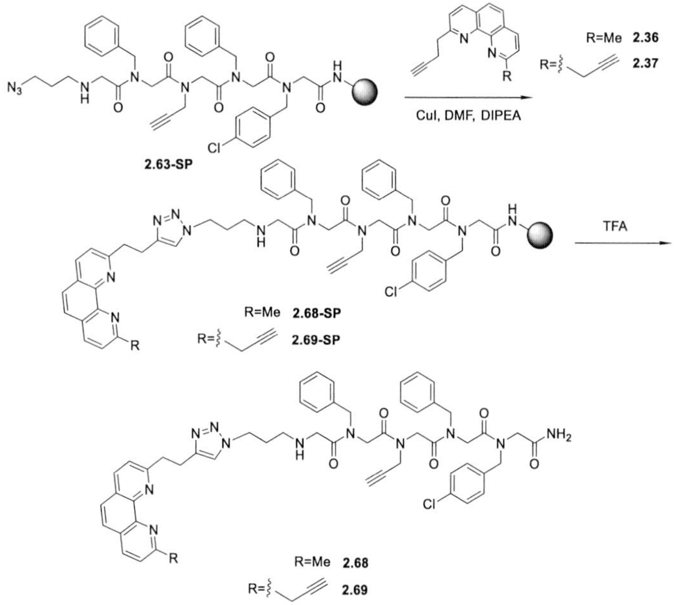

Figure III.2-38 Coupling of peptoid **2.64-SP** with 1,10-phenanthroline derivatives **2.36** and **2.37**, using the CuAAC reaction on solid support to afford peptoids **2.68-SP** and **2.69-SP**, which were subsequently cleaved from the resin to afford peptoids **2.68** and **2.69**.

Fortunately, all the studied reactions turned out to be successful, therefore, RP-HPLC-purified peptoids **2.64-H, 2.65-H, 2.66, 2.67, 2.68, 2.69** were conjugated with Eu. Conjugation with carboxylic acids **2.64-H** and **2.65-H**, using the general scheme $LnCl_3(H_2O)+HL \rightarrow LnL_3(H_2O)_x$, was, however, not successful and only a signal with the mass of the pure peptoid was detected by MALDI-TOF-MS. Unfortunately, the formation of ternary complexes in the case of the peptoids, bearing 1,10-phenanthroline scaffold, was also unsuccessful, because none of these peptoids **2.66** – **2.69** was able to coordinate neither a Eu(III) complex nor just a Eu(III) aqua-ion and hence, no Eu(III)-based isotope pattern was detected by MALDI-TOF-MS. This is probably due to the steric reasons and poor coordination ability of the 2,9-modified 1,10-phenanthroline.

In order to exclude both of these factors, a terpyridine moiety was used for the Eu(III) coordination. A peptoid, bearing terpyridine moiety was synthesized, using SPAAC as demonstrated in Figure III.2-39. The synthetic pathway includes several steps: (i) growing of the peptoid chain of the desired length on solid support (**2.70^{Boc}-SP**); (ii) attachment of the fluorocyclooct-2-yne moiety (**2.71^{Boc}-SP**); (iii) coupling of the cyclooctyne moiety with 4-azido-2,2:6,2-terpyridine (**2.72^{Boc}-SP**); (iv) cleavage of the peptoid from the resin (**2.72**) and (v) conjugation of the obtained peptoid with Eu complex **Eu-2.8** to afford Eu(III) conjugate **Eu-2.73**.

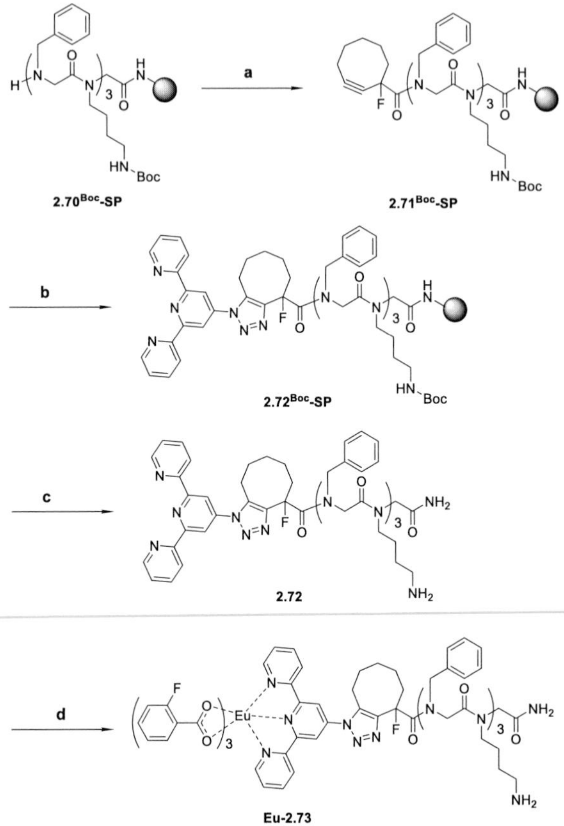

a

2.70^{Boc}-SP

2.71^{Boc}-SP

b

2.72^{Boc}-SP

c

2.72

d

Eu-2.73

Figure III.2-39 Synthetic route to obtain Eu conjugate **Eu-2.73**: a) fluorocyclooct-2-yne carboxylic acid*, HOBt, DIC in DMF, ovrn; b) 4-azido-2,2:6,2-terpyridine in DMF, 16 h; c) 95% TFA in DCM; d) **Eu-2.8**, EtOH, H_2O, reflux.

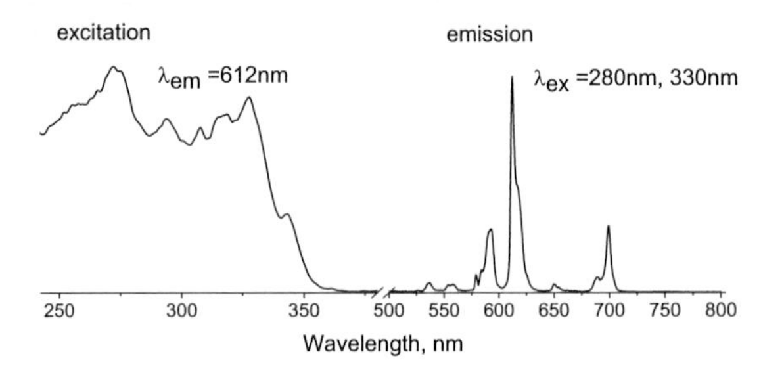

Figure III.2-40 Emission (λ_{em} = 612 nm) and excitation (λ_{ex} = 330 nm) spectra of Eu conjugate **Eu-2.73**, measured for 0.2 mM MeOH solution.

* Synthesized by Dr. S.W. MÜNCH, the group of Prof. BRÄSE

The obtained Eu(III) conjugate indeed exhibited a very intensive luminescence even in highly diluted aqueous solutions (up to 50 μM) which indicates the favorable coordination environment for Eu(III) and high thermodynamic stability of the obtained Eu(III) conjugate (Figure III.2-40). Besides, the variation of the excitation wavelengths in the range 270 – 340 nm did not lead to the appearance of an additional ligand emission band, as it was observed before for conjugates **Eu-2.57** and **Eu-2.60** (Figure III.2-33).

Conjugate **Eu-2.73** was successfully tested in cells. It exhibited moderate luminescence intensity even at very low concentration of 10 μM.

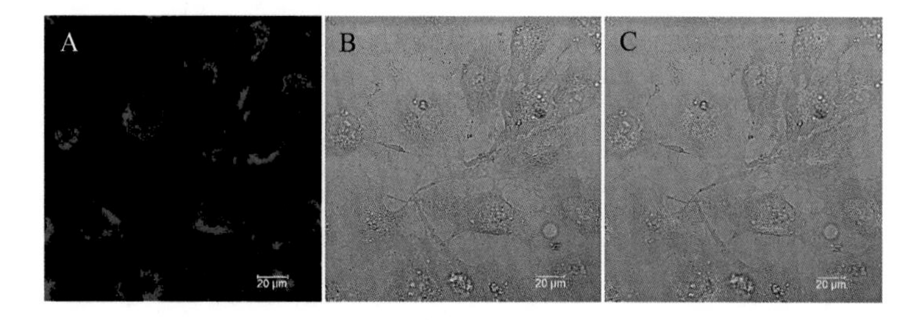

Figure III.2-41 Confocal microscopy images of HeLa cells incubated with 10 μM solution of **Eu-2.73**: A – fluorescence channel, λ_{ex}= 405 nm, laser power = 30%, detection range 580 – 750 nm, B – brightfield channel, C – overlay.

Conclusions

This chapter shows that the coupling of the luminescent Eu(III)-based labels with the peptoids, using solid-phase synthesis was not successful: the final products did not survive the cleavage conditions which is due to the relatively low stability of Ln–O coordination bonds to the acidic medium. However, using already cleaved and purified peptoids for the complexation with Eu(III) was found to be successful, especially, when the peptoid oligomer contains 1,10-phenanthroline or terpyridine moieties which serve as auxiliary ligands and form coordination bonds with the Eu(III) ion. These moieties can be introduced, using SPAAC or CuAAC coupling reactions as well as the direct conjugation, if the amino-modified 1,10-phenanthroline scaffold is used.

To improve the ability of Eu(III) ions to form coordination bonds with peptoids, the latter might contain a chelating macrocycle, so that Eu(III) ion is stabilized in the structure by coordination with several donor atoms. Such conjugates may further possess not only intense Eu(III)-based luminescence but also improved stability due to the chelating effect. This approach may be applied, using cyclic peptoids, bearing a polydentate chelating moiety, to ensure the coordination with the Eu(III) ion. This story will be discussed in the next chapter.

III.2.3 Synthesis and properties of lanthanide conjugates with cyclic peptoids

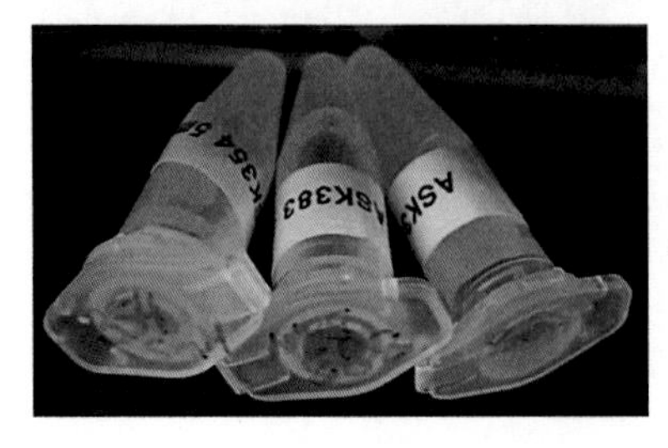

This chapter represents the synthesis and properties of a novel class of compounds, namely the heterocycle-containing cyclic peptoids which can be further used for the design of a targeted luminescence. These peptoids, bearing several carboxylic functionalities, may serve as promising candidates for Eu(III) complexation, featuring the heterocycle moiety responsible for sensitization of Eu(III) emission and the carboxylic functionalities in the peptoid side chains for the strong Eu(III) coordination. Coordination bonds between the Eu(III) ion and the carboxylic groups together with electrostatic interaction between the Eu^{3+} *ion and the lone pairs of nitrogen atoms in the cycle may ensure the formation of stable and water-soluble Eu(III) complexes. These complexes may find various applications in the field of targeted delivery, since they combine the features of Eu(III)-based luminescent labels and peptoid transporting properties. In this work a library containing 9 cyclic peptoids was synthesized and Eu(III) complexes with them were obtained.*

As has been discussed previously, *N*-substituted glycines or peptoids represent a promising class of biocompatible compounds recapitulating the functions and capabilities of natural polypeptides while retaining *in vivo* metabolic stability and remarkable structural diversity.[115] One of the most intriguing features of peptoids is their capacity to generate cyclic structures which have mostly been studied for their targeting feature.[236] Recent research demonstrates the ability of cyclic peptoids as peptidomimetics to coordinate different metal ions,[83,119,120] among them only a few examples of the peptoid coordination with lanthanides are known.[119]

The research work described in this chapter aims to contribute to this poorly investigated field by developing luminescent lanthanide complexes with cyclic peptoids which can serve as bioimaging probes. Due to the presence of several donor atoms, the lanthanide ion can easily be stabilized inside the cyclic peptoid, forming a stable and water-soluble complex exhibiting lanthanide-based luminescence. Such a compound might be very promising, because the combination of these properties is difficult to achieve for many lanthanide complexes.[33]

Due to the poor absorption of Ln(III) f-f levels, it is crucial that the ligand contains fragments with a high molar extinction coefficient to ensure higher energy absorption. Besides, such a fragment should be in a close proximity to the lanthanide ion to be able to transfer the excitation to it. For these purposes heterocyclic moieties are known to be very efficient and the complexes with

1,10-phenanthroline and 2,2'-bipyridine facilitate Eu(III) sensitization, which was also shown in Chapter III.1.2 as well as in literature.[181,237,238]

In this work, the features of heterocycles and peptoids have been combined to obtain thermodynamically stable, water-soluble and highly luminescent compounds for biological applications.

The heterocyclic species was introduced to the backbones of the peptoids. Herein the standard solid-phase peptoid synthesis was adapted for the straightforward preparation of the library of cyclic peptoids featuring cyclometalated Eu(III) complexes. According to the current state of the art, such molecules have never been synthesized before, while only a few attempts to incorporate metal-coordinated heterocycles in peptides (not peptoids) have been made so far.[239]

Synthesis

Cyclic peptoids were synthesized, using the solid-phase synthesis, as described in Chapter III.2.2. However, in this case 2-chlorotrityl chloride resin was used, which allowed to obtain the peptoid with a free carboxylic acid function after the cleavage, ensuring further cyclization of the peptoid (Figure III.2-42) performing a peptide coupling.

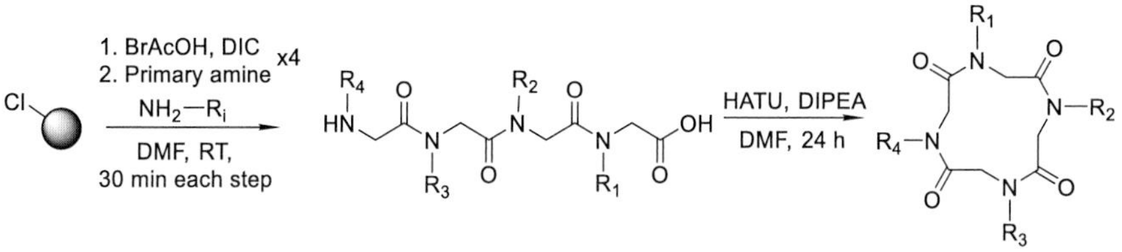

Figure III.2-42 General scheme of the synthesis of cyclic peptoids.

As depicted in Figure III.2-43 the first step of the synthesis of these peptoids on the 2-chlorotrityl chloride resin was an SN_1 reaction with bromoacetic acid and DIPEA in DCM. Subsequent substitution of the bromide with an amine in pDMF completed the first monomer. To introduce the heterocycle into the peptoid backbone, its bromomethyl derivatives were used. In the same way the next amine was coupled with the second bromomethyl end of the heterocycle. The remaining monomers were introduced using the previously described submonomer method, *i.e.* the alteration of addition (bromoacetic acid and DIC) and substitution (replacement of -Br with amine) reactions. As a result, the peptoid with the desired length was obtained (Figure III.2-43) and cleaved from solid support, using relatively mild cleavage conditions (20% hexafluoroisopropanol (HFIP) in DCM).

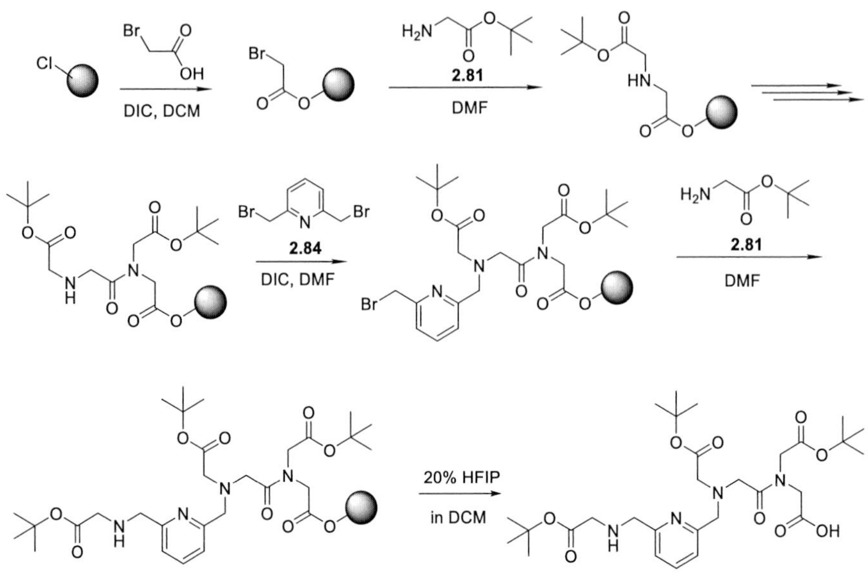

Figure III.2-43 Solid-phase synthesis of the heterocycle-containing linear peptoid on the example of ***H*-(*N*1cx^{tBu})-Pyr-(*N*1cx^{tBu})$_2$-*OH* (2.90)**.

Because the used heterocycles have two bromomethyl functionalities, a side reaction of dimerization was unavoidable in most cases, namely the reaction of the bis(bromomethyl)-functionalized heterocycle with two secondary amines, as demonstrated in Figure III.2-44. The masses of such dimeric fragments were detected by MALDI-TOF-MS for almost all the compounds which were synthesized this way. However, the desired product could be easily isolated from this mixture by RP-HPLC. It is worth noting that the amount of byproduct did not exceed 20% for almost all the synthesized compounds (calculated by the integration of analytical HPLC peaks). Interestingly, the amount of formed dimers depended on the heterocycle and increased in the row pyridine – 2,2'-bipyridine – 1,10-phenanthroline. The following two dimers were isolated: **2.92-*t*Bu-dimer** and **2.98-*t*Bu-dimer** (Figure III.1-45).

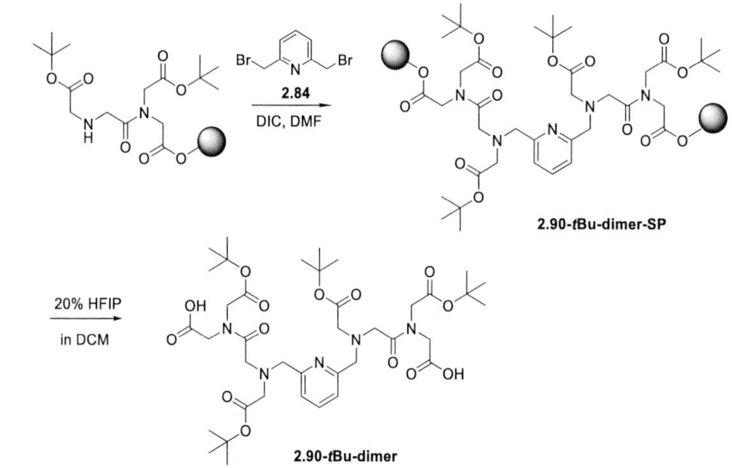

2.84

DIC, DMF

2.90-*t*Bu-dimer-SP

20% HFIP in DCM

2.90-*t*Bu-dimer

Figure III.2-44 Scheme of the side reaction during the peptoid synthesis.

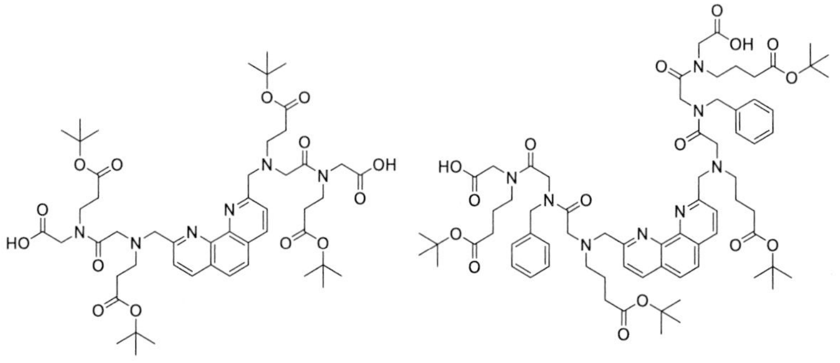

2.92-*t*Bu-dimer

2.98-*t*Bu-dimer

Figure III.2-45 Dimeric byproducts, isolated during the synthesis of the library of heterocycle-containing peptoids.

After the linear peptoids of desirable size bearing a heterocycle moiety had been synthesized, the intramolecular cyclization reaction was performed, using HATU as a coupling agent (Figure III.2-46). The *t*Bu-protecting groups were afterwards removed using the cleavage cocktail TFA/water/TIPS (95/2.5/2.5, volumetric), carboxylic groups were deprotonated and the complexation with the Eu(III) ion was performed, using an aqueous solution of Eu(III) chloride under basic conditions. The general scheme for the synthesis of Eu conjugates with cyclic peptoids is demonstrated in Figure III.2-47. As this synthetic route includes a multistep synthesis of linear peptoid (up to 11 steps), intramolecular cyclization reaction and multiple purification steps, the overall yields were often even less than 1% which includes the losses at the purification stages. The optimization of the method used for RP-HPLC purification might potentially increase the overall yield.

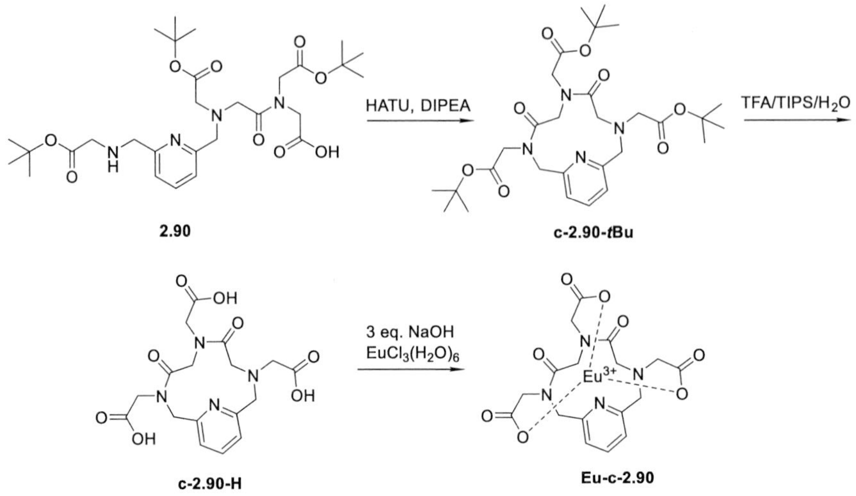

Figure III.2-46 Cyclization of the heterocycle-contained linear peptoid and its subsequent complexation with the Eu(III) ion on the example of peptoid **2.90**.

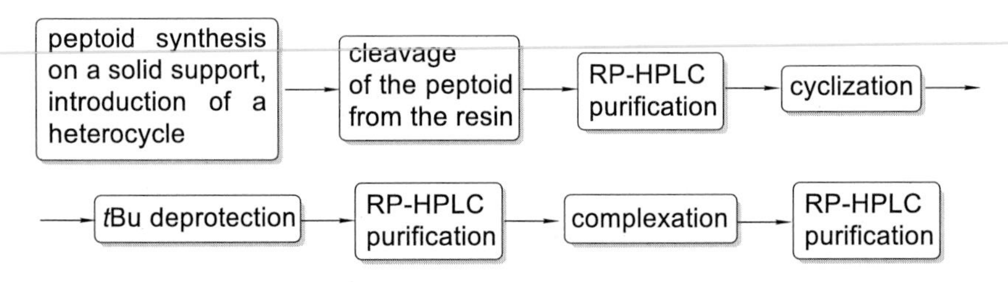

Figure III.2-47 General scheme of the synthesis of Eu conjugates with cyclic peptoids, performed in this work.

During the peptoid synthesis, *t*Bu-protected α-, β- and γ- amino acids were used to utilize carboxylic groups for the Eu(III) coordination after deprotection and deprotonation. In addition to the commercially available *tert*-butyl glycinate **2.81**, *t*Bu-protected β- and γ- amino acids were synthesized in this work in three steps (Figure III.2-48), by exploration of the incorporation of different protection groups (PGs). The synthetic route starts with the Cbz-protection of amino groups (Figure III.2-48a) of the corresponding amino acids: β-alanine (**2.75**) or γ-aminobutyric acid (**2.76**). The carboxylic functionality of the Cbz-protected amino acids **2.77** and **2.78** was then protected with the *t*Bu-PG[240] to afford compounds **2.79** and **2.80**. Finally, deprotection of Cbz-PG afforded *t*Bu-protected α-, β- and γ- amino acids **2.81**, **2.82** and **2.83**.

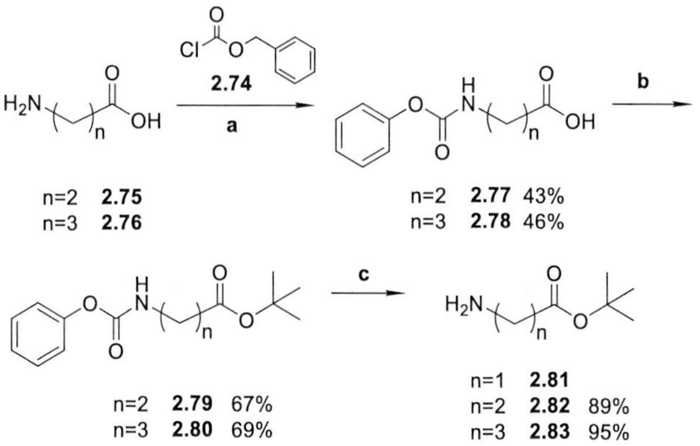

Figure III.2-48 Synthesis of *t*Bu-protected α-, β- and γ- amino acids: a) 1. Cbz-Cl (**2.74**), NaOH, 0 °C – rt, ovrn, 2. HCl; b) *t*BuOH, DMAP, DIC, dry DCM, Ar atmosphere; c) Pd/C, H_2 (balloon), MeOH.

To apply the synthetic route described above, the heterocycle scaffold, containing two bromomethyl functional groups should be afforded. Three types of bis(bromomethyl)-functionalized heterocycles were used in this work (Figure III.2-49), based on

(i) pyridine scaffold **2.84**;

(ii) 2,2'-bipyridine scaffold **2.85**;

(iii) 1,10-phenanthroline scaffold **2.86**.

Complementing the commercially available compound **2.84** (2,6-bis(bromomethyl)pyridine), the other two heterocycles **2.85** and **2.86** were synthesized.

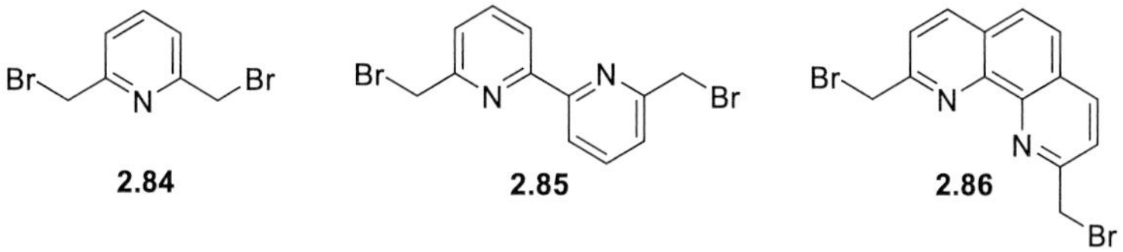

Figure III.2-49 Heterocycles used as building blocks for the peptoid synthesis.

The synthesis of bis(bromomethyl)-functionalized heterocycles was optimized on the example of 2,9-bis(bromomethyl)-1,10-phenanthroline **2.86**. To afford this product, two synthetic routes have been proposed. The first pathway contained three steps,[241] namely (i) oxidation of an aromatic-bound methyl group of neocuproine **2.35** to aldehyde **2.87**, using selenium dioxide in 1,4-dioxane; (ii) reduction of dialdehyde **2.87** to diol **2.88**, using sodium borohydride in ethanol and (iii) conversion of diol **2.88** to the corresponding halide **2.86** (Figure III.2-50). Though each step gave a moderate reaction yield, the total yield over three steps turned out to be only 26%.

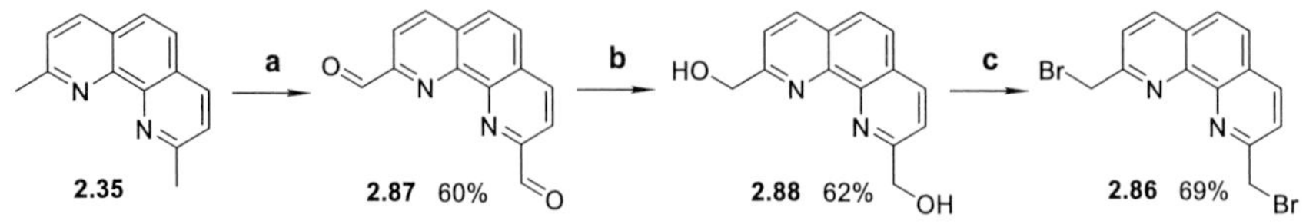

Figure III.2-50 Synthesis of compound **2.86**: a) SeO_2, 1,4-dioxane, H_2O, 1 h, reflux; b) $NaBH_4$, EtOH, 1 h, reflux; c) HBr (aq.), 1 h, reflux.

The second strategy implies one-step radical reaction of neocuproine **2.35** with NBS[242] and afforded product **2.86** in 34% yield after purification by flash column chromatography and subsequent recrystallization from ethanol (Figure III.2-51).

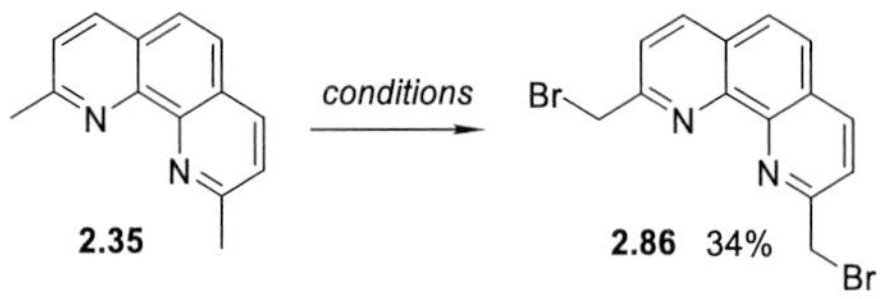

Figure III.2-51 Alternative synthesis of compound **2.86**; conditions: NBS, DBPO in acetonitrile, 24 h, reflux.

Given the superior reaction yield accompanied by the reduction in the number of steps from three to one, the second route was selected for the synthesis of compound **2.85**, but starting with 6,6'-dimethyl 2,2'-bipyridine **2.89**. Interestingly, the use of DCM as a solvent, as described by ZHAO *et al.*,[243] did not afforded product **2.85** and starting material **2.89** was fully recovered. However, when carbon tetrachloride was used, product **2.85** was obtained in 41% yield (Figure III.2-52).

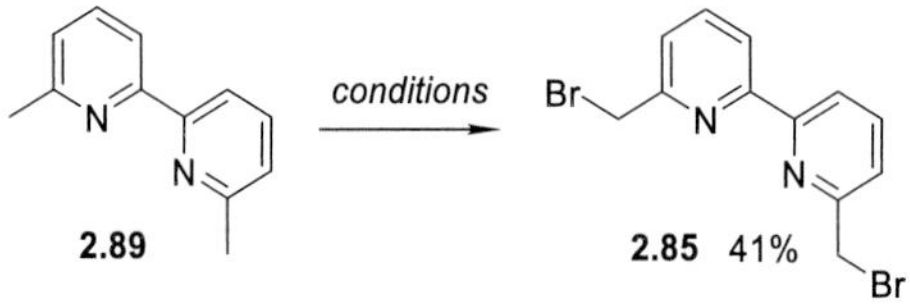

Figure III.2-52 Synthesis of compound **2.85**; conditions: NBS, DBPO in CCl_4, 24 h, reflux.

Using the solid-phase synthesis and the established in this work approach to introduce heterocycles into the peptoid backbone (Figure III.2-43, Figure III.2-46) the library of 9 peptoids **c-2.90** – **c-2.98** have been synthesized (Figure III.2-53), considering three main parameters:

i) the size of the cycle;

ii) the length of the side chain (α-, β- and γ- amino acids);

iii) the type of the heterocycle (pyridine, 2,2'-bipyridine or 1,10-phenanthroline).

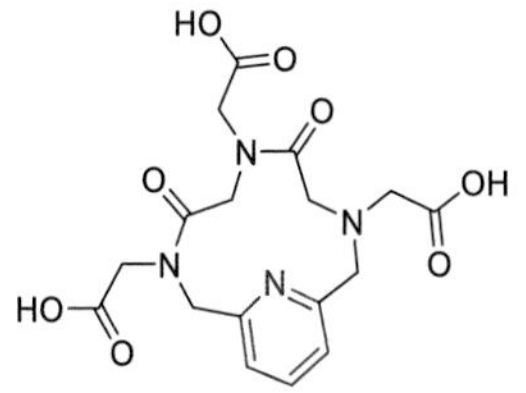

c-2.90 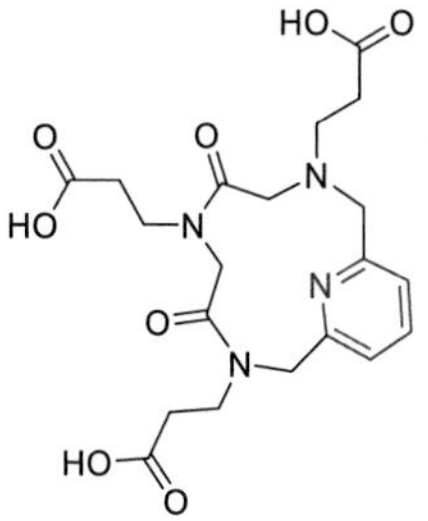 **c-2.91** 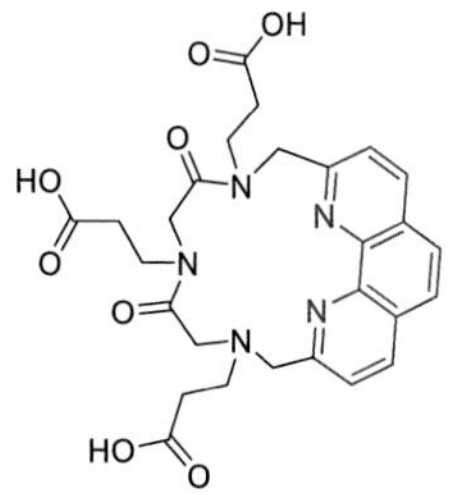 **c-2.92**

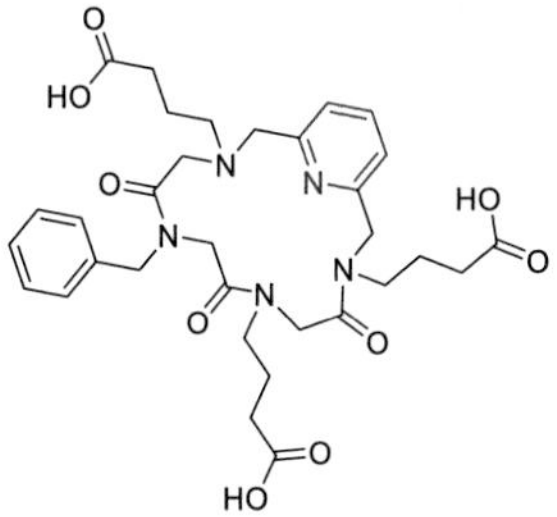

c-2.93 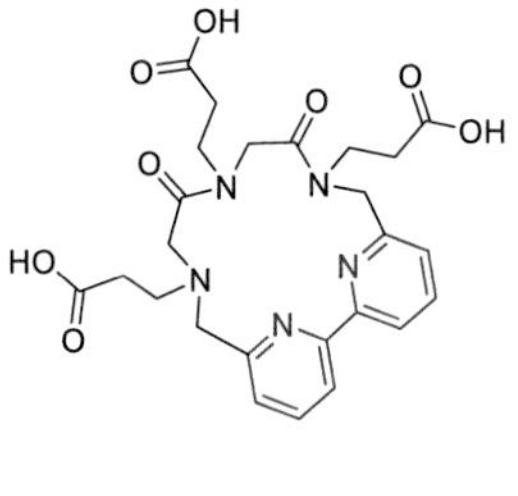 **c-2.94** 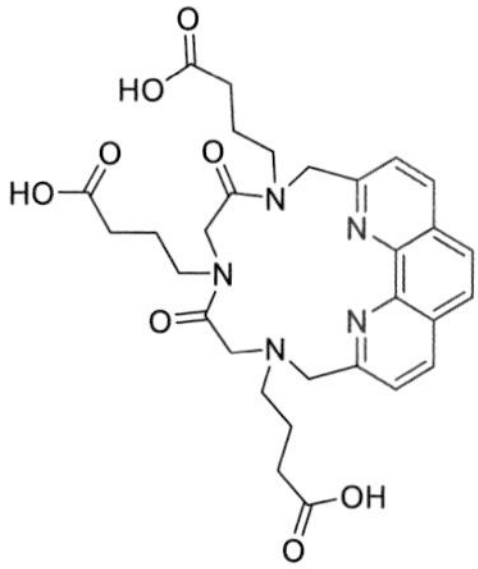 **c-2.95**

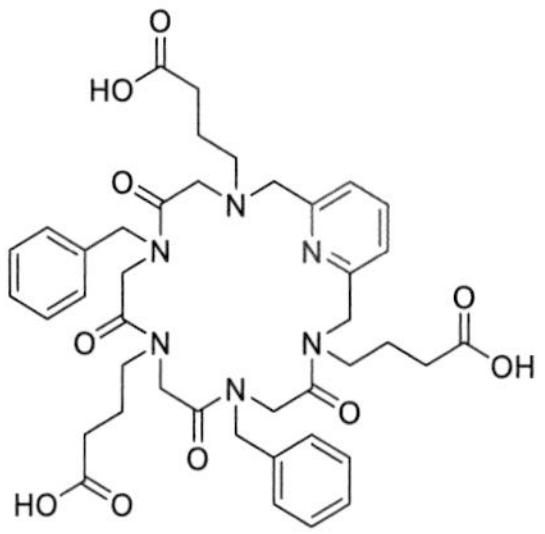

c-2.96 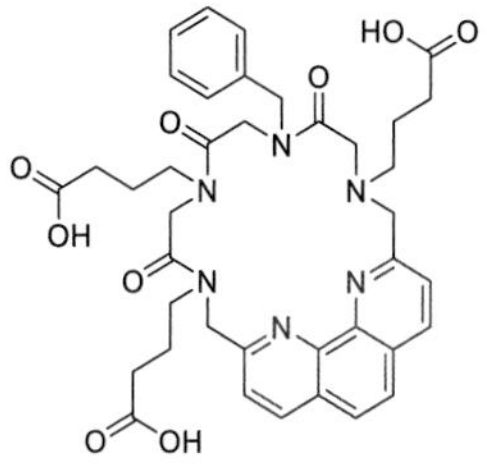 **c-2.97** 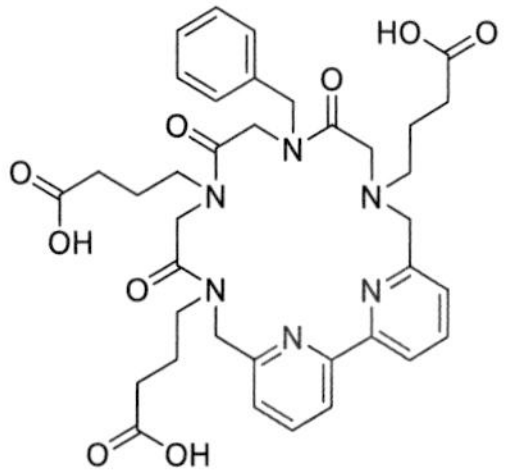 **c-2.98**

Figure III.2-53 Library of heterocycle-containing cyclic peptoids, synthesized in this work. Heterocycles incorporated into the peptoid backbones are highlighted in blue.

In all described peptoids nitrogen atoms are spaced apart by two carbon atoms, therefore, one can estimate the size of the cycle by the amount of nitrogen atoms in the backbone. In this sense the synthesized library contained *4N*-cycles (Figure III.2-53, **c-2.90** and **c-2.91**), *5N*-cycles (Figure III.2-53, **c-2.92** – **c-2.95**) and *6N*-cycles (Figure III.2-53, **c-2.96** – **c-2.98**). The synthesized peptoids contained three side-chains, bearing carboxylic group (Figure III.2-53), which is meant to coordinate Eu^{3+}. When larger cyclic peptoids were synthesized, benzylamine was used as a submonomer in addition to the amino acids (Figure III.2-53, compounds **c-2.93**, **c-2.96** – **c-2.98**).

Geometry optimization

In order to determine the feasibility of the synthesized cyclic peptoids to coordinate Eu(III) and sensitize its luminescence, the geometry optimization for several peptoids was performed, using the TMOLEX program package*. Eu^{3+} ionic radii depend on the coordination number and vary in the range 0.947 Å – 1.12 Å. Therefore, by estimating the geometry of the cavity inside the cyclic peptoid, one can predict if this cavity is able to carry an Eu(III) ion. Besides, the length of the peptoid side chains may also play a role: the optimal length ensures that carboxylic groups are able to stabilize the Eu(III) ion in the cycle and also affect the shape of the cavity inside the cycle. Because the electrostatic forces between positively charged Eu(III) and lone pairs on nitrogen atoms in the peptoid backbone are quite weak, Eu(III) coordination sphere might be saturated by coordination with multiple peptoid moieties to form a coordination polymer, which would make the complex poorly soluble. Hence, the formation of such coordination polymers should be avoided.

To study the impact of the side chain length, the optimized geometries were compared for the pair of cyclic peptoids **c-2.90** and **c-2.91**. These two molecules are *4N*-cyclic peptoids, both containing a pyridine scaffold in their backbones, but possessing different side chains: a glycine moiety was used for the synthesis of **c-2.90**, while β-alanine was used for the synthesis of **c-2.91** (Figure III.2-54a,b). The simulated geometries of these two compounds are indeed different. Compound **c-2.91** has a slightly larger size of the cycle: the distances between the opposite N atoms in **c-2.90** are equal to 3.938 Å and 4.390 Å (Figure III.2-54a), while in **c-2.91** they are equal to 3.977 Å and 4.918 Å (Figure III.2-54b), respectively. However, considering the size of Eu(III) ion and the typical Ln–N distances (2.30 – 2.80 Å for *4N*-macrocycles, according to the CSD analysis), these cavities are rather small to hold Eu(III) ion. Nonetheless, these compounds may still coordinate Eu(III) in an out-plane coordination, where Eu(III) will be stabilized by four nitrogen atoms from the cycle and three carboxylic groups, though not directly inside the cavity. This behaviour might be expected for Eu(III) complex with **c-2.90** and very unlikely for the Eu(III) complex with **c-2.91**, since in the latter case such long side chains may favor the formation of coordination polymers (Figure III.2-54b).

In the case of *5N*-cycles the size of the cavity is significantly larger, possessing an "oval" configuration for compound **c-2.92** with the smallest distance between non-neighboring nitrogen atoms being 3.599 Å and the largest one being 6.828 Å (Figure III.2-54c). As in the previous case an increase of the length of the side chains affects the cavity size and shape. In the case of

* Optimization parameters: functional B3-LYP, basis set def-TZVP, symmetry group C1.

compound **c-2.95** it is less prolonged: the smallest distance between the non-neighboring nitrogen atoms is 3.946 Å and the largest one is 6.018 Å (Figure III.2-54d).

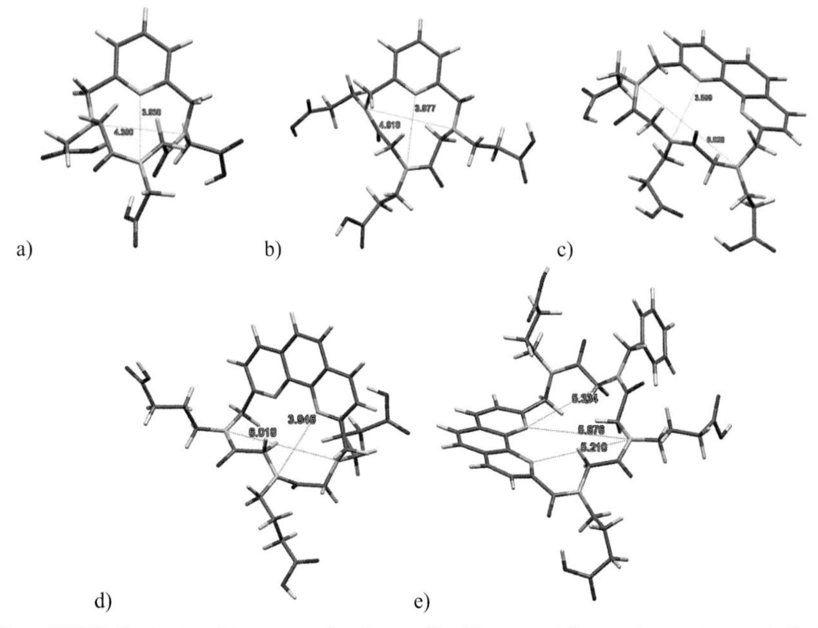

Figure III.2-54 Simulated model structures of cyclic peptoids with measured distances between N atoms inside the cycle: a) **c-2.90**; b) **c-2.91**; c) **c-2.92**; d) **c-2.95**; e) **c-2.97**. Atom assignment: C-grey, O-red, N-blue, H-white.

Both of these compounds, **c-2.92** and **c-2.95**, may coordinate Eu(III) ion, however, not completely on the imaginary plane, formed by nitrogen atoms inside the cavity. This fact can be shown by circumcircling the triangle which is formed by two nitrogen atoms of 1,10-phenanthroline moiety and the opposite nitrogen atom. One can calculate the optimal Ln–N distance for the Ln(III) ion to be located inside this triangle which is equal to 2.040 Å for **c-2.92** and 2.160 Å for **c-2.95** (radius of this circle). These values are still below the range for typical Ln–N distances (2.30 – 2.80 Å). However, such an "oval" configuration for *5N*-cycles is due to the 1,10-phenanthroline moiety which is relatively large and non-flexible, therefore, one would expect *5N*-cycles with 2,2'-bipyridine to be more suitable for complexation. An increase in the cavity size in *6N*-cycles leads to a more favorable geometry which seems to be appropriate for carrying the lanthanide ion, as was calculated for compound **c-2.97** (Figure III.2-54e). The similar cavity size was obtained for Gd-binding cyclic hexapeptoids by De Cola *et al.*,[119] where hexa- and three-carboxylated cyclic hexapeptoids with β-alanine as the side chains featured favorable for the Gd(III) geometry and formed a stable Gd(III) complex.

Complexation, photophysical properties and cellular luminescence

The peptoid library, synthesized in this work, was used for the complexation with Eu(III), generating a library of Eu(III)-conjugates (Figure III.2-55).

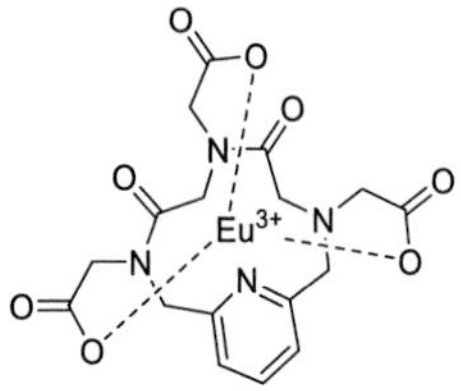

Eu-c-2.90 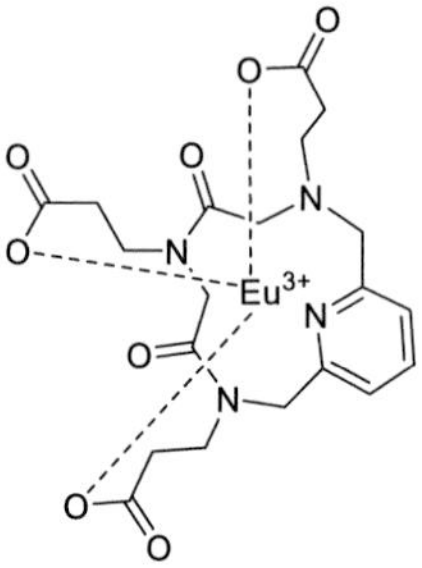 **Eu-c-2.91** 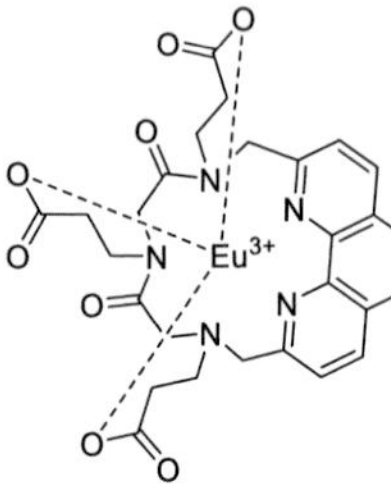 **Eu-c-2.92**

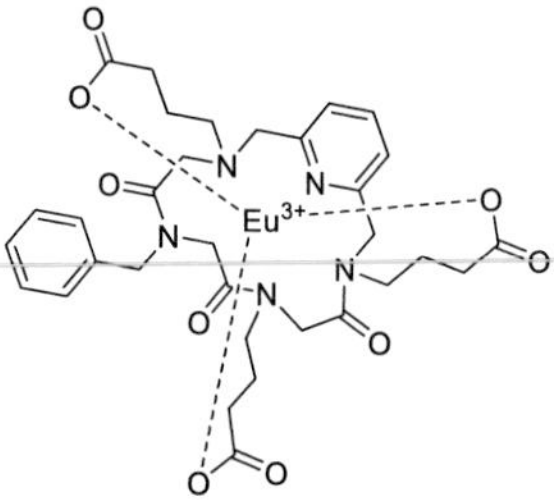

Eu-c-2.93 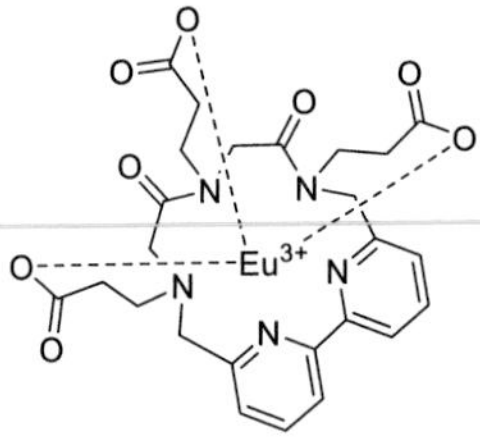 **Eu-c-2.94** 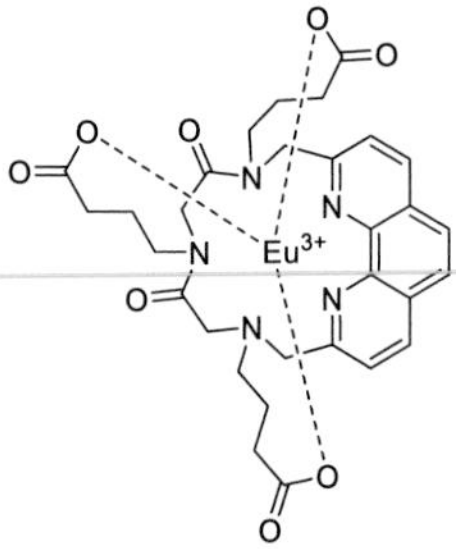 **Eu-c-2.95**

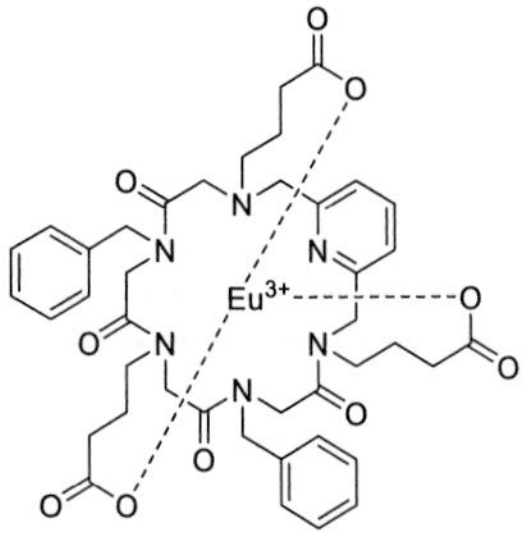

Eu-c-2.96 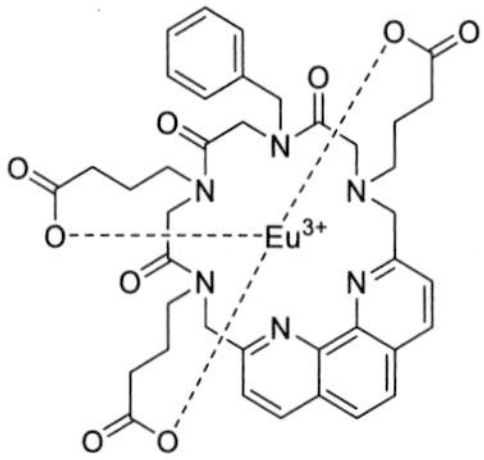 **Eu-c-2.97** 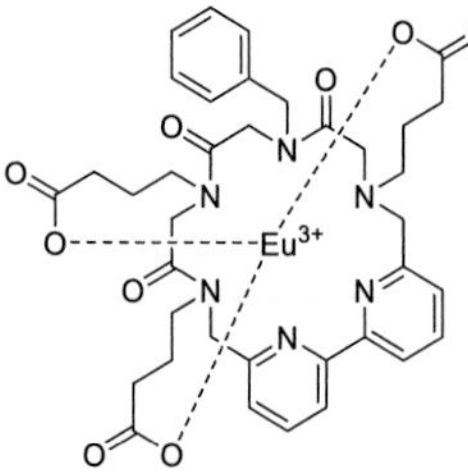 **Eu-c-2.98**

Figure III.2-55 Library of Eu(III) complexes with heterocycle-containing peptoids, synthesized in this work.

Surprisingly, in all cases the obtained Eu(III) complexes turned out to be water-soluble, doubting the formation of Eu(III) coordination polymers. Additionally, no Eu(III)-based emission was detected for compounds **Eu-c-2.91**, **Eu-c-2.92** and **Eu-c-2.93**, instead, these complexes exhibited weak greenish-blue emission under UV lamp (λ_{ex} = 254 nm). Since the ligands themselves were not luminescent, the appearance of weak emission after the complexation may be associated with the ligand phosphorescence due to the introduction of a heavy metal. These results are in line with the geometry optimization study, showing that the side chains of compound **c-2.91** are too long and the size of the cavity is too small (Figure III.2-54b). In this case the complex is still formed, however, due to the unfavorable ligand geometry Eu(III) coordination sphere is probably saturated with quenching solvent molecules, which would explain the absence of Eu(III)-based luminescence. The absence of luminescence in the case of **Eu-c-2.92** is also predicted by the geometry optimization study: due to the rigidity of the 1,10-phenanthroline scaffold and the size of the cycle (*5N*) being too small, peptoid **c-2.92** possesses an oval configuration which won't be able to carry the Ln(III) ion (Figure III.2-54c), whose coordination sphere will be then saturated by solvent molecules. The absence of luminescence in the case of compound **Eu-c-2.93** is rather surprising but could be probably explained by similar reasons.

Other Eu(III) complexes indeed possess Eu-based luminescence (Figure III.2-56), where no any additional luminescence bands were observed. Besides, the ratio between the integral intensities of the 0–1 and 0–2 bands indicates the coordination with the organic ligand (when compared to the Eu(III) aqua-ion luminescence, Figure III.1.12b,d).

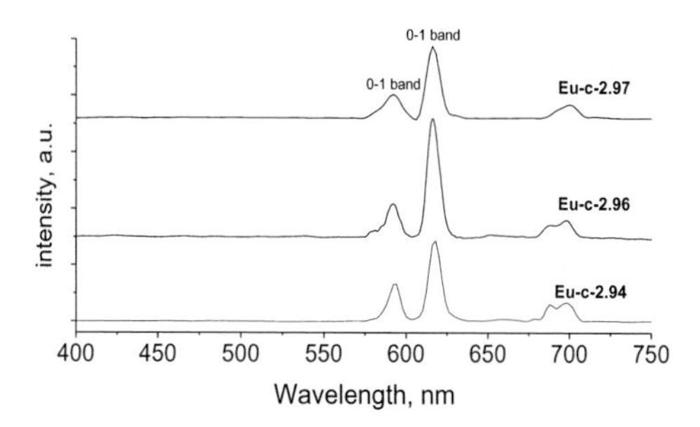

Figure III.2-56 Selected luminescence spectra of Eu-based cyclic peptoids in aqueous solutions, concentration 0.1 mM, λ_{ex} = 320 nm.

The synthesized peptoid library was tested in cells. For this purpose, an MTT assay was firstly carried out in order to determine the safe concentrations. Noteworthy, the safe concentration for all the tested compounds turned out to be relatively high (in the range 1 – 3 mM) which makes these compounds non-toxic in a wide interval of concentrations. After the safe concentration of each peptoid was determined, the cells were treated with the investigated probes at different concentrations below the safe level*. Indeed, the Eu(III)-conjugates possessed a high cellular permeability, being highly emissive even in a very low concentration range (Figure III.2-57).

* MTT testes and confocal microscopy were performed by M. Sc. A.S. MESHKOV within the GRK2039 collaboration, the group of Prof. U. SCHEPERS, Institute of Toxicology and Genetics, Karlsruhe Institute of Technology, Karlsruhe, Germany.

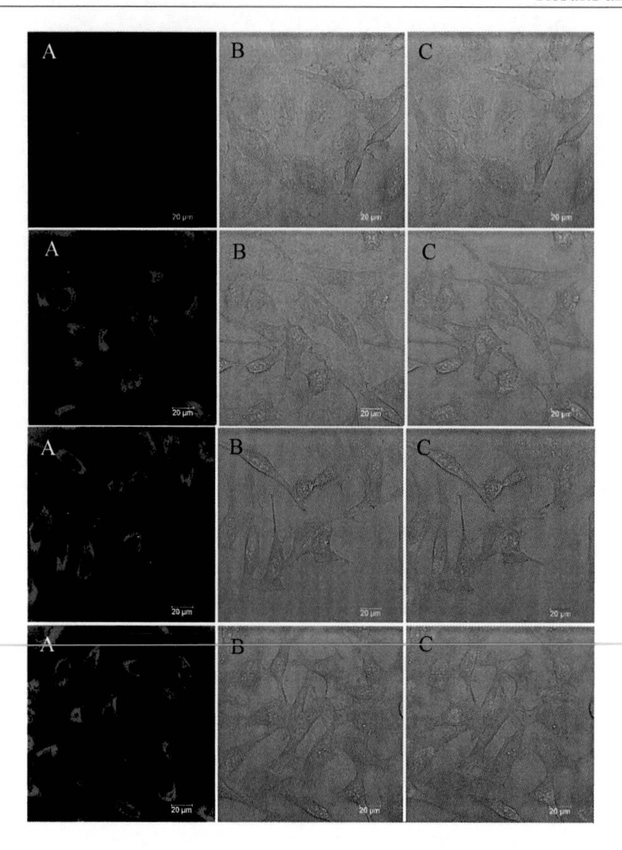

Figure III.2-57 Confocal microscopy images of HeLa cells, incubated with Eu-conjugates. A – fluorescence channel, λ_{ex}= 405 nm, laser power = 30%, detection range 580 – 750 nm, B – brightfield channel, C – overlay. 1st row: control; 2nd row: **Eu-c-2.95**, conc. 10 μM; 3rd row: **Eu-c-2.96,** conc. 5 μM; 4th row: **Eu-c-2.94**, conc. 10 μM.

Conclusions

In this chapter a novel synthetic route to afford heterocycle-containing cyclic peptoids was developed. This route is based on the solid-phase peptoid synthesis, where the heterocycle is introduced, using its bromomethyl derivative. Following this method a library, containing 9 novel cyclic peptoids, was synthesized. The geometry optimization study revealed that compounds **c-2.91** and **c-2.92** are more likely to exhibit no Eu(III)-based luminescence due to their geometric features. The study of photophysical data proofs this assumption: the complexes with these two ligands, together with compound **Eu-c-2.93**, indeed turned out to be non-luminescent. Other Ln(III)-based cyclic peptoids exhibited moderate luminescence and Eu(III)-coordinated peptoids were successfully tested in HeLa cells showing an intensive luminescence even at a low concentration of 5 μM and proved to be promising for bioimaging.

IV. Conclusion and outlook

IV.1 Lanthanide complexes

IV.1.1 Lanthanide fluorobenzoates with different fluorination degrees

This chapter describes the synthesis of 84 lanthanide fluorobenzoates with 9 different ligands and their structure-property relationship. The structural study started with the analysis of different lanthanide pentafluorobenzoates, using lanthanides from La to Lu (excluding Pm) and was based on **14 powder patterns and 14 single-crystal structures** of these complexes. Five different structural types were identified, where the formation of each structural type is determined by both, crystallization conditions and the Ln(III) ion. When crystallized at 80 °C, lanthanide pentafluorobenzoates form coordination polymers with the amount of coordinated water molecules ≤ 2. The crystal structure of the most common phase $Ln(pfb)_3(H_2O)$ was determined from PXRD data and could be observed for Ln = Sm – Er. Slow evaporation of $Ln(pfb)_3$ aqueous solutions exclusively produced crystals with systematically higher water content ($x \geq 4$). Single-crystal XRD studies revealed the reduction of the nuclearity along the lanthanide row, so that the monomeric structures were formed. Some metals demonstrate solvatomorphs of a different hydrate composition and hence, of a different nuclearity within the same batch.

Analyzing other fluorobenzoates, two most thermodynamically stable structures were identified *via* the structural study based on **24 powder patterns and 10 single-crystal structures**. It was shown that the major factor, affecting the crystal structures of partially fluorinated benzoates, is the number of non-hydrogen *ortho*-substituents of the benzene ring (n_{ortho}) which defines the torsion angle between the carboxylic group and the benzene ring. This leads to the differences in the steric hindrance within the complexes of these two groups. Thus, the complexes with $n_{ortho} = 1$, are more flexible and might further possess higher stability in solution. Besides, they were found to be often virtually isomorphous with each other in powders (up to the fluorine atom) and the preferable structural type was found. The complexes with $n_{ortho} = 2$ also tend to form virtually isomorphous powder structures, though different single-crystal structures were found, witnessing their higher tendency to exchange carboxylic ligands with water and hence, lower thermodynamic stability.

Almost all complexes possess a polymeric structure, except the dimeric Eu(III) 2-fluorobenzoate and aforementioned pentafluorobenzoates, possessing a very unusual tendency to crystallize in dimers/monomers.

Despite almost the same value of the ligand triplet level (ca. 20500 cm^{-1}), the luminescence performance turned out to be very different for the complexes. It was shown that the low quantum yield of europium complexes is mostly due to the low efficiency of ligand-to-metal energy transfer and is not related to the quenching *via* coordinated water molecules. Therefore, an introduction of an additional energy transfer step, *i.e.* a Tb^{3+} ion turned out to be beneficial in some cases (sensibilization strategy $L \rightarrow Tb^{3+} \rightarrow Eu^{3+}$). On top of that, the luminescence performance in aqueous solutions was thoroughly studied as well. This study showed that the luminescence intensity is mostly due to the dissociation degree of the complex at a certain concentration.

Another common way to increase the luminescence quantum yield of Eu-based emitters is the introduction of an auxiliary ligand. This strategy was explored choosing fluorobenzoates with the best luminescence performance (Eu complexes with 2-fluorobenzoate, 2,5-difluorobenzoate and 2,6-difluorobenzoate ligands).

IV.1.2 Lanthanide ternary complexes

In this work ternary complexes with the selected fluorobenzoates were synthesized, using 1,10-phenathroline (Phen) and bathophenanthroline (BPhen) as auxiliary ligands. Interestingly, these complexes demonstrated a remarkably high quantum efficiency up to 90% which is very unusual for Eu-based emitters. For several complexes the single-crystal structures were determined and found to be similar for europium complexes within the Phen series (2-fluorobenzoate and 2,5-difluorobenzoate with Phen) and slightly different, comparing to the complex with BPhen (europium 2-fluorobenzoate with BPhen). However, all these compounds possess a dimeric structure with one Phen/BPhen moiety per lanthanide ion which prevents the polymerization of the carboxylate. The study of the coordination environment showed that the polyhedron of these complexes has a favorable geometry, so that the Eu ion is in a low-symmetry environment and its coordination ability is saturated. The study of the photophysical properties revealed that the efficiency of f-f transitions is almost 100% for these ternary complexes. Despite the structural similarity, these complexes showed very different behaviour, when tested as bioprobes and as OLED emitting layers. The ternary complexes with BPhen turned out to be non-permeable in HeLa cells, even at higher incubation times (up to 48 hours), while the complexes with Phen exhibited a moderate cellular luminescence. However, for OLED applications Phen complexes turned out to be inefficient: despite a high photoluminescence quantum yield, these complexes exhibited poor charge transport properties, though an introduction of transporting layers did not lead to any improvement in performance. At the same time, the complexes with BPhen turned out to be very efficient which is due to the improved charge transport properties of BPhen in comparison to Phen. These observations are summarized in Table IV.1-1.

Table IV.1-1 Summary of the application of Eu(III) fluorobenzoates with auxiliary ligands with the formula $EuL_3(Q)$, Where L-fluorobenzoate, Q-auxiliary ligand.

Auxiliary ligand for Eu fluorobenzoates $EuL_3(Q)$	**Q = Phen**	**Q = BPhen**
Structure of an auxiliary ligand	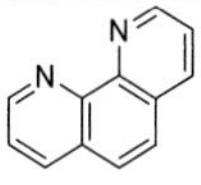	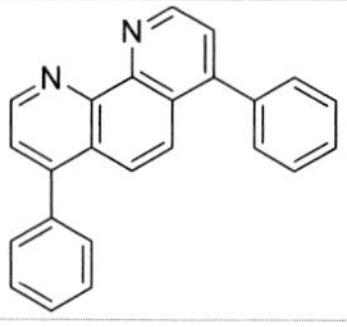
Cell permeability of the complex	+	–
Charge transfer properties of the complex	–	+
Application of the complex with this auxiliary ligand	Cellular imaging	OLED emitting layer

IV.1.3 Lanthanide 9-anthracenates

In this work NIR-emitting lanthanide(III) complexes with anthracene-9-carboxylate were discovered. The bulky ligand anthracene-9-carboxylate with high π-conjugation was used to sensitize NIR Ln-based luminescence (Figure IV.1-1a).

The structural study of these compounds based on the analysis of their powder patterns. Interestingly, the reflections of the *hk*0 zone of a trigonal or hexagonal space group were observed in PXRD patterns for the majority of the $Ln(ant)_3$. These complexes might be called "solid liquid crystals", since they did not crystallize in 3-dimensional crystal structures. Fortunately, europium 9-anthracenate $Eu(ant)_3$ made an exception: its powder pattern, corresponding to a true 3D structure, was indexable and its powder structure was determined (Figure IV.1-1b).

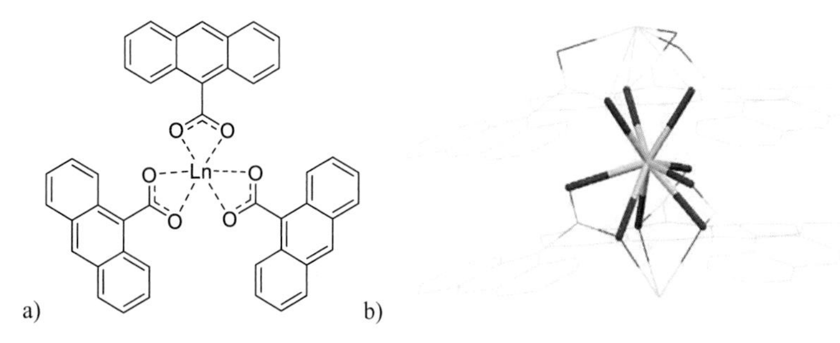

Figure IV.1-1 a) Structural formula of $Ln(ant)_3$; b) molecular structure of $Eu(ant)_3$, showing its polymeric chains; Eu(III) first coordination sphere is highlighted.

The Gd(ant)$_3$ complex was found to be promising for OLED applications due to its room temperature phosphorescence. This work demonstrates a manufacturing process of fully phosphorescent OLED, based on Gd(ant)$_3$.

Lanthanide 9-anthracenates with the NIR-emitting lanthanides Nd^{3+}, Yb^{3+} and Er^{3+} exhibited pure ionic Ln luminescence. The best results were obtained for the Yb^{3+} complex which exhibited the highest luminescence with the quantum yield of 1.5%. Interestingly, this value could be increased up to 2.5% by the introduction of an additional Lu ions to the structure and the resulting formation of the solid solution $Yb_{0.3}Lu_{0.7}$(ant)$_3$. This approach is assumed to dilute the number of emitting Yb(III) ions and hence, decrease the concentration quenching.

The relatively high efficiency of Ln anthracenates as NIR-emitting compounds may be explained by two factors:

i) minimization of the luminescence quenching due to the absence of coordinated solvent molecules or quenching functional groups (Figure IV.1-1b);

ii) efficient energy transfer through the ligand-to-metal charge transfer state (LMCT). This state serves as an additional step between the triplet level of the ligand and the emitting f-level of the Yb^{3+} ion and hence, simplifies the energy transfer process. The host-free OLED, fabricated, using this compound, possessed solely the Yb^{3+} f-f emission band at 1000 nm, with an electroluminescence quantum yield of 0.21%.

The structural features of Ln(ant)$_3$, revealed by powder X-ray data, result in their high electron conductivity through the long chains, caused by stacking interactions, as well as in their morphology, which makes them ready for the fabrication of nanodevices as synthesized. On top of that, Ln(ant)$_3$ possess high thermal stability (up to 350 °C with no phase transition), as well as high stability under electric current (up to 500 mA/cm^2).

The combination of these properties leads to a high electroluminescence efficiency, also in the NIR range, peaking at 0.21% at 14 V for Yb(ant)$_3$-based OLEDs.

IV.1.4 Lanthanide complexes for FRET-assay

The patented[214] synthesis of the commercially available FRET-donor Lumi4Tb® is a multistep process which contains various bottlenecks from a chemical complexity and cost points of view. This work optimized this process, proposing an alternative and innovative syntheses by introducing cheaper and easily available starting materials, as well as reducing reaction times and maintaining high yields. Several steps were optimized in the synthetic route of the cage-type, bicapped macrocyclic ligand, such as:

(i) tosylation of 2-chloroethylamine hydrochloride, where the amounts of the reactants were adjusted, so that the pure product was obtained and no additional purification was required (Figure IV.1-2);

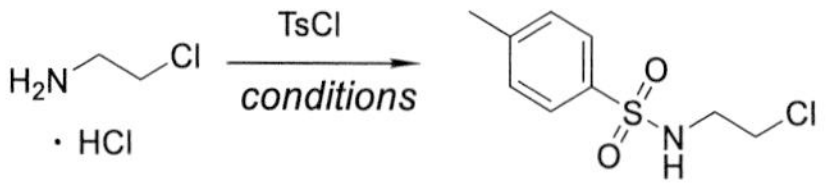

Figure IV.1-2 Optimized tosylation reaction; conditions: amine (1.4 equiv.), TsCl (1.0 equiv.), K_2CO_3 (2.5 equiv.), water, rt.

(ii) synthesis of the "cap" which involved tosylaziridine ring opening followed by the reaction with ethylenediamine, where lower temperatures and increased yields were achieved by using benzene as a solvent instead of the proposed toluene/acetonitrile mixture (Figure IV.1-3);

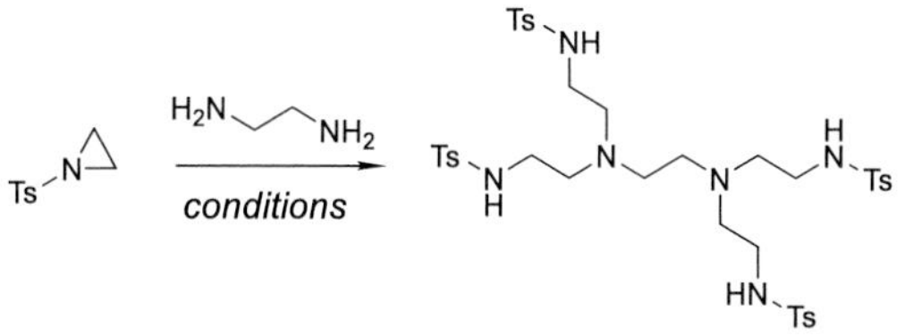

Figure IV.1-3 Optimized synthesis of the "cap"; conditions: benzene, 0 – rt, ovrn.

(iii) synthesis of the monocapped "cage", where more affordable materials have been used (Figure IV.1-4);

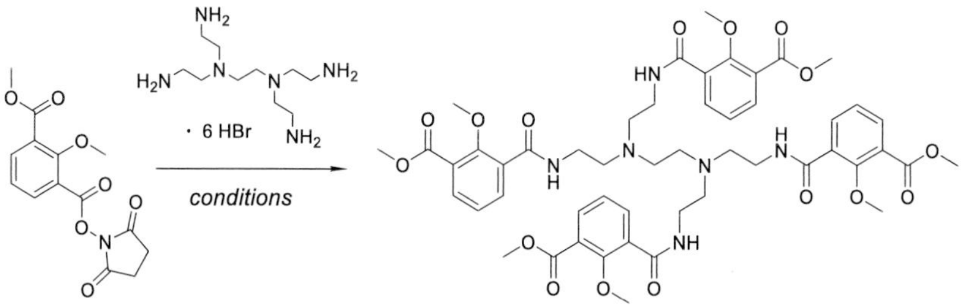

Figure IV.1-4 Synthesis of the monocapped ligand, using a preactivated NHS-ester; conditions: DIPEA, DCM.

(iv) synthesis of the bicapped "cage" which was performed in a shorter time (2 hours instead of 7 days), due to the introduction of a new technique with a parallel portionwise addition of equimolar amounts of reactants to a stirring solution of PyBOP (Figure IV.1-5).

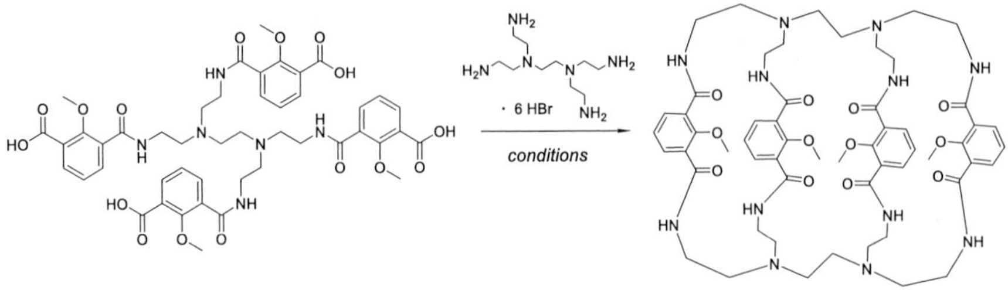

Figure IV.1-5 Coupling of the monocapped acid with PT, using PyBOP as a coupling agent; conditions: PT, PyBOP, DMF, portionwise, 2 h, rt.

The obtained probe possessed a very high luminescence intensity, demonstrating the typical Tb(III) emission profile, similar to those reported by RAYMOND *et al.*[106] On top of that, the study of the single-crystal structure of the bicapped ligand, reported previously, allowed to propose a new way of the linker incorporation, namely, amide alkylation, since four nitrogen atoms of the amides are positioned too far away from the "cage" center and are not able to form coordination bonds with Tb(III). The synthesis of the alkylated "cap" was performed, demonstrating the potential ability of the implementation of the proposed approach (Figure IV.1-6).

In the future, the design of the linkers, alterating their length and purporses, might be of interest, as well as discovering new ways for their introduction.

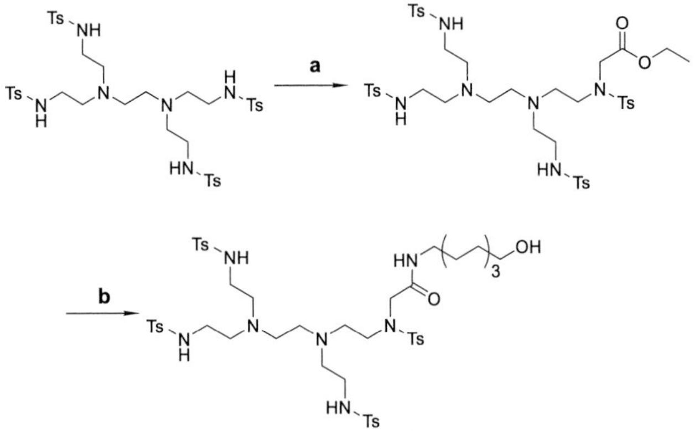

Figure IV.1-6 Incorporation of the linker, using a substitution reaction with subsequent coupling to increase the length of the linker; a) ethyl 2-bromoacetate, acetonitrile, Cs_2CO_3, 4 h, reflux; b) 1. LiOH, MeOH, 1 h, rt, 2. 8-aminooctanol, PyBOP, DCM.

IV.2 Lanthanide conjugates

IV.2.1 Functionalized ligands and lanthanide complexes with them

In this work *p*-substituted fluorobenzoic acid derivatives were synthesized as well as Ln(III) complexes with them. Five single-crystal structures were refined. The study of their geometry showed that Eu(III) 2-fluorobenzoates with $-NH_2$ and $-CH_2NH_2$ groups in *para*-position are isomorphous and reveal the same geometry as unmodified Eu(III) 2-fluorobenzoates, studied in Chapter III.1.1. Besides, I- and Br-substituted benzoates also possess a similar geometry. The investigation of the luminescent properties of these Eu(III) complexes demonstrated that the complexes within these structural groups indeed possess the same emission profile, because of the same coordination environment of the Eu(III) ion. Therefore, luminescence spectroscopy may serve as an indicator for geometry changes and may be used if an additional structural study is difficult.

Several functional groups are found to exhibit luminescence quenching properties. For instance, a *p*-nitro-substituted Eu(III) 2-fluorobenzoate did not exhibit any luminescence in MeOH solutions, the same as *o*-nitro-substituted perfluorinated Eu(III) complex. However, this compound was luminescent in the solid state, though a rather weak emission was detected. This observation may be explained by the large distance between the Eu(III) emitting center and the quenching nitro group in *para*-position of the ligand, so that its quenching influence is less relevant. The higher flexibility of the structures in MeOH solution may be the reason for the luminescence quenching in the Eu(III) complex with the 4-nitro-2-fluorobenzoate ligand.

Azido group was found to be another quencher, when attached directly to the ring. However, the luminescence properties of N_3-modified Eu(III) complexes may be recovered after the conjugation with the corresponding alkyne, using CuAAC cross-coupling, though this affect was observed only in the case, where aliphatic propargylamine was used (Figure IV.2-1). Unfortunately, Eu(III) emission was not recovered, when arylacetylene derivatives were used. Interestingly, if azido group was introduced through a CH_2-spacer, it did not quench the Eu(III)-based luminescence and compound **Eu-2.6** was found to be highly emissive.

These results indicate that *p*-modification of Eu(III) complexes is indeed the most reliable way to introduce a linker into the structure of the complex. The complexes with these linkers may be further used for a coupling with transporters or biologically relevant molecules.

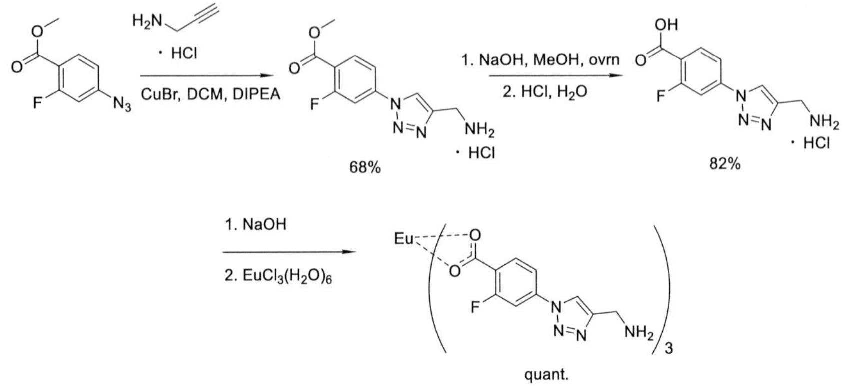

Figure IV.2-1 Synthesis of europium complex with the product of the CuAAC reaction.

IV.2.2 Synthesis and properties of europium conjugates with linear peptoids

This chapter showed that the coupling of the luminescent Eu(III)-based labels with peptoids, using solid-phase synthesis, was not successful: the final products did not survive the cleavage conditions, which is due to the relatively low stability of the Ln-O coordination bonds to the acidic medium. Though, the luminescent molecules were indeed obtained on solid support, the isolation of the product in this case has never been achieved. However, already cleaved and purified peptoids could be successfully conjugated with Eu(III), especially, when the peptoid oligomer contains 1,10-phenanthroline (Figure IV.2-2a) or terpyridine (Figure IV.2-2b) moieties.

a)

b)

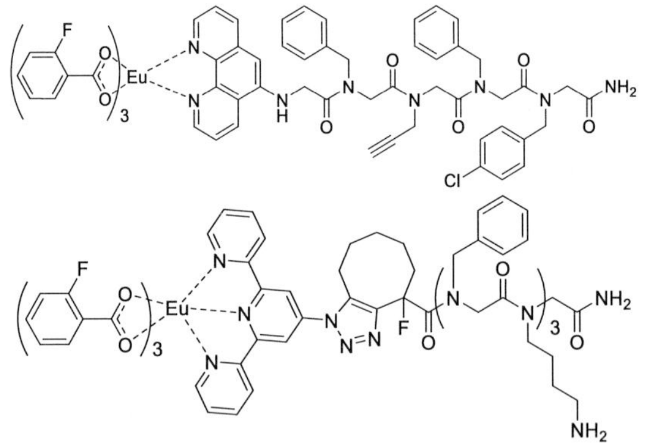

Figure IV.2-2 Successful europium-peptoid conjugates, exhibiting intensive Eu(III)-based luminescence: a) conjugated through an 1,10-phenanthroline scaffold; b) conjugated through a terpyridine scaffold.

In these cases, the peptoids served as auxiliary ligands, equally to the 1,10-phenanthroline ligand in Chapter III.1.2, so that ternary complexes with such peptoids exhibited Eu(III)-based luminescence. The 1,10-phenanthroline derivative was attached to the peptoid directly through amino group during the peptoid solid-phase synthesis. The terpyridine moiety was attached, using the SPAAC reaction. The use of peptoids, bearing carboxylic groups for Eu(III) complexation, was found to be possible, however, only a $EuCl_2$(peptoid) instead of Eu(peptoid)$_3$ complex was obtained, indicating poor coordination properties of COOH-bearing peptoids. Thus, the conjugation of Eu complexes with the peptoids might be done more efficient, when polydentate ligands are used, so that no dissociation would occur.

IV.2.3 Synthesis and properties of lanthanide conjugates with cyclic peptoids

Another way to increase the coordination ability of Eu(III) ions with peptoids is the use of chelating macrocycles, where Eu(III) is stabilized in the structure by coordination with several donor atoms. This may help to obtain very stable luminescent Eu(III) complexes which do not undergo any dissociation. This approach may be applied, using cyclic peptoids as chelating ligands.

In this chapter a novel synthetic route to afford heterocycle-containing cyclic peptoids was developed. This route is based on the solid-phase peptoid synthesis, where the heterocycle is introduced, using its bromomethyl derivative (Figure IV.2-3).

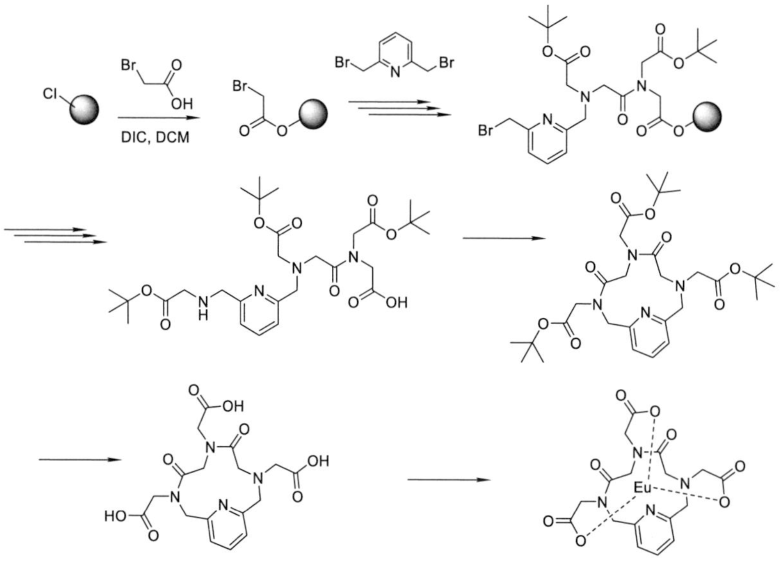

Figure IV.2-3 General scheme for the synthesis of Eu(III) conjugates with cyclic peptoids, bearing heterocycles in their backbones. The scheme includes the following steps: synthesis of the linear peptoid, cyclization, *t*Bu-groups deprotection, complexation.

Following this method, a library, containing nine novel cyclic peptoids, was synthesized. These peptoids were further conjugated with Eu(III), so that the library of Eu(III)-peptoid conjugates was generated (Figure IV.2-4). The geometry optimization study revealed which compounds were more likely to exhibit no Eu(III)-based luminescence due to their geometrical features. The study of photophysical data supported this assumption. Other Ln(III)-based cyclic peptoids exhibited moderate luminescence and Eu(III)-coordinated peptoids were successfully tested in HeLa cells, showing an intensive luminescence even at a low concentration of 5 μM and proved to be promising for bioimaging. In the future, this library might be expanded, particularly towards 6N-cyclic peptoids, alterating the side chains to achieve transporting properties.

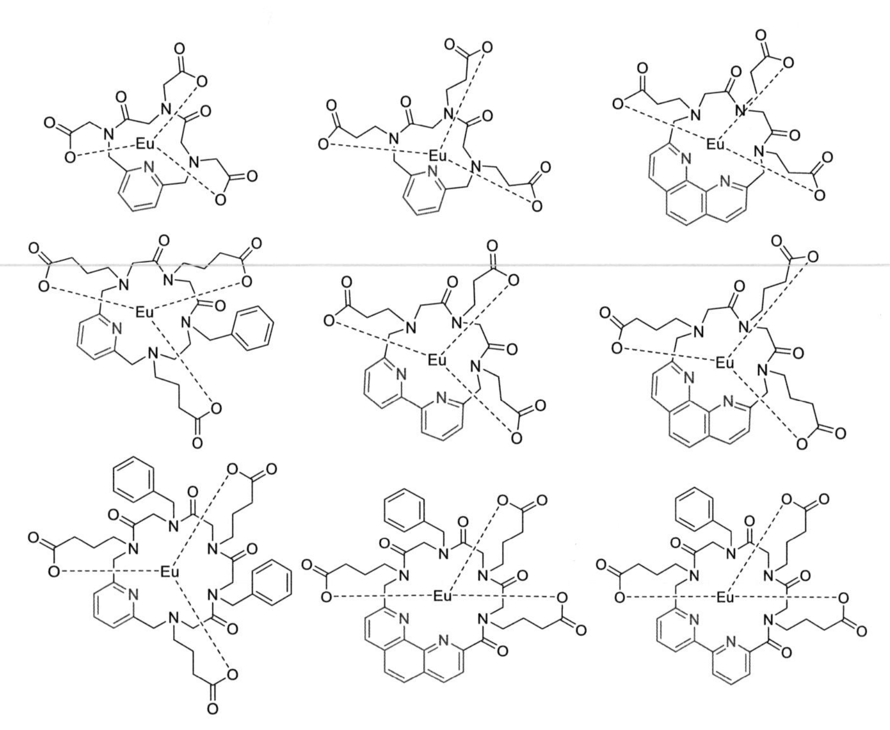

Figure IV.2-4 Library of Eu(III) complexes with heterocycle-containing peptoids, synthesized in this work. Heterocycles incorporated into the peptoid backbones are highlighted in blue.

V. Experimental part

V.1 Miscellaneous

V.1.1 Preparative work

All reactions involving moisture sensitive reactants were carried out under an argon atmosphere, using oven-dried glassware, following the common SCHLENK technique.[244] Reactions have been executed according to SCHLENK-techniques, using argon as an inert gas. Liquids were added *via* plastic syringes and V2-steel cannulas. Solids were added in pulverized form under Ar counterflow. Reactions at 0 °C were cooled with a mixture of ice/water.

All reactions were monitored by thin layer chromatography (TLC) or GCMS.

Solvents were removed at 40 °C with a rotary evaporator. Used solvent mixtures were measured volumetrically (v/v). An UV-lamp as well as Ninhydrin staining was used for detection.

Unless otherwise specified, solutions of inorganic salts are saturated aqueous solutions.

Unless otherwise specified, the crude products, were purified by flash column chromatography following the concepts of STILL *et al.*,[245] using silica gel (SIGMA ALDRICH, pore size 60 Å, particle size 40-63 µm) and sand (calcined and purified with hydrochloric acid) as stationary phase. Solvents were distilled prior to use. *Per analysis* grade solvents (p.a.) were used without further purification. Solvent mixtures were prepared individually in volume ratios and are given as volumetric amounts. The use of a gradient is indicated in the experimental procedures.

Celite® for filtrations was purchased from ALFA AESAR (Celite® 545, treated with Na_2CO_3).

V.1.2 Solid-phase reactions

For the reactions on solid phase, disposable plastic syringes from MULTISYNTHECH of sizes 2 mL, 5 mL or 10 mL were used. The syringes were filled via disposable cannulas from BRAUN.

To shake the resins a rotary shaker IKA-LABORATORY, model KS 501 was used.

V.1.3 Solvents and reagents

Solvents of technical quality have been purified by distillation prior to use. Solvents of the grade p.a. have been purchased (ACROS, FISHER SCIENTIFIC, SIGMA ALDRICH, ROTH, RIEDEL-DE HAËN) and were used without further purification. Absolute solvents have been dried, using the methods listed in Table V.1-1 and were stored under argon afterwards or have been purchased from a commercial supplier (abs. acetonitrile (ACROS, <0.005% water), abs. chloroform (FISCHER, over molecular sieves), abs methanol (FISCHER, <0.005% water), abs. ethanol (ACROS, <0.005% water).

Table V.1-1. Methods for the absolutizing of solvents. All distillations were carried out under argon atmosphere.

Solvent	Method
Dichloromethane	heating to reflux over calcium hydride, distilled over a packed column
Tetrahydrofuran	heating to reflux over sodium metal (benzophenone as an indicator), distilled over a packed column
Diethyl ether	heating to reflux over sodium metal (benzophenone as an indicator) distilled over a packed column
Toluene	heating to reflux over sodium metal (benzophenone as an indicator) distilled over a packed column

Reagents have been purchased from commercial suppliers (Companies: ABCR, ACROS, ALFA AESAR, CARBOLUTION, CHEMPUR, FLUKA, IRIS, MERCK, RIEDEL-DE HAËN, TCI, THERMO FISHER SCIENTIFIC, SIGMA ALDRICH). They have been used without further purification unless stated otherwise.

V.1.4 Analytics and equipment

Nuclear Magnetic Resonance (NMR)

NMR spectra have been recorded, using the following machines:

^{1}H NMR: BRUKER *Avance 300* (300 MHz), BRUKER *Avance 400* (400 MHz), BRUKER *Avance DRX 500* (500 MHz). The chemical shift δ is expressed in parts per million (ppm) where the residual signal of the solvent has been used as reference: chloroform-d_1 (δ = 7.26 ppm), dimethyl sulfoxide-d_6 (δ = 2.50 ppm), deuterium oxide-d_2 (δ = 4.79 ppm), methanol-d_4 (δ = 4.87 ppm), acetonitrile-d_3 (δ = 1.94 ppm).[246] The spectra were analyzed according to first order.

^{13}C NMR: BRUKER *Avance 300* (75 MHz), BRUKER *Avance 400* (101 MHz), BRUKER *Avance DRX 500* (126 MHz). The chemical shift δ is expressed in parts per million (ppm) where the residual signal of the solvent has been used as reference: chloroform-d_1 (δ = 77.0 ppm), dimethyl sulfoxide-d_6 (δ = 39.4 ppm) and dichloromethane-d_2 (δ = 53.8 ppm) methanol-d_4 (δ = 49.0 ppm), acetonitrile-d_3 (δ = 118.3 ppm).[246] The spectra were ^{1}H-decoupled and characterization of the ^{13}C NMR-spectra ensured through the DEPT-technique (DEPT = Distortionless Enhancement by Polarization Transfer) and are stated as follows: DEPT: "+" = primary or secondary carbon atoms (positive DEPT-signal), "–" = secondary carbon atoms (negative DEPT-signal), $C_{quart.}$ = quaternary carbon atoms (no DEPT-signal).

^{19}F NMR: BRUKER *Avance 400* (377 MHz). Chemical shifts in ^{19}F NMR spectra were calculated without reference by the instrument.

All spectra were obtained at room temperature. NMR-solvents were obtained from EURISOTOP and SIGMA ALDRICH: chloroform-d_1, dimethylsulfoxide-d_6, deuterium oxide-d_2, methanol-d_4, acetonitrile-d_3, dichloromethane-d_2. For central symmetrical signals the midpoint and for multiplets the range of the signal region are given. The multiplicities of the signals are abbreviated as follows: s = singlet, d = doublet, t = triplet, q = quartet, hept = heptet, bs = broad singlet, m = multiplet, b = broad (unresolved) and combinations thereof. All coupling constants '*J*' are stated as modulus in Hertz [Hz].

Elemental analysis (EA)

Elemental analysis measurements were performed on a ELEMENTAR *Vario Micro*. As analytical scale the model SARTORIUS *M2P* was used. Notation of Carbon (C), Hydrogen (H) and Nitrogen (N) is given in mass percent. Following abbreviations are used: calc. = expected value (calculated); found = value found in analysis.

Infrared spectroscopy (IR)

IR-spectra were recorded on a BRUKER *Alpha P* and a BRUKER *IFS 88*. Measurements of the samples were conducted via attenuated total reflection (ATR). The intensity of bands (strength of absorption) was described as follows: vs = very strong (0-10% transmission, T), s = strong (10-40% T); m = middle (40-70% T), w = weak (70-90% T), vw = very weak (90-100% T). Position of the absorption bands is given as wavenumber $\tilde{\nu}$ with the unit [cm^{-1}].

Time-resolved Fourier-Transform Infrared spectroscopy (trFTIR)

Time-resolved FTIR (trFTIR) experiments were performed with a BRUKER *VERTEX 80v* spectrometer that was operated in step-scan mode. Signal recording and processing was done by a liquid-nitrogen cooled mercury-cadmium-telluride (MCT) detector (KOLMAR TECH., model KV100-1-B-7/190) with a risetime of 25 ns, connected to a fast preamplifier and a 14-bit transient recorder board (Spectrum Germany, M3I4142, 400 MS/s). For the excitation of the sample a Q-switched Nd:YAG laser (LUMONICS *HY750*, frequency tripled to 355 nm) that generates excitation pulses with a half-width of about 10 ns and a repetition rate of 10 Hz was used. The timing of the laser pulse and the step-scan triggering was performed by a STANFORD RESEARCH SYSTEMS DG535 delay generator. The UV pump beam was adjusted to have a maximum overlap with the spectrometer's IR probe beam. To avoid backscattering of the UV laser radiation into the detector or interferometer compartment, anti-reflection-coated germanium filters were placed inside the sample compartment. The 355 nm excitation laser beam was attenuated to ca. 5 mJ/shot at a diameter of about 9 mm. The temporal resolution of the digitization at the 14-bit transient recorder board was set to 100 ns. The step-scan experiment was started 50 μs before the laser (355 nm) reached the sample. Hence, this time was set as zero point in all spectra. A total number of 1248 co-additions at each interferogram point were recorded. The spectral region was limited by undersampling to 0-1975 cm^{-1} with a spectral resolution of 4 cm^{-1} (resulting in 1110 interferogram points). For the prevention of problems by performing a Fourier transformation, an IR longpass filter (cutoff at 1900 cm^{-1}) was used (i.e. no IR intensity outside the measured region should be observed). After every step-scan measurement, the stability of the sample was checked by a (ground state) FTIR spectrum. *Sample Preparation.* Complex **Eu-1.9-Phen** (approx. 2 mg) was mixed with dry KBr (approx. 200 mg) (stored in a compartment dryer at 70 °C) and grinded to a homogenous mixture. This mixture was filled in an evacuable pellet with a diameter of 13 mm and sintered at a pressure of 0.75 GPa. The concentration of the sample in the pellet has been chosen in a way that the most intense peak at 1610 cm^{-1} had an absorption of ca. 0.65 OD. The sample compartment was evacuated after the sample was fixed in the sample position.

Time-dependent density functional theory (TD-DFT) calculations

The crystal structure was used as initial structure for the **quantum-chemical calculations**. For the optimization, GAUSSIAN 09 or TMOLEX program packages were used. The B3LYP functional of was used with the cc-pVDZ basis set for F, C, H, N and O and the MWB52 ECP for Eu. Frequency calculation was used to demonstrate the existence of a minimum structure where no imaginary frequency is found. Excitations were calculated, using TD-DFT with the same functional and basis set. Since the MWB52 ECP was used for Eu, no f-f excitation could be simulated.

Mass Spectrometry (LC-ELSD-MS, GC-MS, EI-MS, FAB-MS, MALDI-TOF, HRMS)

LC-ELSD-MS was performed on a WATERS 2545 HPLC equipped with a 2998 diode array detector, a WATERS 3100 ESI-MS module, using a XTerraMS C18 5 μm, 4.6 x 50 mm column at a flow rate of 5 mL/min with a linear gradient (95% A: 5% B to 100% B with 90 sec and 30 sec hold at 100% B, solvent A = water + 0.1% formic acid, solvent B = acetonitrile + 0.1% formic acid).

GC-MS (Gas chromatography–mass spectrometry). The measurements have been recorded with an AGILENT TECHNOLOGIES model 6890N (electron impact ionization), equipped with a AGILENT 19091S-433 column (5% phenyl methyl siloxane, 30 m, 0.25 μm) and a 5975B VL MSD detector with turbo pump. As a carrier gas helium was used.

EI-MS and **FAB-MS**. Mass spectra were recorded on a FINNIGAN *MAT 95*. Ionization was achieved through either EI (electron ionization), FAB (fast atom bombardment).

MALDI-TOF (matrix-assisted laser desorption/ionisation - time of flight). Mass spectrometry was performed on the following two devices:

1) BRUKER *Biflex IV* spectrometer, equipped with a nitrogen laser (λ = 337 nm), the software FlexControl version 1.1 and XMASS-XTOF version 5.1.1. The spectrum was shot about 100 to 300 times with a repetition rate of 1-3 Hz and the results were averaged. The BRUKER target plate "MTP 384 massive target T" was used.

2) AXIMA *Confidence* spectrometer from SHIMADZU BIOTECH, equipped with a nitrogen laser (λ = 337 nm) and the associated software from SHIMADZU BIOTECH. BRUKER standard stainless steel sample plate with 386 "spots" was used as a target.

The matrices were purchased from SIGMA ALDRICH and used as acetonitrile/water (1/1) saturated solutions with 0.1% TFA:

1) 1/1 mixture of 2,5-dihydroxybenzoic acid (DHB) and α-cyano-4-hydroxycinnamic acid (CHCA)
2) pure CHCA
3) 2,4,6-trihydroxyacetophenone (THAP) which does not allow any leakage of metal ions[115,247].

The crystallization was carried out in air at room temperature. The protonated molecular ion peak $[M+H]^+$ or pseudomolecular ion peaks with sodium $[M+Na]^+$ or potassium $[M+K]^+$ were reported.

HR-MS (high resolution-mass spectra). The spectra were recorded with an LTQ Orbitrap XL mass spectrometer. The sample (3 mM solution in methanol was ionized by electrospray ionization, both in positive and negative mode. Spray voltages were 4 kV and 2 kV respectively. The following abbreviations were used: calc. = expected value (calculated); found = value found in analysis.

Notation of molecular fragments is given as mass to charge ratio (*m/z*); the intensities of the signals are noted in percent relative to the base signal (100%). As abbreviation for the ionized molecule $[M]^+$ is used. Characteristic fragmentation peaks are given as $[M\text{-fragment}]^+$ and $[\text{fragment}]^+$.

Single-crystal X-ray diffraction (XRD)/ Powder X-ray diffraction (PXRD)

Single-crystal X-ray diffraction studies were carried out, using a BRUKER *APEX Duo CCD* or BRUKER *Nonius Kappa CCD* diffractometer (ω-scans) at 120 K, using Cu-Kα- or Mo-Kα-radiation. All structures were solved by direct methods and then refined least-squarely, using isotropic-anisotropic full-matrix approximation against $|F_{hkl}|^2$. For organic compounds as well as for several lanthanide complexes hydrogen atoms were found from difference Fourier synthesis of electron density, while for other structures (mostly the complexes with earlier Ln, such as La, Sm, Eu, Gd, Tb) H-atoms were calculated, using H-bonding network analysis both, in monomers and in dimers. All hydrogen atoms were refined isotropically, using the riding model with $U_{iso}(H)$ parameters equal to 1.2 $U_{eq}(Ci)$, where $U_{eq}(Ci)$ are respectively the equivalent thermal parameters of the atoms to which corresponding H-atoms are bonded. For heterometallic complexes coordinates and anisotropic displacement parameters of metal atoms were constrained on each other while their populations were fixed and equal to 0.5. For the crystals determined as twins, the indexing was performed, using CELL_NOW routine, structures were solved and initially refined, using only one domain. The calculations performed, using OLEX or SHELX ver. 2014/6.

The powder patterns of all studied compounds were measured on a BRUKER *D8 Advance Vario* diffractometer with Ge(III) monochromator in transmission mode between MYLAR films or on a in reflection mode on Si zero-background holder (BRAGG-BRENTANO geometry, variable slits). Both diffractometers were fitted with LynxEye 1-D positional-sensitive detector. The indexing performed, using the SVD[248] algorithm as implemented in TOPAS 4.2[249] and solved, using the Parallel Tempering algorithm as implemented in FOX[250]. The resulting structures were refined and verified, using a symmetrized modification of MORSE restraints and statistical analysis of bond length distributions at different values of penalty weighting[251]. Still, given the relative complexity of the studied structures, we used periodic DFT calculations to check the refinement results and obtain more reliable complex geometry[51]. The calculation was performed, using PW-PBE approach with GRIMME correction[252] as implemented in VASP[253–255], with a fixed unit cell and 680 eV energy cutoff. The r.m.s deviation of the optimized structures from reasonably restrained geometry as obtained from powder data was 0.16 Å, lower than the cutoff value proposed by VAN DE STREEK[255] for the verification of single-crystal structures. The calculation also unambiguously determined the position of hydrogen atoms of the water molecule. The restraints generated from the calculated structure were used to re-refine the powder data, resulting in slight drop in the R_{wp}. The final structure reported, with r.m.s. bond Δd from restraints of 0.01 Å, demonstrated $R_{wp}/R_{wp}'/R_p/R_p'/R_{Bragg}$ of 1.70/10.14/1.16/11.25/0.5% and r.m.s deviation from the optimized structure of 0.10 Å.

X-ray absorption spectroscopy (XAS)

The local structure of lanthanide complexes has been elucidated, using Ln L_{III}-edge **X-ray absorption spectroscopy (XAS)**. The X-ray absorption spectra were measured at the Structural Materials Science beamline of the Kurchatov Centre for Synchrotron Radiation (National Research Centre “Kurchatov Institute”, Moscow, Russia) in transmission mode, using two ionization chambers filled with appropriate N_2/Ar mixtures. Experimental data reduction and analysis performed, using the IFEFFIT software suite[256] with FEFF[257] *ab initio* photoelectron phase and amplitude functions.

Ultra violet/visible light absorption spectra (UV/Vis)

UV/Vis spectra were recorded on a ANALYTIK JENA *Specord 50/plus* or on a PERKIN-ELMER Lambda 650 spectrometer.

Excitation/Emission Spectra

Excitation and Emission spectra have been recorded on A HORIBA SCIENTIFIC *fluoromax-4* spectrofluorometer or AGILENT *Cary Eclipse* fluorescent spectrometer both equipped with Czerny-Turner-type monochromator and a R928P PMT detector.

Luminescence lifetime measurements

Luminescence lifetime measurements were performed and detected on a HORIBA SCIENTIFIC FLUOROMAX-4 spectrofluorometer, using the time-correlated single photon counting (TCSPC) method with the FM-2013 accessory and a TCSPC hub from HORIBA JOBIN YVON. For this, a NanoLED 370 was used as excitation source ($\lambda = 370$ nm, 1.5 ns pulse). Decay curves were analyzed with the software DAS-6 and DataStation provided by HORIBA JOBIN YVON. The quality of the fit was determined by the χ^2 method of PEARSON.

Photoluminescence quantum yield (PLQY)

For the determination of the **photoluminescence quantum yield** Q_{Ln}^{L}, an absolute PL quantum yield measurement system from HAMAMATSU PHOTONICS was used. The system consisted of a photonic multichannel analyzer PMA-12, a model C99200-02G calibrated integrating sphere and a monochromatic light source L9799-02 (150 W Xe and Hg-Xe lamps). Data analysis was performed with the PLQY measurement software U6039-05, provided by HAMAMATSU PHOTONICS.

Preparative high-performance liquid chromatography (HPLC)

For purification of synthesized peptoids, two **preparative reversed phase chromatography** instruments were used. The first device is a JASCO LC-NetII / ADC HPLC system equipped with two PU-2087 Plus pumps, a CO-2060 Plus thermostat, an MD-2010 Plus diode array detector and a CHF-122SC fraction collector from ADVANTEC. The stationary phase was a preparative C18 column (VDSpher C18-M-SE, C18, 10 µm, 250 × 20 mm from VDS OPTILAB) at a flow rate of 15 mL/min. Purification was carried out at 25 °C with a linear gradient of A: 5% acetonitrile, 0.1% TFA in bidistilled water to B: 95% acetonitrile, 0.1% TFA in bidistilled water in 40 min. The separation of the peptoids was detected at $\lambda = 218$ nm, 254 nm and 560 nm and *via* MALDI-TOF mass spectrometry.

The second HPLC system is a PURIFLASH® 4125 from INTERCHIM, equipped with InterSoft V5.1.08 software and a UV diode array detector (200-600 nm). The stationary phase was a preparative C18 column (VDSpher C18-M-SE, C18, 10 µm, 250 × 20 mm from VDS OPTILAB) at a flow rate of 15 mL / min. The precolumn was a small C18 column (VDSpher, C18, 10 µm, 16 × 40 mm). The purification was carried out at 25 °C with a linear gradient of A: 5% acetonitrile, 0.1% TFA in bidistilled water to B: 100% acetonitrile, 0.1% TFA in 45 min. The separation of the peptoids was detected at $\lambda = 218$ nm, 254 nm and 560 nm and *via* MALDI-TOF mass spectrometry.

Analytical HPLC

The determination of the purity of the compounds was carried out on an AGILENT 1100 series **HPLC system** with a G1322A degasser, a G1311A pump, G1313A autosampler, a G1316A column oven and a G1315B diode array detector. The flow rate was 1 mL / min. The stationary phase used was a VDSpher C18-M-SE (VDS OPTILAB) C18 column (5 µm, 4.0 mm × 250 mm). The runs were carried out with a linear gradient of A: 5% acetonitrile, 0.1% TFA in water to B: 95% acetonitrile, 0.1% TFA in water within 30 min. The purity was determined by integration of the signals at 218 or 256 nm.

MTT assay

The cytotoxicity of the samples was assessed, using an MTT assay. To determine the toxic effect of the probes towards HeLa cells, the CellTiter 96® Non-Radioactive Cell Proliferation Assay (PROMEGA) was used. This assay is based on the intracellular reduction of a tetrazolium salt (yellow) into a formazan product (blue), which only takes place in metabolic active cells. The generated formazan is detectable at wavelengths between 630-750 nm and is a direct measure for the viability of the cells. For this assay, each well of a 96-well plate (CSTAR 3596, 96 Well Cell Culture Cluster, sterile) was seeded with 1×10^4 HeLa cells in 100 µl Dulbecco's modified Eagle's medium (DMEM, high glucose, GIBCO) supplemented with 10% fetal calf serum (FCS, PAA) and 1 U/mL Penicillin/Streptomycin at 37 °C, 5% CO_2 and 95% humidity. After 24 h cells were incubated with the samples at different concentrations. For each concentration 6 wells were prepared and incubated for 72 h. A set of positive (cells treated with 5 µl of 20% triton) and negative (untreated cells) control wells, as well as the test samples, were treated with 15 µl of the dye solution and incubated for 4 h. 100 µl Solubilization solution/stop mix was then added to each well to solubilize the formazan product, according to the manufacturer's manual. After 24 h incubation the absorbance at 595 nm, using a 96-well plate reader (ULTRA MICROPLATE READER ELx808, BIOTEK INSTRUMENTS, INC) was measured. Data were averaged and the multiple determination of each substance and concentration made it possible to calculate the standard deviation.

Confocal microscopy

Two hours after seeding 1×10^4 HeLa cells per well plate were transferred into 8-well ibiTreat chamber slides (IBIDI, Martinsried, Germany) in 0.2 mL the medium. After 24 h, cells were treated with the tested compounds at the desired concentrations. After additional 24 h the cells were washed and then investigated, using confocal microscope LEICA TCS-SPE (DM2500), equipped with ACS APO 63x/1.30 OIL object-glass. Fluorescence excitation was performed by the 488 nm line or 405 nm line of an Ar-ion laser 15%, resolution 8 bit, line average 16, format 1024×1024 pixels, 200 Hz. Fluorescence detection took place either in green channel (450-580 nm) or in the red channel (600-750 nm). Additionally, brightfield images were recorded in a third independent channel. The images were recorded and then analyzed, using LAS-AF 2.0.2.4647 software.

Thin films deposition and characterisation

Thin films tested compounds were deposited, using spin-coating technique on a glass/ITO substrate. In order to obtain high quality thin films, the deposition parameters were optimized, by varying the solvents (ethanol/toluene/ethanol:benzene(1/1)/chloroform), the mode of deposition (solution dropping to the rotating substrate; solution fast pouring on the rotating substrate; solution pouring on the resting substrate, followed by rotation), rotation time (20-60 sec), rotation rate (2500-5000 rpm), the amount (0.2-0.3 mL) and concentration (10-100 g/L) of compound solution. **The thin film thickness** was measured, using a profilometer Talystep (TAYLOR-HOBSON). **The surface morphology and roughness** were characterized with a NT-MDT-INTEGRA AURA AFM in semi-contact mode at ambient atmosphere (25 °C), using MICROMASH-NSC15/AlBS cantilevers. The observed surface structures were analyzed by NOVA-1.0.26 software. The morphology of the films was also monitored by a scanning electron microscope (SEM) SUPRA-50 VP (LEO, Germany). **Transmission spectra** were recorded on a double-beam spectrophotometer with a concave holographic grating Lambda 35 (PERKIN-ELMER).

OLED manufacturing

OLED manufacturing took place in a clean room class 10000 (P.N. Lebedev Physical Institute, Moscow, Russia) in a glovebox with argon atmosphere. The substrates were cleaned by ultrasonication in the following media: NaOH aq. solution, distilled water, acetone and 2-propanol for 16 min each. A 40 nm thin film of a hole-injective layer PEDOT:PSS (poly(3,4- ethylenedioxythiophene):poly(styrenesulfonate)) was first deposited. An aq. solution of PEDOT:PSS (5 mL) was poured onto the preheated (70 °C) patterned ITO glass substrate, after which the substrate was rotated for 60 s at 2000 rpm. Finally, the deposited film was annealed in air at 80 °C for 60 min. As hole-transport layer PVK solution was spin-coated from toluene (c = 5 g/L). The emission layer was spin-coated from toluene solution (c = 5 mM) on the heterostructure. The electron-transport/hole-blocking layer (TPBi) was thermally evaporated (Univex-300, LEYBOLD HERAEUS) under a pressure below 106 mm Hg. The thickness (~ 30 nm) was controlled by a quartz indicator. The contacts were attached to the electrodes and the device was sealed with epoxy resin (NORLAND OPTICAL ADHESIVE). Electroluminescence spectra were measured on a PICOQUANT time-correlated single photon counting system used as a conventional spectrofluorimeter. Spectral resolution was 4 nm.

Solubility measurements

The solubility of the synthesized compounds was measured as follows. A suspension of each compound in a desired solvent was refluxed during several hours. After cooling down and complete precipitation 5 mL clear solution was placed in a vessel with a known mass and the solution was evaporated to dryness. The mass change of the vessel corresponded to the quantity of the dissolved product in 5 mL solvent.

Thin Layer Chromatography (TLC):

All reactions were monitored by thin layer chromatography (TLC). The TLC plates were purchased from MERCK (silica gel 60 on aluminum plate, fluorescence indicator F254, 0.25 mm layer thickness). Detection was carried out under UV-light at $\lambda = 254$ nm and $\lambda = 366$ nm.

Lyophilisator

The Christ ALPHA 1-2 LD plus model was used as a freeze-drying system.

Analytical scales

Used machine: SARTORIUS Basic.

V.2 Synthesis and Characterization

V.2.1 General procedures

General Procedure for the synthesis of non-soluble lanthanide carboxylates (**GP1a**).[33,258]

A suspension of HL (3.00 equiv.) in water is treated with 1 M aq. solution of NaOH at 60 °C until the pH is slightly acidic (pH = 5 – 6) and the solution becomes clear. Then an aq. solution of

$LnCl_3 \cdot 6\ H_2O$ (1.00 equiv.) is added following by an immediate precipitation of a product. The reaction mixture is stirred for additional 3 h at room temperature. The crude product is filtered, washed with ethanol and dried in vacuo. The product is then purified (if necessary) by means of recrystallization.

General Procedure for the synthesis of water-soluble lanthanide carboxylates (**GP1b**).[139]

An excess of concentrated aq. solution of NH_3 (10.0 equiv.) is added to an aq. solution of $LnCl_3 \cdot 6\ H_2O$ (1.00 equiv.) in water (approx. 50.0 mL per 2.00 mmol of $LnCl_3 \cdot 6\ H_2O$). The mixture is stirred for 30 min, the precipitate of $Ln(OH)_3$ is centrifuged and washed with water until the pH of the washing solution becomes neutral. A small excess of freshly prepared $Ln(OH)_3$ is placed into a beaker and the solution of HL (2.70 equiv.) in an acetone-methanol mixture (3/1) is added. The reaction mixture is stirred for 1 h at 60 °C and the unreacted components are separated by filtration followed by the evaporation of the clear solution to dryness. The obtained solid powder of the lanthanide complexes is then recrystallized from water. Subsequent evaporation procedure at 80 °C during 1 h leads to the precipitation of microcrystalline powder of the general formula $Ln(L)_3(H_2O)_x$, where $x = 0 - 2$. The slow evaporation of the aq. solution of the corresponding powder at room temperature gives prismatic and needle-shaped single crystals with significantly higher water content ($x = 5 - 6$).

General Procedure for the synthesis of lanthanide ternary complexes (**GP2**).[184,259]

The solution of an auxillary ligand in ethanol (1.00 equiv.) is added to a solution of lanthanide carboxylate (1.00 eqiuv.) in ethanol or ethanol/water (1/1) and the reaction mixture is refluxed for 2 h. The reaction mixture is then allowed to cool to room temperature. The precipitate is recovered by filtration, washed with cold ethanol and dried in vacuo. When no precipitate is formed, ethanol is removed under reduced pressure and the rest is lyophilized.

General Procedure for the protection of carboxylic acids with a methyl ester group (**GP3a**)[260].

To a stirred solution of a benzoic acid derivative (1.00 equiv.) in methanol (approx. 50 mL per 1 g of starting material), concentrated H_2SO_4 (3.00 equiv.) is added and the mixture is refluxed for 48 h. After concentration in vacuo, the residue is triturated with water. The resulting precipitate is collected and dried in vacuo.

General Procedure for the protection of carboxylic acids with *tert*-butyl ester group (**GP3b**).[261]

The carboxylic acid (1.00 equiv.) is solubilized in dry DCM under Ar. Then dry *t*BuOH (3.00 equiv.) with DMAP (0.20 equiv.) and DIC (1.15 equiv.) are added. The reaction mixture is stirred for 3 days under Ar. The precipitate is then filtered off and washed with DCM. Afterwards the filtrate is washed with water (3 × 300 mL), 5% $NaHCO_3$ solution (3 × 300 mL) and water again. The organic phase is dried over Na_2SO_4, the volatiles are removed under reduced pressure and the residue is then purified *via* flash column chromatography.

General Procedure for the deprotection of methyl/ethyl esters (**GP4a**).[262]

Benzoic acid derivative methyl ester (1.00 equiv.) is dissolved in methanol (approx. 100 mL per 1.00 g) and treated with a solution of NaOH (2.20 equiv.), then stirred for 2.5 h. A solution of 1 M HCl is added to acidify the reaction mixture. The carboxylic acid is collected by filtration, then dried in vacuo.

General Procedure for the deprotection of a tert-butyl esters of peptoids (**GP4b**).[263,264]

The peptoid protected with *t*Bu is treated with a deprotection cocktail of TFA, triisopropylsilane (TIPS) and water (95/2.5/2.5) for 4 h. The volatiles are then removed under airflow and reduced pressure, the crude product is solubilized in an CH_3CN/H_2O mixture and purified by means of preparative HPLC.

General Procedure for the protection of amines with a carboxybenzyl protecting group (**GP5**).[265]

To a solution of a primary amine (1.00 equiv.) in 2 M NaOH (1.15 equiv.) a suspension of benzyl chloroformate (1.08 equiv.) in 2 M NaOH (1.15 equiv.) is added slowly at 0 °C. The reaction is stirred for 40 min at 0 °C and then overnight at room temperature. The crude product is washed with diethyl ether (3 × 50 mL), acidified to pH = 3 with concentrated HCl and cooled in an ice bath. The white precipitate is filtered, washed with cold 0.1 M HCl and dried in vacuo.

General Procedure for the deprotection of carboxybenzyl protecting group (**GP6**).[266]

The solution of a benzylated secondary amine (1.00 equiv.) in methanol (approx. 100 mL per 10.00 g) is treated with Pd/C (0.30 equiv.). The reaction flask is flushed with H_2 and then fitted with a balloon filled with H_2. The mixture is stirred under balloon pressure of H_2 until TLC showed the reaction to be complete (usually 3 – 4 h). After that the catalyst is filtered through a Celite® pad and washed with methanol. The solvent is then evaporated under reduced pressure.

General Procedure for the synthesis of methyl azidofluorobenzoates (**GP7**).[267]

An aminofluorobenzoate (1.00 equiv.) is stirred in 50% HCl (approx. 20.0 mL per 10 mmol). The solution/suspension is cooled to 0 °C and a solution of sodium nitrite (1.30 equiv.) in water (approx. 5.0 mL per 7.50 mmol) is added dropwise over a period of 15 min while stirring. After

15 min of further stirring at 0 – 5 °C, a solution of sodium azide (1.50 equiv.) in water (approx. 5.0 mL per 9.0 mmol) is added at such a rate that the temperature of the reaction mixture does not exceed 15 °C till the end of addition. The solid product, that precipitates after 10 – 15 min is filtered, washed with ice water and dried in vacuo. In case of azides that are oils, the reaction mixture was extracted with DCM three times. The combined organic layers were dried over Na_2SO_4, filtered and evaporated under reduced pressure.

General Procedure for the Sonogashira coupling of terminal alkynes with aryl halides (**GP8**).

To a solution of a halide (8.58 mmol, 1.00 equiv.) in 15 mL 1,4-dioxane or dry THF (~ 15 mL per 8 mmol of the starting material) and triethylamine (~ 15 mL per 8 mmol of the starting material), trimethylsilylacetylene (1.30 equiv.) is added. As catalysts dichlorobis(triphenylphosphine)-palladium(II) or tetrakis(triphenylphosphine)palladium(0) (0.05 equiv.) and copper(I) iodide (0.03 equiv.) are added to the reaction mixture which is then stirred at 60 °C or refluxed for 24 h. After allowing to cool down to room temperature, the reaction mixture is filtered through Celite® and extracted with DCM or EtOAc. The organic layer is washed with water and saturated brine and dried over Na_2SO_4. After filtration, the filtrate was concentrated under reduced pressure and the obtained crude product is purified by silica gel chromatography or distillation.

General Procedure for the deprotection of trimethylsilyl group (**GP9**).

To a solution of 1.00 g of a trimethylsilyl protected alkyl (1.00 equiv.) in methanol (~ 10 mL per 1.0 mmol of the starting material) NaOH (2.00 equiv.) is added and allowed to reflux for 4 h. The reaction mixture is concentrated and the residual liquid is treated with 5% aq. $NaHCO_3$ and DCM. The organic layer is separated and the aqueous layer is extracted twice with DCM. Combined organic layers are dried over Na_2SO_4 and evaporated.

General Procedure for the CuAAC reaction (**GP10a**).[268]

To a solution of an azide (1.00 equiv.) in dry DCM (approx. 50.0 mL per 5.00 mmol) triethylamine (3.60 equiv.), copper (I) bromide (5.00 mol%) and the corresponding acetylene (1.15 equiv.) are added and stirred for 4 h. The reaction mixture is then filtered, the solvent is evaporated and the product is recrystallized from ethanol.

General Procedure for the CuAAC reaction on solid support (**GP10b**).[117]

The resin (1.00 equiv.) is swollen in DMF for 2 h. Then CuI (0.60 equiv.) together with the solution of an azide (1.05 equiv.) and DIPEA (2.40 equiv.) in pDMF is added to the resin and the mixture is shaken at room temperature for 16 h. Afterwards the liquid is filtered off and the resin is washed with THF, methanol, DMF and DCM until the filtrate is colorless (using 3 × 3.00 mL for each washing step). The product is cleaved from the resin according to **GP12a** or **GP12b.**

General Procedure for the synthesis of linear peptoids on solid support with rink amid resin (**GP11a**).

Step 1 – Swelling. 1.00 equiv. of dry rink amide aminomethyl resin (0.61 mmol/g) is swollen in a plastic fritted-syringe with twice its volume of DMF for 2 h. The solvent is then removed.

Step 2 – Fmoc deprotection. The resin is treated with a solution of 20% piperidine in pDMF (approx. 0.500 – 1.00 mL/100 mg resin) for 5 min at room temperature. Then the solution is filtered off and the procedure is repeated two more times. Afterwards the resin is washed with DMF (3 ×) and pDMF (1 ×).

Step 3 – Acetylation. 1.2 M solution of bromoacetic acid (8.00 equiv.) and DIC (8.00 equiv.) in pDMF (approx. 1 M) is added to the resin (1.00 equiv.). The mixture is stirred for 30 min. Then the resin is washed with DMF (4 ×) and pDMF (1 ×).

Step 4 – Substitution. 1.0 M solution of the desired amine (8.00 equiv.) in pDMF is added to the resin (1.00 equiv.). The mixture is stirred for 45 min. Then the resin is washed with DMF (4 ×) and pDMF (1 ×).

The step of acetylation is carried out in alternation with the substitution reaction until the desired peptoid is synthesized. After the last substitution reaction the resin is washed with pDMF (2 ×) and DCM (3 ×).

General Procedure for the synthesis of linear peptoids on solid support with 2-chlortritylchlorid resin (**GP11b**).

Step 1 – Swelling. 1.00 equiv. of dry 2-chlortritylchlorid resin (0.61 mmol/g) is swollen in a plastic fritted-syringe with twice its volume of DCM for 2 h. The solvent is then removed.

Step 2 – 1st acetylation. The resin is treated with 1.2 M solution of bromoacetic acid (5.00 equiv.) in DCM and DIPEA (5.00 equiv.). The mixture is stirred for 2 h. Then the solution is filtered off and the resin is washed with DCM (3 ×), DMF (2 ×) and pDMF (1 ×).

Step 3 – Substitution. 1.0 M solution of the desired amine (5.00 equiv.) in pDMF is added to the resin (1.00 equiv.). The mixture is stirred for 45 min. Then the resin is washed with DMF (4 ×) and pDMF (1 ×).

Step 3 – 2nd and subsequent acetylation. 1.2 M solution of bromoacetic acid (5.00 equiv.) and DIC (5.00 equiv.) in pDMF (approx. 1.0 M) is added to the resin (1.00 equiv.). The mixture is stirred for 30 min. Then the resin is washed with DMF (4 ×) and pDMF (1 ×).

The step of acetylation is carried out in alternation with the substitution reaction until the desired peptoid is synthesized. After the last substitution reaction the resin is washed with pDMF (2 ×) and DCM (3 ×).

General Procedure for the cleavage and isolation of linear peptoids on solid support with rink amid resin (**GP12a**).

Step 1 – Cleavage. Linear peptoid products are cleaved from solid support by treatment with 95% TFA in DCM (approx. 1.00 mL/100 mg resin) for 1.5-4 h at room temperature. The solution is filtered off and the next cleavage mixture is added to the resin and let shaken for additional 2 h. The resin is washed with the same volume of DCM (3 ×) and methanol (3 ×). The cleavage cocktail is evaporated under airflow, the remains are combined with washes and the solvent is removed under reduced pressure.

Step 2 – Purification and isolation. The crude product is dissolved in approx. 5.00 mL acetonitrile and passed through a syringe filter. The crude product is purified by means of HPLC and lyophilized.

General Procedure for the cleavage and isolation of linear peptoids on solid support with 2-chlortritylchlorid (**GP12b**).

Step 1 – Cleavage. 20% HFIP solution in DCM is added to the resin (approx. 1 mL/100 mg resin) and let shaken at room temperature for 2 h. The solution is collected, the resin is washed with DCM (2 ×). New cleavage mixture is added to the resin and let shaken for additional 2 h at room temperature. After filtering off the solution, the resin is washed with the same volume of DCM (3 ×) and methanol (3 ×). The cleavage solutions and washes are combined and the solvents are removed under reduced pressure.

Step 2 – Purification and isolation. The crude product is dissolved in approx. 5.00 mL acetonitrile and passed through a syringe filter. The crude product is purified by means of HPLC and lyophilized.

General Procedure for the cyclisation of the linear peptoids (**GP13**).

To a solution of the crude linear peptoid (1.00 equiv.) in pDMF (100.0 mL per 100 mg peptoid), a diluted solution of DIPEA (6.20 equiv.) and HATU (4.00 equiv.) in pDMF (100-150 mL per

100 mg peptoid) is added slowly under Ar, using a dropping funnel. The resulting mixture is stirred for 3 days under Ar at room temperature. Afterwards, the solvent is evaporated under reduced pressure and the residue is taken up in EtOAc (approx. ⅕ of the volume used for cyclization). The organic layer is washed with an equal amount of water and the aqueous layer is extracted with EtOAc (3 ×). The combined organic layers are dried over Na_2SO_4, filtered and concentrated under reduced pressure. The crude product is purified *via* preparative HPLC.

V.2.2 Synthesis and characterisation Chapter III.1 Lanthanide complexes

V.2.2.1 Lanthanide fluorobenzoates with different fluorination degrees.

Lanthanide 2,3,4,5,6-tetrafluorobenzoates $Ln(L_I)_3(H_2O)_n$, *n=0 – 2 (Ln-1.1)*

According to **GP1b** an excess of a concentrated aq. solution of NH_3 (1.30 mL, 20.0 equiv.) was added to an aq. solution of $LnCl_3 \cdot x\ H_2O$ (1.00 mmol, 1.00 equiv.) in 25 mL water. The mixture was stirred for 30 min, the precipitate of $Ln(OH)_3$ was centrifuged and washed with water until the pH of the washing solution became neutral. A small excess of freshly prepared $Ln(OH)_3$ was placed into a beaker and a solution of pentafluorobenzoic acid HL_I (573 mg, 2.70 mmol, 2.70 equiv.) in 20 mL acetone/methanol (3/1 v/v) was added. The reaction mixture was stirred for 1 h at 60 °C and the unreacted components were separated by filtration followed by the evaporation of the clear solution to dryness. The obtained powder of the lanthanide complex was then recrystallized from water. Subsequent drying at 80 °C in vacuo for 1 h led to the formation of a microcrystalline powder of general formula $Ln(L_I)_3(H_2O)_n$ (n depends on Ln used).

lanthanum 2,3,4,5,6-pentafluorobenzoate dihydrate $La(L_I)_3(H_2O)_2$ (La-1.1)

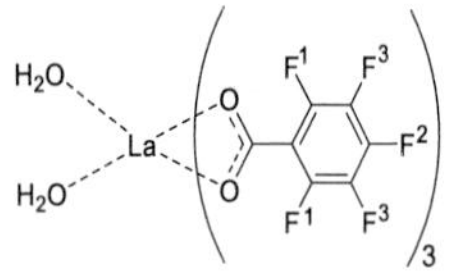

For the synthesis 373 mg of $LaCl_3 \cdot 7\ H_2O$ were used. 570 mg (89%) of an off-white solid of general formula $La(L_I)_3(H_2O)_2$ were obtained.

^{19}F NMR (377 MHz, D_2O): δ (ppm) = -138.91 (s, 2F, 2 × F_1), -154.68 (s, 1F, F_2), -162.03 (s, 2F, 2 × F_3). – **IR** (ATR): $\tilde{\nu}$ (cm^{-1}) = 3655.3 (w), 3340.0 (m), 1644.8 (m), 1571.8 (w), 1522.3 (w), 1492.9 (vw), 1376.1 (w), 1291.1 (m), 1119.7 (w), 1101.9 (vw), 986.6 (w), 937.6 (m), 832.2 (m), 816.4 (w), 764.2 (w), 506.6 (vw). – **EA**: calcd for $La(L_I)_3(H_2O)_2$,% C 31.21, H 0.50, found C 31.11, H 0.47.

praseodymium 2,3,4,5,6-pentafluorobenzoate dihydrate $Pr(L_1)_3(H_2O)_2$ (Pr-1.1)

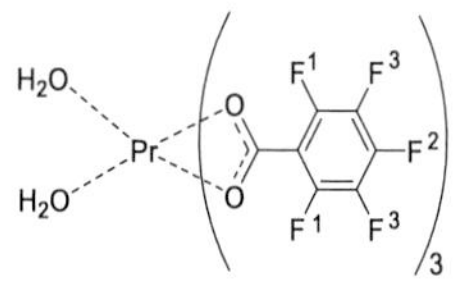

For the synthesis 373 mg of $PrCl_3 \cdot 7\ H_2O$ were used. 570 mg (89%) of an off-white solid of general formula $Pr(L_1)_3(H_2O)_2$ were obtained.

^{19}F NMR (377 MHz, MeOD-d_4): δ (ppm) = -141.14 (s, 2F, 2 × F_1), -157.14 (s, 1F, F_2), -163.82 (s, 2F, 2 × F_3). – **EA**: calcd for $Pr(L_1)_3(H_2O)_2$,% C 31.21, H 0.50, found C 31.33, H 0.72.

neodymium 2,3,4,5,6-pentafluorobenzoate dihydrate $Nd(L_1)_3(H_2O)_2$ (Nd-1.1)

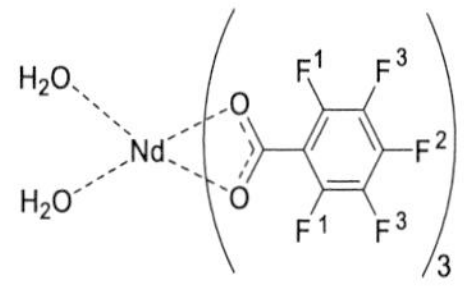

For the synthesis 359 mg of $NdCl_3 \cdot 6\ H_2O$ were used. 664 mg (91%) of an off-white solid of general formula $Nd(L_1)_3(H_2O)_2$ were obtained.

^{19}F NMR (377 MHz, MeOD-d_4): δ (ppm) = -143.16 (s, 2F, 2 × F_1), -157.66 (s, 1F, F_2), -164.59 (s, 2F, 2 × F_3). – **^{19}F NMR** (377 MHz, D_2O): δ (ppm) = -139.03 (s, 2F, 2 × F_1), -154.69 (s, 1F, F_2), -162.35 (s, 2F, 2 × F_3). – **IR** (ATR): $\tilde{\nu}$ (cm^{-1}) = 3653.5 (w), 1647.1 (m), 1577.4 (w), 1524.9 (m), 1492.6 (w), 1430.5 (m), 1398.3 (w), 1375.6 (w), 1295.9 (w), 1139.5 (w), 1116.8 (vw), 992.6 (w), 935.3 (m), 887.9 (w), 830.9 (vw), 768.5 (w), 741.4 (vw), 706.4 (w), 676.5 (vw), 651.8 (w). – **EA**: calcd for $Nd(L_1)_3(H_2O)_2$,% C 31.01, H 0.50, found C 30.80, H 0.62.

samarium 2,3,4,5,6-pentafluorobenzoate monohydrate $Sm(L_1)_3(H_2O)$ (Sm-1.1)

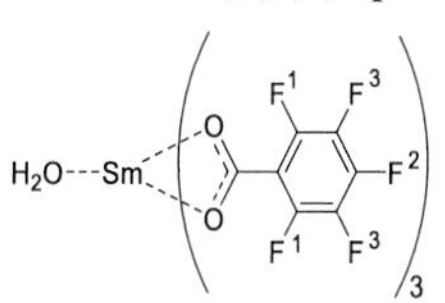

For the synthesis 365 mg of $SmCl_3 \cdot 6\ H_2O$ were used. 679 mg (94%) of an off-white solid of general formula $Sm(L_1)_3(H_2O)$ were obtained.

^{19}F NMR (377 MHz, MeOD-d_4): δ (ppm) = -142.90 (s, 2F, 2 × F_1), -157.01 (s, 1F, F_2), -164.93 (s, 2F, 2 × F_3). – **EA**: calcd for $Sm(L_1)_3(H_2O)$,% C 31.47, H 0.25, found C 31.23, H 0.30.

europium 2,3,4,5,6-pentafluorobenzoate monohydrate $Eu(L_1)_3(H_2O)$ (Eu-1.1)

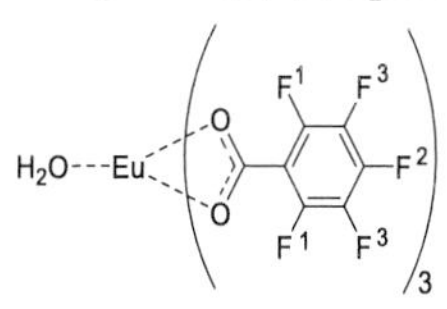

For the synthesis 366 mg of $EuCl_3 \cdot 6\ H_2O$ were used. 651 mg (90%) of an off-white solid of general formula $Eu(L_1)_3(H_2O)$ were obtained.

^{19}F NMR (377 MHz, MeOD-d_4): δ (ppm) = -145.68 (s, 2F, 2 × F_1), -159.37 (s, 1F, F_2), -167.71 (s, 2F, 2 × F_3). – **^{19}F NMR** (377 MHz, D_2O): δ (ppm) = -139.11 (s, 2F, 2 × F_1), -154.74 (s, 1F, F_2), -162.45 (s, 2F, 2 × F_3). – **IR** (ATR): $\tilde{\nu}$ (cm^{-1}) = 3632.8 (vw), 1725.2 (vw), 1701.5 (vw), 1651.5 (m), 1629.9 (w), 1595.7 (w), 1524.2 (w), 1489.9 (m), 1428.9 (w), 1397.2 (m), 1376.0 (w), 1325.7 (w), 1296.9 (vw), 1295.6 (vw), 1238.5 (w), 1111.5 (vw), 992.6 (m), 944.8 (m),

931.7 (w), 916.0 (m), 823.4 (w), 770.5 (w), 741.2 (vw). – **EA**: calcd for $Eu(L_1)_3(H_2O)$,% C 31.40, H 0.25, found C 30.89, H 0.30.

gadolinium 2,3,4,5,6-pentafluorobenzoate monohydrate $Gd(L_1)_3(H_2O)$ (Gd-1.1)

For the synthesis 372 mg of $GdCl_3 \cdot 6\ H_2O$ were used. 640 mg (88%) of an off-white solid of general formula $Gd(L_1)_3(H_2O)$ were obtained.

^{19}F NMR (377 MHz, MeOD-d_4): δ (ppm) = -140.93 (s, 2F, 2 × F_1), -157.53 (s, 1F, F_2), -166.68 (s, 2F, 2 × F_3). – **^{19}F NMR** (377 MHz, D_2O): δ (ppm) = -139.07 (s, 2F, 2 × F_1), -154.67 (s, 1F, F_2), -162.40 (s, 2F, 2 × F_3). – **IR** (ATR): $\tilde{\nu}$ (cm^{-1}) = 3633.9 (vw), 1650.5 (m), 1633.2 (m), 1591.4 (w), 1522.4 (m), 1487.7 (w), 1426.2 (vw), 1401.4 (m), 1379.3 (w), 1332.4 (w), 1290.6 (w), 1142.8 (vw), 1118.4 (w), 990.2 (w), 943.3 (m), 933.5 (m), 829.8 (w), 768.3 (vw), 743.9 (vw), 704.4 (w), 650.0 (vw). – **EA**: calcd for $Gd(L_1)_3(H_2O)$,% C 31.20, H 0.25, found C 30.95, H 0.27.

terbium 2,3,4,5,6-pentafluorobenzoate monohydrate $Tb(L_1)_3(H_2O)$ (Tb-1.1)

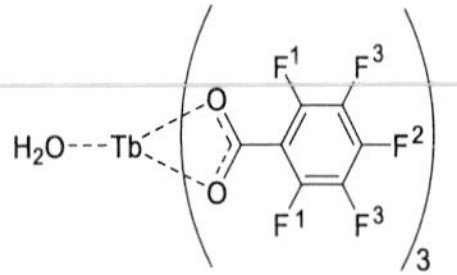

For the synthesis 373 mg of $TbCl_3 \cdot 6\ H_2O$ were used. 649 mg (89%) of an off-white solid of general formula $Tb(L_1)_3(H_2O)$ were obtained.

^{19}F NMR (377 MHz, MeOD-d_4): δ (ppm) = -129.62 (s, 2F, 2 × F_1), -153.62 (s, 1F, F_2), -161.80 (s, 2F, 2 × F_3). – **^{19}F NMR** (377 MHz, D_2O): δ (ppm) = -139.14 (s, 2F, 2 × F_1), -154.77 (s, 1F, F_2), -162.48 (s, 2F, 2 × F_3). – **IR** (ATR): $\tilde{\nu}$ (cm^{-1}) = 3635.5 (vw), 1649.8 (m), 1630.1 (w), 1597.8 (w), 1521.5 (m), 1491.9 (vw), 1429.9 (m), 1407.8 (w), 1378.1 (w), 1336.3 (m), 1306.7 (w), 1291.6 (w), 1259.6 (w), 1146.3 (vw), 1115.0 (m), 993.0 (vw), 946.0 (m), 936.2 (w), 830.3 (m), 773.4 (w), 743.8 (w), 721.7 (w), 706.7 (w), 652.6 (vw). – **EA**: calcd for $Tb(L_1)_3(H_2O)$,% C 31.13, H 0.25, found C 31.02, H 0.30.

terbium/europium 2,3,4,5,6-pentafluorobenzoate monohydrate $Tb_{0.5}Eu_{0.5}(L_1)_3(H_2O)$ (TbEu-1.1)

For the synthesis 187 mg of $TbCl_3 \cdot 6\ H_2O$ (0.50 mmol, 0.50 equiv.) and 183 mg $EuCl_3 \cdot 6\ H_2O$ (0.50 mmol, 0.50 equiv.) were used. 653 mg (90%) of an off-white solid of general formula $Tb_{0.5}Eu_{0.5}(L_1)_3(H_2O)$ were obtained.

^{19}F NMR (377 MHz, MeOD-d_4): δ (ppm) = -135.74 (s, 2F, 2 × F_1), -155.50 (s, 1F, F_2), -163.44 (s, 2F, 2 × F_3). – **IR** (ATR): $\tilde{\nu}$ (cm^{-1}) = 3603.9 (vw), 3412.1 (vw), 1632.6 (m), 1581.0 (m), 1485.1 (w), 1459.5 (m), 1402.2 (m), 1380.1 (w), 1299.2 (w), 1240.5 (w), 1194.8 (w), 1019.4 (vw), 948.5 (m),

815.2 (w), 779.3 (w), 709.4 (w), 624.5 (m), 602.7 (w), 577.1 (w), 515.7 (w), 413.1 (vw). – **EA**: calcd for $Tb_{0.5}Eu_{0.5}(L_1)_3(H_2O)$,% C 31.27, H 0.25, found C 31.37, H 0.20.

dysprosium 2,3,4,5,6-pentafluorobenzoate monohydrate $Dy(L_1)_3(H_2O)$ (Dy-1.1)

For the synthesis 377 mg of $DyCl_3{\cdot}6\ H_2O$ were used. 681 mg (93%) of an off-white solid of general formula $Dy(L_1)_3(H_2O)$ were obtained.

^{19}F NMR (377 MHz, MeOD-d_4): δ (ppm) = -115.56 (s, 2F, 2 × F_1), -149.58 (s, 1F, F_2), -156.48 (s, 2F, 2 × F_3). – **EA**: calcd for $Dy(L_1)_3(H_2O)$,% C 31.00, H 0.25, found C 31.10, H 0.22.

holmium 2,3,4,5,6-pentafluorobenzoate monohydrate $Ho(L_1)_3(H_2O)$ (Ho-1.1)

For the synthesis 379 mg of $HoCl_3{\cdot}6\ H_2O$ were used. 676 mg (92%) of an off-white solid of general formula $Ho(L_1)_3(H_2O)$ were obtained.

^{19}F NMR (377 MHz, MeOD-d_4): δ (ppm) = -129.93 (s, 2F, 2 × F_1), -153.26 (s, 1F, F_2), -161.10 (s, 2F, 2 × F_3). – **EA**: calcd for $Ho(L_1)_3(H_2O)$,% C 30.91, H 0.25, found C 30.50, H 0.32.

erbium 2,3,4,5,6-pentafluorobenzoate monohydrate $Er(L_1)_3(H_2O)$ (Er-1.1)

For the synthesis 381 mg of $ErCl_3{\cdot}6\ H_2O$ were used. 663 mg (90%) of an off-white solid of general formula $Er(L_1)_3(H_2O)$ were obtained.

^{19}F NMR (377 MHz, MeOD-d_4): δ (ppm) = -145.00 (s, 2F, 2 × F_1), -158.25 (s, 1F, F_2), -166.57 (s, 2F, 2 × F_3). – **^{19}F NMR** (377 MHz, D_2O): δ (ppm) = -139.07 (s, 2F, 2 × F_1), -154.63 (s, 1F, F_2), -162.34 (s, 2F, 2 × F_3). – **IR** (ATR): $\tilde{\nu}$ (cm^{-1}) = 3633.1 (vw), 1651.2 (m), 1593.8 (w), 1519.2 (m), 1489.3 (w), 1431.8 (m), 1404.3 (w), 1384.4 (vw), 1307.4 (w), 1292.3 (w), 1137.9 (vw), 1120.3 (vw), 988.3 (vw), 950.8 (m), 935.7 (w), 828.8 (w), 771.3 (w), 746.2 (vw), 703.9 (w), 691.6 (w), 656.5 (w). – **EA**: calcd for $Er(L_1)_3(H_2O)$,% C 30.82, H 0.25, found C 31.13, H 0.48.

thulium 2,3,4,5,6-pentafluorobenzoate $Tm(L_1)_3$ (Tm-1.1)

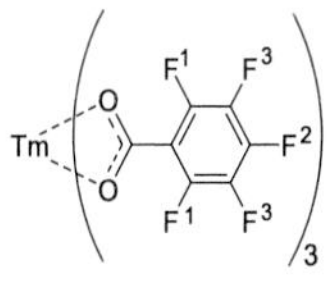

For the synthesis 383 mg of $TmCl_3{\cdot}6\ H_2O$ were used. 650 mg (90%) of an off-white solid of general formula $Tm(L_1)_3$ were obtained.

^{19}F NMR (377 MHz, MeOD-d_4): δ (ppm) = -149.09 (s, 2F, 2 × F_1), -159.77 (s, 1F, F_2), -168.20 (s, 2F, 2 × F_3). – **EA**: calcd for $Tm(L_1)_3$,% C 31.44, found C 31.76.

ytterbium 2,3,4,5,6-pentafluorobenzoate $Yb(L_1)_3$ (Yb-1.1)

For the synthesis 388 mg of $YbCl_3 \cdot 6\ H_2O$ were used. 701 mg (88%) of an off-white solid of general formula $Yb(L_1)_3$ were obtained.

^{19}F NMR (377 MHz, MeOD-d_4): δ (ppm) = -145.99 (s, 2F, 2 × F_1), -158.21 (s, 1F, F_2), -166.30 (s, 2F, 2 × F_3). – **^{19}F NMR** (377 MHz, D_2O): δ (ppm) = -139.06 (s, 2F, 2 × F_1), -154.70 (s, 1F, F_2), -162.40 (s, 2F, 2 × F_3). – **IR** (ATR): $\tilde{\nu}$ (cm^{-1}) = 1685.4 (m), 1653.3 (w), 1616.0 (w), 1583.9 (m), 1526.8 (m), 1487.2 (w), 1405.5 (m), 1296.5 (w), 1261.9 (w), 1147.8 (w), 1115.7 (w), 994.4 (vw), 939.8 (m), 823.1 (w), 776.3 (w), 748.9 (w). – **EA**: calcd for $Yb(L_1)_3$,% C 31.28, found C 31.33.

lutetium 2,3,4,5,6-pentafluorobenzoate $Lu(L_1)_3$ (Lu-1.1)

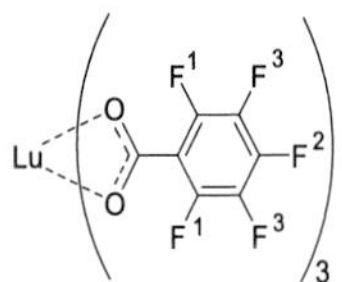

For the synthesis 389 mg of $LuCl_3 \cdot 6\ H_2O$ was used. 701 mg (88%) of an off-white solid of general formula $Lu(L_1)_3$ were obtained.

^{19}F NMR (377 MHz, MeOD-d_4): δ (ppm) = -143.17 (s, 2F, 2 × F_1), -157.14 (s, 1F, F_2), -165.11 (s, 2F, 2 × F_3). – **^{19}F NMR** (377 MHz, D_2O): δ (ppm) = -139.05 (s, 2F, 2 × F_1), -154.68 (s, 1F, F_2), -162.39 (s, 2F, 2 × F_3). – **IR** (ATR): $\tilde{\nu}$ (cm^{-1}) = 1653.6 (m), 1607.5 (w), 1523.3 (m), 1488.4 (vw), 1422.4 (m), 1405.9 (w),1393.0 (vw), 1297.7 (vw), 1259.2 (m), 1140.1 (w), 1112.6 (w), 1066.8 (w), 996.2 (m), 934.7 (m), 876.1 (w), 830.0 (m), 776.9 (w), 744.1 (m), 688.9 (w). – **EA**: calcd for $Lu(L_1)_3$,% C 31.21, found C 31.01.

Lanthanide 2,3,4,5-tetrafluorobenzoates $Ln(L_2)_3(H_2O)_n$, n=0 – 2 (Ln-1.2)

According to **GP1b** an excess of concentrated aq. solution of NH_3 (1.30 mL, 20.0 equiv.) was added to an aq. solution of $LnCl_3 \cdot x\ H_2O$ (1.00 mmol, 1.00 equiv.) in 25 mL water. The mixture was stirred for 30 min, the precipitate of $Ln(OH)_3$ was centrifugated and washed with water until the pH of the washing solution became neutral. A small excess of freshly prepared $Ln(OH)_3$ was placed into a beaker and a solution of 524 mg of 2,3,4,5-tetrafluorobenzoic acid HL_2 (2.70 mmol, 2.70 equiv.) in 20 mL acetone/methanol (3/1) was added. The reaction mixture was stirred for 1 h at 60 °C and the unreacted components were separated by filtration followed by the evaporation of the clear solution to dryness. The obtained solid powder of the lanthanide complex was then recrystallized from water. Subsequent drying at 80 °C in vacuo for 1 h led to the formation of a microcrystalline powder of general formula $Ln(L_2)_3(H_2O)_n$ (n depends on Ln used).

lanthanum 2,3,4,5-tetrafluorobenzoate dihydrate $La(L_2)_3(H_2O)_2$ (La-1.2)

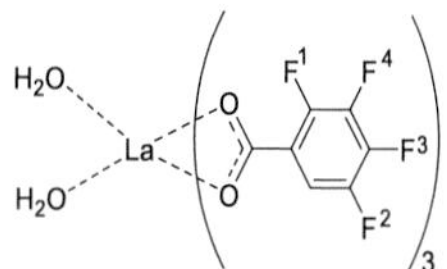

For the synthesis 373 mg of $LaCl_3{\cdot}7\ H_2O$ were used. 590 mg (87%) of an off-white solid of general formula $La(L_2)_3(H_2O)_2$ were obtained.

^{1}H NMR (400 MHz, DMSO-d_6): δ (ppm) = 7.58 – 7.63 (m, 1H, H_{Ar}). – **^{19}F NMR** (377 MHz, DMSO-d_6): δ (ppm) = -139.85 – -140.08 (m, 1F, F_1), -140.80 – -140.59 (m, 1F, F_2), -154.98 (bs, 1F, F_3), -156.61 (t, J = 22.5 Hz, 1F, F_4). – **IR** (ATR): $\tilde{\nu}$ (cm^{-1}) = 3665.7 (vw), 3084.6 (vw), 1695.8 (m), 1571.8 (w), 1529.3 (m), 1488.9 (w), 1396.1 (m), 1291.1 (w), 1269.1 (vw), 1195.7 (m), 1101.9 (w), 986.6 (w), 927.6 (w), 892.2 (vw), 816.4 (m), 744.5 (w), 716.9 (w), 698.4 (m), 634.0 (w), 506.6 (w). – **EA**: calcd for $La(L_2)_3(H_2O)_2$,% C 33.45, H 0.94, found C 33.38, H 0.85.

neodymium 2,3,4,5-tetrafluorobenzoate dihydrate $Nd(L_2)_3(H_2O)_2$ (Nd-1.2)

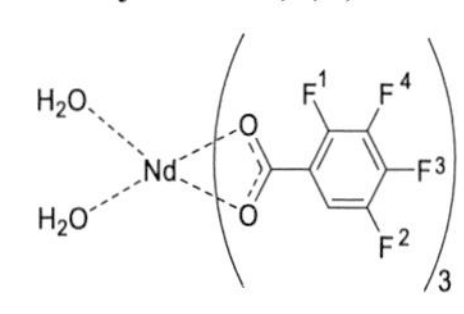

For the synthesis 359 mg of $NdCl_3{\cdot}6\ H_2O$ were used. 615 mg (90%) of an off-white solid of general formula $Nd(L_2)_3(H_2O)_2$ were obtained.

^{1}H NMR (400 MHz, DMSO-d_6): δ (ppm) = 8.14 (bs, 1H, H_{Ar}). – **^{19}F NMR** (377 MHz, DMSO-d_6): δ (ppm) = -134.52 (bs, 1F, F_1), -135.24 (p, J = 10.1 Hz, F_2), -148.72 (bs, 1F, F_3), -151.02 (t, J = 21.5 Hz, 1F, F_4). – **IR** (ATR): $\tilde{\nu}$ (cm^{-1}) = 3655.3 (vw), 3340.6 (vw), 1644.8 (m), 1571.8 (w), 1522.3 (m), 1492.9 (w), 1376.1 (m), 1291.1 (w), 1119.7 (m), 1101.9 (w), 986.6 (w), 937.6 (m), 832.2 (w), 816.4 (vw), 764.2 (m), 506.6 (w). – **EA**: calcd for $La(L_2)_3(H_2O)_2$,% C 33.21, H 0.93, found C 33.30, H 0.78.

europium 2,3,4,5-tetrafluorobenzoate monohydrate $Eu(L_2)_3(H_2O)$ (Eu-1.2)

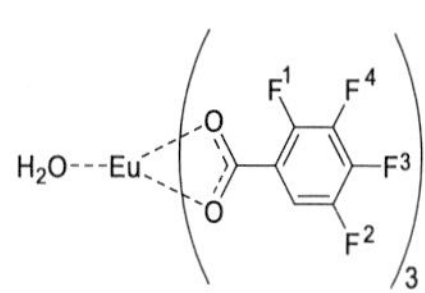

For the synthesis 366 mg of $EuCl_3{\cdot}6\ H_2O$ were used. 634 mg (94%) of an off-white solid of general formula $Eu(L_2)_3(H_2O)$ were obtained.

^{1}H NMR (400 MHz, DMSO-d_6): δ (ppm) = 6.62 (bs, 1H, H_{Ar}). – **^{19}F NMR** (377 MHz, DMSO-d_6): δ (ppm) = -138.89 (bs, 1F, F_1), -141.71 – -141.39 (m, 1F, F_2), -154.02 (bs, 1F, F_3), -157.24 (t, J = 22.0 Hz, 1F, F_4). – **IR** (ATR): $\tilde{\nu}$ (cm^{-1}) = 3660.2 (vw), 3085.5 (vw), 1676.0 (vw), 1569.6 (w), 15217 (w), 1474.4 (w), 1391.5 (m), 1291.0 (w), 1264.5 (w), 1208.7 (vw), 1101.9 (w), 1034.4 (w), 914.6 (w), 887.0 (vw), 798.6 (w), 771.1 (w), 751.3 (w), 712.9 (w), 695.7 (w), 627.7 (vw), 523.0 (vw), 456.9 (vw), 386.5 (w). – **EA**: calcd for $Eu(L_2)_3(H_2O)$,% C 33.67, H 0.67, found C 32.56, H 0.92.

gadolinium 2,3,4,5-tetrafluorobenzoate monohydrate $Gd(L_2)_3(H_2O)$ (Gd-1.2)

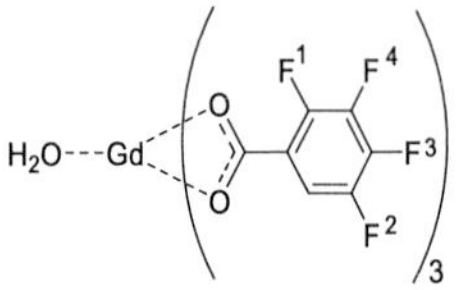

For the synthesis 372 mg of $GdCl_3 \cdot 6\ H_2O$ were used. 640 mg (88%) of an off-white solid of general formula $Gd(L_2)_3(H_2O)$ were obtained.

^{1}H NMR (400 MHz, DMSO-d_6): δ (ppm) = 8.20 (bs, 1H, H_{Ar}). – **^{19}F NMR** (377 MHz, DMSO-d_6): δ (ppm) = -138.28 (bs, 2F, F_1, F_2), -148.30 – -148.91 (m, 1F, F_3), -153.06 (bs, 1F, F_4). – **IR** (ATR): $\tilde{\nu}$ (cm^{-1}) = 3636.2 (vw), 3065.5 (vw), 1656.0 (m), 1559.0 (w), 1534.5 (w), 1474.4 (vw), 1381.4 (m), 1290.3 (m), 1267.1 (w), 1209.1 (w), 1104.9 (vw), 1038.4 (w), 914.4 (m), 886.8 (w), 793.6 (w), 775.3 (m), 752.7 (w), 694.0 (vw), 622.4 (m), 513.0 (w).– **EA**: calcd for $Gd(L_2)_3(H_2O)$,% C 33.43, H 0.67, found C 32.54, H 0.87.

terbium 2,3,4,5-tetrafluorobenzoate monohydrate $Tb(L_2)_3(H_2O)$ (Tb-1.2)

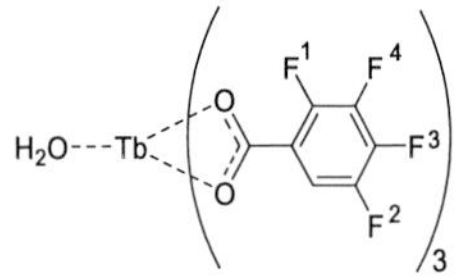

For the synthesis 373 mg of $TbCl_3 \cdot 6\ H_2O$ were used. 524 mg (77%) of an off-white solid of general formula $Tb(L_2)_3(H_2O)$ were obtained.

^{1}H NMR (400 MHz, DMSO-d_6): δ (ppm) = 5.36 (bs, 1H, H_{Ar}). – **^{19}F NMR** (377 MHz, DMSO-d_6): δ (ppm) = -121.55 (bs, 1F, F_1), -136.47 (bs, 1F, F_2), -147.70 (bs, 1F, F_4), -149.97 (bs, 1F, F_3). – **IR** (ATR): $\tilde{\nu}$ (cm^{-1}) = 3452.9 (vw), 1564.9 (m), 1535.5 (w), 1465.6 (m), 1417.5 (w), 1278.5 (m), 1188.9 (w), 1105.4 (w), 1039.0 (m), 921.4 (vw), 881.0 (w), 792.4 (w), 771.9 (vw). – **EA**: calcd for $Tb(L_2)_3(H_2O)$,% C 33.36, H 0.67, found C 32.93, H 0.60.

terbium/europium 2,3,4,5-tetrafluorobenzoate monohydrate $Tb_{0.5}Eu_{0.5}(L_2)_3(H_2O)$ ((TbEu)-1.2)

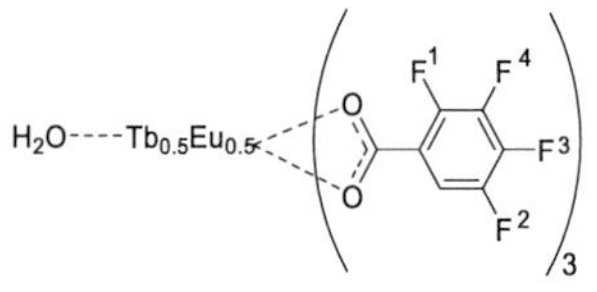

For the synthesis 187 mg of $TbCl_3 \cdot 6\ H_2O$ (0.50 mmol, 0.50 equiv.) and 183 mg $EuCl_3 \cdot 6\ H_2O$ (0.50 mmol, 0.50 equiv.) were used. 603 mg (90%) of an off-white solid of general formula $Tb_{0.5}Eu_{0.5}(L_2)_3(H_2O)$ were obtained.

^{1}H NMR (400 MHz, DMSO-d_6): δ (ppm) = 8.20 (bs, 1H, H_{Ar}). – **^{19}F NMR** (377 MHz, DMSO-d_6): δ (ppm) = -139.20 (bs, 1F, F_1), -142.32 (bs, 1F, F_2), -153.02 – -154.16 (m, 1F, F_3), -156.15 (bs, 1F, F_4). – **IR** (ATR): $\tilde{\nu}$ (cm^{-1}) = 1566.4 (w), 1524.4 (w), 1476.8 (vw), 1396.6 (w), 1267.0 (vw), 1202.0 (vw), 1107.9 (vw), 1036.3 (w), 996.6 (vw), 920.0 (vw), 889.0 (vw), 798.9 (vw), 771.6 (vw), 752.1 (vw), 711.6 (vw), 696.2 (vw), 626.6 (vw), 522.4 (vw), 459.2 (vw), 392.4 (vw). – **EA**: calcd for $Tb_{0.5}Eu_{0.5}(L_2)_3(H_2O)$,% C 33.51, H 0.67, found C 34.14, H 0.84.

erbium 2,3,4,5-tetrafluorobenzoate monohydrate $Er(L_2)_3(H_2O)$ (Er-1.2)

For the synthesis 381 mg of $ErCl_3 \cdot 6\ H_2O$ were used. 612 mg (89%) of an off-white solid of general formula $Er(L_2)_3(H_2O)$ were obtained.

^{1}H NMR (400 MHz, DMSO-d_6): δ (ppm) = 5.25 (bs, 1H, H_{Ar}). – **^{19}F NMR** (377 MHz, DMSO-d_6): δ (ppm) = -136.84 (bs, 1F, F_1), -138.53 (bs, 1F, F_2), -149.52 (bs, 1F, F_3), -153.43 (t, J = 23.0 Hz, 1F, F_4). – **IR** (ATR): $\tilde{\nu}$ (cm^{-1}) = 3542.1 (vw), 1544.9 (m), 1533.5 (w), 1425.6 (w), 1399.5 (w), 1288.5 (w), 1190.9 (m), 1115.6 (w), 1028.8 (w), 924.7 (m), 890.8 (w), 791.4 (m), 778.9 (w), 719.1 (vw), 697.5 (w), 629.4 (w), 526.5 (vw), 464.7 (vw), 391.3 (vw). – **EA**: calcd for $Er(L_2)_3(H_2O)$,% C 32.99, H 0.66, found C 32.42, H 0.86.

ytterbium 2,3,4,5-tetrafluorobenzoate $Yb(L_2)_3$ (Yb2)

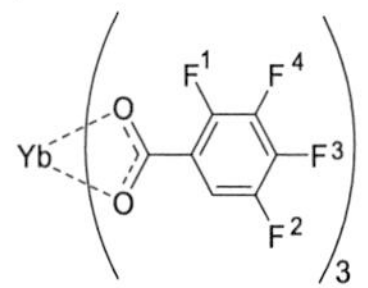

For the synthesis 388 mg of $YbCl_3 \cdot 6\ H_2O$ were used. 701 mg (88%) of an off-white solid of general formula $Yb(L_2)_3$ were obtained.

^{1}H NMR (400 MHz, DMSO-d_6): δ (ppm) = 4.53 (bs, 1H, H_{Ar}). – **^{19}F NMR** (377 MHz, DMSO-d_6): δ (ppm) = -136.11 (bs, 1F, F_1), -138.11 (bs, 1F, F_2), -149.64 (bs, 1F, F_3), -152.73 (t, J = 22.0 Hz, 1F, F_4). – **IR** (ATR): $\tilde{\nu}$ (cm^{-1}) = 1692.6 (m), 1564.5 (w), 1523.1 (w), 1482.6 (vw), 1399.3 (w), 1289.8 (w), 1269.0 (m), 1190.1 (w), 1038.5 (vw), 924.6 (m), 890.3 (w), 795.2 (vw), 781.2 (w), 717.4 (vw), 697.5 (vw), 640.4 (w), 523.5 (m), 461.7 (w), 392.3 (w). – **EA**: calcd for $Yb(L_2)_3$,% C 33.53, H 0.40 found C 32.17, H 0.44.

lutetium 2,3,4,5-tetrafluorobenzoate $Lu(L_2)_3$ (Lu-1.2)

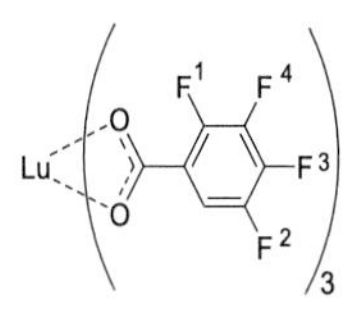

For the synthesis 389 mg of $LuCl_3 \cdot 6\ H_2O$ were used. 611 mg (90%) of an off-white solid of general formula $Lu(L_2)_3$ were obtained.

^{1}H NMR (400 MHz, DMSO-d_6): δ (ppm) = 7.86 (s, 1H, H_{Ar}). – **^{19}F NMR** (377 MHz, DMSO-d_6): δ (ppm) = -139.08 – -139.42 (m, 1F, F_1), -140.55 – -140.80 (m, 1F, F_2), -154.01 (bs, 1F, F_3), -156.55 (t, J = 21.5 Hz, 1F, F_4). – **IR** (ATR): $\tilde{\nu}$ (cm^{-1}) = 1695.3 (vw), 1562.5 (m), 1527.1 (m), 1478.6 (w), 1405.6 (m), 1290.1 (w), 1269.6 (w), 1196.0 (w), 1107.6 (w), 1037.4 (m), 923.0 (m), 889.8 (w), 799.5 (m), 752.2 (m), 726.0 (w), 710.8 (w), 694.6 (m), 626.9 (w), 522.2 (w), 485.6 (w), 461.6 (w), 397.7 (w), 3669.3 (vw). – **EA**: calcd for Lu $(L_2)_3$,% C 33.44, H 0.40, found C 32.96, H 0.87.

Lanthanide 2,3,4,5-tetrafluoro-6-nitrobenzoates dihydrates $Ln(L_3)_3(H_2O)_2$ ***(Ln-1.3)***

According to **GP1a**, a suspension of 717 mg of 2,3,4,5-tetrafluoro-6-nitrobenzoic acid HL_3 (3.00 mmol, 3.00 equiv.) in water was treated with 1 M aq. solution of NaOH at 60 °C until the pH was slightly acidic (pH = 5 – 6, approx. 3 mL) and the solution became clear. Then an aq. solution of $LnCl_3 \cdot x\ H_2O$ (1.00 mmol, 1.00 equiv.) was added following by an immediate precipitation of the product. The reaction mixture was stirred for additional 3 h at rt, the crude product was filtered, washed with ethanol and dried in vacuo to afford a pale-yellow solid.

lanthanum 2,3,4,5-tetrafluoro-6-nitrobenzoate dihydrate $La(L_3)_3(H_2O)_2$ (La-1.3)

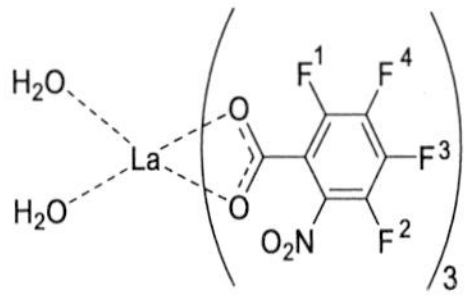

For the synthesis 373 mg of $LaCl_3 \cdot 7\ H_2O$ were used. 836 mg (94%) of a pale-yellow solid of general formula $La(L_3)_3(H_2O)_2$ were obtained.

^{19}F NMR (377 MHz, DMSO-d_6): δ (ppm) = -137.40 – -137.56 (m, 1F, F_1), -142.66 – -142.76 (m, 1F, F_2), -145.20 – -145.34 (m, 1F, F_3), -152.18 – -152.28 (m, 1F, F_4). – **IR** (ATR): $\tilde{\nu}$ (cm^{-1}) = 3609.4 (vw), 1699.8 (vw), 1608.3 (w), 1537.7 (m), 1519.3 (w), 1478.1 (w), 1396.0 (w), 1281.8 (m), 1269.1 (w), 1126.9 (vw), 983.1 (m), 925.9 (w), 837.1 (w), 764.5 (m), 706.9 (w), 643.1 (vw), 536.3 (w), 485.3 (vw), 445.9 (vw), 385.4 (vw). – **EA**: calcd for $La(L_3)_3(H_2O)_2$,% C 28.37, H 0.45, N 4.73, found C 29.11, H 0.27, N 4.90.

neodymium 2,3,4,5-tetrafluoro-6-nitrobenzoate dihydrate $Nd(L_3)_3(H_2O)_2$ (Nd-1.3)

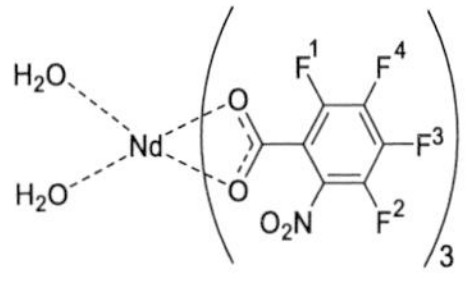

For the synthesis 359 mg of $NdCl_3 \cdot 6\ H_2O$ were used. 850 mg (95%) of a pale-yellow solid of general formula $Nd(L_3)_3(H_2O)_2$ were obtained.

^{19}F NMR (377 MHz, DMSO-d_6): δ (ppm) = -137.30 – -137.45 (m, 1F, F_1), -142.60 – -142.66 (m, 1F, F_2), -145.18 – -145.26 (m, 1F, F_3), -152.16 – -152.24 (m, 1F, F_4). – **IR** (ATR): $\tilde{\nu}$ (cm^{-1}) = 3602.4 (vw), 1709.2 (vw), 1609.7 (w), 1538.6 (m), 1514.5 (w), 1476.9 (w), 1395.0 (m), 1351.8 (w), 1285.1 (w), 1127.4 (w), 1076.2 (w), 953.5 (m), 927.2 (w), 837.2 (m), 768.8 (w), 744.2 (w), 702.7 (w), 640.4 (m), 615.8 (w), 536.6 (vw), 485.5 (vw), 446.5 (vw), 385.7 (vw). – **EA**: calcd for $Nd(L_3)_3(H_2O)_2$,% C 28.20, H 0.45, N 4.70, found C 29.21, H 0.27, N 4.84.

europium 2,3,4,5-tetrafluoro-6-nitrobenzoate dihydrate $Eu(L_3)_3(H_2O)_2$ (Eu-1.3)

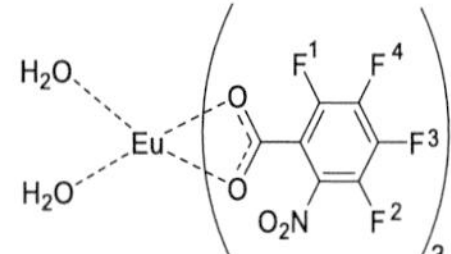

For the synthesis 366 mg of $EuCl_3 \cdot 6\ H_2O$ were used. 840 mg (93%) of a pale-yellow solid of general formula $Eu(L_3)_3(H_2O)_2$ were obtained.

^{19}F NMR (377 MHz, DMSO-d_6): δ (ppm) = -137.35 – -137.43 (m, 1F, F_1), -142.58 – -142.65 (m, 1F, F_2), -145.23 – -145.33 (m, 1F, F_3), -152.20 – -152.26 (m, 1F, F_4). – **IR** (ATR): $\tilde{\nu}$ (cm^{-1}) = 3601.8 (vw), 1715.7 (vw), 1613.4 (m), 1538.8 (m), 1515.4 (w), 1476.4 (m), 1396.3 (m), 1352.8 (m), 1284.5 (w), 1126.8 (w), 1077.2 (m), 954.7 (w), 927.5 (vw), 837.1 (w), 769.3 (m), 743.9 (m), 703.4 (w), 640.6 (vw), 536.0 (w), 485.7 (w), 447.2 (w), 387.0 (w). – **EA**: calcd for $Eu(L_3)_3(H_2O)_2$,% C 27.96, H 0.45, N 4.66, found C 28.53, H 0.21, N 4.96.

gadolinium 2,3,4,5-tetrafluoro-6-nitrobenzoate dihydrate $Gd(L_3)_3(H_2O)_2$ (Gd-1.3)

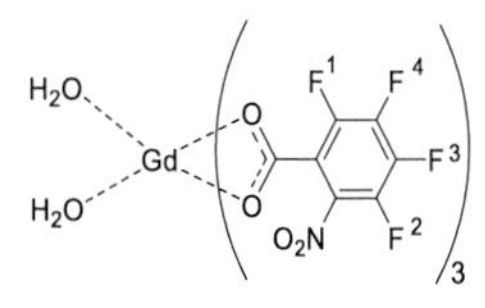

For the synthesis 372 mg of $GdCl_3 \cdot 6\ H_2O$ were used. 830 mg (91%) of a pale-yellow solid of general formula $Gd(L_3)_3(H_2O)_2$ were obtained.

^{19}F NMR (377 MHz, DMSO-d_6): δ (ppm) = -136.61 (bs, 1F, F_1), -141.89 (bs, 1F, F_2), -144.51 (bs, 1F, F_3), -151.48 (bs, 1F, F_4). – **IR** (ATR): $\tilde{\nu}$ (cm^{-1}) = 3604.1 (vw), 1718.4 (w), 1614.2 (m), 1538.3 (m), 1515.5 (m), 1476.2 (m), 1396.9 (m), 1351.9 (m), 1284.7 (m), 1127.2 (m), 1076.9 (m), 954.8 (m), 927.6 (vw), 837.2 (w), 769.1 (m), 745.1 (m), 703.4 (w), 640.3 (vw), 536.3 (w), 485.6 (w), 447.1 (w), 386.9 (w). – **EA**: calcd for $Gd(L_3)_3(H_2O)_2$,% C 27.79, H 0.44, N 4.63, found C 28.71, H 0.30, N 4.96.

terbium 2,3,4,5-tetrafluoro-6-nitrobenzoate dihydrate $Tb(L_3)_3(H_2O)_2$ (Tb-1.3)

For the synthesis 373 mg of $TbCl_3 \cdot 6\ H_2O$ were used. 825 mg (91%) of a pale-yellow solid of general formula $Tb(L_3)_3(H_2O)_2$ were obtained.

^{19}F NMR (377 MHz, DMSO-d_6): δ (ppm) = -137.41 (bs, 1F, F_1), -142.61 (bs, 1F, F_2), -144.28 (bs, 1F, F_3), -152.24 (bs, 1F, F_4). – **IR** (ATR): $\tilde{\nu}$ (cm^{-1}) = 3595.2 (vw), 1719.1 (vw), 1617.9 (w), 1537.9 (m), 1515.6 (w), 1475.7 (m), 1396.6 (m), 1351.4 (m), 1283.3 (w), 1125.6 (w), 1076.5 (m), 954.4 (w), 927.4 (w), 836.9 (w), 769.1 (m), 743.3 (m), 702.7 (w), 640.1 (vw). – **EA**: calcd for $Tb(L_3)_3(H_2O)_2$,% C 27.74, H 0.44, N 4.62, found C 28.65, H 0.24, N 4.78.

terbium/europium 2,3,4,5-tetrafluoro-6-nitrobenzoate dihydrate $Tb_{0.5}Eu_{0.5}(L_3)_3(H_2O)_2$ (TbEu-1.3)

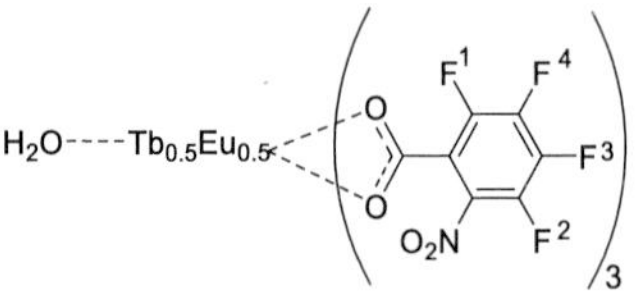

For the synthesis 187 mg of $TbCl_3·6\,H_2O$ (0.50 mmol, 0.50 equiv.) and 183 mg of $EuCl_3·6\,H_2O$ (0.50 mmol, 0.50 equiv.) were used. 846 mg (93%) of a pale-yellow solid of general formula $Tb_{0.5}Eu_{0.5}(L_3)_3(H_2O)_2$ were obtained.

^{19}F NMR (377 MHz, DMSO-d_6): δ (ppm) = -137.35 – -137.43 (m, 1F, F_1), -142.56 – -142.63 (m, 1F, F_2), -145.25 – -145.33 (m, 1F, F_3), -152.17 – -152.25 (m, 1F, F_4). – **IR** (ATR): $\tilde{\nu}$ (cm^{-1}) = 3598.0 (vw), 1702.1 (vw), 1613.6 (w), 1536.8 (m), 1514.4 (w), 1476.8 (vw), 1396.6 (m), 1276.3 (w), 1282.0 (w), 1127.9 (w), 1076.3 (w), 994.6 (w), 928.0 (m), 839.0 (vw), 768.9 (m), 745.1 (w), 703.6 (w), 643.2 (vw), 532.4 (w), 489.2 (w), 382.4 (w). – **EA**: calcd for $Tb_{0.5}Eu_{0.5}(L_3)_3(H_2O)_2$,% C 27.96, H 0.45, N 4.66, found C 27.95, H 0.19, N 4.85.

erbium 2,3,4,5-tetrafluoro-6-nitrobenzoate dihydrate $Er(L_3)_3(H_2O)_2$ (Er-1.3)

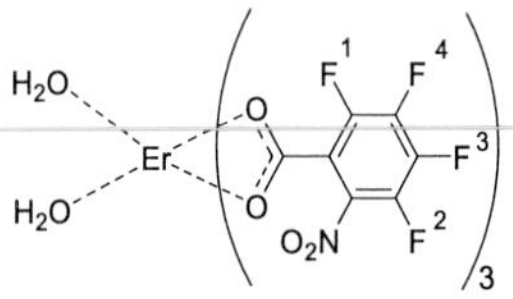

For the synthesis 381 mg of $ErCl_3·6\,H_2O$ were used. 841 mg (92%) of a pale-yellow solid of general formula $Er(L_3)_3(H_2O)_2$ were obtained.

^{19}F NMR (377 MHz, DMSO-d_6): δ (ppm) = -137.45 (bs, 1F, F_1), -142.61 (bs, 1F, F_2), -145.48 (bs, 1F, F_3), -152.33 (bs, 1F, F_4). – **IR** (ATR): $\tilde{\nu}$ (cm^{-1}) = 3596.3 (vw), 1724.8 (w), 1627.9 (m), 1538.4 (m), 1515.8 (m), 1476.0 (m), 1398.6 (m), 1351.2 (m), 1284.0 (m), 1126.3 (m), 1077.1 (m), 955.3 (m), 927.7 (w), 837.0 (w), 769.3 (m), 744.5 (m), 703.1 (m), 640.4 (w), 536.4 (w), 486.1 (m), 448.3 (w), 388.0 (w). – **EA**: calcd for $Er(L_3)_3(H_2O)_2$,% C 27.49, H 0.44, N 4.58, found C 28.65, H 0.24, N 4.78.

ytterbium 2,3,4,5-tetrafluoro-6-nitrobenzoate dihydrate $Yb(L_3)_3(H_2O)_2$ (Yb-1.3)

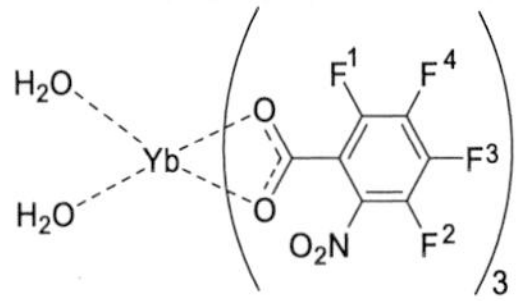

For the synthesis 388 mg of $YbCl_3·6\,H_2O$ were used. 825 mg (89%) of a pale-yellow solid of general formula $Yb(L_3)_3(H_2O)_2$ were obtained.

^{19}F NMR (377 MHz, DMSO-d_6): δ (ppm) = -137.38 – -137.46 (m, 1F, F_1), -142.58 – -142.65 (m, 1F, F_2), -145.20 – -145.28 (m, 1F, F_3), -152.24 – -152.30 (m, 1F, F_4). – **IR** (ATR): $\tilde{\nu}$ (cm^{-1}) = 3594.4 (vw), 1727.7 (vw), 1631.0 (m), 1539.1 (m), 1515.8 (m), 1475.8 (m), 1400.0 (m), 1351.2 (m), 1284.4 (m), 1126.9 (m), 1077.4 (m), 955.7 (m), 928.1 (w), 837.0 (w), 769.4 (m), 745.3 (m), 703.3 (w), 640.3 (w), 536.8 (w), 486.3 (w), 448.6 (w), 388.2 (w). – **EA**: calcd for $Yb(L_3)_3(H_2O)_2$,% C 27.96, H 0.45, N 4.66, found C 28.94, H 0.23, N 4.77.

lutetium 2,3,4,5-tetrafluoro-6-nitrobenzoate dihydrate $Lu(L_3)_3(H_2O)_2$ (Lu-1.3)

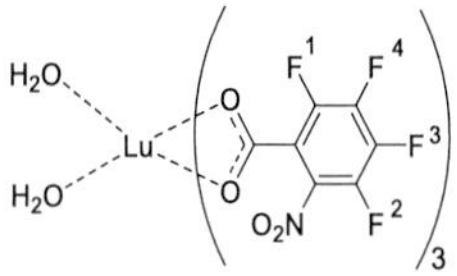

For the synthesis 389 mg of $LuCl_3 \cdot 6\ H_2O$ were used. 818 mg (88%) of a pale-yellow solid of general formula $Lu(L_3)_3(H_2O)_2$ were obtained.

^{19}F NMR (377 MHz, DMSO-d_6): δ (ppm) = -137.38 – -137.44 (m, 1F, F_1), -142.58 – -142.66 (m, 1F, F_2), -145.24 – -145.32 (m, 1F, F_3), -152.20 – -152.28 (m, 1F, F_4). – **IR** (ATR): $\tilde{\nu}$ (cm^{-1}) = 3589.5 (vw), 1729.4 (vw), 1630.5 (w), 1539.9 (w), 1516.2 (w), 1475.5 (w), 1399.9 (m), 1350.6 (m), 1283.4 (w), 1126.0 (w), 1076.7 (w), 955.3 (w), 927.7 (vw), 836.8 (w), 769.1 (w), 745.1 (m), 702.5 (w), 640.3 (vw), 536.7 (w), 485.1 (w), 447.7 (vw), 386.2 (w). – **EA**: calcd for $Lu(L_3)_3(H_2O)_2$,% C 27.26, H 0.44, N 4.54, found C 28.51, H 0.25, N 4.52.

Lanthanide 2,3,6-trifluorobenzoates monohydrates $Ln(L_4)_3(H_2O)$ (Ln-1.4)

According to **GP1b** an excess of concentrated aq. solution of NH_3 (1.30 mL, 20.0 equiv.) was added to an aq. solution of $LnCl_3 \cdot x\ H_2O$ (1.00 mmol, 1.00 equiv.) in 25 mL water. The mixture was stirred for 30 min, the precipitate of $Ln(OH)_3$ was centrifuged and washed with water until the pH of the washing solution became neutral. A small excess of freshly prepared $Ln(OH)_3$ was placed into a beaker and a solution of 475 mg of 2,3,6-trifluorobenzoic acid HL_4 (2.70 mmol, 2.70 equiv.) in 20 mL acetone/methanol (3/1) was added. The reaction mixture was stirred for 1 h at 60 °C and the unreacted components were separated by filtration followed by the evaporation of the clear solution to dryness. The obtained solid powder of the lanthanide complexes was then recrystallized from water. Subsequent drying at 80 °C in vacuo for 1 h led to the formation of a microcrystalline off-white powder of general formula $Ln(L_4)_3(H_2O)$.

lanthanum 2,3,6-trifluorobenzoate monohydrate $La(L_4)_3(H_2O)$ (La-1.4)

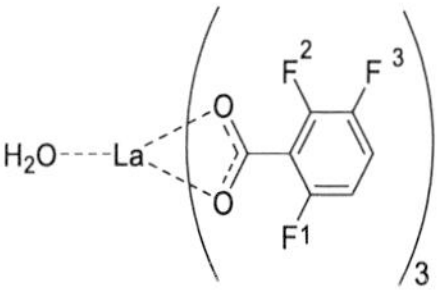

For the synthesis 373 mg of $LaCl_3 \cdot 7\ H_2O$ were used. 513 mg (88%) of a pale-yellow solid of general formula $La(L_4)_3(H_2O)$ were obtained.

^{1}H NMR (400 MHz, DMSO-d_6): δ (ppm) = 7.43 – 7.25 (m, 1H, H_{Ar}), 7.08 – 6.96 (m, 1H, H_{Ar}). – **^{19}F NMR** (377 MHz, DMSO-d_6): δ (ppm) = -118.26 (bs, 1F, F_1), -137.97 (bs, 1F, F_2), -143.38 – -143.50 (m, 1F, F_3). – **IR** (ATR): $\tilde{\nu}$ (cm^{-1}) = 3612.4 (vw), 3433.1 (vw), 1637.2 (w), 1574.7 (m), 1485.7 (w), 1458.3 (m), 1406.9 (w), 1379.1 (m), 1297.4 (w), 1241.2 (w), 1194.5 (w), 1020.9 (vw), 947.3 (m), 811.5 (w), 780.1 (w), 733.2 (w), 709.8 (w), 628.1 (w), 602.5 (vw), 578.3 (w), 513.8 (w), 456.9 (vw), 414.7 (vw). – **EA**: calcd for $La(L_4)_3(H_2O)$,% C 36.97, H 1.18, found C 35.88, H 1.15.

neodymium 2,3,6-trifluorobenzoate monohydrate $Nd(L_4)_3(H_2O)$ (Nd-1.4)

For the synthesis 359 mg of $NdCl_3 \cdot 6\ H_2O$ were used. 540 mg (87%) of a pale-yellow solid of general formula $Nd(L_4)_3(H_2O)$ were obtained.

^{1}H NMR (400 MHz, DMSO-d_6): δ (ppm) = 7.48 – 7.30 (m, 1H, H_{Ar}), 7.24 – 7.10 (m, 1H, H_{Ar}). – **^{19}F NMR** (377 MHz, DMSO-d_6): δ (ppm) = -118.23 (bs, 1F, F_1), -138.10 – -137.76 (m, 1F, F_2), -142.68 – -143.00 (m, 1F, F_3). – **IR** (ATR): $\tilde{\nu}$ (cm^{-1}) = 3608.7 (vw), 3436.5 (vw), 3087.8 (vw), 1697.3 (vw), 1629.6 (w), 1579.8 (m), 1485.1 (m), 1457.8 (w), 1401.0 (m), 1379.6 (m), 1316.4 (w), 1298.5 (w), 1241.5 (w), 1194.7 (w), 1139.1 (vw), 1020.0 (m), 947.3 (w), 816.5 (m), 780.3 (m), 733.0 (vw), 709.2 (w), 624.7 (w), 602.6 (w), 578.2 (w), 513.6 (w), 456.1 (vw), 414.2 (w). – **EA**: calcd for $Nd(L_4)_3(H_2O)$,% C 36.69, H 1.17, found C 36.30, H 1.18.

europium 2,3,6-trifluorobenzoate monohydrate $Eu(L_4)_3(H_2O)$ (Eu-1.4)

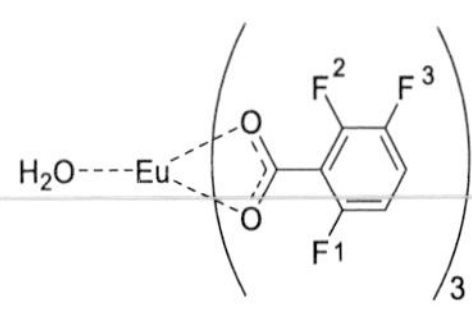

For the synthesis 366 mg of $EuCl_3 \cdot 6\ H_2O$ were used. 543 mg (87%) of a pale-yellow solid of general formula $Eu(L_4)_3(H_2O)$ were obtained.

^{1}H NMR (400 MHz, DMSO-d_6): δ (ppm) = 7.13 – 7.32 (m, 1H, H_{Ar}), 7.90 – 7.67 (m, 1H, H_{Ar}). – **^{19}F NMR** (377 MHz, DMSO-d_6): δ (ppm) = -118.01 (bs, 1F, F_1), -138.05 – -137.72 (m, 1F, F_2), -143.93 – -143.67 (m, 1F, F_3). – **IR** (ATR): $\tilde{\nu}$ (cm^{-1}) = 3117.9 (w), 3031.9 (w), 1632.3 (w), 1581.7 (w), 1485.6 (w), 1458.6 (w), 1399.9 (m), 1240.4 (w), 1019.5 (w), 948.1 (w), 816.0 (m), 779.6 (w), 709.3 (w), 624.4 (w), 602.7 (w), 577.3 (w), 514.8 (w), 414.6 (w). – **EA**: calcd for $Eu(L_4)_3(H_2O)$,% C 36.28, H 1.16, found C 37.01, H 1.20.

gadolinium 2,3,6-trifluorobenzoate monohydrate $Gd(L_4)_3(H_2O)$ (Gd-1.4)

For the synthesis 372 mg of $GdCl_3 \cdot 6\ H_2O$ were used. 538 mg (85%) of a pale-yellow solid of general formula $Gd(L_4)_3(H_2O)$ were obtained.

^{1}H NMR (400 MHz, DMSO-d_6): δ (ppm) = 7.68 (bs, 1H, H_{Ar}), 7.43 (bs, 1H, H_{Ar}). – **IR** (ATR): $\tilde{\nu}$ (cm^{-1}) = 3603.1 (vw), 3444.3 (vw), 1632.7 (w), 1581.0 (w), 1485.6 (w), 1459.3 (w), 1403.4 (w), 1382.0 (w), 1299.0 (vw), 1240.3 (w), 1194.6 (vw), 1019.5 (w), 948.4 (w), 815.8 (w), 779.9 (w), 733.5 (vw), 709.7 (w), 624.4 (vw), 602.8 (w), 577.5 (w), 515.6 (vw), 453.6 (vw), 414.4 (w). – **EA**: calcd for $Gd(L_4)_3(H_2O)$,% C 36.01, H 1.15, found C 36.30, H 1.18.

terbium 2,3,6-trifluorobenzoate monohydrate $Tb(L_4)_3(H_2O)$ (Tb-1.4)

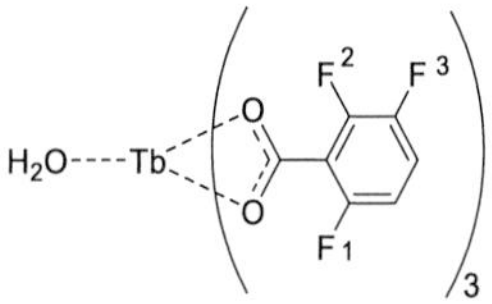

For the synthesis 373 mg of $TbCl_3 \cdot 6\ H_2O$ were used. 542 mg (86%) of a pale-yellow solid of general formula $Tb(L_4)_3(H_2O)$ were obtained.

^{1}H NMR (400 MHz, DMSO-d_6): δ (ppm) = 7.99 (bs, 1H, H_{Ar}), 7.57 (bs, 1H, H_{Ar}). – **^{19}F NMR** (377 MHz, DMSO-d_6): δ (ppm) = -114.40 (bs, 1F, F_1), -137.40 (bs, 1F, F_2), -143.64 (bs, 1F, F_3). – **IR** (ATR): $\tilde{\nu}$ (cm^{-1}) = 3603.7 (vw), 3433.3 (vw), 1632.9 (w), 1581.6 (w), 1485.0 (w), 1459.6 (w), 1403.2 (w), 1381.6 (w), 1299.1 (w), 1239.8 (w), 1194.2 (w), 1019.4 (w), 948.2 (w), 815.5 (w), 779.6 (w), 734.0 (vw), 223.5 (w), 602.8 (w), 577.4 (w), 516.9 (w), 413.9 (vw). – **EA**: calcd for $Tb(L_4)_3(H_2O)$,% C 35.92, H 1.15, found C 35.02, H 1.40.

terbium/europium 2,3,6-trifluorobenzoate monohydrate $Tb_{0.5}Eu_{0.5}(L_4)_3(H_2O)$ (TbEu-1.4)

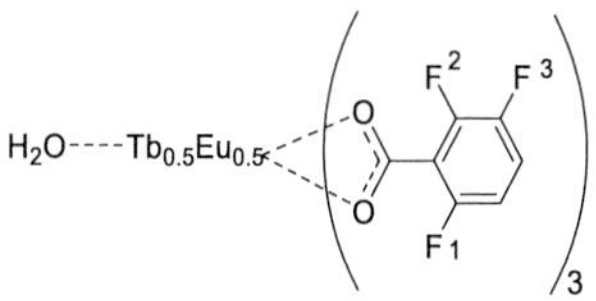

For the synthesis 187 mg of $TbCl_3 \cdot 6\ H_2O$ (0.50 mmol, 0.50 equiv.) and 183 mg $EuCl_3 \cdot 6\ H_2O$ (0.50 mmol, 0.50 equiv.) were used. 532 mg (85%) of a pale-yellow solid of general formula $Tb_{0.5}Eu_{0.5}(L_4)_3(H_2O)$ were obtained.

^{1}H NMR (400 MHz, DMSO-d_6): δ (ppm) = 7.39 – 7.23 (m, 1H, H_{Ar}), 7.10 – 6.95 (m, 1H, H_{Ar}). – **^{19}F NMR** (377 MHz, DMSO-d_6): δ (ppm) = -116.18 (bs, 1F, F_1), -137.72 (bs, 1F, F_2), -143.78 (bs, 1F, F_3). – **IR** (ATR): $\tilde{\nu}$ (cm^{-1}) = 3077.2 (vw), 1624.5 (w), 1547.2 (w), 1512.6 (w), 1436.8 (w), 1383.4 (w), 1327.8 (w), 1293.3 (w), 1199.7 (w), 1147.5 (w), 1078.3 (vw), 901.6 (vw), 853.1 (w), 786.4 (w), 731.5 (w), 629.7 (w), 602.9 (w), 452.2 (vw), 392.2 (vw). – **EA**: calcd for $Tb_{0.5}Eu_{0.5}(L_4)_3(H_2O)$,% C 36.12, H 1.15, found C 35.84, H 1.28.

erbium 2,3,6-trifluorobenzoate monohydrate $Er(L_4)_3(H_2O)$ (Er-1.4)

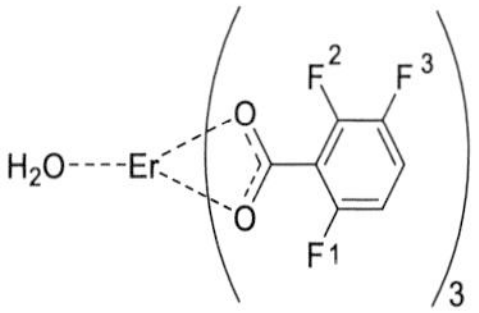

For the synthesis 381 mg of $ErCl_3 \cdot 6\ H_2O$ were used. 544 mg (85%) of a pale-yellow solid of general formula $Er(L_4)_3(H_2O)$ were obtained.

^{1}H NMR (400 MHz, DMSO-d_6): δ (ppm) = 7.20 (bs, 1H, H_{Ar}), 6.88 (bs, 1H, H_{Ar}). – **^{19}F NMR** (377 MHz, DMSO-d_6): δ (ppm) = -116.79 (bs, 1F, F_1), -136.55 (bs, 1F, F_2), -142.69 (bs, 1F, F_3). – **IR** (ATR): $\tilde{\nu}$ (cm^{-1}) = 3086.6 (vw), 1601.5 (w), 1557.7 (m), 1486.1 (w), 1430.8 (w), 1390.3 (w), 1298.1 (w), 1252.6 (m), 1188.6 (vw), 1120.0 (w), 1080.0 (w), 944.5 (w), 899.6 (w), 819.8 (m), 762.2 (w), 670.8 (w), 574.0 (w), 542.2 (vw), 467.3 (vw), 411.2 (w).– **EA**: calcd for $Er(L_4)_3(H_2O)$,% C 35.50, H 1.13, found C 35.98, H 1.10.

ytterbium 2,3,6-trifluorobenzoate monohydrate $Yb(L_4)_3(H_2O)$ (Yb-1.4)

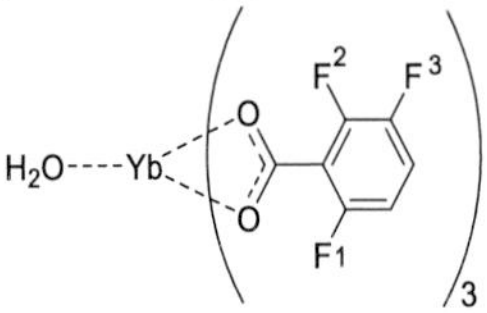

For the synthesis 388 mg of $YbCl_3 \cdot 6\ H_2O$ were used. 561 mg (87%) of a pale-yellow solid of general formula $Yb(L_4)_3(H_2O)$ were obtained.

^{1}H NMR (400 MHz, DMSO-d_6): δ (ppm) = 6.97 – 6.79 (m, 1H, H_{Ar}), 6.57 – 6.39 (m, 1H, H_{Ar}). – **^{19}F NMR** (377 MHz, DMSO-d_6): δ (ppm) = -118.65 (bs, 1F, F_1), -137.93 (bs, 1F, F_2), -145.02 (bs, 1F, F_3). – **IR** (ATR): $\tilde{\nu}$ (cm^{-1}) = 3603.2 (vw), 3086.6 (vw), 1634.5 (w), 1575.7 (m), 1473.1 (w), 1420.8 (w), 1388.3 (m), 1298.7 (w), 1244.6 (w), 1198.6 (vw), 1120.4 (w), 1080.0 (w), 948.5 (m), 819.5 (w), 762.3 (vw), 670.1 (w), 584.3 (w), 542.6 (w), 466.7 (vw), 411.3 (vw).– **EA**: calcd for $Yb(L_4)_3(H_2O)$,% C 35.21, H 1.13, found C 35.87, H 1.21.

lutetium 2,3,6-trifluorobenzoate monohydrate $Lu(L_4)_3(H_2O)$ (Lu-1.4)

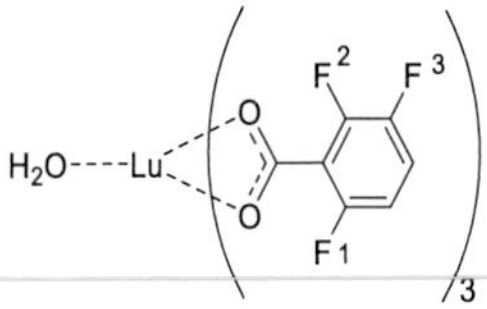

For the synthesis 389 mg of $LuCl_3 \cdot 6\ H_2O$ were used. 569 mg (88%) of a pale-yellow solid of general formula $Lu(L_4)_3(H_2O)$ were obtained.

^{1}H NMR (400 MHz, DMSO-d_6): δ (ppm) = 7.41 – 7.25 (m, 1H, H_{Ar}), 7.06 – 6.91 (m, 1H, H_{Ar}). – **^{19}F NMR** (377 MHz, DMSO-d_6): δ (ppm) = -118.28 (bs, 1F, F_1), -137.98 (bs, 1F, F_2), -143.37 (bs, 1F, F_3). – **IR** (ATR): $\tilde{\nu}$ (cm^{-1}) = 3611.1 (vw), 3044.6 (vw), 1638.1 (w), 1576.4 (m), 1478.1 (w), 1429.8 (vw), 1387.4 (vw), 1292.5 (w), 1245.8 (w), 1194.9 (w), 1123.9 (w), 1081.6 (vw), 944.8 (m), 814.9 (w), 762.6 (w), 673.5 (w), 587.1 (vw), 548.4 (w), 468.2 (w), 412.5 (vw).– **EA**: calcd for $Lu(L_4)_3(H_2O)$,% C 35.12, H 1.12, found C 35.54, H 1.18.

Lanthanide 2,4,5-trifluorobenzoates dihydrates $Ln(L_5)_3(H_2O)_2$ (Ln-1.5)

According to **GP1a**, a suspension of 528 mg of 2,4,5-trifluorobenzoic acid HL_5 (3.00 mmol, 3.00 equiv.) in water was treated with 1 M aq. solution of NaOH at 60 °C until the pH was slightly acidic (pH = 5 – 6, approx. 3 mL) and the solution became clear. Then an aq. solution of $LnCl_3 \cdot x\ H_2O$ (1.00 mmol, 1.00 equiv.) was added following by an immediate precipitation of the product. The reaction mixture was stirred for additional 3 h at rt, the crude product was filtered, washed with ethanol and dried in vacuo to afford the product as an off-white solid.

lanthanum 2,4,5-trifluorobenzoate dihydrate $La(L_5)_3(H_2O)_2$ (La-1.5)

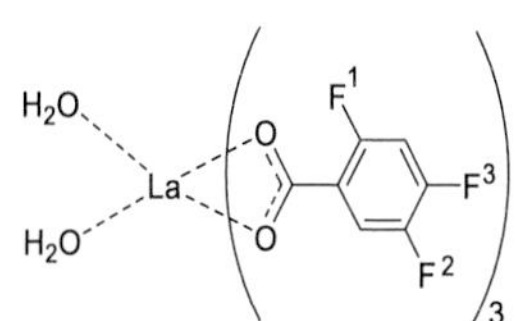

For the synthesis 373 mg of $LaCl_3 \cdot 7\ H_2O$ were used. 632 mg (90%) of an off-white solid of general formula $La(L_5)_3(H_2O)_2$ were obtained.

^{1}H NMR (400 MHz, DMSO-d_6): δ (ppm) = 7.89 – 7.78 (m, 1H, H_{Ar}), 7.59 – 7.45 (m, 1H, H_{Ar}). – **^{19}F NMR** (377 MHz, DMSO-d_6): δ (ppm) = -111.03 (bs, 1F, F_1), -128.70 (bs, 1F, F_2), -142.76 (bs, 1F, F_3). – **IR** (ATR): $\tilde{\nu}$ (cm^{-1}) = 3365.1 (vw), 3073.1 (vw), 1632.3 (vw), 1544.7 (m), 1435.3 (w), 1389.0 (w), 1297.4 (m), 1196.1 (vw), 1080.6 (w), 917.3 (w), 811.2 (w), 790.4 (m), 731.5 (w), 638.2 (vw), 604.7 (w), 476.9 (vw), 454.7 (vw), 401.6 (vw). – **EA**: calcd for $La(L_5)_3(H_2O)_2$,% C 36.02, H 1.44, found C 35.93, H 1.58.

neodymium 2,4,5-trifluorobenzoate dihydrate $Nd(L_5)_3(H_2O)_2$ (Nd-1.5)

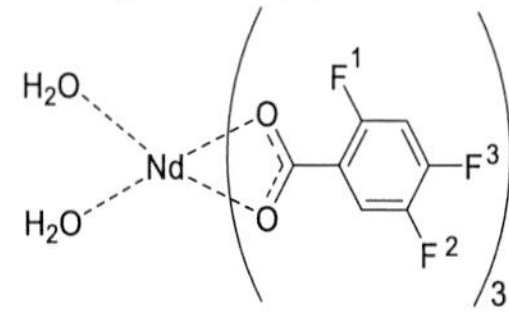

For the synthesis 359 mg of $NdCl_3 \cdot 6\ H_2O$ were used. 615 mg (87%) of an off-white solid of general formula $Nd(L_5)_3(H_2O)_2$ were obtained.

^{19}F NMR (377 MHz, DMSO-d_6): δ (ppm) = -112.71 (bs, 1F, F_1), -130.79 (bs, 1F, F_2), -143.74 (bs, 1F, F_3). – **IR** (ATR): $\tilde{\nu}$ (cm^{-1}) = 3537.2 (vw), 3083.0 (vw), 1624.8 (w), 1560.5 (m), 1512.9 (m), 1438.8 (m), 1378.7 (m), 1328.2 (w), 1293.1 (w), 1200.0 (m), 1147.7 (w), 1076.3 (vw), 901.8 (w), 853.7 (w), 800.6 (vw), 778.2 (w), 731.6 (w), 626.2 (w), 600.9 (w), 477.6 (vw), 451.2 (w). – **EA**: calcd for $Nd(L_5)_3(H_2O)_2$,% C 35.75, H 1.43, found C 35.51, H 1.53.

europium 2,4,5-trifluorobenzoate dihydrate $Eu(L_5)_3(H_2O)_2$ (Eu-1.5)

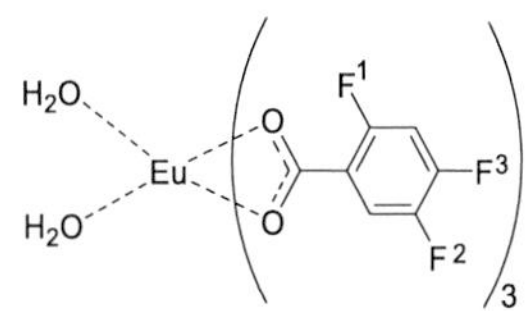

For the synthesis 366 mg of $EuCl_3 \cdot 6\ H_2O$ were used. 651 mg (91%) of an off-white solid of general formula $Eu(L_5)_3(H_2O)_2$ were obtained.

^{1}H NMR (400 MHz, DMSO-d_6): δ (ppm) = 7.10 (bs, 1H, H_{Ar}), 6.81 (bs, 1H, H_{Ar}). – **^{19}F NMR** (377 MHz, DMSO-d_6): δ (ppm) = -112.13 (bs, 1F, F_1), -130.72 (bs, 1F, F_2), -144.30 (bs, 1F, F_3). – **IR** (ATR): $\tilde{\nu}$ (cm^{-1}) = 3344.5 (vw), 3075.1 (vw), 1624.8. (w), 1611.6 (w), 1543.3 (m), 1432.9 (m), 1384.9 (m), 1322.6 (w), 1204.0 (w), 1197.3 (m), 1148.5 (w), 1078.8 (w), 901.1 (w), 871.2 (w), 853.3 (w), 811.7 (w), 790.1 (m), 730.8 (w), 633.2 (w), 605.0 (m), 475.7 (vw), 442.5 (w), 386.7 (w). – **EA**: calcd for $Eu(L_5)_3(H_2O)_2$,% C 35.36, H 1.41, found C 35.01, H 1.44.

gadolinium 2,4,5-trifluorobenzoate dihydrate $Gd(L_5)_3(H_2O)_2$ (Gd-1.5)

For the synthesis 372 mg of $GdCl_3 \cdot 6\ H_2O$ were used. 660 mg (92%) of an off-white solid of general formula $Gd(L_5)_3(H_2O)_2$ were obtained.

^{19}F NMR (377 MHz, DMSO-d_6): δ (ppm) = -90.13 (bs, 1F, F_1), -130.76 (bs, 1F, F_2), -145.57 (bs, 1F, F_3). – **IR** (ATR): $\tilde{\nu}$ (cm^{-1}) = 3554.4 (vw), 1615.1 (w), 1540.1 (m), 1434.6 (m), 1391.8 (m), 1326.7 (w), 1294.5 (w), 1199.0 (m), 1147.0 (w), 1083.8 (vw), 914.4 (vw), 896.1 (vw), 854.4 (m), 802.4 (w), 784.7 (m), 733.7 (w), 687.3 (vw), 628.0 (vw), 605.0 (w), 480.5 (w), 452.1 (w), 381.8 (vw). – **EA**: calcd for $Gd(L_5)_3(H_2O)_2$,% C 35.10, H 1.40, found C 35.93, H 1.58.

terbium 2,4,5-trifluorobenzoate dihydrate $Tb(L_5)_3(H_2O)_2$ (Tb-1.5)

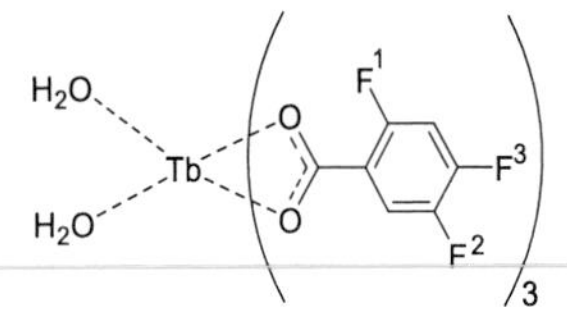

For the synthesis 373 mg of $TbCl_3 \cdot 6\ H_2O$ were used. 664 mg (92%) of an off-white solid of general formula $Tb(L_5)_3(H_2O)_2$ were obtained.

^{19}F NMR (377 MHz, DMSO-d_6): δ (ppm) = -95.17 (bs, 1F, F_1), -126.73 (bs, 1F, F_2), -142.58 (bs, 1F, F_3). – **IR** (ATR): $\tilde{\nu}$ (cm^{-1}) = 3351.2 (w), 3075.9 (w), 1625.3 (w), 1611.7 (w), 1545.1 (s), 1514.9 (m), 1435.6 (m), 1386.3 (m), 1322.9 (w), 1294.5 (m), 1197.7 (m), 1148.7 (m), 1079.7 (w), 920.4 (vw), 900.7 (w), 873.0 (w), 853.2 (m), 812.5 (m), 790.7 (s), 731.4 (m), 633.4 (m), 605.3 (m), 477.0 (w), 443.0 (w), 388.4 (w). – **EA**: calcd for $Tb(L_5)_3(H_2O)_2$,% C 35.88, H 1.63, found C 35.93, H 1.58.

terbium/europium 2,4,5-trifluorobenzoate dihydrate $Tb_{0.5}Eu_{0.5}(L_5)_3(H_2O)_2$ (TbEu-1.5)

For the synthesis 187 mg of $TbCl_3 \cdot 6\ H_2O$ (0.50 mmol, 0.50 equiv.) and 183 mg $EuCl_3 \cdot 6\ H_2O$ (0.50 mmol, 0.50 equiv.) were used. 637 mg (85%) of an off-white solid of general formula $Tb_{0.5}Eu_{0.5}(L_5)_3(H_2O)_2$ were obtained.

^{1}H NMR (400 MHz, DMSO-d_6): δ (ppm) = 7.90 – 7.76 (m, 1H, H_{Ar}), 7.59 – 5.43 (m, 1H, H_{Ar}). – **^{19}F NMR** (377 MHz, DMSO-d_6): δ (ppm) = -111.03 (bs, 1F, F_1), -128.70 (bs, 1F, F_2), -142.76 (bs, 1F, F_3). – **IR** (ATR): $\tilde{\nu}$ (cm^{-1}) = 3077.2 (vw), 1624.5 (w), 1547.2 (w), 1512.6 (w), 1436.8 (w), 1383.4 (w), 1327.8 (w), 1293.3 (w), 1199.7 (w), 1147.5 (w), 1078.3 (vw), 901.6 (vw), 853.1 (w), 786.4 (w), 731.5 (w), 629.7 (w), 602.9 (w), 452.2 (vw), 392.2 (vw). – **EA**: calcd for $Tb_{0.5}Eu_{0.5}(L_5)_3(H_2O)_2$,% C 35.17, H 1.40, found C 35.40, H 1.38.

erbium 2,4,5-trifluorobenzoate dihydrate Er(L5)3(H2O)2 (Er-1.5)

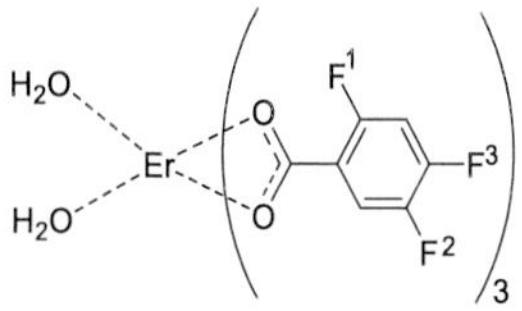

For the synthesis 381 mg of $ErCl_3·6\ H_2O$ were used. 659 mg (89%) of an off-white solid of general formula $Er(L_5)_3(H_2O)_2$ were obtained.

^{1}H NMR (400 MHz, DMSO-d_6): δ (ppm) = 5.74 (bs, 2H, 2 × H_{Ar}). – **^{19}F NMR** (377 MHz, DMSO-d_6): δ (ppm) = -117.99 (bs, 1F, F_1), -132.05 (bs, 1F, F_2), -145.06 (bs, 1F, F_3). – **IR** (ATR): $\tilde{\nu}$ (cm^{-1}) = 3076.5 (w), 1612.6 (w), 1549.7 (m), 1513.4 (m), 1437.2 (m), 1381.5 (m), 1328.5 (w), 1293.5 (w), 1200.2 (m), 1148.1 (m), 1078.7 (vw), 902.3 (w), 871.9 (w), 853.3 (m), 813.2 (w), 789.4 (m), 731.4 (w), 632.7 (w), 603.0 (m), 476.5 (vw), 453.9 (w), 390.8 (w). – **EA**: calcd for $Er(L_5)_3(H_2O)_2$,% C 34.62, H 1.38, found C 35.01, H 1.44.

ytterbium 2,4,5-trifluorobenzoate dihydrate Yb(L5)3(H2O)2 (Yb-1.5)

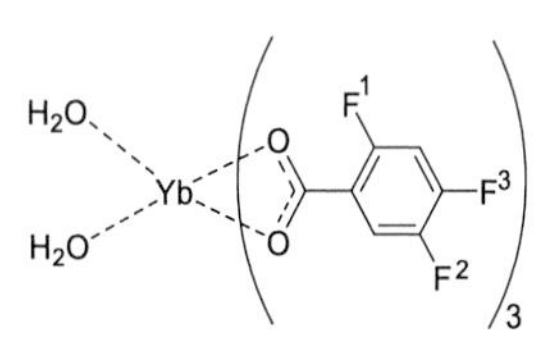

For the synthesis 388 mg of $YbCl_3·6\ H_2O$ were used. 648 mg (88%) of an off-white solid of general formula $Yb(L_5)_3(H_2O)_2$ were obtained.

^{1}H NMR (400 MHz, DMSO-d_6): δ (ppm) = 6.17 (bs, 2H, 2 × H_{Ar}). – **^{19}F NMR** (377 MHz, DMSO-d_6): δ (ppm) = -116.96 (bs, 1F, F_1), -132.07 (bs, 1F, F_2), -144.31 (bs, 1F, F_3). – **IR** (ATR): $\tilde{\nu}$ (cm^{-1}) = 3082.9 (vw), 1614.8 (m), 1550.3 (m), 1514.8 (w), 1430.6 (w), 1381.9 (m), 1329.5 (w), 1293.0 (w), 1202.7 (w), 1147.3 (w), 1083.6 (vw), 904.5 (w), 852.4 (w), 814.2 (w), 801.8 (w), 788.1 (m), 734.0 (w), 664.4 (vw), 634.1 (w), 605.3 (w), 465.0 (w), 405.5 (w). – **EA**: calcd for $Yb(L_5)_3(H_2O)_2$,% C 34.35, H 1.37, found C 34.44, H 1.51.

lutetium 2,4,5-trifluorobenzoate dihydrate Lu(L5)3(H2O)2 (Lu-1.5)

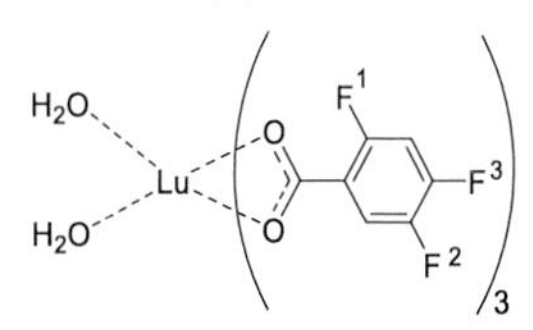

For the synthesis 389 mg of $LuCl_3·6\ H_2O$ were used. 622 mg (84%) of an off-white solid of general formula $Lu(L_5)_3(H_2O)_2$ were obtained.

^{1}H NMR (400 MHz, DMSO-d_6): δ (ppm) = 7.69 (bs, 1H, H_{Ar}), 7.41 (bs, 1H, H_{Ar}). – **^{19}F NMR** (377 MHz, DMSO-d_6): δ (ppm) = -112.68 (bs, 1F, F_1), -131.61 (bs, 1F, F_2), -143.99 (bs, 1F, F_3). – **IR** (ATR): $\tilde{\nu}$ (cm^{-1}) = 3374.9 (vw), 3077.5 (vw), 1616.0 (w), 1553.0 (m), 1515.9 (w), 1433.7 (m), 1385.8 (m), 1329.1 (w), 1293.5 (w), 1201.6 (m), 1146.9 (w), 1082.5 (vw), 907.3 (w), 852.0 (w), 800.2 (w), 784.0 (m), 733.4 (w), 662.6 (vw), 635.0 (w), 606.7 (w), 490.7 (vw), 462.9 (w), 389.6 (w). – **EA**: calcd for $Lu(L_5)_3(H_2O)_2$,% C 34.26, H 1.37, found C 34.67, H 1.43.

Lanthanide 2,4-difluorobenzoates dihydrates $Ln(L_6)_3(H_2O)_2$ (Ln-1.6)

According to **GP1a**, a suspension of 474 mg of 2,4-difluorobenzoic acid HL_6 (3.00 mmol, 3.00 equiv.) in water was treated with 1 M aq. solution of NaOH at 60 °C until the pH was slightly acidic (pH = 5–6, approx. 3 mL) and the solution became clear. Then an aq. solution of $LnCl_3 \cdot x\ H_2O$ (1.00 mmol, 1.00 equiv.) was added following by an immediate precipitation of the product. The reaction mixture was stirred for additional 3 h at rt, the crude product was filtered, washed with ethanol and dried in vacuo to afford the product as an off-white solid.

lanthanum 2,4-difluorobenzoate dihydrate $La(L_6)_3(H_2O)_2$ (La-1.6)

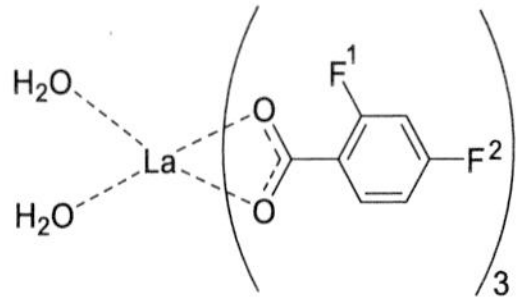

For the synthesis 373 mg of $LaCl_3 \cdot 7\ H_2O$ were used. 542 mg (84%) of an off-white solid of general formula $La(L_6)_3(H_2O)_2$ were obtained.

^{1}H NMR (400 MHz, DMSO-d_6): δ (ppm) = 7.89 – 8.78 (m, 1H, H_{Ar}), 7.22 – 7.10 (m, 1H, H_{Ar}), 7.08 – 6.98 (m, 1H, H_{Ar}). – **^{19}F NMR** (377 MHz, DMSO-d_6): δ (ppm) = -107.59 (bs, 1F, F_2), -107.97 (bs, 1F, F_1). – **IR** (ATR): $\tilde{\nu}$ (cm^{-1}) = 3482.4 (vw), 1601.9 (m), 1528.5 (m), 1428.7 (w), 1389.9 (m), 1266.1 (m), 1137.8 (w), 1091.3 (w), 971.9 (w), 847.3 (m), 782.4 (m), 734.3 (w), 683.0 (w), 610.1 (w), 596.8 (w), 436.9 (w). – **EA**: calcd for $La(L_6)_3(H_2O)_2$,% C 39.03, H 2.03, found C 38.29, H 2.15.

neodymium 2,4-difluorobenzoate dihydrate $Nd(L_6)_3(H_2O)_2$ (Nd-1.6)

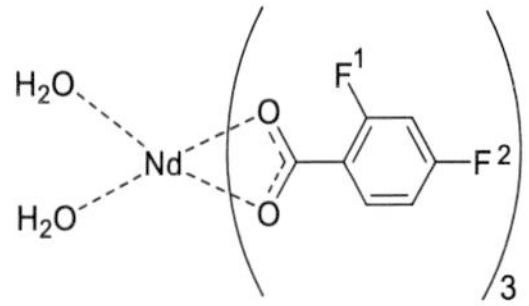

For the synthesis 359 mg of $NdCl_3 \cdot 6\ H_2O$ were used. 561 mg (86%) of an off-white solid of general formula $Nd(L_6)_3(H_2O)_2$ were obtained.

^{1}H NMR (400 MHz, DMSO-d_6): δ (ppm) = 7.62 (bs, 3H, 3 × H_{Ar}). – **^{19}F NMR** (377 MHz, DMSO-d_6): δ (ppm) = -107.27 (bs, 2F, $F_{1,2}$). – **IR** (ATR): $\tilde{\nu}$ (cm^{-1}) = 3543.8 (vw), 1613.3 (m), 1530.5 (m), 1503.8 (m), 1428.6 (m), 1394.5 (m), 1267.2 (m), 1240.8 (w), 1139.5 (m), 1091.4 (m), 974.2 (m), 846.6 (m), 782.2 (m), 734.4 (w), 683.2 (w), 599.1 (m), 574.3 (w), 519.4 (w), 440.9 (m), 383.3 (vw). – **EA**: calcd for $Nd(L_6)_3(H_2O)_2$,% C 38.71, H 2.01, found C 39.03, H 2.25.

europium 2,4-difluorobenzoate dihydrate $Eu(L_6)_3(H_2O)_2$ (Eu-1.6)

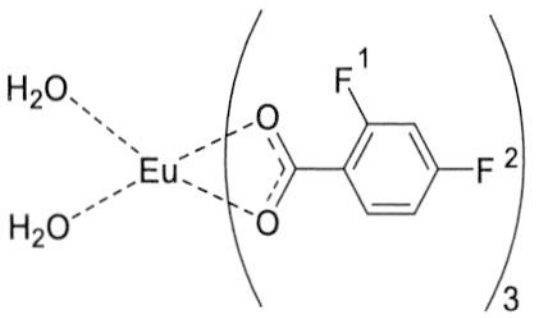

For the synthesis 366 mg of $EuCl_3 \cdot 6\,H_2O$ were used. 587 mg (89%) of an off-white solid of general formula $Eu(L_6)_3(H_2O)_2$ were obtained.

^{1}H NMR (400 MHz, DMSO-d_6): δ (ppm) = 6.57 (m, 3H, 3 × HAr). – **^{13}C NMR** (101 MHz, DMSO-d_6): δ (ppm) = 162.41 ($C_{quart.}$, 3 × *C*OO), 161.66 ($C_{quart.}$, 3 × C_{Ar}F), 133.50 ($C_{quart.}$, 3 × C_{Ar}F), 110.00 ($C_{quart.}$, 3 × C_{Ar}), 109.78 (+, 3 × C_{Ar}H), 102.39 (+, 3 × C_{Ar}H), 101.88 (+, 3 × C_{Ar}H). – **^{19}F NMR** (377 MHz, DMSO-d_6): δ (ppm) = -107.02 (bs, 2F, $F_{1,2}$). – **IR** (ATR): $\tilde{\nu}$ (cm^{-1}) = 3085.7 (vw), 1601.4 (w), 1534.6 (w), 1501.0 (w), 1425.9 (w), 1388.1 (w), 1268.9 (w), 1137.9 (w), 1092.4 (w), 971.3 (w), 849.1 (w), 780.0 (w), 735.3 (w), 684.1 (w), 612.2 (w), 519.9 (vw), 444.8 (vw), 395.9 (w). – **EA**: calcd for $Eu(L_6)_3(H_2O)_2$,% C 38.26, H 1.99, found C 38.09, H 2.31.

gadolinium 2,4-difluorobenzoate dihydrate $Gd(L_6)_3(H_2O)_2$ (Gd-1.6)

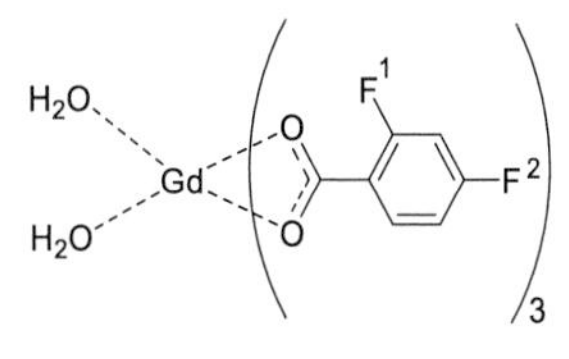

For the synthesis 372 mg of $GdCl_3 \cdot 6\,H_2O$ were used. 582 mg (86%) of an off-white solid of general formula $Gd(L_6)_3(H_2O)_2$ were obtained.

^{19}F NMR (377 MHz, DMSO-d_6): δ (ppm) = -106.69 (bs, 2F, $F_{1,2}$). – **IR** (ATR): $\tilde{\nu}$ (cm^{-1}) = 3233.6 (w), 1600.6 (m), 1536.5 (m), 1500.7 (w), 1423.3 (m), 1395.0 (m), 1268.0 (m), 1239.1 (w), 1137.4 (w), 1092.0 (m), 969.7 (m), 847.2 (m), 776.1 (m), 731.2 (w), 680.8 (w), 618.6 (m), 604.7 (m), 590.9 (m), 517.8 (w), 439.9 (w), 391.7 (w). – **EA**: calcd for $Gd(L_6)_3(H_2O)_2$,% C 37.95, H 1.97, found C 38.01, H 2.14.

terbium 2,4-difluorobenzoate dihydrate $Tb(L_6)_3(H_2O)_2$ (Tb-1.6)

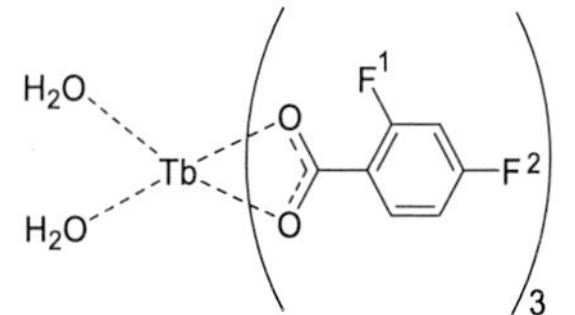

For the synthesis 373 mg of $TbCl_3 \cdot 6\,H_2O$ were used. 560 mg (84%) of an off-white solid of general formula $Tb(L_6)_3(H_2O)_2$ were obtained.

^{19}F NMR (377 MHz, DMSO-d_6): δ (ppm) = -94.32 (bs, 1F, F_1), -103.86 (bs, 1F, F_2). – **IR** (ATR): $\tilde{\nu}$ (cm^{-1}) = 3631.3 (vw), 3331.0 (w), 1649.1 (w), 1589.6 (m), 1521.0 (w), 1488.1 (m), 1426.0 (m), 1379.1 (m), 1292.6 (w), 1114.4 (m), 990.1 (m), 943.9 (w), 934.0 (w), 827.4 (w), 771.2 (m), 743.8 (m), 704.0 (w), 623.6 (vw), 582.9 (vw), 506.7 (w), 474.6 (vw), 381.7 (w). – **EA**: calcd for $Tb(L_6)_3(H_2O)_2$,% C 37.86, H 1.97, found C 38.15, H 2.30.

terbium/europium 2,4-difluorobenzoate dihydrate $Tb_{0.5}Eu_{0.5}(L_6)_3(H_2O)_2$ (TbEu-1.6)

For the synthesis 187 mg of $TbCl_3 \cdot 6\,H_2O$ (0.50 mmol, 0.50 equiv.) and 183 mg $EuCl_3 \cdot 6\,H_2O$ (0.50 mmol, 0.50 equiv.) were used. 563 mg (85%) of an off-white solid of general formula $Tb_{0.5}Eu_{0.5}(L_6)_3(H_2O)_2$ were obtained.

^{19}F NMR (377 MHz, DMSO-d_6): δ (ppm) = -96.88 (bs, 1F, F_1), -105.12 (bs, 1F, F_2). – **IR** (ATR): $\tilde{\nu}$ (cm^{-1}) = 3084.2 (vw), 1600.8 (m), 1539.3 (w), 1502.3 (w), 1430.2 (w), 1392.3 (m), 1268.2 (m), 1236.7 (w), 1137.9 (w), 1093.1 (w), 971.6 (w), 848.8 (m), 779.7 (m), 734.9 (w), 683.3 (w), 609.0 (m), 520.0 (vw), 456.7 (w), 393.2 (w). – **EA**: calcd for $Tb_{0.5}Eu_{0.5}(L_6)_3(H_2O)_2$,% C 38.06, H 1.98, found C 38.38, H 1.87.

erbium 2,4-difluorobenzoate dihydrate $Er(L_6)_3(H_2O)_2$ (Er-1.6)

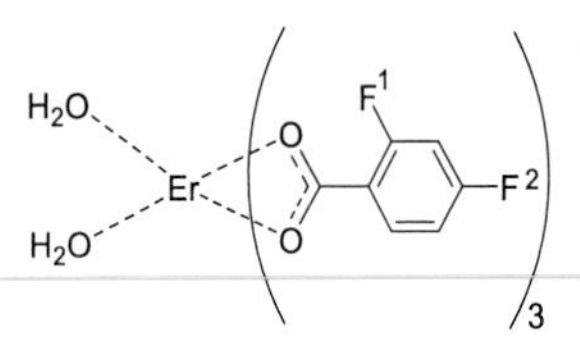

For the synthesis 381 mg of $ErCl_3 \cdot 6\,H_2O$ were used. 601 mg (89%) of an off-white solid of general formula $Er(L_6)_3(H_2O)_2$ were obtained.

^{1}H NMR (400 MHz, DMSO-d_6): δ (ppm) = 6.13 (bs, 2H, 2 × H_{Ar}), 5.42 (bs, 1H, H_{Ar}). – **^{19}F NMR** (377 MHz, DMSO-d_6): δ (ppm) = -107.33 (bs, 1F, F_1), -112.52 (bs, 1F, F_2). – **IR** (ATR): $\tilde{\nu}$ (cm^{-1}) = 3232.5 (vw), 1600.2 (m), 1539.7 (w), 1501.3 (w), 1425.7 (w), 1396.5 (m), 1268.6 (m), 1241.0 (w), 1137.9 (w), 1092.0 (w), 970.1 (w), 846.5 (m), 775.6 (w), 731.9 (w), 681.2 (w), 619.6 (m), 604.6 (w), 591.2 (m), 518.1 (w), 441.6 (w), 395 (vw). – **EA**: calcd for $Er(L_6)_3(H_2O)_2$,% C 37.39, H 1.94, found C 37.95, H 1.74.

ytterbium 2,4-difluorobenzoate dihydrate $Yb(L_6)_3(H_2O)_2$ (Yb-1.6)

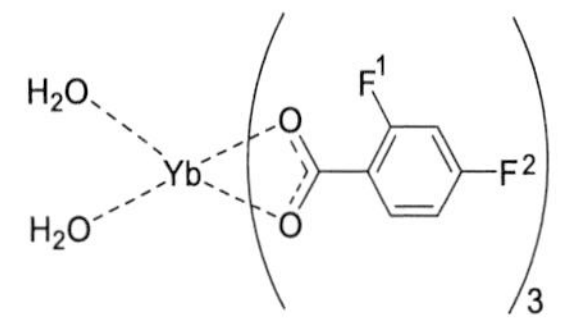

For the synthesis 388 mg of $YbCl_3 \cdot 6\,H_2O$ were used. 598 mg (88%) of an off-white solid of general formula $Yb(L_6)_3(H_2O)_2$ were obtained.

^{1}H NMR (400 MHz, DMSO-d_6): δ (ppm) = 6.29 (bs, 2H, 2 × H_{Ar}), 5.75 (bs, 1H, H_{Ar}). – **^{19}F NMR** (377 MHz, DMSO-d_6): δ (ppm) = -107.46 (bs, 1F, F_1), -111.76 (bs, 1F, F_2). – **IR** (ATR): $\tilde{\nu}$ (cm^{-1}) = 3233.6 (w), 1598.7 (m), 1539.4 (m), 1501.5 (w), 1426.6 (m), 1397.1 (m), 1268.8 (m), 1241.8 (w), 1138.0 (w), 1091.9 (m), 970.2 (m), 855.5 (w), 846.1 (m), 775.4 (m), 732.2 (w), 681.4 (w), 619.8 (m), 604.2 (m), 591.3 (m), 518.4 (w), 442.2 (w), 396.1 (vw). – **EA**: calcd for $Yb(L_6)_3(H_2O)_2$,% C 37.07, H 1.93, found C 37.46, H 1.78.

lutetium 2,4-difluorobenzoate dihydrate $Lu(L_6)_3(H_2O)_2$ (Lu-1.6)

For the synthesis 389 mg of $LuCl_3 \cdot 6\,H_2O$ were used. 570 mg (83%) of an off-white solid of general formula $Lu(L_6)_3(H_2O)_2$ were obtained.

^{1}H NMR (400 MHz, DMSO-d_6): δ (ppm) = 7.95 – 7.83 (m, 1H, H_{Ar}), 7.21 – 7.12 (m, 1H, H_{Ar}), 7.08 – 6.96 (m, 1H, H_{Ar}). – **^{19}F NMR** (377 MHz, DMSO-d_6): δ (ppm) = -106.69 (bs, 1F, F_1), -107.02 (bs, 1F, F_2). – **IR** (ATR): $\tilde{\nu}$ (cm^{-1}) = 3344.1 (w), 1606.2 (m), 1555.4 (m), 1501.2 (w), 1428.6 (m), 1395.8 (m), 1272.4 (m), 1240.6 (w), 1137.3 (w), 1091.2 (m), 975.1 (m), 864.7 (w), 846.1 (m), 781.7 (m), 732.1 (w), 679.4 (w), 625.7 (m), 608.7 (m), 596.2 (m), 520.1 (w), 441.1 (w), 391.2 (vw). – **EA**: calcd for $Lu(L_6)_3(H_2O)_2$,% C 36.97, H 1.92, found C 37.29, H 2.15.

Lanthanide 2,5-difluorobenzoates dihydrates $Ln(L_7)_3(H_2O)_2$ (Ln-1.7)

According to **GP1a**, a suspension of 474 mg of 2,5-difluorobenzoic acid HL_7 (3.00 mmol, 3.00 equiv.) in water was treated with 1 M aq. solution of NaOH at 60 °C until the pH was slightly acidic (pH = 5 – 6, approx. 3 mL) and the solution became clear. Then an aq. solution of $LnCl_3 \cdot x\,H_2O$ (1.00 mmol, 1.00 equiv.) was added following by an immediate precipitation of the product. The reaction mixture was stirred for additional 3 h at rt, the crude product was filtered, washed with ethanol and dried in vacuo to afford the product as an off-white solid.

lanthanum 2,5-difluorobenzoate dihydrate $La(L_7)_3(H_2O)_2$ (La-1.7)

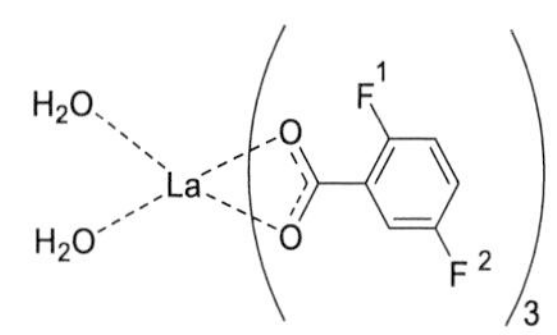

For the synthesis 373 mg of $LaCl_3 \cdot 7\,H_2O$ were used. 551 mg (85%) of an off-white solid of general formula $La(L_7)_3(H_2O)_2$ were obtained.

^{1}H NMR (400 MHz, DMSO-d_6): δ (ppm) = 7.25 – 7.18 (m, 1H, H_{Ar}), 7.03 (bs, 2H, 2 × H_{Ar}). – **^{19}F NMR** (377 MHz, DMSO-d_6): δ (ppm) = -118.60 (bs, 1F, F_1), -120.07 (bs, 1F, F_2). – **IR** (ATR): $\tilde{\nu}$ (cm^{-1}) = 3392.7 (vw), 1599.5 (w), 1559.0 (w), 1429.8 (w), 1386.9 (w), 1256.1 (vw), 1187.1 (w), 1122.3 (vw), 894.4 (vw), 847.2 (vw), 820.0 (w), 764.2 (w), 680.1 (vw), 543.3 (w), 417.0 (vw). – **EA**: calcd for $La(L_7)_3(H_2O)_2$,% C 39.03, H 2.03, found C 39.14, H 2.10.

neodymium 2,5-difluorobenzoate dihydrate $Nd(L_7)_3(H_2O)_2$ (Nd-1.7)

For the synthesis 359 mg of $NdCl_3 \cdot 6\ H_2O$ were used. 591 mg (91%) of an off-white solid of general formula $Nd(L_7)_3(H_2O)_2$ were obtained.

^{1}H NMR (400 MHz, DMSO-d_6): δ (ppm) = 8.15 (bs, 1H, H_{Ar}), 7.45 (bs, 2H, 2 × H_{Ar}). – **^{19}F NMR** (377 MHz, DMSO-d_6): δ (ppm) = -118.88 (bs, 1F, F_1), -119.25 (bs, 1F, F_2). – **IR** (ATR): $\tilde{\nu}$ (cm^{-1}) = 3317.1 (vw), 1600.8 (w), 1559.1 (w), 1484.1 (w), 1429.2 (w), 1387.1 (w), 1254.3 (vw), 1188.9 (w), 1118.4 (vw), 941.1 (vw), 895.5 (vw), 819.6 (w), 762.2 (w), 674.7 (vw), 563.1 (w), 540.9 (w), 411.1 (w). – **EA**: calcd for $Nd(L_7)_3(H_2O)_2$,% C 38.71, H 2.01, found C 38.89, H 2.22.

europium 2,5-difluorobenzoate dihydrate $Eu(L_7)_3(H_2O)_2$ (Eu-1.7)

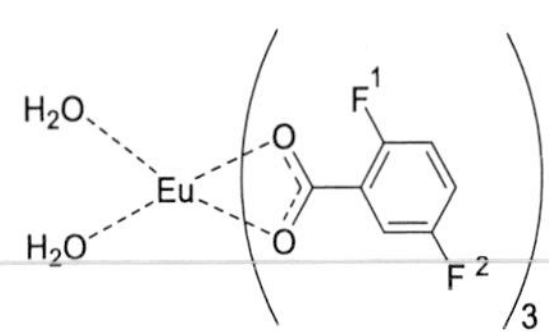

For the synthesis 366 mg of $EuCl_3 \cdot 6\ H_2O$ were used. 594 mg (90%) of an off-white solid of general formula $Eu(L_7)_3(H_2O)_2$ were obtained.

^{1}H NMR (400 MHz, DMSO-d_6): δ (ppm) = 6.98 – 6.90 (m, 1H, H_{Ar}), 6.79 – 6.67 (m, 1H, H_{Ar}), 6.48 – 6.39 (m, 1H, H_{Ar}). – **^{13}C NMR** (101 MHz, DMSO-d_6): δ (ppm) = 156.76 ($C_{quart.}$, 6 × C_{Ar}F, 3 × COO, 3 × C_{Ar}, m), 119.06 (+, 3 × C_{Ar}H, dd, J = 24.3, 7.8 Hz), 118.14 (+, 3 × C_{Ar}H, d, J = 24.1 Hz), 115.82 (+, 3 × C_{Ar}H, dd, J = 25.6, 8.2 Hz). – **^{19}F NMR** (377 MHz, DMSO-d_6): δ (ppm) = -118.63 (bs, 1F, F_1), -120.57 (bs, 1F, F_2). – **IR** (ATR): $\tilde{\nu}$ (cm^{-1}) = 3085.7 (vw), 1600.6 (w), 1540.2 (m), 1486.4 (w), 1432.0 (m), 1374.9 (m), 1244.5 (w), 1191.1 (m), 1121.7 (w), 1083.3 (vw), 944.7 (vw), 890.2 (vw), 822.3 (m), 761.7 (m), 671.9 (w), 584.5 (w), 542.1 (w), 468.5 (w), 416.0 (w). – **EA**: calcd for $Eu(L_7)_3(H_2O)_2$,% C 38.26, H 1.99, found C 38.63, H 2.42.

gadolinium 2,5-difluorobenzoate dihydrate $Gd(L_7)_3(H_2O)_2$ (Gd-1.7)

For the synthesis 372 mg of $GdCl_3 \cdot 6\ H_2O$ were used. 583 mg (88%) of an off-white solid of general formula $Gd(L_7)_3(H_2O)_2$ were obtained.

IR (ATR): $\tilde{\nu}$ (cm^{-1}) = 3352.0 (vw), 1600.7 (vw), 1559.7 (w), 1484.9 (w), 1431.6 (w), 1388.9 (w), 1245.4 (vw), 1189.2 (w), 1119.3 (vw), 942.2 (vw), 895.7 (vw), 820.9 (w), 762.3 (w), 672.7 (vw), 585.9 (vw), 539.6 (vw), 465.7 (vw), 418.4 (vw). – **EA**: calcd for $Gd(L_7)_3(H_2O)_2$,% C 37.95, H 1.97, found C 38.13, H 2.19.

terbium 2,5-difluorobenzoate dihydrate $Tb(L_7)_3(H_2O)_2$ (Tb-1.7)

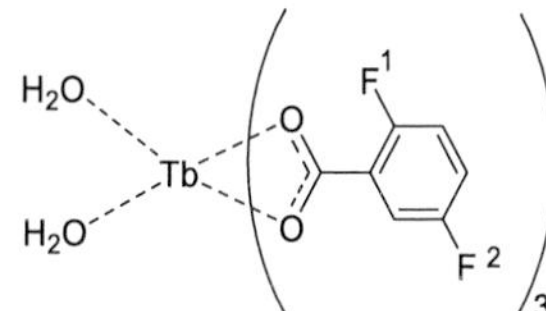

For the synthesis 373 mg of $TbCl_3 \cdot 6\ H_2O$ were used. 574 mg (86%) of an off-white solid of general formula $Tb(L_7)_3(H_2O)_2$ were obtained.

IR (ATR): $\tilde{\nu}$ (cm^{-1}) = 3345.0 (vw), 1600.3 (w), 1557.6 (w), 1485.1 (w), 1433.5 (w), 1389.7 (w), 1245.9 (vw), 1189.1 (w), 1118.4 (vw), 942.3 (vw), 895.8 (vw), 820.4 (w), 762.4 (w), 673.9 (vw), 538.3 (vw), 466.9 (vw), 411.8 (vw). – **EA**: calcd for $Tb(L_7)_3(H_2O)_2$,% C 37.86, H 1.97, found C 38.21, H 2.01.

terbium/europium 2,5-difluorobenzoate dihydrate $Tb_{0.5}Eu_{0.5}(L_7)_3(H_2O)_2$ (TbEu-1.7)

For the synthesis 187 mg of $TbCl_3 \cdot 6\ H_2O$ (0.50 mmol, 0.50 equiv.) and 183 mg $EuCl_3 \cdot 6\ H_2O$ (0.50 mmol, 0.50 equiv.) were used. 550 mg (83%) of an off-white solid of general formula $Tb_{0.5}Eu_{0.5}(L_7)_3(H_2O)_2$ were obtained.

IR (ATR): $\tilde{\nu}$ (cm^{-1}) = 3392.9 (w), 1599.4 (w), 1557.1 (m), 1485.1 (m), 1435.0 (m), 1388.0 (m), 1298.9 (w), 1245.9 (w), 1190.3 (m), 1118.5 (w), 1082.2 (vw), 943.4 (vw), 895.9 (vw), 821.1 (m), 763.5 (m), 676.4 (w), 581.5 (w), 543.1 (w), 465.9 (w), 416.4 (w). – **EA**: calcd for $Tb_{0.5}Eu_{0.5}(L_7)_3(H_2O)_2$,% C 38.06, H 1.98, found C 38.13, H 2.03.

erbium 2,5-difluorobenzoate dihydrate $Er(L_7)_3(H_2O)_2$ (Er-1.7)

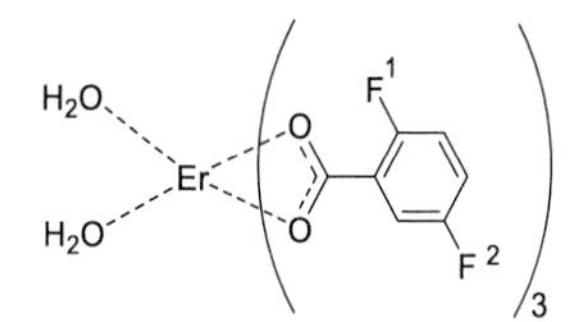

For the synthesis 381 mg of $ErCl_3 \cdot 6\ H_2O$ were used. 601 mg (89%) of an off-white solid of general formula $Er(L_7)_3(H_2O)_2$ were obtained.

^{1}H NMR (400 MHz, DMSO-d_6): δ (ppm) = 6.50 (bs, 2H, 2 × H_{Ar}), 5.71 (bs, 1H, H_{Ar}). – **^{19}F NMR** (377 MHz, DMSO-d_6): δ (ppm) = -120.55 (bs, 1F, F_1), -121.77 (bs, 1F, F_2). – **IR** (ATR): $\tilde{\nu}$ (cm^{-1}) = 3351.1 (vw), 1600.3 (m), 1558.4 (w), 1485.5 (w), 1433.8 (w), 1389.9 (m), 1246.3 (m), 1182.4 (w), 1242.8 (w), 1183.9 (w), 1111.3 (vw), 946.4 (w), 895.9 (m), 823.2 (m), 767.8 (w), 669.8 (w), 534.3 (w), 468.9 (w), 411.5 (vw). – **EA**: calcd for $Er(L_7)_3(H_2O)_2$,% C 37.39, H 1.94, found C 38.05, H 1.90.

ytterbium 2,5-difluorobenzoate dihydrate $Yb(L_7)_3(H_2O)_2$ (Yb-1.7)

For the synthesis 388 mg of $YbCl_3 \cdot 6\,H_2O$ were used. 558 mg (82%) of an off-white solid of general formula $Yb(L_7)_3(H_2O)_2$ were obtained.

^{1}H NMR (400 MHz, DMSO-d_6): δ (ppm) = 6.67 (bs, 2H, 2 × H_{Ar}), 5.95 (bs, 1H, H_{Ar}). – **^{19}F NMR** (377 MHz, DMSO-d_6): δ (ppm) = -119.90 (bs, 1F, F_1), -121.04 (bs, 1F, F_2). – **IR** (ATR): $\tilde{\nu}$ (cm^{-1}) = 3374.2 (w), 1609.2 (m), 1554.8 (m), 1483.8 (m), 1434.3 (m), 1386.7 (m), 1248.4 (w), 1185.8 (w), 1239.1 (w), 1188.6 (w), 1113.6 (w), 942.5 (m), 893.6 (vw), 821.4 (w), 763.8 (m), 678.2 (w), 535.4 (m), 464.4 (w), 413.7 (vw). – **EA**: calcd for $Yb(L_7)_3(H_2O)_2$,% C 37.07, H 1.93, found C 37.71, H 2.13.

lutetium 2,5-difluorobenzoate dihydrate $Lu(L_7)_3(H_2O)_2$ (Lu-1.7)

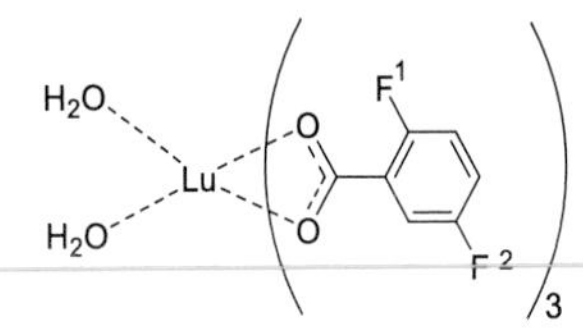

For the synthesis 389 mg of $LuCl_3 \cdot 6\,H_2O$ were used. 573 mg (84%) of an off-white solid of general formula $Lu(L_7)_3(H_2O)_2$ were obtained.

^{1}H NMR (400 MHz, DMSO-d_6): δ (ppm) = 7.23 (bs, 2H, 2 × H_{Ar}), 7.00 (bs, 1H, H_{Ar}). – **^{19}F NMR** (377 MHz, DMSO-d_6): δ (ppm) = -118.60 (bs, 1F, F_1), -119.70 (bs, 1F, F_2). – **IR** (ATR): $\tilde{\nu}$ (cm^{-1}) = 3299.5 (vw), 1602.3 (w), 1559.9 (m), 1486.7 (w), 1433.1 (m), 1391.9 (m), 1298.0 (vw), 1252.8 (w), 1189.1 (w), 1121.2 (vw), 1080.6 (vw), 944.3 (vw), 899.4 (vw), 820.4 (m), 762.8 (m), 671.0 (w), 583.5 (vw), 542.7 (w), 467.8 (vw), 414.5 (w), 382.9 (vw). – **EA**: calcd for $Lu(L_7)_3(H_2O)_2$,% C 36.97, H 1.92, found C 37.29, H 2.15.

Lanthanide 2-chloro-6-fluorobenzoates monohydrates $Ln(L_8)_3(H_2O)$ (Ln-1.8)

According to **GP1b** an excess of concentrated aq. solution of NH_3 (1.30 mL, 20.00 equiv.) was added to an aq. solution of $LnCl_3 \cdot x\,H_2O$ (1.00 mmol, 1.00 equiv.) in 25 mL water. The mixture was stirred for 30 min, the precipitate of $Ln(OH)_3$ was centrifugated and washed with water until the pH of the washing solution became neutral. A small excess of freshly prepared $Ln(OH)_3$ was placed into a beaker and the solution of 471 mg of 2-chloro-6-fluorobenzoic acid HL_8 (2.70 mmol, 2.70 equiv.) in 20 mL acetone/methanol (3/1) was added. The reaction mixture was stirred for 1 h at 60 °C and the unreacted components were separated by filtration followed by the evaporation of the clear solution to dryness. The obtained solid powder of the lanthanide complexes was then recrystallized from water. Subsequent drying at 80 °C in vacuo for 1 h led to the formation of a microcrystalline off-white powder of general formula $Ln(L_8)_3(H_2O)$.

lanthanum 2-chloro-6-fluorobenzoate monohydrate La(L8)3(H2O) (La-1.8)

For the synthesis 373 mg of $LaCl_3 \cdot 7\ H_2O$ were used. 512 mg (84%) of an off-white solid of general formula $La(L_8)_3(H_2O)_2$ were obtained.

^{1}H NMR (400 MHz, DMSO-d_6): δ (ppm) = 7.28 (td, J = 8.1 Hz, J = 6.0 Hz, 1H, H_{Ar}), 7.21 (dd, J = 8.1 Hz, J = 1.2 Hz, 1H, H_{Ar}), 7.18 – 7.09 (m, 1H, H_{Ar}). – **^{19}F NMR** (377 MHz, DMSO-d_6): δ (ppm) = -113.92 (s, 1F, F). – **IR** (ATR): $\tilde{\nu}$ (cm^{-1}) = 3313.5 (w), 1574.8 (m), 1553.6 (m), 1458.3 (m), 1403.6 (m), 1365.2 (m), 1302.4 (w), 1238.7 (w), 1189.9 (w), 904.5 (m), 877.2 (w), 825.3 (vw), 775.3 (w), 747.2 (vw), 736.0 (w), 693.6 (w), 586.2 (w), 558.0 (m), 483.4 (w), 422.1 (w). – **EA**: calcd for $La(L_8)_3(H_2O)$,% C 37.23, H 1.64, found C 38.16, H 1.99.

neodymium 2-chloro-6-fluorobenzoate monohydrate Nd(L8)3(H2O) (Nd-1.8)

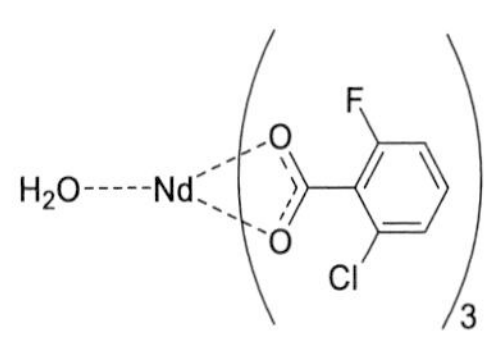

For the synthesis 359 mg of $NdCl_3 \cdot 6\ H_2O$ were used. 495 mg (81%) of an off-white solid of general formula $Nd(L_7)_3(H_2O)_2$ were obtained.

^{1}H NMR (400 MHz, DMSO-d_6): δ (ppm) = 7.34 (bs, 2H, 2 × H_{Ar}), 7.32 – 7.25 (m, 1H, H_{Ar}). – **^{19}F NMR** (377 MHz, DMSO-d_6): δ (ppm) = -113.91 (s, 1F, F). – **IR** (ATR): $\tilde{\nu}$ (cm^{-1}) = 3246.5 (w), 1695.1 (vw), 1578.1 (m), 1535.3 (m), 1445.5 (m), 1398.6 (m), 1242.7 (m), 1186.6 (w), 1058.1 (w), 899.2 (m), 855.6 (m), 782.3 (w), 770.0 (m), 703.2 (w), 580.5 (vw), 552.0 (m), 484.4 (w), 432.8 (w). – **EA**: calcd for $Nd(L_8)_3(H_2O)$,% C 36.94, H 1.62, found C 37.34, H 2.05.

europium 2-chloro-6-fluorobenzoate monohydrate Eu(L8)3(H2O) (Eu-1.8)

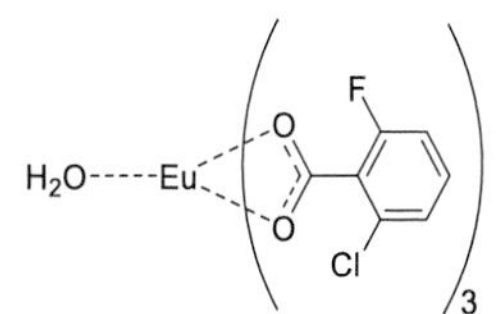

For the synthesis 366 mg of $EuCl_3 \cdot 6\ H_2O$ were used. 510 mg (82%) of an off-white solid of general formula $Eu(L_8)_3(H_2O)_2$ were obtained.

^{1}H NMR (400 MHz, DMSO-d_6): δ (ppm) = 7.16 (td, J = 8.1 Hz, J = 5.7 Hz, 1H, H_{Ar}), 6.91 (d, J = 8.0 Hz, 1H, H_{Ar}), 6.81 (t, J = 8.4 Hz, 1H, H_{Ar}). – **^{19}F NMR** (377 MHz, DMSO-d_6): δ (ppm) = -113.75 (s, 1F, F). – **IR** (ATR): $\tilde{\nu}$ (cm^{-1}) = 3626.2 (vw), 3317.6 (vw), 1582.7 (m), 1561.8 (m), 1446.0 (m), 1399.8 (m), 1243.0 (w), 1186.8 (w), 1136.3 (vw), 1058.7 (vw), 900.8 (m), 861.0 (w), 770.0 (w), 723.8 (w), 701.0 (w), 553.3 (m), 487.0 (w), 438.8 (vw). – **EA**: calcd for $Eu(L_8)_3(H_2O)$,% C 36.52, H 1.61, found C 37.48, H 2.09.

gadolinium 2-chloro-6-fluorobenzoate monohydrate $Gd(L_8)_3(H_2O)$ (Gd-1.8)

For the synthesis 372 mg of $GdCl_3 \cdot 6\ H_2O$ were used. 551 mg (88%) of an off-white solid of general formula $Gd(L_8)_3(H_2O)_2$ were obtained.

IR (ATR): $\tilde{\nu}$ (cm^{-1}) = 3627.8 (vw), 2916.3 (vw), 2660.7 (vw), 1696.8 (w), 1576.6 (m), 1445.8 (m), 1406.6 (m), 1300.5 (w), 1275.5 (w), 1242.8 (w), 1186.4 (w), 1129.3 (w), 1058.0 (vw), 898.3 (m), 871.5 (w), 859.5 (w), 809.0 (w), 791.1 (w), 770.7 (w), 726.2 (w), 693.9 (w), 581.7 (w), 552.1 (w), 487.4 (w), 442.4 (vw), 416.2 (w). – **EA**: calcd for $Gd(L_8)_3(H_2O)$,% C 36.24, H 1.59, found C 36.14, H 1.95.

terbium 2-chloro-6-fluorobenzoate monohydrate $Tb(L_8)_3(H_2O)$ (Tb-1.8)

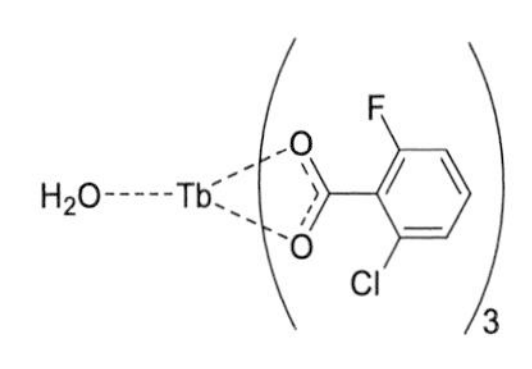

For the synthesis 373 mg of $TbCl_3 \cdot 6\ H_2O$ were used. 546 mg (87%) of an off-white solid of general formula $Tb(L_8)_3(H_2O)$ were obtained.

^{1}H NMR (400 MHz, DMSO-d_6): δ (ppm) = 8.00 (bs, 1H, H_{Ar}), 7.71 (bs, 1H, H_{Ar}), 7.30 (bs, 1H, H_{Ar}). – **^{19}F NMR** (377 MHz, DMSO-d_6): δ (ppm) = -111.89 (s, 1F, *F*). – **IR** (ATR): $\tilde{\nu}$ (cm^{-1}) = 3625.3 (vw), 3301.4 (vw), 1697.1 (vw), 1566.7 (m), 1445.4 (m), 1407.2 (m), 1301.9 (w), 1242.9 (w), 1186.9 (w), 898.5 (m), 871.9 (w), 858.3 (w), 809.4 (w), 783.7 (w), 770.8 (w), 725.8 (w), 693.5 (w), 577.0 (vw), 553.6 (w), 487.4 (w), 415.4 (vw). – **EA**: calcd for $Tb(L_8)_3(H_2O)$,% C 36.16, H 1.59, found C 36.16, H 1.68.

terbium/europium 2-chloro-6-fluorobenzoate monohydrate $Tb_{0.5}Eu_{0.5}(L_8)_3(H_2O)$ (TbEu-1.8)

For the synthesis 187 mg of $TbCl_3 \cdot 6\ H_2O$ (0.50 mmol, 0.50 equiv.) and 183 mg $EuCl_3 \cdot 6\ H_2O$ (0.50 mmol, 0.50 equiv.) were used. 537 mg (86%) of an off-white solid of general formula $Tb_{0.5}Eu_{0.5}(L_8)_3(H_2O)$ were obtained.

^{1}H NMR (400 MHz, DMSO-d_6): δ (ppm) = 7.78 (bs, 1H, H_{Ar}), 7.34 (bs, 1H, H_{Ar}), 7.00 (bs, 1H, H_{Ar}). – **^{19}F NMR** (377 MHz, DMSO-d_6): δ (ppm) = -113.52 (s, 1F, *F*). – **IR** (ATR): $\tilde{\nu}$ (cm^{-1}) = 3317.8 (vw), 1700.5 (vw), 1579.5 (w), 1446.3 (w), 1408.1 (w), 1302.9 (w), 1244.8 (w), 1188.0 (vw), 1134.2 (vw), 1059.2 (vw), 900.3 (w), 868.5 (w), 807.0 (vw), 771.3 (w), 727.5 (w), 697.8 (w), 552.8 (w), 486.2 (w), 420.4 (vw). – **EA**: calcd for $Tb_{0.5}Eu_{0.5}(L_8)_3(H_2O)$,% C 36.34, H 1.60, found C 37.01, H 1.85.

erbium 2-chloro-6-fluorobenzoate monohydrate Er(L8)3(H2O) (Er-1.8)

For the synthesis 381 mg of $ErCl_3 \cdot 6\ H_2O$ were used. 515 mg (81%) of an off-white solid of general formula $Er(L_8)_3(H_2O)$ were obtained.

^{1}H NMR (400 MHz, DMSO-d_6): δ (ppm) = 7.42 (bs, 1H, H_{Ar}), 7.14 (bs, 1H, H_{Ar}), 7.01 (bs, 1H, H_{Ar}). – **^{19}F NMR** (377 MHz, DMSO-d_6): δ (ppm) = -113.29 (s, 1F, *F*). – **IR** (ATR): $\tilde{\nu}$ (cm^{-1}) = 3367.7 (vw), 1579.7 (w), 1537.0 (w), 1480.9 (w), 1449.8 (w), 1411.6 (m), 1359.3 (w), 1247.8 (w), 1228.5 (w), 1192.4 (w), 895.1 (w), 879.0 (w), 828.9 (w), 809.7 (w), 773.3 (w), 732.8 (w), 703.1 (w), 581.6 (w), 555.1 (w), 479.5 (w), 463.1 (w), 432.2 (vw), 414.3 (vw), 382.1 (vw), 354.2 (vw). – **EA**: calcd for $Er(L_8)_3(H_2O)$,% C 35.73, H 1.57, found C 35.16, H 2.03.

ytterbium 2-chloro-6-fluorobenzoate monohydrate Yb(L8)3(H2O)2 (Yb-1.8)

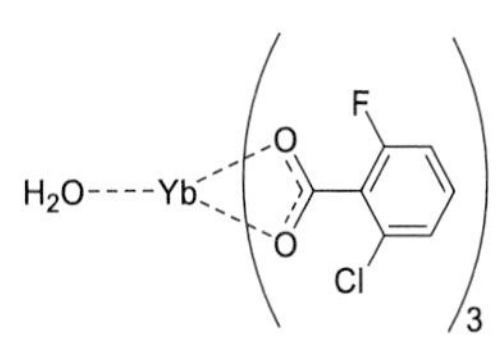

For the synthesis 388 mg of $YbCl_3 \cdot 6\ H_2O$ were used. 532 mg (83%) of an off-white solid of general formula $Yb(L_8)_3(H_2O)$ were obtained.

^{1}H NMR (400 MHz, DMSO-d_6): δ (ppm) = 6.78 (bs, 1H, H_{Ar}), 6.71 – 6.64 (m, 1H, H_{Ar}), 7.01 (bs, 1H, H_{Ar}). – **^{19}F NMR** (377 MHz, DMSO-d_6): δ (ppm) = -115.16 (s, 1F, *F*). – **IR** (ATR): $\tilde{\nu}$ (cm^{-1}) = 3110.1 (vw), 1653.7 (w), 1607.9 (m), 1586.2 (w), 1551.3 (m), 1474.8 (w), 1450.9 (w), 1418.3 (m), 1387.6 (w), 1278.5 (w), 1248.1 (w), 1191.3 (w), 1154.2 (w), 1134.6 (vw), 1056.1 (w), 971.1 (w), 876.6 (w), 846.6 (w), 811.7 (w), 792.3 (w), 774.8 (w), 731.4 (w), 709.7 (w), 694.5 (w), 659.9 (vw). – **EA**: calcd for $Yb(L_8)_3(H_2O)$,% C 35.44, H 1.56, found C 36.13, H 1.88.

lutetium 2-chloro-6-fluorobenzoate monohydrate Lu(L8)3(H2O)2 (Lu-1.8)

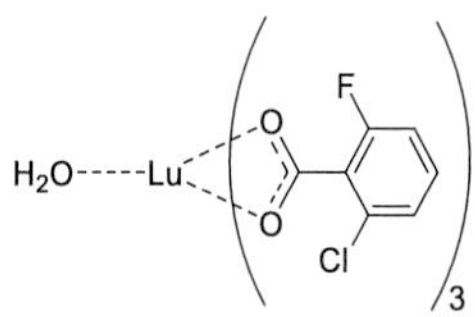

For the synthesis 389 mg of $LuCl_3 \cdot 6\ H_2O$ were used. 514 mg (80%) of an off-white solid of general formula $Lu(L_8)_3(H_2O)_2$ were obtained.

^{1}H NMR (400 MHz, DMSO-d_6): δ (ppm) = 7.29 (td, J = 8.1 Hz, J = 6.0 Hz, 1H, H_{Ar}), 7.20 (d, J = 8.0 Hz, 1H, H_{Ar}), 7.17 – 7.09 (m, 1H, H_{Ar}). – **^{19}F NMR** (377 MHz, DMSO-d_6): δ (ppm) = -113.51 (s, 1F, *F*). – **IR** (ATR): $\tilde{\nu}$ (cm^{-1}) = 3353.3 (w), 1579.4 (m), 1551.0 (m), 1448.7 (m), 1413.7 (m), 1361.6 (m), 1302.0 (w), 1247.2 (w), 1189.9 (w), 901.9 (m), 879.6 (w), 843.5 (vw), 829.5 (vw), 808.5 (w), 786.0 (w), 772.3 (w), 732.2 (w), 698.2 (w), 582.6 (w), 550.8 (m), 489.6 (w), 427.6 (w). – **EA**: calcd for $Lu(L_8)_3(H_2O)$,% C 35.35, H 1.55, found C 34.89, H 1.62.

Lanthanide 2-fluorobenzoates dihydrates $Ln(L_7)_3(H_2O)_2$ (Ln-1.9)

According to **GP1a**, a suspension of 420 mg of 2-fluorobenzoic acid HL_9 (3.00 mmol, 3.00 equiv.) in water was treated with 1 M aq. solution of NaOH at 60 °C until the pH was slightly acidic (pH = 5 – 6, approx. 3 mL) and the solution became clear. Then an aq. solution of $LnCl_3 \cdot x\ H_2O$ (1.00 mmol, 1.00 equiv.) was added following by an immediate precipitation of the product. The reaction mixture was stirred for additional 3 h at rt, the crude product was filtered, washed with ethanol and dried in vacuo to afford the product as an off-white solid.

lanthanum 2-fluorobenzoate dihydrate $La(L_9)_3(H_2O)_2$ (La-1.9)

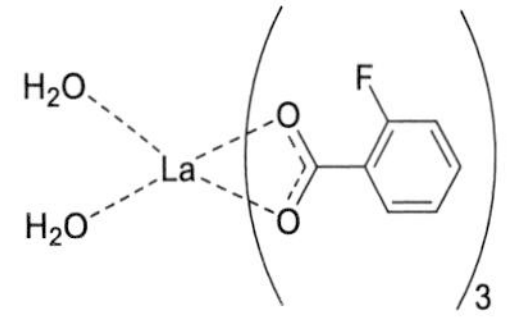

For the synthesis 373 mg of $LaCl_3 \cdot 7\ H_2O$ were used. 480 mg (90%) of an off-white solid of general formula $La(L_7)_3(H_2O)_2$ were obtained.

^{1}H NMR (400 MHz, DMSO-d_6): δ (ppm) = 7.78 (bs, 2H, 2 × H_{Ar}), 7.48 (bs, 1H, H_{Ar}), 7.21 (bs, 1H, H_{Ar}). – **^{19}F NMR** (377 MHz, DMSO-d_6): δ (ppm) = -112.14 (bs, 1F, *F*). – **IR** (ATR): $\tilde{\nu}$ (cm^{-1}) = 3420.9 (vw), 1670.7 (w), 1601.7 (w), 1561.5 (w), 1530.8 (w), 1485.5 (w), 1452.6 (w), 1387.2 (m), 1297.2 (w), 1220.5 (w), 1159.6 (w), 1135.0 (w), 1096.2 (w), 1031.8 (vw), 871.3 (w), 844.0 (w), 812.0 (vw), 794.5 (vw), 751.5 (m), 693.4 (w), 659.5 (w), 567.0 (w), 544.1 (w), 515.9 (w), 422.6 (w). – **EA**: calcd for $La(L_9)_3(H_2O)_2$,% C 42.59, H 2.72, found C 41.87, H 3.05.

neodymium 2-fluorobenzoate dihydrate $Nd(L_9)_3(H_2O)_2$ (Nd-1.9)

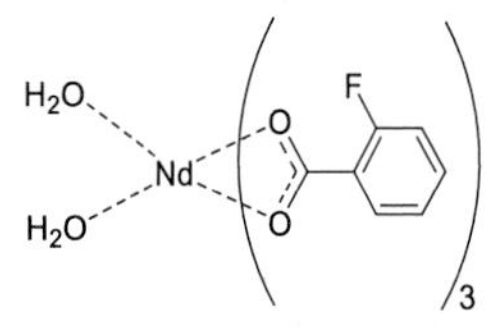

For the synthesis 359 mg of $NdCl_3 \cdot 6\ H_2O$ were used. 526 mg (88%) of an off-white solid of general formula $Nd(L_9)_3(H_2O)_2$ were obtained.

^{1}H NMR (400 MHz, DMSO-d_6): δ (ppm) = 8.72 (bs, 1H, H_{Ar}), 7.76 (dd, J = 8.1 Hz, J = 4.3 Hz, 1H, H_{Ar}), 7.55 (q, J = 8.8 Hz, J = 6.6 Hz, 2H, 2 × H_{Ar}). – **^{19}F NMR** (377 MHz, DMSO-d_6): δ (ppm) = -111.99 (bs, 1F, *F*). – **IR** (ATR): $\tilde{\nu}$ (cm^{-1}) = 3418.2 (vw), 1669.7 (w), 1601.1 (m), 1562.0 (w), 1528.5 (w), 1486.2 (w), 1451.7 (w), 1386.6 (m), 1294.9 (w), 1216.9 (w), 1159.3 (w), 1134.2 (w), 1096.4 (w), 1031.4 (vw), 872.8 (w), 844.0 (w), 810.7 (w), 750.9 (m), 692.0 (w), 651.4 (w), 566.5 (w), 544.1 (w), 516.5 (w), 448.1 (w), 397.6 (w). – **EA**: calcd for $Nd(L_9)_3(H_2O)_2$,% C 42.21, H 2.70, found C 42.01, H 3.02.

europium 2-fluorobenzoate dihydrate $Eu(L_9)_3(H_2O)_2$ (Eu-1.9)

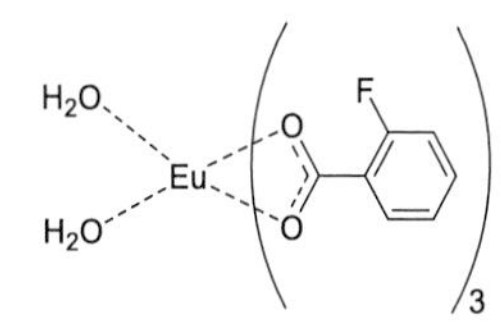

For the synthesis 366 mg of $EuCl_3 \cdot 6\ H_2O$ were used. 557 mg (92%) of an off-white solid of general formula $Eu(L_9)_3(H_2O)_2$ were obtained.

^{1}H NMR (400 MHz, DMSO-d_6): δ (ppm) = 7.22 (bs, 2H, 2 × H_{Ar}), 6.81 (bs, 2H, 2 × H_{Ar}). – **^{19}F NMR** (377 MHz, DMSO-d_6): δ (ppm) = -112.14 (bs, 1F, *F*). – **IR** (ATR): $\tilde{\nu}$ (cm^{-1}) = 3415.0 (vw), 1672.0 (w), 1595.2 (m), 1531.3 (w), 1486.6 (w), 1451.6 (w), 1385.6 (m), 1294.5 (w), 1215.8 (w), 1159.4 (w), 1133.9 (w), 1096.4 (w), 1031.2 (vw), 874.0 (w), 843.8 (w), 810.5 (w), 750.5 (m), 691.3 (w), 651.3 (w), 566.8 (w), 543.8 (w), 516.5 (w), 450.1 (w), 427.7 (vw), 398.2 (w). – **EA**: calcd for $Eu(L_9)_3(H_2O)_2$,% C 41.67, H 2.66, found C 42.31, H 2.95.

gadolinium 2-fluorobenzoate dihydrate $Gd(L_9)_3(H_2O)_2$ (Gd-1.9)

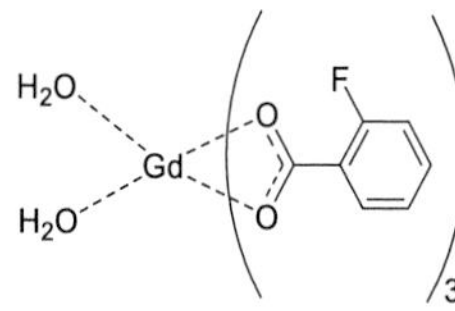

For the synthesis 372 mg of $GdCl_3 \cdot 6\ H_2O$ were used. 531 mg (87%) of an off-white solid of general formula $Gd(L_9)_3(H_2O)_2$ were obtained.

IR (ATR): $\tilde{\nu}$ (cm^{-1}) = 3420.9 (vw), 1672.4 (w), 1604.1 (m), 1531.8 (m), 1486.8 (w), 1451.5 (m), 1386.2 (m), 1295.0 (m), 1216.0 (m), 1159.5 (w), 1133.9 (w), 1096.7 (w), 1031.3 (w), 875.2 (m), 843.8 (w), 811.2 (w), 792.0 (w), 750.3 (s), 691.7 (w), 651.5 (m), 567.1 (m), 544.1 (m), 516.9 (m), 450.7 (w), 420.8 (w), 398.7 (w). – **EA**: calcd for $Gd(L_9)_3(H_2O)_2$,% C 41.31, H 2.64, found C 41.73, H 3.15.

terbium 2-fluorobenzoate dihydrate $Tb(L_9)_3(H_2O)_2$ (Tb-1.9)

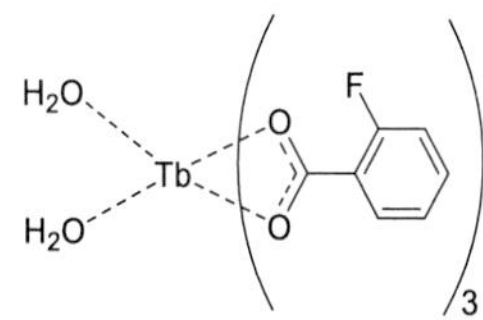

For the synthesis 373 mg of $TbCl_3 \cdot 6\ H_2O$ were used. 545 mg (89%) of an off-white solid of general formula $Tb(L_9)_3(H_2O)_2$ were obtained.

^{19}F NMR (377 MHz, DMSO-d_6): δ (ppm) = -94.28 (bs, 1F, *F*). – **IR** (ATR): $\tilde{\nu}$ (cm^{-1}) = 3413.7 (vw), 1674.3 (w), 1605.2 (m), 1531.9 (w), 1486.9 (w), 1451.8 (w), 1386.4 (m), 1295.0 (w), 1216.2 (w), 1159.6 (w), 1133.8 (w), 1096.8 (w), 1031.3 (vw), 875.9 (w), 843.9 (w), 811.1 (w), 792.1 (w), 750.6 (m), 692.0 (w), 651.8 (w), 567.2 (w), 544.4 (w), 517.0 (w), 451.4 (w), 427.9 (w), 399.0 (w). – **EA**: calcd for $Tb(L_9)_3(H_2O)_2$,% C 41.20, H 2.63, found C 41.87, H 3.05.

terbium/europium 2-fluorobenzoate dihydrate $Tb_{0.5}Eu_{0.5}(L_9)_3(H_2O)_2$ (TbEu-1.9)

For the synthesis 187 mg of $TbCl_3 \cdot 6\,H_2O$ (0.50 mmol, 0.50 equiv.) and 183 mg $EuCl_3 \cdot 6\,H_2O$ (0.50 mmol, 0.50 equiv.) were used. 548 mg (90%) of an off-white solid of general formula $Tb_{0.5}Eu_{0.5}(L_9)_3(H_2O)_2$ were obtained.

IR (ATR): $\tilde{\nu}$ (cm^{-1}) = 3407.5 (vw), 1672.7 (w), 1604.5 (m), 1531.9 (w), 1486.9 (w), 1451.6 (w), 1386.1 (m), 1294.9 (w), 1216.1 (w), 1159.5 (w), 1133.9 (w), 1096.7 (w), 1031.5 (vw), 875.0 (w), 843.7 (w), 810.9 (w), 750.4 (m), 691.7 (w), 651.5 (w), 566.8 (w), 543.3 (w), 516.7 (w), 450.7 (w), 398.6 (w). – **EA**: calcd for $Tb_{0.5}Eu_{0.5}(L_9)_3(H_2O)_2$,% C 41.44, H 2.64, found C 42.01, H 2.85.

erbium 2-fluorobenzoate dihydrate $Er(L_9)_3(H_2O)_2$ (Er-1.9)

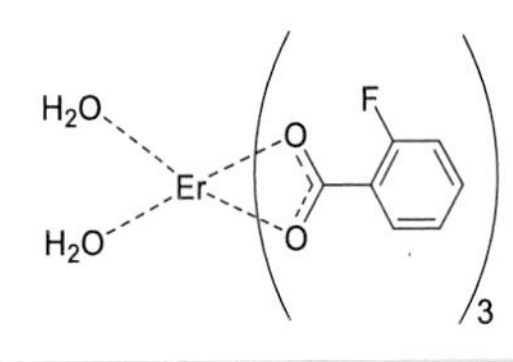

For the synthesis 381 mg of $ErCl_3 \cdot 6\,H_2O$ were used. 546 mg (88%) of an off-white solid of general formula $Er(L_9)_3(H_2O)_2$ were obtained.

IR (ATR): $\tilde{\nu}$ (cm^{-1}) = 3418.4 (vw), 1678.0 (w), 1608.6 (m), 1535.8 (m), 1487.3 (w), 1452.8 (m), 1386.2 (m), 1295.2 (m), 1216.1 (w), 1159.6 (w), 1133.5 (w), 1097.0 (w), 1031.4 (vw), 877.3 (w), 843.7 (w), 811.1 (w), 750.2 (m), 691.7 (w), 651.9 (m), 567.4 (w), 544.3 (w), 516.7 (w), 452.8 (w), 399.7 (w). – **EA**: calcd for $Er(L_9)_3(H_2O)_2$,% C 40.64, H 2.60, found C 41.37, H 2.91.

ytterbium 2-fluorobenzoate dihydrate $Yb(L_9)_3(H_2O)_2$ (Yb-1.9)

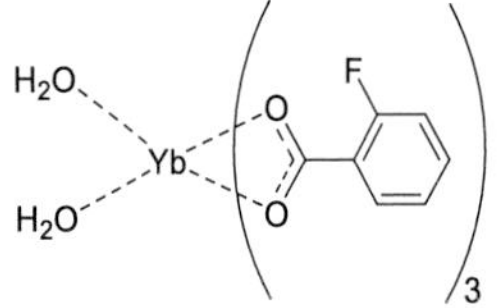

For the synthesis 388 mg of $YbCl_3 \cdot 6\,H_2O$ were used. 545 mg (87%) of an off-white solid of general formula $Yb(L_9)_3(H_2O)_2$ were obtained.

IR (ATR): $\tilde{\nu}$ (cm^{-1}) = 3418.3 (vw), 1681.8 (w), 1609.5 (m), 1537.2 (w), 1487.4 (w), 1453.3 (m), 1386.9 (m), 1294.7 (m), 1216.4 (w), 1159.8 (w), 1133.1 (w), 1097.1 (w), 1031.4 (vw), 877.9 (w), 843.4 (w), 809.4 (w), 750.2 (m), 691.9 (w), 629.4 (m), 544.5 (w), 516.6 (w), 453.0 (w). – **EA**: calcd for $Yb(L_9)_3(H_2O)_2$,% C 40.27, H 2.57, found C 40.43, H 3.11.

lutetium 2-fluorobenzoate dihydrate $Lu(L_9)_3(H_2O)_2$ (Lu-1.9)

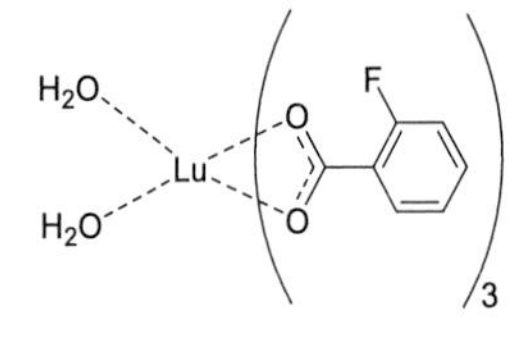

For the synthesis 389 mg of $LuCl_3 \cdot 6\,H_2O$ were used. 559 mg (89%) of an off-white solid of general formula $Lu(L_9)_3(H_2O)_2$ were obtained.

^{1}H NMR (400 MHz, DMSO-d_6): δ (ppm) = 7.82 (td, J = 7.6 Hz, J = 2.0 Hz, 1H, H_{Ar}), 7.45 (tdd, J = 7.5 Hz, J = 4.9 Hz, J = 2.0 Hz, 1H, H_{Ar}), 7.21 – 7.09 (m, 2H, 2 × H_{Ar}). – **^{19}F NMR** (377 MHz, DMSO-d_6): δ (ppm) = -112.28 (bs, 1F, F). – **IR** (ATR): $\tilde{\nu}$ (cm^{-1}) = 3420.9 (vw), 1671.3 (vw), 1610.1 (w), 1592.9 (w), 1535.8 (w), 1485.1 (w), 1453.5 (w), 1409.9 (m), 1263.9 (w), 1225.4 (w), 1164.9 (w), 1149.2 (w), 1096.0 (w), 1031.8 (vw), 864.1 (w), 798.6 (w), 753.5 (m), 695.9 (w), 649.8 (w), 583.8 (w), 546.2 (w), 521.7 (w), 445.0 (w), 387.9 (w). – **EA**: calcd for $Lu(L_9)_3(H_2O)_2$,% C 40.14, H 2.57, found C 40.64, H 2.70.

V.2.2.2 Lanthanide ternary complexes

$Eu(L_1)_3$(Phen) (HL_1=2-fluorobenzoic acid, Phen= 1,10-phenanthroline) (Eu-1.9-Phen)

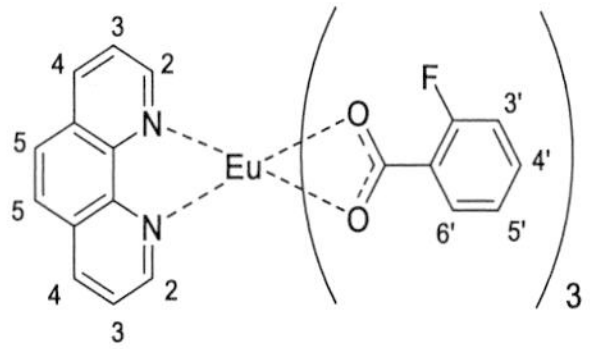

According to **GP2**, a solution of 68.0 mg of 1,10-phenanthroline **2.32** (0.38 mmol, 1.00 equiv.) in 20 mL ethanol was added to a solution of 229 mg of tris-europium-2-fluorobenzoate dihydrate **Eu-1.9** (0.38 mmol, 1.00 eqiuv.) in 30 mL ethanol and the reaction mixture was refluxed for 2 h and then allowed to cool to room temperature. The precipitate was recovered by filtration, washed with cold ethanol and dried in vacuo to afford 230 mg (81%) of a pale-pink solid.

^{1}H NMR (300 MHz, DMSO-d_6): δ (ppm) = 9.10 (bs, 2H, 2 × H_2), 8.50 (d, J = 8.0 Hz, 2H, 2 × H_4), 8.00 (s, 2H, 2 × H_5), 7.77 (bs, 2H, 2 × H_3), 7.04 (bs, 3H, 3 × $H_{4'}$), 6.58 (bs, 9H, 3 × $H_{6',3',5'}$). – **^{13}C NMR** (101 MHz, DMSO-d_6): δ (ppm) = 161.52 ($C_{quart.}$, COO), 158.99 ($C_{quart.}$, 3 × C_{Ar}F), 149.89 (+, 2 × C_2H), 145.67 ($C_{quart.}$, 2 × CN), 136.26 (+, 2 × C_4H), 131.98 (+, 3 × $C_{4'}$H), 131.24 (+, 3 × $C_{6'}$H), 128.46 (+, 2 × C_5H, 2 × $C_{quart.}$, C_{Phen}), 126.44 (+, 2 × C_3H), 122.78 (+, 3 × $C_{5'}$H), 114.24 (+, 3 × $C_{3'}$H), 114.02 ($C_{quart.}$, 3 × C_{Ar}COO). – **^{19}F NMR** (377 MHz, DMSO-d_6): δ (ppm) = -116.64 (s, 1F). – **IR** (ATR): $\tilde{\nu}$ (cm^{-1}) = 1606.9 (m), 1540.6 (w), 1514.6 (w), 1484.9 (w), 1451.3 (w), 1400.2 (m), 1260.1 (vw), 1222.2 (w), 1143.8 (vw), 1093.0 (vw), 1032.7 (vw), 864.3 (w), 853.1 (w), 807.1 (vw), 790.8 (vw), 751.2 (m), 731.8 (m), 722.1 (w), 651.7 (w), 636.9 (vw), 543.2 (vw), 459.4 (vw), 409.4 (vw). – **EA**: calcd for $C_{33}H_{20}F_3N_2O_6Eu$,% C 52.88, H 2.69, N 3.74, found C 52.97, H 2.61, N 3.82.

Eu(L$_2$)$_3$(Phen) (HL$_2$=2,5-difluorobenzoic acid, Phen= 1,10-phenanthroline) (Eu-1.7-Phen)

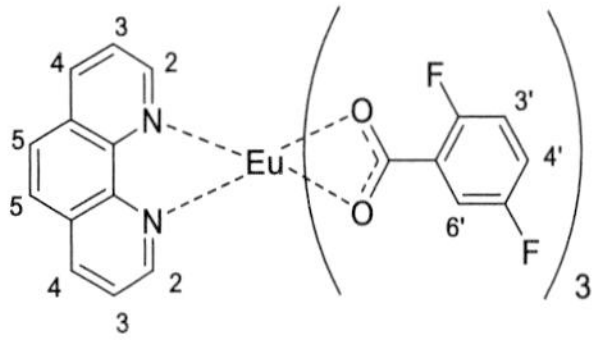

According to **GP2**, a solution of 68.0 mg of 1,10-phenanthroline (0.38 mmol, 1.00 equiv.) in 20 mL ethanol was added to a solution of 235 mg of tris-europium-2,5-difluorobenzoate dihydrate **Eu-1.7** (0.38 mmol, 1.00 eqiuv.) in 30 mL ethanol and the reaction mixture was refluxed for 2 h and then allowed to cool to room temperature. The precipitate was recovered by filtration, washed with cold ethanol and dried in vacuo to afford 247 mg (81%) of a pale-pink solid.

^{1}H NMR (300 MHz, DMSO-d_6): δ (ppm) = 9.09 (bs, 2H, 2 × H_2), 8.51 (bs, 2H, 2 × H_4), 8.01 (s, 2H, 2 × H_5), 7.76 (bs, 2H, 2 × H_3), 6.92 (bs, 3H, 3 × $H_{6'}$), 6.66 (bs, 3H, 3 × $H_{3'}$), 6.20 (bs, 3H, 3 × $H_{4'}$). – **^{19}F NMR** (377 MHz, DMSO-d_6): δ (ppm) = -122.91 (s, 1F), -124.80 (d, J = 18.6 Hz, 1F). – **IR** (ATR): $\tilde{\nu}$ (cm^{-1}) = 1614.0 (m), 1583.1 (m), 1551.1 (m), 1515.7 (w), 1487.7 (m), 1423.2 (m), 1383.0 (s), 1245.0 (m), 1191.9 (m), 1122.0 (w), 941.5 (vw), 890.6 (vw), 853.0 (m), 820.3 (s), 794.1 (w), 756.4 (m), 731.2 (m), 722.1 (m), 697.4 (w), 667.5 (w), 636.9 (vw), 609.5 (vw), 592.6 (vw), 547.3 (vw), 524.6 (w), 468.0 (vw), 415.1 (m). – **EA**: calcd for $C_{33}H_{17}F_6N_2O_6Eu$,% C 49.33, H 2.13, N 3.49, found C 49.37, H 1.98, N 3.58.

Eu(L$_3$)$_3$(Phen) (HL$_3$=2,4-difluorobenzoic acid, Phen= 1,10-phenanthroline) (Eu-1.6-Phen).

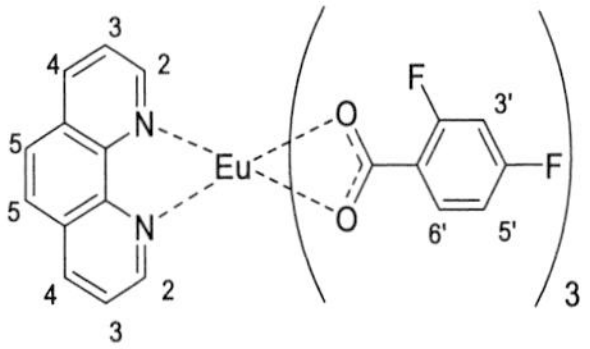

According to **GP2**, a solution of 68.0 mg of 1,10-phenanthroline (0.38 mmol, 1.00 equiv.) in 20 mL ethanol was added to a solution of 235 mg of tris-europium-2,4-difluorobenzoate dihydrate **Eu-1.6** (0.38 mmol, 1.00 eqiuv.) in 30 mL ethanol and the reaction mixture was refluxed for 2 h and then allowed to cool to room temperature. The precipitate was recovered by filtration, washed with cold ethanol and dried in vacuo to afford 244 mg (80%) of a pale-pink solid.

^{1}H NMR (400 MHz, DMSO-d_6): δ (ppm) = 9.08 (bs, 2H, H_2), 8.53 (bs, 2H, H_4), 8.03 (s, 2H, H_5), 7.73 (s, 2H, H_3), 6.47 (d, J = 46.3 Hz, 9H, $H_{3',4',6'}$). – **^{13}C NMR** (101 MHz, DMSO-d_6): δ (ppm) = 164.1 ($C_{quart.}$, COO), 162.3 ($C_{quart.}$, 3 × C_{Ar}F), 159.7 ($C_{quart.}$, 3 × C_{Ar}F), 150.1 (+, 2 × C_2H), 145.5 ($C_{quart.}$, 2 × CN), 136.2 (+, 2 × C_4H), 133.2 (+, 3 × $C_{6'}$H), 128.4 (+, 2 × C_5H, $C_{quart.}$, 2 × C_{Phen}), 126.5 (+, 2 × C_3H), 123.2 (+, 3 × $C_{5'}$H), 109.8 (+, 3 × $C_{3'}$H), 102.0 ($C_{quart.}$, 3 × C_{Ar}COO). – **^{19}F NMR** (377 MHz, DMSO-d_6): δ (ppm) = -111.45 (s, 2F). – **IR** (ATR): $\tilde{\nu}$ (cm^{-1}) = 1605.8 (m), 1548.7 (m), 1518.4 (vw), 1502.0 (w), 1424.4 (m), 1390.0 (s), 1268.6 (m), 1135.4 (m), 1091.5 (m), 972.2 (m), 864.5 (w),

846.9 (m), 780.4 (m), 731.1 (m), 700.1 (vw), 620.1 (m), 606.3 (m), 589.6 (m), 520.6 (vw), 458.5 (vw), 417.4 (vw), 390.6 (w). – **EA**: calcd for $C_{33}H_{17}F_6N_2O_6Eu$,% C 49.33, H 2.13, N 3.49, found C 49.42, H 2.01, N 3.61.

Eu(L$_1$)$_3$(BPhen) (HL$_1$=2-fluorobenzoic acid, BPhen= bathophenanthroline) (Eu-1.9-BPhen).

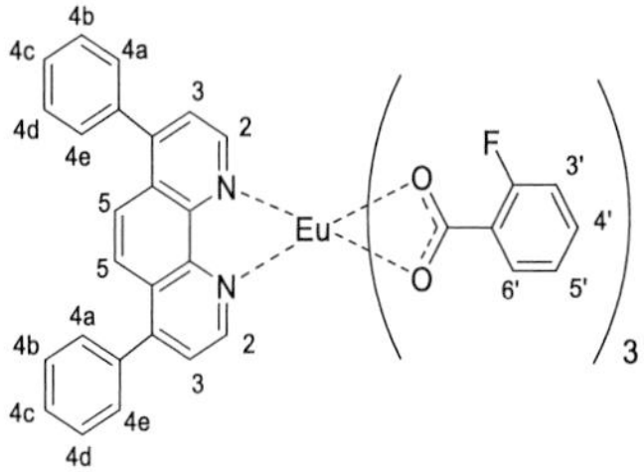

According to **GP2**, a solution of 125 mg of bathophenanthroline (0.38 mmol, 1.00 equiv.) in 20 mL ethanol was added to a solution of 229 mg of tris-europium-2-fluorobenzoate dihydrate **Eu-1.9** (0.38 mmol, 1.00 eqiuv.) in 30 mL ethanol and the reaction mixture was refluxed for 2 h and then allowed to cool to room temperature. The precipitate was recovered by filtration, washed with cold ethanol and dried in vacuo to afford 264 mg (77%) of a pale-pink solid.

^{1}H NMR (300 MHz, DMSO-d_6): δ (ppm) = 9.17 (d, J = 4.7 Hz, 2H, 2 × H_2), 7.86 (s, 2H, 2 × H_5), 7.73 (d, J = 4.5 Hz, 2H, 2 × H_3), 7.60 (bs, 10H, 2 × $H_{4a\text{-}e}$), 7.09 (s, 3H, 3 × $H_{6'}$), 6.63 (s, 9H, 3 × $H_{3',4',5'}$). – **^{13}C NMR** (101 MHz, DMSO-d_6): δ (ppm) = 161.6 (C$_{quart.}$, *C*OOH), 159.1 (C$_{quart.}$, C_{Ar}), 132.1 (+, C_{Ar}H), 132.0 (C$_{quart.}$, C_{Ar}), 131.4 (+, C_{Ar}H), 128.9 (+, C_{Ar}H), 122.8 (+, C_{Ar}H), 123.2 (C$_{quart.}$, C_{Ar}), 114.7 (C$_{quart.}$, C_{Ar}). – **^{19}F NMR** (377 MHz, DMSO-d_6): δ (ppm) = -116.55 (bs, 1F). – **IR** (ATR): $\tilde{\nu}$ (cm^{-1}) = 1608.1 (m), 1553.4 (m), 1489.1 (vw), 1383.5 (m), 1262.8 (m), 1221.8 (vw), 1148.8 (w), 1092.3 (w), 970.2 (w), 925.0 (vw), 841.3 (m), 765.4 (m), 750.1 (vw), 701.7 (m), 610.8 (m), 543.5 (w), 492.1 (vw), 469.5 (vw). – **EA**: calcd for $C_{45}H_{28}F_3N_2O_6Eu$,% C 59.94, H 3.13, N 3.11, found C 59.11, H 3.27, N 3.09.

Eu(L$_2$)$_3$(BPhen) (HL$_2$=2,5-difluorobenzoic acid, BPhen= bathophenanthroline) (Eu-1.7-BPhen).

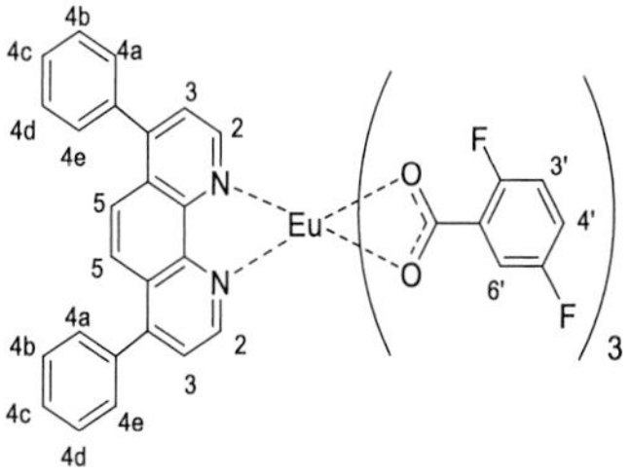

A solution of 125 mg of bathophenanthroline (0.38 mmol, 1.00 equiv.) in 20 mL ethanol was added to a solution of 235 mg of tris-europium-2,5-difluorobenzoate dihydrate **Eu-1.7** (0.38 mmol, 1.00 eqiuv.) in 30 mL ethanol and the reaction mixture was refluxed for 2 h and then allowed to cool to room temperature. The precipitate was recovered by filtration, washed with cold ethanol and dried in vacuo to afford 269 mg (74%) of a pale-pink solid.

^{1}H NMR (300 MHz, DMSO-d_6): δ (ppm) = 9.18 (d, J = 4.5 Hz, 2H, 2 × H_2), 7.87 (s, 2H, 2 × H_5), 7.73 (d, J = 4.5 Hz, 2H, 2 × H_3), 7.59 (bs, 10H, 2 × $H_{4a\text{-}e}$), 6.95 (s, 9H, 3 × $H_{6',4',3'}$). – **^{13}C NMR** (101 MHz,

DMSO-d_6): δ (ppm) = 157.6 ($C_{quart.}$, 3 ×COO), 156.93 ($C_{quart.}$, 3 × C_{Ar}) , 155.2 ($C_{quart.}$, 5 × C_{Ar}), 154.5 (+, 2 × C_{Ar}H), 137.0 ($C_{quart.}$, 2 × C_{Ar}), 130.2 (+, 6 × C_{Ar}H), 129.0 (+, 7 × C_{Ar}H), 123.3 (+, 2 × C_{Ar}H, $C_{quart.}$, 2 × C_{Ar}), 118.2 ($C_{quart.}$, 3 × C_{Ar}COO, d, J = 17.9 Hz), 117.3 (+, 6 × C_{Ar}H, d, J = 24.0 Hz), 115.2 (+, 2 × C_{Ar}H, dd, J = 25.7, 8.2 Hz). – ^{19}**F NMR** (377 MHz, DMSO-d_6): δ (ppm) = -122.72 (bs, 1F), -124.84 (d, J = 18.6 Hz, 1F). – **IR** (ATR): $\tilde{\nu}$ (cm^{-1}) = 1607.6 (m), 1552.2 (m), 1507.4 (vw), 1486.6 (vw), 1380.8 (m), 1269.4 (m), 1189.3 (vw), 1072.4 (w), 1019.2 (w) 970.3 (vw), 923.6 (vw), 841.3 (w), 810.6 (w), 765.2 (m), 749.2 (vw), 701.6 (m), 618.4 (m), 593.2 (w), 541.6 (w), 495.7 (vw), 469.0 (vw). – **EA**: calcd for $C_{45}H_{25}F_6N_2O_6Eu$,% C 56.56, H 2.64, N 2.93, found C 56.51, H 2.77, N 2.84.

Eu(L_3)$_3$(BPhen) (HL_3=2,4-difluorobenzoic acid, BPhen= bathophenanthroline) (Eu-1.6-BPhen)

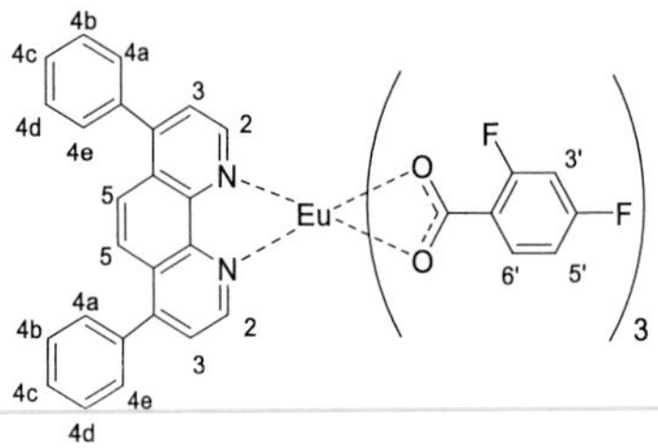

A solution of 125 mg of bathophenanthroline (0.38 mmol, 1.00 equiv.) in 20 mL ethanol was added to a solution of 235 mg of tris-europium-2,4-difluorobenzoate dihydrate **Eu-1.6** (0.38 mmol, 1.00 eqiuv.) in 30 mL ethanol and the reaction mixture was refluxed for 2 h and then allowed to cool to room temperature. The precipitate was recovered by filtration, washed with cold ethanol and dried in vacuo to afford 276 mg (76%) of a pale-pink solid.

^{1}**H NMR** (300 MHz, DMSO-d_6): δ (ppm) = 9.18 (bs, 2H, 2 × H_2),), 7.97 – 7.37 (m, 17H, 2 × H_5, 2 × H_3, 2 × $H_{4a\text{-}e}$, 3 × $H_{6'}$), 6.62 (d, J = 35.6 Hz, 6H, 3 × $H_{3',5'}$). – ^{13}**C NMR** (101 MHz, DMSO-d_6): δ (ppm) = 161.6 ($C_{quart.}$, COOH), 149.7 ($C_{quart.}$, C_{Ar}), 133.1 (+, C_{Ar}H), 129.7 ($C_{quart.}$, C_{Ar}), 128.9 (+, C_{Ar}H), 123.7 (+, C_{Ar}H), 110.05 (+, C_{Ar}H), 102.21 (+, C_{Ar}H), 101.95 (+, C_{Ar}H). – ^{19}**F NMR** (377 MHz, DMSO-d_6): δ (ppm) = -111.59 (bs, 2F). – **IR** (ATR): $\tilde{\nu}$ (cm^{-1}) = 1608.1 (m), 1553.8 (m), 1499.3 (vw), 1384.7 (m), 1269.4 (m), 1234.8 (vw), 1138.8 (w), 1091.1 (w), 970.3 (w), 924.1 (vw), 841.3 (m), 784.5 (w), 765.9 (m), 750.2 (vw), 740.0 (w), 701.6 (m), 608.3 (m), 573.5 (w), 543.4 (w), 492.3 (vw), 469.7 (vw). – **EA**: calcd for $C_{45}H_{25}F_6N_2O_6Eu$,% C 56.56, H 2.64, N 2.93, found C 57.02, H 2.71, N 2.96.

V.2.2.3 Lanthanide 9-anthracenates

According to **GP1a**, a suspension of 667 mg of anthracene-9-carboxylic acid Hant (3.00 mmol, 3.00 equiv.) in water was treated with 1 M aq. solution of NaOH at 60 °C until the pH was slightly acidic (pH = 5–6, approx. 3 mL) and the solution became clear. Then an aqueous solution of $LnCl_3 \cdot x\ H_2O$ (1.00 mmol, 1.00 equiv.) was added following by an immediate precipitation of the product. The reaction mixture was stirred for additional 3 h at rt, the crude product was filtered, washed with ethanol and dried in vacuo to afford the product as a pale-yellow solid.

lanthanum anthracene-9-carboxylate La(ant)$_3$ (La-1.10)

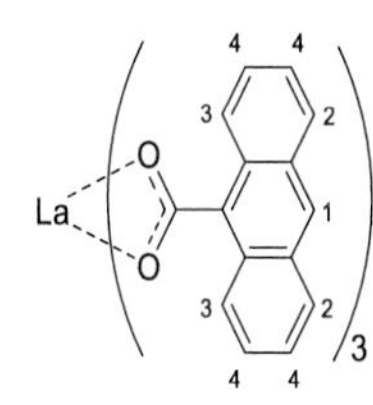

For the synthesis 373 mg of $LaCl_3 \cdot 7\ H_2O$ were used. 730 mg (91%) of a pale-yellow solid of general formula La(ant)$_3$ were obtained.

^{1}H NMR (300 MHz, DMSO-*d$_6$*): δ (ppm) = 8.62 (d, J = 8.6 Hz, 2H, 2 × H_2), 8.55 (bs, 1H, H_1), 8.07 (d, J = 8.5 Hz, 2H, 2 × H_3), 7.48 – 7.32 (m, J = 8.5 Hz, J = 6.2 Hz, 4H, 4 × H_4). – **^{13}C NMR** (101 MHz, DMSO-*d$_6$*): δ (ppm) = 169.3 ($C_{quart.}$, COOH), 130.9 ($C_{quart.}$, 3 × C_{Ar}), 129.5 ($C_{quart.}$, 2 × C_{Ar}), 127.8 (+, 4 × C_{Ar}H), 126.5 (+, 5 × C_{Ar}H). – **EA**: calcd for La(ant)$_3$,% C 67.34, H 3.39, found C 64.11, H 3.38. – **MS** (MALDI-TOF, THAP): m/z = 803.1 $[M+H]^+$.

praseodimium anthracene-9-carboxylate Pr(ant)$_3$ (Pr-1.10)

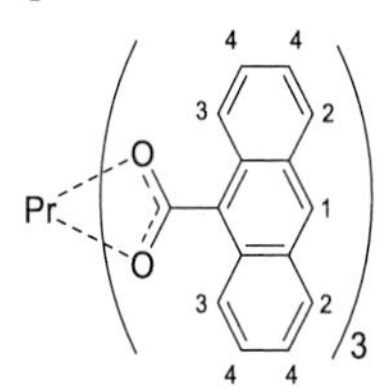

For the synthesis 373 mg of $PrCl_3 \cdot 7\ H_2O$ were used. 716 mg (89%) of a pale-yellow solid of general formula Pr(ant)$_3$ were obtained.

^{1}H NMR (300 MHz, DMSO-*d$_6$*): δ (ppm) = 11.55 (bs, 2H, 2 × H_3), 9.45 (bs, 1H, H_1), 8.78 (bs, 2H, 2 × H_2), 7.86 (d, J = 66.1 Hz, 4H, 4 × H_4). – **EA**: calcd for Pr(ant)$_3$,% C 64.42, H 3.29, found C 65.16, H 3.04. – **MS** (MALDI-TOF, THAP): m/z = 805.6 $[M+H]^+$.

neodymium anthracene-9-carboxylate Nd(ant)$_3$ (Nd-1.10)

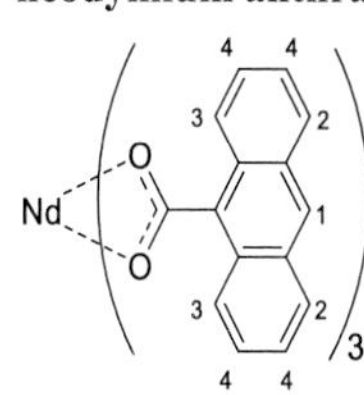

For the synthesis 359 mg of $NdCl_3 \cdot 6\ H_2O$ were used. 751 mg (93%) of a pale-yellow solid of general formula Nd(ant)$_3$ were obtained.

^{1}H NMR (300 MHz, DMSO-*d$_6$*): δ (ppm) = 9.17 (bs, 2H, 2 × H_3), 8.73 (bs, 1H, H_1), 8.27 (d, J = 8.6 Hz, 2H, 2 × H_2), 7.66 – 7.32 (m, 4H, 4 × H_4). – **EA**: calcd for Nd(ant)$_3$,% C 66.90, H 3.37, found C 66.58, H 3.16. – **MS** (MALDI-TOF, THAP): m/z = 806.7 $[M+H]^+$.

europium anthracene-9-carboxylate Eu(ant)$_3$ (Eu-1.10)

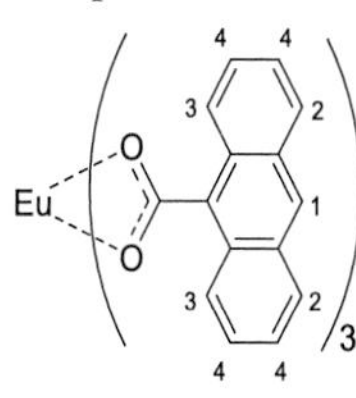

For the synthesis 366 mg of $EuCl_3 \cdot 6\ H_2O$ were used. 726 mg (89%) of a pale-yellow solid of general formula Eu(ant)$_3$ were obtained.

^{1}H NMR (300 MHz, DMSO-*d$_6$*): δ (ppm) = 8.43 (bs, 1H, H_1), 7.81 (d, J = 8.5 Hz, 2H, 2 × H_2), 7.44 – 7.17 (m, 6H, 2 × H_3, 4 × H_4). – **EA**: calcd for

Eu(ant)$_3$,% C 66.26, H 3.34, found C 65.18, H 3.05. – **MS** (MALDI-TOF, THAP): m/z = 817.2 $[M+H]^+$.

gadolinium anthracene-9-carboxylate Gd(ant)$_3$ (Gd-1.10)

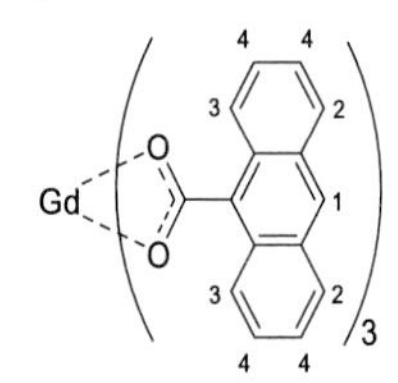

For the synthesis 372 mg of $GdCl_3 \cdot 6\ H_2O$ were used. 739 mg (90%) of a pale-yellow solid of general formula Gd(ant)$_3$ were obtained.

EA: calcd for Gd(ant)$_3$,% C 65.84, H 3.32, found C 65.69, H 3.06. – **MS** (MALDI-TOF, THAP): m/z = 822.2 $[M+H]^+$.

terbium anthracene-9-carboxylate Tb(ant)$_3$ (Tb-1.10)

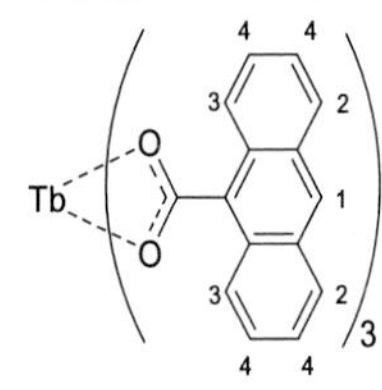

For the synthesis 373 mg of $TbCl_3 \cdot 6\ H_2O$ were used. 740 mg (90%) of a pale-yellow solid of general formula Tb(ant)$_3$ were obtained.

^{1}H NMR (300 MHz, DMSO-d_6): δ (ppm) = 11.65 (bs, 1H, H_1), 10.14 (bs, 2H, 2 × H_2), 8.61 (bs, 2H, 2 × H_3), 7.59 (bs, 4H, 4 × H_4). – **EA**: calcd for Tb(ant)$_3$,% C 65.70, H 3.31, found C 64.44, H 3.13. – **MS** (MALDI-TOF, THAP): m/z = 823.2 $[M+H]^+$.

dysprosium anthracene-9-carboxylate Dy(ant)$_3$ (Dy-1.10)

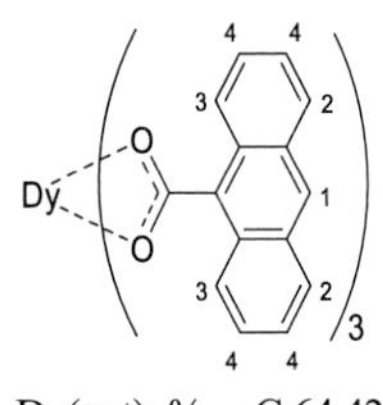

For the synthesis 377 mg of $DyCl_3 \cdot 6\ H_2O$ were used. 727 mg (88%) of a pale-yellow solid of general formula Dy(ant)$_3$ were obtained.

^{1}H NMR (300 MHz, DMSO-d_6): δ (ppm) = 14.78 (bs, 1H, H_1), 12.62 (bs, 2H, 2 × H_2), 10.44 (bs, 4H, 4 × H_4), 9.83 (bs, 2H, 2 × H_3). – **EA**: calcd for Dy(ant)$_3$,% C 64.42, H 3.29, found C 65.16, H 3.04. – **MS** (MALDI-TOF, THAP): m/z = 827.3 $[M+H]^+$.

erbium anthracene-9-carboxylate Er(ant)$_3$ (Er-1.10)

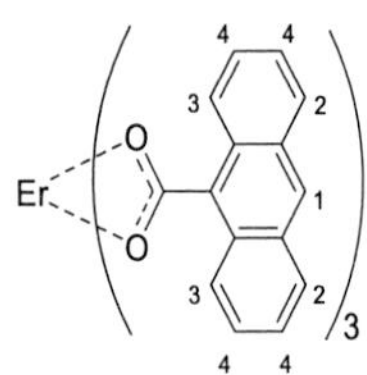

For the synthesis 381 mg of $ErCl_3 \cdot 6\ H_2O$ were used. 764 mg (92%) of a pale-yellow solid of general formula Er(ant)$_3$ were obtained.

^{1}H NMR (300 MHz, DMSO-d_6): δ (ppm) = 8.13 – 6.51 (bs, 9H, 9 × H_{Ar}). – **EA**: calcd for Er(L$_9$)$_3$(H$_2$O)$_2$,% C 65.04, H 3.28, found C 64.78, H 3.05. – **MS** (MALDI-TOF, THAP): m/z = 830.1 $[M+H]^+$.

ytterbium anthracene-9-carboxylate Yb(ant)3 (Yb-1.10)

For the synthesis 388 mg of $YbCl_3 \cdot 6\ H_2O$ were used. 736 mg (88%) of a pale-yellow solid of general formula Yb(ant)3 were obtained.

^{1}H NMR (300 MHz, DMSO-d_6): δ (ppm) = 8.01 – 6.85 (m, 9H, 9 × H_{Ar}). – **EA**: calcd for Yb(ant)3,% C 65.59, H 3.25, found C 65.36, H 2.92. – **MS** (MALDI-TOF, THAP): m/z = 838.2 $[M+H]^+$.

lutetium anthracene-9-carboxylate Lu(ant)3 (Lu-1.10)

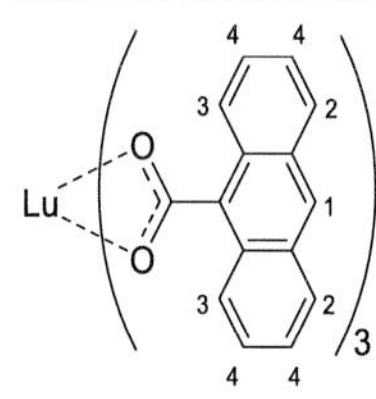

For the synthesis 389 mg of $LuCl_3 \cdot 6\ H_2O$ were used. 763 mg (91%) of a pale-yellow solid of general formula Lu(ant)3 were obtained.

^{1}H NMR (300 MHz, DMSO-d_6): δ (ppm) = 8.55 (s, 3H, H_1, 2 × H_2), 8.06 (d, J = 8.5 Hz, 2H, 2 × H_3), 7.53 – 7.16 (m, 4H, 4 × H_4). – **EA**: calcd for Lu(ant)3,% C 64.45, H 3.25, found C 64.13, H 2.95. – **MS** (MALDI-TOF, THAP): m/z = 839.3 $[M+H]^+$.

V.2.2.4 Lanthanide complexes for FRET-assay

***N*-(2-chloroethyl)-4-methylbenzenesulfonamide (1.13)**

To a solution of 5.75 g of 2-chloroethylamine hydrochloride **1.11** (49.6 mmol, 1.40 equiv.) and 12.2 g of K_2CO_3 (88.5 mmol, 2.50 equiv.) in distilled water (70 mL), 6.75 g of *p*-toluenesulfonyl chloride **1.12** (35.4 mmol, 1.00 equiv.) were added slowly at rt and left stirring overnight. The resulting precipitate was collected by suction filtration, washed with water and dried in vacuo to afford 8.08 g (98%) of a white solid. Analytical data are consistent with the literature.[269]

***R*f** (Hex/EtOAc 5:1): 0.65. – **^{1}H NMR** (400 MHz, $CDCl_3$): δ (ppm) = 7.76 (dd, J = 8.3 Hz, J = 1.9 Hz, 2H, 2 × H_{Ar}), 7.33 (d, J = 7.9 Hz, 2H, 2 × H_{Ar}), 4.90 (bs, J = 6.5 Hz, 1H, NH), 3.55 (td, J = 5.8 Hz, J = 1.8 Hz, 2H, CH_2Cl), 3.30 (q, J = 6.0 Hz, 2H, CH_2N), 2.44 (s, 3H, CH_3). – **^{13}C NMR** (101 MHz, $CDCl_3$): δ (ppm) = 143.9 ($C_{quart.}$, $C_{Ar}CH_3$), 136.9 ($C_{quart.}$, $C_{Ar}S$), 130.3 (+, 2 × $C_{Ar}H$), 127.1 (+, 2 × $C_{Ar}H$), 44.7 (–, NHCH_2), 43.6 (–, ClCH_2), 21.56 (+, CH_3). **LC-ELSD-MS**: t_{Ret} = 1.02 min.

1-tosylaziridine (1.14)

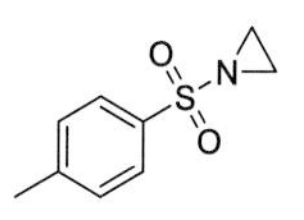

To 12 mL of a stirred 1.4 M solution of NaOH (5.00 equiv., 685 mg) in a salt/ice bath, 1.00 g of N-(2-chloroethyl)-4-methylbenzenesulfonamide **1.13** (4.28 mmol, 1.00 equiv.) was slowly added and stirring was continued for 3.5 h.

The precipitate was then allowed to settle overnight at 0 °C. The product was collected, washed with ice-cold distilled water and dried in vacuo to afford 801 mg (95%) of a white solid. Analytical data are consistent with the literature.[213]

^{1}H NMR (400 MHz, $CDCl_3$): δ (ppm) = 7.84 (d, J = 7.9 Hz, 2H, 2 × H_{Ar}), 7.35 (d, J = 7.9 Hz, 2H, 2 × H_{Ar}), 2.45 (s, 3H, CH_3), 2.37 (s, 4H, 2 × CH_2). – **^{13}C NMR** (101 MHz, $CDCl_3$): δ (ppm) = 144.8 ($C_{quart.}$, $C_{Ar}CH_3$), 135.0 ($C_{quart.}$, $C_{Ar}S$), 130.0 (+, $C_{Ar}H$), 129.9 (+, $C_{Ar}H$), 128.2 (+, $C_{Ar}H$), 127.8 (+, $C_{Ar}H$), 27.58 (–, 2 × CH_2), 21.8 (+, CH_3). **LC-ELSD-MS**: t_{Ret} = 0.94 min, m/z (ESI+, %) = 156.28 (100) $[Ts]^+$, 198.28 (30) $[M+H]^+$.

N,N',N'',N'''-((ethane-1,2-diylbis(azanetriyl))tetrakis(ethane-2,1-diyl))tetrakis(4-methylbenzenesulfonamide) (1.16)

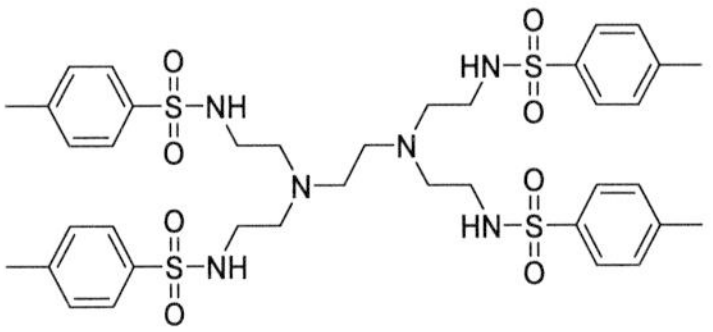

Under Ar atmosphere, 424 mg of tosylaziridine **1.14** (2.15 mmol, 5.00 equiv.) were dissolved in 3 mL of dry benzene at rt. A solution of 26.0 mg of ethylenediamine **1.15** (0.43 mmol, 1.00 equiv.) in 1 µL of benzene was added dropwise over a period of 1 h. The mixture was then heated at 30 °C overnight with stirring. The precipitate was allowed to settle in the fridge for 2 h and was collected as a white powder which was washed with acetonitrile and dried in vacuo to afford 241 mg (66%) of a white solid.

^{1}H NMR (400 MHz, $CDCl_3$): δ (ppm) = 7.77 (d, J = 7.9 Hz, 8H, 8 × H_{Ar}), 7.28 (s, 8H, 8 × H_{Ar}), 5.79 (d, J = 7.3 Hz, 4H, 4 × NHTs), 3.07 (bs, 12H, 6 ×CH_2), 2.56 (bs, 8H, 4 ×CH_2), 2.40 (s, 12H, 4 ×CH_3). – **^{13}C NMR** (101 MHz, $CDCl_3$): δ (ppm) = 143.9 ($C_{quart.}$, 4 × C_{Ar}), 137.1 ($C_{quart.}$, 4 × C_{Ar}), 130.2 (+, 8 × $C_{Ar}H$), 127.7 (+, 8 × $C_{Ar}H$), 54.0 (–, 4 × NCH_2CH_2NH), 53.4 (–, NCH_2CH_2N), 40.4 (–, 4 × NCH_2CH_2NH), 22.0 (+, 4 × CH_3). – **LC-ELSD-MS**: t_{Ret} = 1.02 min, m/z (ESI+, %) = 849.43 (100) $[M+H]^+$, 438.33 (55) $[M–2 × (CH_2)_2NH\text{-}Ts]^+$, m/z (ESI–, %) = 847.19 (100).

N1,N1'-(ethane-1,2-diyl)bis(N1-(2-aminoethyl)ethane-1,2-diamine) hydrobromide salt (1.17)

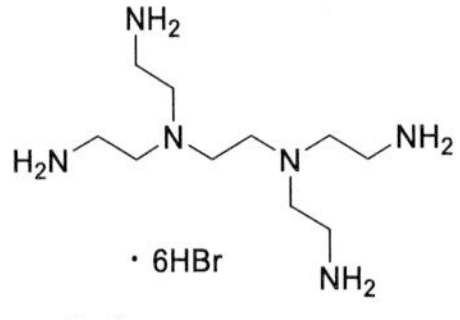

1.50 g of **1.16** (1.77 mmol, 1.00 equiv.) were dissolved in 20 mL of a 33% solution of HBr in acetic acid and stirred under reflux for 48 h. As it was warmed, the reaction mixture became more and more red until it was a deep brick red. After 48 h the reaction mixture was cooled to room temperature and then placed in an ice bath. 20 mL of a diethyl ether/ethanol mixture (1/1) were added slowly to the reaction mixture while it was stirred in the ice bath. The mixture was stirred in the ice bath for an additional 1 h. The resulting precipitate was collected, washed with the diethyl ether/ethanol mixture and dried in vacuo to afford 1.14 g (90%) of an off-white solid.

^{1}H NMR (400 MHz, D_2O): δ (ppm) = 3.08 (t, *J* = 6.5 Hz, 8H, 4 × $NH_3^+CH_2CH_2$), 2.84 (t, *J* = 6.6 Hz, 8H, 4 × $NH_3^+CH_2CH_2$), 2.75 (s, 4H, NCH_2CH_2N). – **^{13}C NMR** (101 MHz, D_2O): δ (ppm) = 50.3 (–, 4 × $NH_3^+CH_2CH_2$), 49.1 (–, NCH_2CH_2N), 36.6 (–, 4 × $NH_3^+CH_2CH_2$). **LC-ELSD-MS**: t_{Ret} = 0.15 min, *m/z* (ESI–, %) = 391.05 (50), 393.05 (100), 395.05 (50) $[C_{10}H_{28}N_6{\cdot}2HBr{-}H]^-$.

2-methoxyisophthalic acid (1.19)

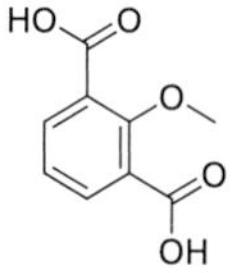

A suspension of 5.00 g of 1,3-dimethylanisole **1.18** (36.7 mmol, 1.00 equiv.), 38.3 g of potassium permanganate (242 mmol, 6.60 equiv.) and 6.59 g of potassium hydroxide (118 mmol, 3.20 equiv.) in 200 mL water was stirred at 80 °C for 10 h and then cooled to rt. The solid was filtered off and the filtrate was acidified with concentrated hydrochloric acid to pH 2. The resulting precipitate was filtered, washed with water and dried in vacuo to afford 3.67 g (51%) of the desired product as a white solid. Analytical data are consistent with the literature.[219]

***R*f** (DCM/MeOH 5:1): 0.47. – **^{1}H NMR** (400 MHz, DMSO-*d6*): δ (ppm) = 13.11 (s, 2H, 2 × COO*H*), 7.81 (d, *J* = 7.7 Hz, 2H, 2 × H_{Ar}), 7.26 (t, *J* = 7.7 Hz, 1H, H_{Ar}), 3.81 (s, 3H, CH_3). – **^{13}C NMR** (101 MHz, DMSO-*d6*): δ (ppm) = 167.5 ($C_{quart.}$, 2 × *C*OOH), 158.2 ($C_{quart.}$, $C_{Ar}OCH_3$), 134.0 (+, 2 × $C_{Ar}H$), 128.2 ($C_{quart.}$, 2 × $C_{Ar}COOH$), 124.1 (+, $C_{Ar}H$), 63.4 (+, OCH_3). **LC-ELSD-MS**: t_{Ret} = 0.59 min, *m/z* (ESI–, %) = 194.92 (100) $[M{-}H]^-$, 391.09 (79) $[2M{-}H]^-$.

dimethyl 2-methoxyisophthalate (1.20)

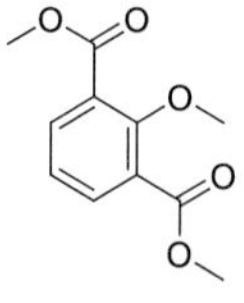

According to **GP3a**, to a stirred solution of 2.50 g of 2-methoxyisophthalic acid **1.19** (12.7 mmol, 1.00 equiv.) in 250 mL methanol, 2.05 mL concentrated H_2SO_4 (38.2 mmol, 3.00 equiv.) were added and the mixture was refluxed for 48 h. After concentration in vacuo, the residue was triturated with water. The resulting precipitate was collected and dried in vacuo to afford 2.59 g (91%) of a colorless oil.

***R*f** (Hex/EtoAc 4:1): 0.90. – **^{1}H NMR** (400 MHz, DMSO-*d6*): δ (ppm) = 7.86 (dd, *J* = 7.7 Hz, *J* = 4.1 Hz, 2H, 2 × H_{Ar}), 7.14 (td, *J* = 7.8 Hz, *J* = 4.2 Hz, 1H, H_{Ar}), 3.91 – 3.82 (bs, 9H, 3 × CH_3). – **^{13}C NMR** (101 MHz, DMSO-*d6*): δ (ppm) = 166.0 ($C_{quart.}$, 2 × *C*OOCH$_3$), 159.5 ($C_{quart.}$, $C_{Ar}OCH_3$), 134.8 (+, 2 × $C_{Ar}H$), 126.4 ($C_{quart.}$, 2 × $C_{Ar}COCH_3$), 123.3 (+, $C_{Ar}H$), 63.5 (+, OCH_3), 52.2 (+, 2 × COOCH_3). – **LC-ELSD-MS**: t_{Ret} = 0.95 min, *m/z* (ESI+, %) = 225.41 (10) $[M{+}H]^+$, 197.08 (65) $[M{-}2CH_3{+}H]^+$, 211.04 (100) $[M{-}CH_3{+}2H]^+$.

2-methoxy-3-(methoxycarbonyl)benzoic acid (1.21)

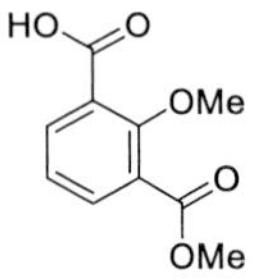

According to **GP4a**, 122 mg of dimethyl 2-methoxyisophthalate **1.20** (1.00 equiv., 544 µmol) were dissolved in 2.00 mL methanol and treated with 544 µL of a 1.00 M NaOH solution (1.00 equiv.), then heated at 50 °C for 4 h and afterwards concentrated in vacuo. The solid residue was dissolved in water (5.0 mL) and washed twice with EtOAc to remove unreacted starting material. Then the aqueous layer was acidified with 6 M HCl solution and the product was extracted with DCM (3 × 20 mL). The organic layers were combined, washed with water and brine and dried over Na_2SO_4. The solvent was evaporated to afford 40.0 mg (35%) of a white solid.

^{1}H NMR (400 MHz, $CDCl_3$): δ (ppm) = 10.85 (s, 1H, COO*H*), 8.34 (d, *J* = 7.4 Hz, 1H, H_{Ar}), 8.11 (dd, *J* = 7.8 Hz, *J* = 2.6 Hz, 1H, H_{Ar}), 7.35 (dd, *J* = 8.9 Hz, *J* = 6.9 Hz, 1H, H_{Ar}), 4.07 (s, 3H, $COOCH_3$), 3.98 (s, 3H, OCH_3). – **^{13}C NMR** (101 MHz, DMSO-d_6): δ (ppm) = 164.8 ($C_{quart.}$, *C*OOH), 164.4 ($C_{quart.}$, *C*$OOCH_3$), 159.0 ($C_{quart.}$, $C_{Ar}OCH_3$), 137.1 (+, 2 × C_{Ar}H), 125.9 ($C_{quart.}$, $C_{Ar}COOCH_3$), 124.8 ($C_{quart.}$, C_{Ar}COOH), 124.4 (+,C_{Ar}H), 64.4 (+, OCH_3), 52.6 (+, $COOCH_3$). – **LC-ELSD-MS**: t_{Ret} = 0.76 min, *m/z* (ESI+, %) = 197.08 (65) $[M–CH_3+H]^+$, 211.04 (100) $[M–CH_3+2H]^+$.

1-(2,5-dioxopyrrolidin-1-yl) 3-methyl 2-methoxyisophthalate (1.22)

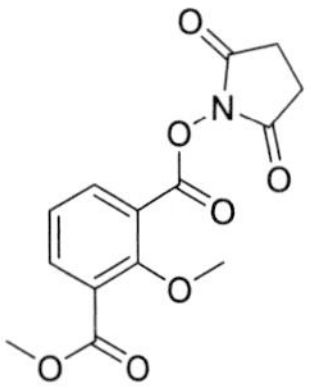

To a stirred solution of 100 mg of **1.21** (0.48 mmol, 1.00 equiv.) and 82.1 mg of NHS (0.71 mmol, 1.50 equiv.) in 5 mL DCM, 148 mg of EDC (0.95 mmol, 2.00 equiv.) and 1.0 mL DIPEA (5.71 mmol, 12.0 equiv.) were added. The reaction was monitored by LC-MS and additional 82.1 mg of NHS (0.71 mmol, 1.50 equiv.), 148 mg of EDC (0.95 mmol, 2.00 equiv.) and 2.0 mL DIPEA (11.4 mmol, 24.0 equiv.) were added to push the reaction towards completion. The mixture was left stirring overnight. The product was washed with 1 M HCl, $NaHCO_3$ and brine. The organic layer was dried over Na_2SO_4 and the solvent was removed under reduced pressure to afford a pale-yellow oil. This oil was recrystallized from MeOH to afford 391 mg (86%) of an off-white solid.

^{1}H NMR (400 MHz, $CDCl_3$): δ (ppm) = 8.15 (dd, *J* = 7.8 Hz, *J* = 1.6 Hz, 1H, H_{Ar}), 8.06 (dd, *J* = 7.9 Hz, *J* = 1.6 Hz, 1H, H_{Ar}), 7.30 (d, *J* = 7.8 Hz, 1H, H_{Ar}), 3.99 (s, 3H, $COOCH_3$), 3.95 (s, 3H, OCH_3), 2.92 (s, 4H, 2 × CH_2). – **^{13}C NMR** (101 MHz, $CDCl_3$): δ (ppm) = 169.2 ($C_{quart.}$, 2 × N*C*O), 166.2 ($C_{quart.}$, *C*$OOCH_3$), 161.2 ($C_{quart.}$, *C*OON), 159.2 ($C_{quart.}$, $C_{Ar}OCH_3$), 137.3 (+, C_{Ar}H), 135.9 (+, C_{Ar}H), 123.9 (+, C_{Ar}H), 112.8 ($C_{quart.}$, 2 × C_{Ar}COO), 64.3 (+, OCH_3), 52.7 (+, $COOCH_3$), 25.9 (–, 2 × CH_2). – **LC-ELSD-MS**: t_{Ret} = 0.84 min, *m/z* (ESI+, %) = 308.37 (65) $[M+H]^+$. *m/z* (ESI–, %) = 338.13 (100) $[M–CH_3+HCOOH]^-$.

tetramethyl 3,3',3'',3'''-(5,8-bis(2-formamidoethyl)-2,5,8,11-tetraazadodecanedioyl) tetrakis(2-methoxybenzoate) (1.24)

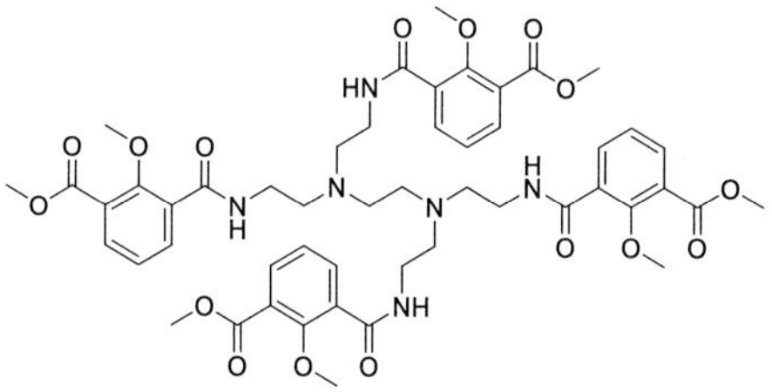

Variant A. To a suspension of 58.0 mg of **1.17** (0.0800 µmol, 1.00 equiv.) in 3.0 mL of DCM, 76.4 mg of **1.21** (0.350 µmol, 4.50 equiv.) were added, following by the addition of 210 mg of PyBOP (0.400 µmol, 5.00 equiv.) and 0.5 mL DIPEA (3.23 mmol, 40.0 equiv.). The reaction mixture was stirred overnight. The product was washed with 1 M HCl, $NaHCO_3$ and brine, dried over Na_2SO_4 and purified *via* flash column chromatography to afford 21 mg (27%) of a white solid.

Variant B. To a solution of 87.7 mg of the NHS-ester **1.22** (286 mmol, 4.10 equiv.) in 3.0 mL of DCM, the solution of 50.0 mg of amine hydrobromide salt **1.17** (69.7 mmol, 1.00 equiv.) and 121 µL of DIPEA (697 µmol, 10.0 equiv.) in 2 mL DCM was added and the mixture was stirred for 3 h at rt. The solvent was removed under reduced pressure and the resulting residue was purified *via* flash column chromatography on silica, using DCM and CH_3OH (95:5) as the eluent to afford 53.7 mg (77%) of a white foam.

^{1}H NMR (400 MHz, MeOD-d_4): δ (ppm) = 7.80 (ddd, J = 16.2, 7.8, 1.8 Hz, 8H, 8 × H_{Ar}), 7.19 (t, J = 7.7 Hz, 4H, 4 × H_{Ar}), 3.86 (s, 12H, 4 × $C_{Ar}OCH_3$), 3.81 (s, 12H, 4 × $COOCH_3$), 3.54 (t, J = 6.4 Hz, 8H, 4 × $NHCH_2CH_2N$), 2.88 (d, J = 5.7 Hz, 8H, 4 × $NHCH_2CH_2N$), 2.67 (s, 4H, NCH_2CH_2N). – **^{13}C NMR** (101 MHz, MeOD-d_4): δ (ppm) = 174.9 ($C_{quart.}$, 4 × $COOCH_3$), 168.2 ($C_{quart.}$, 4 × NHCO), 167.4 ($C_{quart.}$, 4 × $C_{Ar}OCH_3$), 135.1 (+, 4 × $C_{Ar}H$), 134.9 (+, 4 × $C_{Ar}H$), 130.9 (+, 4 × $C_{Ar}H$), 126.8 ($C_{quart.}$, 4 × $C_{Ar}CONH$), 125.1 ($C_{quart.}$, 4 × $C_{Ar}COOCH_3$), 64.1 (+, 4 × OCH_3), 54.8 (–, 4 × NCH_2CH_2NH), 54.4 (–, NCH_2CH_2N), 52.9 (+, 4 × $COOCH_3$), 38.8 (–, 4 × NCH_2CH_2NH). – **LC-ELSD-MS**: t_{Ret} = 0.88 min, t_{Ret} (5–95% acetonitrile + 0.1% formic acid in 10 min) = 4.99 min, m/z (ESI+, %) = 1001.56 (100), 1002.56 (55), 1003.56 (20) $[M+H]^+$, m/z (ESI–, %) = 999.32 (65) $[M-H]^-$, 1045.32 (100), 1046.40 (50), 1047.38 (15) $[M+HCOO^-]^-$.

3,3',3'',3'''-(5,8-bis(2-formamidoethyl)-2,5,8,11-tetraazadodecanedioyl)tetrakis(2-methoxybenzoic acid) (1.25)

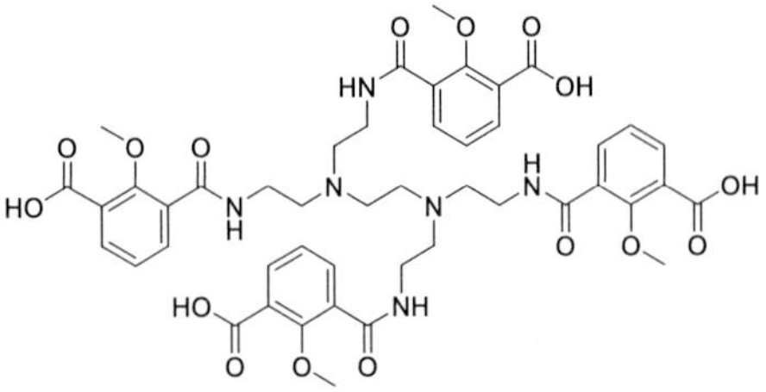

According to **GP4a**, 88.0 mg of **1.24** (1.00 equiv., 88.0 µmol) were dissolved in 1.00 mL methanol and treated with 0.50 mL of a 1.78 M solution of KOH in methanol (10.0 equiv., 0.88 mmol), refluxed for 4 h and acidified with 6 M HCl solution. The solvent was then removed under reduced pressure and the product was purified, using a reverse-phase column chromatography to afford 74.8 mg (91%) of a white foam.

^{1}H NMR (400 MHz, MeOD-d_4): δ (ppm) = 7.78 – 7.63 (m, 8H, 8 × H_{Ar}), 7.08 (t, J = 7.7 Hz, 4H, 4 × H_{Ar}), 3.77 (s, 12H, 4 × $C_{Ar}OCH_3$), 3.62 (t, J = 6.2 Hz, 8H, 4 × $NHCH_2CH_2N$), 3.25 – 3.22 (m, 4H, NCH_2CH_2N), 3.15 (t, J = 6.3 Hz, 8H, 4 × $NHCH_2CH_2N$). – **^{13}C NMR** (101 MHz, MeOD-d_4): δ (ppm) = 170.2 ($C_{quart.}$, 4 × COOH), 169.0 ($C_{quart.}$, 4 × NHCO), 158.6 ($C_{quart.}$, 4 × $C_{Ar}OCH_3$), 135.0 ($C_{quart.}$, 4 × C_{Ar}), 134.2 (+, 4 × $C_{Ar}H$), 133.9 (+, 4 × $C_{Ar}H$) 129.6 ($C_{quart.}$, 4 × C_{Ar}), 129.4 ($C_{quart.}$, 4 × C_{Ar}), 124.9 (+, 4 × $C_{Ar}H$), 63.8 (+, 4 × OCH_3), 54.8 (–, 4 × NCH_2CH_2NH), 52.1 (–, NCH_2CH_2N), 37.7 (–, 4 × NCH_2CH_2NH). – **LC-ELSD-MS**: t_{Ret} = 0.62 min, m/z (ESI+, %) = 945.55 (100), 944.31 (50), 945.27 (15) $[M+H]^+$. m/z (ESI–, %) = 943.31 (100), 944.31 (50), 945.27 (10) $[M-H]^-$.

3,3',3'',3'''-(5,8-bis(2-formamidoethyl)-2,5,8,11-tetraazadodecanedioyl)tetrakis(2-hydroxybenzoic acid) (1.28)

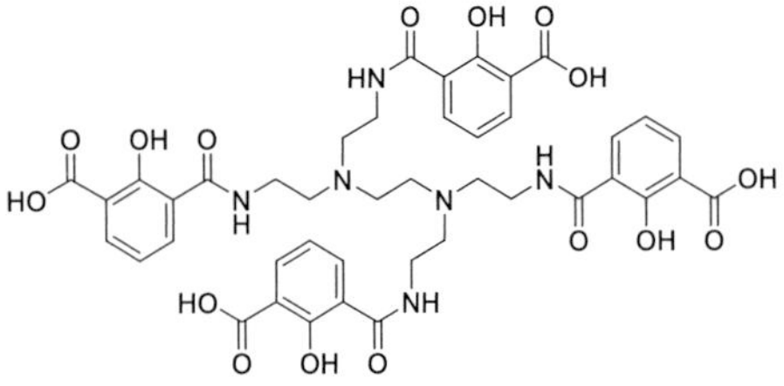

A mixture of 5.00 mg of **1.24** (4.99 µmol, 1.00 equiv.) and 33.4 mg of LiI (0.25 mmol, 50.0 equiv.) in 2.0 mL of 2,6-lutidine was refluxed at 145 °C for 4 h. 0.500 mL of concentrated HCl and 1.00 mL of H_2O were added to the ice-cooled solution which was then left at room temperature for an additional 30 min. The reaction mixture was evaporated to dryness and the product was purified by reverse phase chromatography to afford 3.09 mg (69%) of a white solid.

LC-ELSD-MS: t_{Ret} = 1.03 min, m/z (ESI+, %) = 889.54 (100), 890.57 (50), 891.56 (15) $[M+H]^+$. m/z (ESI–, %) = 887.30 (100), 888.36 (80) $[M-H]^-$.

48,49,63,64-tetramethoxy-1,4,7,15,18,21,24,32,37,45,52,60-dodecaazaheptacyclo [19.13.13.13^{4,18}.1^{9,13}.1^{26,30}.1^{39,43}.1^{54,58}]tetrahexaconta-9,11,13(63),26,28,30(49),39,41,43(48), 54,56,58(64)-dodecaene-8,14,25,31,38,44,53,59-octone (1.26)

129 mg of PyBOP (248 µmol, 5.00 equiv.) were dissolved in 50 mL DMF and left stirring for 10 minutes. Meanwhile, 46.9 mg of **1.25** (49.6 µmol, 1.00 equiv.) and 35.6 mg of **1.17** (49.6 µmol, 1.00 equiv.) with 50 µL of DIPEA (248 µmol, 5.00 equiv.) were dissolved in 5.00 mL and 4.95 mL DMF respectively, ensuring that the molar concentration of both solutions is the same. These solutions were added to a stirred solution of PyBOP portionwise, adding a 0.50 mL portion of each solution simultaneously every 5 min (10 portions in total). After 10 portions had been added, the reaction mixture was left stirring for 1 h and then the solvent was evaporated under reduced pressure. The crude product was purified *via* reverse phase column chromatography to afford 22.6 mg (41%) of a white solid.

^{1}H NMR (400 MHz, MeOD-d_4): δ (ppm) = 8.36 (bs, 8H, N*H*), 7.56 (d, J = 7.7 Hz, 4H, 4 × H_{Ar}), 7.26 – 6.93 (m, 8H, 8 × H_{Ar}), 3.81 (s, 12H, 4 × $C_{Ar}OCH_3$), 3.77 – 3.51 (m, 24H, 12 × CH_2), 3.28 – 3.11 (m, 8H, 4 × CH_2), 2.66 – 2.49 (m, 8H, 4 × CH_2). – **LC-ELSD-MS**: t_{Ret} = 0.47 min, *m/z* (ESI+, %) = 553.54 (100), 554.07 (65), 554.53 (15) $[M+2H]^{2+}$, 1105.80 (20), 1106.77 (10), 1107.74 (5) $[M+H]^+$.

48,49,63,64-tetrahydroxy-1,4,7,15,18,21,24,32,37,45,52,60-dodecaazaheptacyclo [19.13.13.13^{4,18}.1^{9,13}.1^{26,30}.1^{39,43}.1^{54,58}]tetrahexaconta-9,11,13(63),26,28,30(49),39,41,43(48), 54,56,58(64)-dodecaene-8,14,25,31,38,44,53,59-octone (1.27)

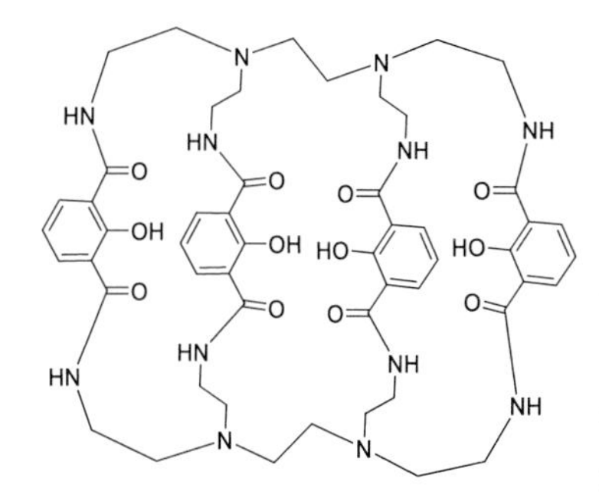

A mixture of 7.10 mg of **1.26** (6.40 µmol, 1.00 equiv.) and 43 mg of LiI (0.32 mmol, 50.0 equiv.) in 1.0 mL of 2,6-lutidine was refluxed at 145 °C for 4 h. 0.500 mL of concentrated HCl and 1.00 mL of H_2O were added to the ice-cooled solution which was then left at room temperature for an additional 30 min. The reaction mixture was evaporated to dryness and the product was purified by reverse phase chromatography to afford 1.10 mg (16%) of a white solid.

LC-ELSD-MS: t_{Ret} = 0.59 min, *m/z* (ESI+, %) = 525.55 (100), 525.95 (50), 526.48 (20) $[M+2H]^{2+}$, 1049.83 (60), 1050.70 (15), 1051.74 (5) $[M+H]^{+}$; *m/z* (ESI–, %) = 1047.51 (100), 1048.54 (60), 1049.60 (15) $[M–H]^{-}$.

Tb(III) complex with 48,49,63,64-tetrahydroxy-1,4,7,15,18,21,24,32,37,45,52,60-dodecaazaheptacyclo[19.13.13.13^{4,18}.1^{9,13}.1^{26,30}.1^{39,43}.1^{54,58}]tetrahexaconta-9,11,13(63),26,28,30(49),39,41,43(48),54,56,58(64)-dodecaene-8,14,25,31,38,44,53,59-octone (Tb-1.27)

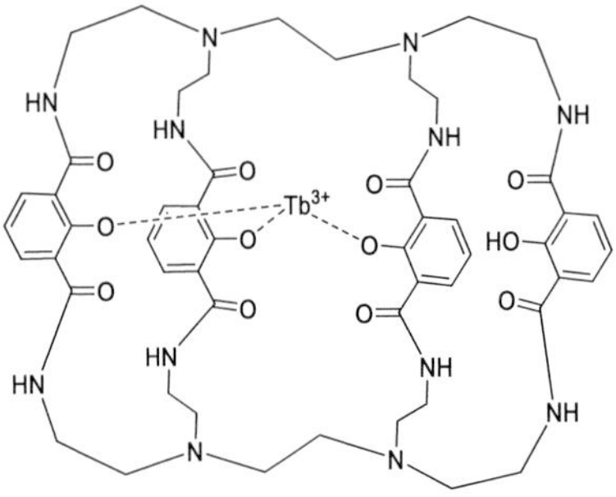

To a solution of 1.10 mg of **1.27** (1.05 µmol, 1.00 equiv.) in MeOH (1.00 mL), a 0.0248 M solution of Tb(III) chloride hexahydrate (99.99%, 0.360 mg, 0.953 µmol) in MeOH was added with stirring. The mixture was heated to gentle reflux. After the solution was heated at 60 °C for 2 h, the solvent was removed under reduced pressure to afford 1.12 mg (97%) of a white powder.

LC-ELSD-MS: t_{Ret} = 0.16 min, purity 85%, *m/z* (ESI+, %) = 1205.90 (100) $[M+H]^{+}$; *m/z* (ESI–, %) = 1203.54 (100) $[M–H]^{-}$.

ethyl *N*-(2-((2-(bis(2-((4-methylphenyl)sulfonamido)ethyl)amino)ethyl)(2-((4-methylphenyl)sulfonamido)ethyl)amino)ethyl)-*N*-tosylglycinate (1.31)

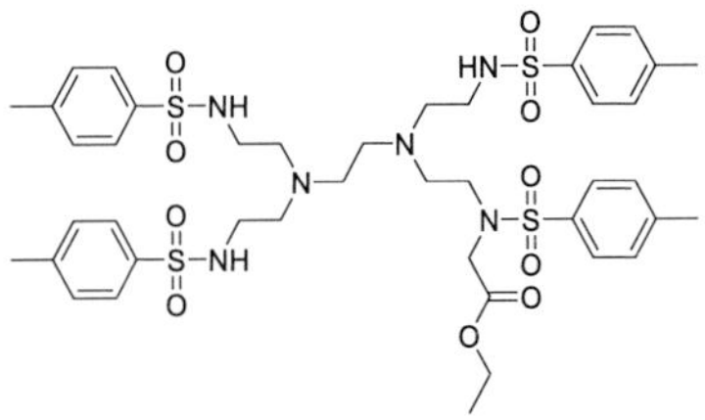

To a suspesion of p-toluenesulfonamide (154 mg, 0.180 mmol, 1.00 equiv.) and Cs_2CO_3 (58.9 mg, 0.180 mmol, 1.00 equiv.) in acetonitrile (10 mL), 20 µL ethyl 2-bromoacetate (0.180 mmol, 1.00 equiv.) were added and the mixture was refluxed for 4 h. The suspension was centrifuged and starting material **1.16** was recovered as an unreacted solid (82 mg). The filtrate was evaporated to dryness and the product was purified by flash column chromatography to afford 5 mg (6%) of a yellow oil.

^{1}H NMR (400 MHz, MeOD-d_4): δ (ppm) = 8.55 (bs, 3H, 3 × N*H*), 7.71 (d, *J* = 7.8 Hz, 8H, 8 × H_{Ar}), 7.36 (dd, *J* = 8.2, 3.3 Hz, 8H, 8 × H_{Ar}), 4.13–3.97 (m, 4H, NCH_2COOEt, $COOCH_2CH_3$), 3.23 (t, *J* = 7.2 Hz, 2H, CH_2NCH_2COOEt), 2.83 (dq, *J* = 12.2, 6.3 Hz, 6H, 3 × $NHCH_2CH_2N$), 2.69 – 2.62 (m, 6H, 3 × $NHCH_2CH_2N$), 2.60 (t, *J* = 7.1 Hz, 2H, $NCH_2CH_2NCH_2COOEt$), 2.41 (s, 12H, 4 × $C_{Ar}CH_3$), 2.37 (s, 4H, NCH_2CH_2N), 1.16 (t, *J* = 7.1 Hz, 3H, $COOCH_2CH_3$). – **^{13}C NMR** (101 MHz, MeOD-d_4): δ (ppm) = 170.5 ($C_{quart.}$, *C*OOEt), 144.9 ($C_{quart.}$, 4 × C_{Ar}), 139.0 ($C_{quart.}$, 4 × C_{Ar}), 131.1 (+, 8 × C_{Ar}H),

128.8 (+, 8 × C_{Ar}H), 62.6 (–, COOCH_2CH_3), 55.2 (–, 3 × NCH_2CH_2NH), 50.1 (–, NCH_2CH_2N), 47.7 (–, NCH_2COOEt), 45.3 (–, NCH_2CH_2NCH_2COOEt), 41.7 (–, 3 × N$CH_2$$CH_2$NH), 21.7 (+, 4 × $C_{Ar}CH_3$), 14.6 (+, COOCH_2CH_3). – **LC-ELSD-MS**: t_{Ret} = 6.84 min (5–95% acetonitrile + 0.1% formic acid in 10 min), *m/z* (ESI+, %) = 935.60 (100), 936.62 (50), 937.65 (20) $[M+H]^+$, 1870.61 (20) $[2M+2H]^+$; *m/z* (ESI–, %) = 933.39 (100), 934.38 (70) $[M–H]^-$.

V.2.3 Synthesis and characterisation Chapter III.2 Lanthanide conjugates

V.2.3.1 Functionalized ligands and lanthanide complexes with them

2-fluoro-4-nitrotoluene (2.22) and 2-fluoro-5-nitrotoluene (2.23)

A stirred solution of 10.0 g of 2-fluorotoluene (45.0 mmol, 110 mL, 1.00 equiv.) in 30 mL H_2SO_4 was cooled to 0 ° C. Then 24 mL concentrated HNO_3 were added dropwise so that the temperature did not exceed 50–60 °C. The mixture was stirred for additional 4 h and then poured onto crushed ice. The pH was adjusted to ~ 7, using NaOH (2 M). The mixture was extracted with EtOAc and washed with water and brine. The combined organic layers were dried over Na_2SO_4 and evaporated to dryness. Two isomers were isolated *via* flash column chromatography (CH/EtOAc 8:1). Analytical data are consistent with the literature.[270]

2.22 brown solid, 3.02 g (30%)

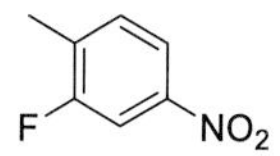

^{1}H NMR (300 MHz, $CDCl_3$): δ (ppm) = 7.98 – 7.91 (m, 2H, $H_{3,6}$), 7.17 – 7.11 (m, 1H, H_5), 2.37 (s, 3H, CH_3). – **^{19}F NMR** (377 MHz, $CDCl_3$): δ (ppm) = -107.02 (s, 1F). – **IR** (ATR): $\tilde{\nu}$ (cm^{-1}) = 3112.0 (m), 3088.4 (m), 2967.7 (s), 2929.1 (s), 2855.9 (s), 1667.3 (s), 1631 (vs), 1491.2 (s), 1457.4 (s), 1431.3 (m), 1361.0 (vs), 1318.0 (s), 1289.3 (m), 1270.9 (s), 1233.2 (s), 1198.8 (s), 1190.0 (s), 1141.8 (m), 1128.5 (m), 1078.0 (s), 1055.6 (w), 1004.3 (w), 925.7 (s), 884.5 (s), 827.0 (m), 817.1 (s), 751.1 (s), 644.9 (w), 539.4 (m), 443.5 (w). – **EA**: calcd for $C_7H_6FNO_2$,% C 54.20, H 3.90, N 9.03, found C 53.14, H 3.74, N 8.89.

2.23 yellow solid, 1.82 g, (18%)

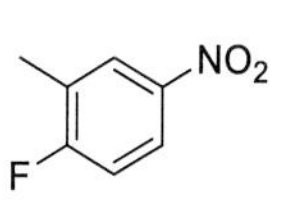

^{1}H NMR (300 MHz, $CDCl_3$): δ (ppm) = 7.96 – 8.01 (m, 2H, $H_{6,4}$), 7.36 – 7.42 (m, 1H, H_3), 2.21 (s, 3H, CH_3). – **^{19}F NMR** (377 MHz, $CDCl_3$): δ (ppm) = -113.40 (s, 1F). – **IR** (ATR): $\tilde{\nu}$ (cm^{-1}) = 3083.0 (m), 2957.7 (s), 2928.6 (vs), 2866.6 (s), 2729.1 (w), 1896.3 (vw), 1583.0 (m), 1515.8 (s), 1489.7 (s), 1382.0 (w), 1341.0 (s), 1315.1 (s), 1277.3 (m), 1240.6 (s), 1183.5 (m), 1118.8 (m), 1088.0 (s), 923.7 (m), 904.5 (s), 827.0 (s), 810.9 (m),

758.5 (m), 744.3 (s), 696.7 (w), 634.0 (s), 533.4 (m), 440.5 (m), 421.0 (m). – **EA**: calcd for $C_7H_6FNO_2$,% C 54.20, H 3.90, N 9.03, found C 54.11, H 4.01, N 8.95.

2-fluoro-4-nitrobenzoic acid (2.1-H)

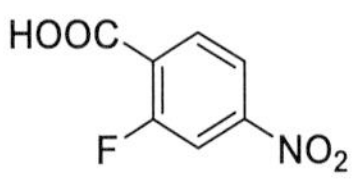

A stirred suspension of 2.55 g of 2-fluoro-4-nitrotoluene **2.22** (16.4 mmol, 1.00 equiv.) and 9.00 g of potassium dichromate (30.5 mmol, 1.00 equiv.) in 15 mLwater was cooled to 0 °C. Then 26.5 mL concentrated H_2SO_4 were added dropwise, so that the temperature did not exceed 50–60°C. After the addition was complete (15 min), the reaction was stirred for additional 15 min and then heated to 65 °C for 3 h. The reaction mixture was allowed to cool to room temperature and poured onto 100 mL crushed ice. The resulting mixture was stirred for 15 min and filtered. The solid was added to 30 mL of a 2 M Na_2CO_3 solution and the resulting suspension was filtered. The filtrate was cooled to 0 °C and 21 mL of a 25% aq. HCl was added dropwise. The mixture was extracted with EtOAc and washed with water and brine. The combined organic layers were dried over Na_2SO_4 and evaporated to dryness to afford 2.12 g (70%) of a pale-yellow solid. Analytical data are consistent with the literature.[271]

^{1}H NMR (400 MHz, $CDCl_3$): δ (ppm) = 13.95 (s, 1H, COO*H*), 8.19 (dd, *J* = 10.2 Hz, *J* = 2.1 Hz, 1H, H_3), 8.16 – 8.05 (m, 2H, $H_{5,6}$). – **^{13}C NMR** (75 MHz, $CDCl_3$): δ (ppm) = 163.8 ($C_{quart.}$, *C*OOH), 161.7 ($C_{quart.}$, C_{Ar}), 150.5 ($C_{quart.}$, C_{Ar}), 133.2 (+, C_{Ar}H), 125.5 ($C_{quart.}$, C_{Ar}), 119.4 (+, C_{Ar}H), 112.9 (+, C_{Ar}H). – **^{19}F NMR** (377 MHz, $CDCl_3$): δ (ppm) = -111.45 (s, 1H). – **IR** (ATR): $\tilde{\nu}$ (cm^{-1}) = 3111.8 (w), 1706.9 (w), 1609.4 (w), 1514.9 (s), 1415.2 (m), 1381.1 (w), 1345.9 (s), 1315.9 (m), 1268.7 (m), 1227.9 (s), 1191.0 (m), 1139.3 (m), 1125.0 (m), 1071.5 (m), 933.2 (m), 879.5 (s), 829.6 (w), 813.4 (s), 760.9 (w), 738.1 (s), 637.9 (w), 540.7 (w), 441.0 (m), 417.1 (w).

methyl 2-fluoro-4-nitrobenzoate (2.1-Me)

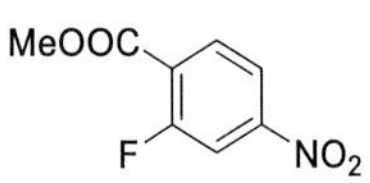

According to **GP3a**, 1.00 mL concentrated H_2SO_4 was added to a stirred solution of 2.00 g of 2-fluoro-4-nitrobenzoic acid 2.1-H (10.8 mmol, 1.00 equiv.) in 100 mL methanol and the mixture was refluxed for 48 h. After concentration in vacuo, the residue was triturated with water. The resulting precipitate was collected and dried in vacuo to afford 1.87 g (87%) of a yellow solid.

^{1}H NMR (400 MHz, $CDCl_3$): δ (ppm) = 8.24 – 7.89 (m, 3H, 3 × H_{Ar}), 3.99 (s, 3H, CH_3). – **^{13}C NMR** (101 MHz, $CDCl_3$): δ (ppm) = 156.9 ($C_{quart.}$, *C*OOMe), 136.7 ($C_{quart.}$, C_{Ar}), 132.9 ($C_{quart.}$, C_{Ar}), 132.3 (+, C_{Ar}H), 124.9 ($C_{quart.}$, *C*COOMe), 121.1 (+, C_{Ar}H), 111.62 (+, C_{Ar}H), 45.88 (+, CH_3). – **IR** (ATR): $\tilde{\nu}$ (cm^{-1}) = 3610.9 (w), 3108.7 (m), 3034.3 (m), 3025.2 (m), 3016.3 (m), 2882.5 (m), 2667.3 (w),

2528.7 (w), 1716.6 (s), 1602.5 (m), 1589.2 (m), 1538.1 (vs), 1474,1 (m), 1409.5 (m), 1343.8 (vs), 1286.3 (s), 1284.6 (m), 1150.2 (m), 1131.1 (w), 1108.0 (m), 1048.5 (m), 901.3 (s), 851.6 (m), 674.3 (w), 636.2 (w). – **EA**: calcd for $C_8H_6FNO_4$,% C 45.42, H 2.18, N 7.57, found C 45.32, H 2.21, N 7.50.

methyl 4-amino-2-fluorobenzoate (2.2-Me)

A solution of 1.80 g of methyl 2-fluoro-4-nitrobenzoate **2.1-Me** (9.00 mmol, 1.00 equiv.) in 15 mL methanol and 5 mL THF was hydrogenated at room temperature overnight in the presence of 300 mg of Raney-Ni (89%, 50 mol%, 4.50 mmol, 0.50 equiv.). After filtration through Celite®, the mixture was concentrated in vacuo to afford 1.50 g (quant.) of a brown solid.

^{1}H NMR (400 MHz, DMSO-d_6): δ (ppm) = 7.58 (t, J = 8.7 Hz, 1H, H_6), 6.40 (dd, J = 8.7 Hz, J = 2.2 Hz, 1H, H_3), 6.32 (d, J = 2.1 Hz, 1H, H_5), 6.27 (bs, 2H, NH_2), 3.72 (s, 3H, CH_3). – **^{13}C NMR** (101 MHz, DMSO-d_6): δ (ppm) = 164.6 ($C_{quart.}$, *C*OOMe), 162.1 ($C_{quart.}$, C_{Ar}), 155.5 ($C_{quart.}$, C_{Ar}), 133.2 (+, C_{Ar}H), 109.3 (+, C_{Ar}H), 103.4 ($C_{quart.}$, C_{Ar}), 99.5 (+, C_{Ar}H), 51.2 (+, CH_3). – **^{19}F NMR** (377 MHz, DMSO-d_6): δ (ppm) = -113.03 (s, 1F). – **EA**: calcd for $C_8H_8FNO_2$,% C 56.80, H 4.77, N 8.28, found C 56.61, H 4.73, N 8.34.

4-amino-2-fluorobenzoic acid (2.2-H)

According to **GP4a**, 1.50 g of methyl 4-amino-2-fluorobenzoate **2.2-Me** (8.87 mmol, 1.00 equiv.) were dissolved in 100 mL methanol and treated with ~ 20 mL 1 M NaOH solution (19.5 mmol, 2.20 equiv.), then stirred for 2.5 h. A solution of 1 M HCl was added to acidify the reaction mixture. The carboxylic acid was collected by filtration, then dried in vacuo to afford 1.24 g (90%) of a yellow solid.

^{1}H NMR (400 MHz, DMSO-d_6): δ (ppm) = 12.90 (s, 1H, COO*H*), 7.56 (t, J = 8.7 Hz, 1H, H_6), 6.37 (dd, J = 8.6 Hz, J = 2.1 Hz, 1H, H_3), 6.27 (dd, J = 14.1 Hz, J = 2.0 Hz, 1H, H_5), 6.20 (s, 2H, NH_2). – **^{13}C NMR** (101 MHz, DMSO-d_6): δ (ppm) = 165.2 ($C_{quart.}$, *C*OOH), 162.3 ($C_{quart.}$, C_{Ar}), 155.2 ($C_{quart.}$, C_{Ar}), 133.5 (+, C_{Ar}H), 109.2 ($C_{quart.}$, C_{Ar}), 104.6 (+, C_{Ar}H), 99.8 (+, C_{Ar}H). – **^{19}F NMR** (377 MHz, DMSO-d_6): δ (ppm) = -113.17 (s, 1F). – **EA**: calcd for $C_7H_6FNO_2$,% C 54.20, H 3.90, N 9.03, found C 54.90, H 3.83, N 9.12.

europium 4-amino-2-fluorobenzoate dihydrate (Eu-2.2)

For the synthesis 366 mg of $EuCl_3 \cdot 6\,H_2O$ and 465 mg of 4-amino-2-fluorobenzoic acid 2.2-H were used. 547 mg (84%) of an off-white solid of general formula $Eu(L)_3(H_2O)_2$ were obtained, where L=4-amino-2-fluorobenzoate.

^{1}H NMR (300 MHz, DMSO-d_6): δ (ppm) = 5.66 – 5.10 (m, 9H, 3 × 3 × H_{Ar}). – **^{19}F NMR** (377 MHz, DMSO-d_6): δ (ppm) = -109.72 (s, 1F, *F*). – **IR** (ATR): $\tilde{\nu}$ (cm^{-1}) = 3330.3 (w), 1613.6 (m), 1588.3 (m), 1550.8 (w), 1510.0 (m), 1444.0 (w), 1401.4 (m), 1381.1 (m), 1312.5 (m), 1239.6 (w), 1171.8 (m), 1112.0 (w), 964.7 (w), 844.6 (m), 785.8 (m), 688.6 (w), 620.7 (m), 602.8 (m), 524.1 (w), 431.3 (w). – **EA**: calcd for $Eu(L)_3(H_2O)_2$, where L = 4-amino-2-fluorobenzoate, % C 38.78, H 2.94, N 6.46, found C 38.91, H 3.01, N 6.55

methyl 4-azido-2-fluorobenzoate (2.3-Me)

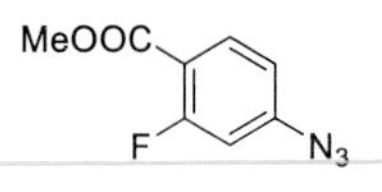

According to **GP7**, 1.00 g of methyl 4-amino-2-fluorobenzoate **2.2-Me** (5.92 mmol, 1.00 equiv.) was stirred in 10 mL of 50% HCl. The suspension was cooled to 0 °C and a solution of 531 mg of sodium nitrite (7.70 mmol, 1.30 equiv.) in 5 mL water was added dropwise over a period of 15 min while stirring. After 15 min of further stirring at 0–5 °C, a solution of 578 mg of sodium azide (8.89 mmol, 1.50 equiv.) in 5 mLwater was added at such a rate that the temperature of the reaction mixture did not exceed 15 °C until the end of the addition. The solid product, that precipitated after 10–15 min was filtered, washed with ice water and dried in vacuo to afford 1.00 g (87%) of a pale-yellow crystalline solid.

^{1}H NMR (400 MHz, $CDCl_3$): δ (ppm) = 8.04 (t, *J* = 8.2 Hz, 1H, H_6), 6.91 (dd, *J* = 8.6 Hz, *J* = 2.2 Hz, 1H, H_3), 6.82 (dd, *J* = 11.3 Hz, *J* = 2.2 Hz, 1H, H_5), 3.92 (s, 3H, CH_3). – **^{13}C NMR** (101 MHz, $CDCl_3$): δ (ppm) = 167.2 ($C_{quart.}$, *C*OOMe), 165.3 ($C_{quart.}$, C_{Ar}), 138.1 ($C_{quart.}$, C_{Ar}), 134.8 (+, C_{Ar}H), 115.3 (+, C_{Ar}H), 108.4 ($C_{quart.}$, C_{Ar}), 108.2 (+, C_{Ar}H), 54.2 (+, CH_3). – **^{19}F NMR** (377 MHz, $CDCl_3$): δ (ppm) = -109.24 (s, 1F). – **IR** (ATR): $\tilde{\nu}$ (cm^{-1}) = 2953.1 (w), 2119.4 (m), 1704.2 (m), 1608.0 (m), 1575.7 (m), 1501.0 (w), 1435.9 (m), 1281.9 (m), 1211.7 (m), 1120.5 (m), 966.5 (w), 943.7 (m), 868.3 (m), 828.6 (m), 765.1 (m), 676.8 (w), 607.0 (w), 592.9 (w), 558.5 (w), 529.2 (w), 484.0 (w), 456.9 (w), 408.2 (w). – **EA**: calcd for $C_8H_6FN_3O_2$,% C 49.24, H 3.10, N 21.53, found C 49.45, H 3.02, N 21.21.

4-azido-2-fluorobenzoic acid (2.3-H)

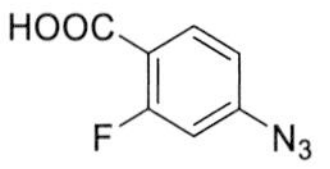

According to **GP4a**, 1.00 g of methyl 4-azido-2-fluorobenzoate **2.3-Me** (5.13 mmol, 1.00 equiv.) were dissolved in 70 mL methanol and treated with ~ 14 mL 1 M NaOH solution (11.3 mmol, 2.20 equiv.), then stirred for 2.5 h. A solution of 1 M HCl was added to acidify the reaction mixture. The carboxylic acid was collected by filtration, then dried in vacuo to afford 0.93 g (92%) of a grey solid.

^{1}H NMR (400 MHz, $CDCl_3$): δ (ppm) = 12.91 (s, 1H, COO*H*), 7.94 – 7.87 (m, 1H, H_6), 7.16 – 7.04 (m, 2H, $H_{3,5}$).– **^{13}C NMR** (101 MHz, $CDCl_3$): δ (ppm) = 161.0 ($C_{quart.}$, *C*OOH), 159.8 ($C_{quart.}$, C_{Ar}), 137.8 (+, C_{Ar}H), 127.6 ($C_{quart.}$, C_{Ar}), 118.0 (+, C_{Ar}H), 117.4 ($C_{quart.}$, C_{Ar}), 115.7 (+, C_{Ar}H). – **^{19}F NMR** (377 MHz, $CDCl_3$): δ (ppm) = -111.84 (s, 1F). – **EA**: calcd for $C_7H_4FN_3O_2$,% C 46.42, H 2.23, N 23.20, found C 45.99, H 2.02, N 23.21.

methyl 2-fluoro-4-methylbenzoate (2.4-Me)

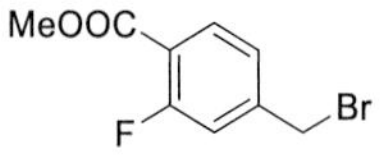

To a solution of 3.00 g of methyl 4-methylbenzoate **2.12-Me** (17.8 mmol, 1.00 equiv.) in dry $CHCl_3$ (60 mL), 3.91 g of NBS (21.8 mmol, 1.23 equiv.) and 58.0 mg of AIBN (0.02 mmol, 0.01 equiv.) were added. The reaction was refluxed for 12 h under Ar atmosphere. The solid was filtered off and the filtrate was evaporated in vacuo to afford 3.50 g (80%) of a pale-yellow oil.

^{1}H NMR (300 MHz, $CDCl_3$): δ (ppm) = 7.93 – 7.79 (m, 1H, H_6), 7.31 – 7.06 (m, 2H, $H_{3,5}$), 4.37 (d, J = 4.0 Hz, 2H, CH_2Br), 3.86 (s, 3H, CH_3). – **^{13}C NMR** (75 MHz, $CDCl_3$): δ (ppm) = 165.4 ($C_{quart.}$, *C*OOMe), 158.5 ($C_{quart.}$, +, C_{Ar}, C_{Ar}H), 138.9 ($C_{quart.}$, C_{Ar}), 133.7 (+, C_{Ar}H), 126.3 (+, C_{Ar}H), 117.4 ($C_{quart.}$, C_{Ar}), 56.0 (+, CH_3), 29.6 (–, CH_2).

4-(aminomethyl)-2-fluorobenzoic acid (2.5-H)

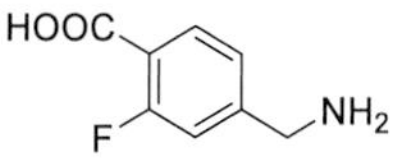

2.24 g (13.6 mmol, 1.00 equiv.) of 4-cyano-2-fluorobenzoic acid **2.13-H** were stirred at room temperature overnight under a hydrogen atmosphere in the presence of 300 mg of Pd/C (10 mol%). The mixture was then filtered through Celite® and the volatiles were removed under reduced pressure to afford 2.27 g (quant.) of a white solid. Analytical data are consistent with the literature.[272]

^{1}H NMR (300 MHz, D_2O): δ (ppm) = 7.94 – 7.84 (m, 1H, H_6), 7.57 – 7.43 (m, 2H, $H_{5,3}$), 4.47 (s, 2H, CH_2). – **^{13}C NMR** (75 MHz, D_2O): δ (ppm) = 172.4 ($C_{quart.}$, *C*OOH), 160.5 ($C_{quart.}$, C_{Ar}), 136.6 ($C_{quart.}$, C_{Ar}), 131.3 (+, C_{Ar}H), 124.9 (+, C_{Ar}H), 118.2 ($C_{quart.}$, C_{Ar}), 117.2 (+, C_{Ar}H), 43.0 (–, CH_2). – **^{19}F NMR** (377 MHz, D_2O): δ (ppm) = -114.81 (dd, J = 10.8 Hz, J = 7.7 Hz). – **IR** (ATR): $\tilde{\nu}$ (cm^{-1}) = 3440.1 (w),

3238.0 (m), 3046.2 (m), 2925.6 (vs), 2904.3 (s), 2841.2 (m), 2643.1 (s), 2446.6 (w), 2360.3 (m), 2230.4 (vw), 2137.1 (w), 1928.0 (m), 1628.6 (s), 1587.3 (s), 1541.1 (vs), 1500.4 (s), 1464.8 (m), 1424.7 (m), 1391.2 (m), 1372.2 (vs), 1301.6 (s), 1268.5 (w), 1241.4 (w), 1171.9 (w), 1162.0 (w), 1124.5 (vw), 1093.1 (m), 988.0 (w), 947.3 (vw), 905.9 (m), 882.7 (w), 854.4 (vw), 823.4 (w), 794.7 (m), 735.2 (m), 699.8 (vw), 645.0 (m), 617.6 (m), 594.2 (w), 585.3 (w), 540.7 (w), 522.1 (w), 508.8 (vw), 498.3 (w), 486.1 (w), 469.5 (vw), 421.1 (w). – **EA**: calcd for $C_8H_8FNO_2$,% C 56.80, H 4.77, N 8.28, found C 56.91, H 4.97, N 8.34.

methyl 4-(aminomethyl)-2-fluorobenzoate (2.5-Me)

MeOOC F NH_2

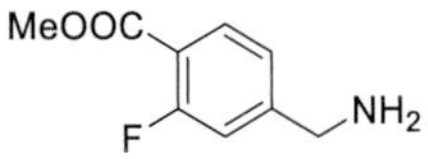

According to **GP3a**, 1.00 mL concentrated H_2SO_4 was added to a stirred solution of 1.82 g of 4-(aminomethyl)-2-fluorobenzoic acid **2.5-H**) (10.8 mmol, 1.00 equiv.) in 100 mL methanol and the mixture was refluxed for 48 h. After concentration in vacuo, the residue was triturated with water. The resulting precipitate was collected and dried in vacuo to give 1.72 g (87%) of a pale-yellow solid.

^{1}H NMR (300 MHz, D_2O): δ (ppm) = 7.99 (t, J = 8.0 Hz, 1H, H_6), 7.39-7.32 (m, 2H, $H_{3,5}$), 4.27 (s, 2H, CH_2), 3.94 (s, 3H, CH_3). – **^{13}C NMR** (75 MHz, D_2O): δ (ppm) = 166.3 ($C_{quart.}$, *COOMe*), 162.7 ($C_{quart.}$, C_{Ar}), 140.1 ($C_{quart.}$, C_{Ar}), 132.6 (+, C_{Ar}H), 124.5 (+, C_{Ar}H), 118.3 (+, C_{Ar}H), 117.3 ($C_{quart.}$, C_{Ar}), 52.9 (+, CH_3), 42.1 (–, CH_2). – **EA**: calcd for $C_9H_{10}FNO_2$,% C 59.01, H 5.50, N 7.65, found C 59.11, H 5.44, N 7.70.

methyl 4-(azidomethyl)-2-fluorobenzoate (2.6-Me)

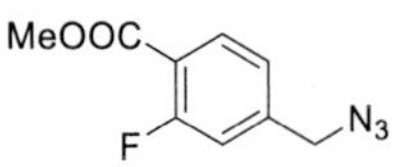

MeOOC F N_3

Variant A. According to **GP7**, 1.11 g of methyl 4-(aminomethyl)-2-fluorobenzoate **2.5-Me** (5.92 mmol, 1.00 equiv.) were stirred in 10 mL of 50% HCl. The solution/suspension was cooled to 0 °C and a solution of 531 mg of sodium nitrite (7.70 mmol, 1.30 equiv.) in 5 mLwater was added dropwise over a period of 15 min with stirring. After 15 min of further stirring at 0–5 °C, a solution of 578 mg of sodium azide (8.89 mmol, 1.50 equiv.) in 5 mL water was added at such a rate that the temperature of the reaction mixture did not exceed 15 °C towards the end of the addition. The solid product, that precipitated after 10–15 min was filtered, washed with ice water and dried in vacuo to afford 1.03 g (82%) of a pale-yellow solid.

Variant B. To a stirred solution of 2.00 g of methyl 4-(bromomethyl)benzoate (8.73 mmol, 1.00 equiv.) in 40 mL acetone/H_2O (3/1), 1.14 g of NaN_3 (17.5 mmol, 2.00 equiv.) were added. The resulting suspension was stirred for 30 min at 25 °C and the solvent was evaporated. The residue was purified, using flash column chromatography (DCM/CH 1:1) to afford 1.60 g (88%) of a white solid.

^{1}H NMR (300 MHz, $CDCl_3$): δ (ppm) = 8.11 – 7.85 (m, 1H, H_6), 7.12 – 7.01 (m, 2H, $H_{3,5}$), 4.42 (s, 2H, CH_2), 3.94 (s, 3H, CH_3).– **^{13}C NMR** (75 MHz, $CDCl_3$): δ (ppm) = 164.1 ($C_{quart.}$, $COOCH_3$, 159.8 ($C_{quart.}$, C_{Ar}), 137.5 ($C_{quart.}$, C_{Ar}), 136.2 (+, C_{Ar}H), 127.8 (+, C_{Ar}H), 117.1 ($C_{quart.}$, C_{Ar}), 114.7 (+, C_{Ar}H), 52.7 (+, CH_3), 49.3 (–, CH_2). – **IR** (ATR): $\tilde{\nu}$ (cm^{-1}) = 3049.9 (w), 2958.4 (w), 2191.2 (vw), 2102.6 (s), 1712.9 (s), 1622.4 (m), 1572.7 (w), 1505.4 (m), 1418.1 (s), 1346.1 (m), 1305.0 (m), 1260.3 (s), 1188.7 (m), 1141.8 (s), 1083.1 (s), 982.4 (w), 959.0 (m), 920.4 (s), 878.9 (m), 828.2 (m), 816.2 (m), 770.7 (s), 756.6 (s), 687.0 (m), 609.6 (w), 558.1 (m), 447.0 (w), 422.8 (m). – **EA**: calcd for $C_9H_8FN_3O_2$,% C 51.68, H 3.86, N 20.09, found C 51.45, H 3.02, N 21.21.

4-(azidomethyl)-2-fluorobenzoic acid (2.6-H)

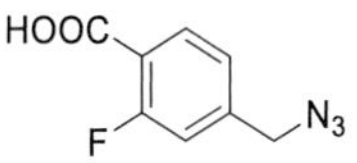

According to **GP4a**, 900 mg of methyl 4-(azidomethyl)-2-fluorobenzoate **2.6-Me** (4.30 mmol, 1.00 equiv.) were dissolved in 60 mL methanol and treated with ~ 10 mL 1 M NaOH solution (9.46 mmol, 2.20 equiv.), then stirred for 2.5 h. A solution of 1 M HCl was added to acidify the reaction mixture. The carboxylic acid was collected by filtration, then dried in vacuo to afford 761 mg (90%) of a pale-yellow solid.

^{1}H NMR (400 MHz, $CDCl_3$): δ (ppm) = 8.05 (t, J = 7.7 Hz, 1H, H_6), 7.23 – 7.13 (m, 2H, $H_{3,5}$), 4.45 (s, 2H, CH_2). – **^{13}C NMR** (75 MHz, $CDCl_3$): δ (ppm) = 168.4 ($C_{quart.}$, COOH), 164.5 ($C_{quart.}$, C_{Ar}), 137.5 ($C_{quart.}$, C_{Ar}), 133.8 (+, C_{Ar}H), 123.8 (+, C_{Ar}H), 117.5 ($C_{quart.}$, C_{Ar}), 116.7 (+, C_{Ar}H), 54.1 (–, CH_2). – **^{19}F NMR** (377 MHz, $CDCl_3$): δ (ppm) = -111.41 (s, 1F). – **IR** (ATR): $\tilde{\nu}$ (cm^{-1}) = 2553.7 (w), 2196.1 (w), 2106.4 (m), 1679.9 (s), 1622.0 (m), 1573.3 (m), 1502.5 (w), 1416.5 (m), 1341.1 (w), 1287.3 (m), 1216.5 (m), 1156.1 (m), 1090.5 (m), 966.6 (m), 913.3 (m), 868.1 (m), 834.6 (m), 791.7 (m), 772.9 (m), 745.6 (m), 682.6 (w), 656.4 (w), 627.9 (w), 582.8 (m), 523.8 (w), 448.4 (w), 423.5 (m). – **EA**: calcd for $C_8H_6FN_3O_2$,% C 49.24, H 3.10, N 9.74, found C 49.42, H 3.08, N 9.68.

europium 4-(azidomethyl)-2-fluorobenzoate dihydrate (Eu-2.6)

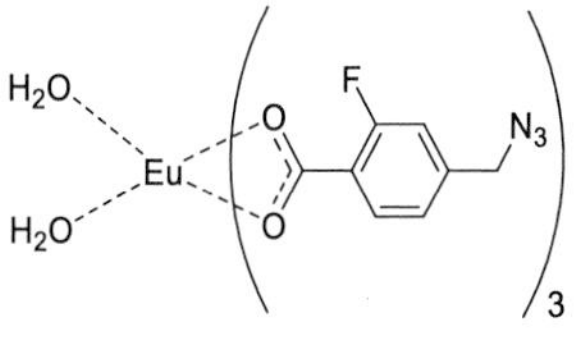

For the synthesis 366 mg of $EuCl_3 \cdot 6\,H_2O$ and 939 mg of 4-amino-2,5-difluorobenzoic acid **2.6-H** were used. 601 mg (78%) of an off-white solid of general formula $Eu(L)_3(H_2O)_2$ were obtained, where L=4-(azidomethyl)-2-fluorobenzoate.

^{1}H NMR (300 MHz, DMSO-d_6): δ (ppm) = 7.01 – 6.17 (m, 9H, 3 × 3 × H_{Ar}), 4.31 (s, 6H, 3 × CH_2). – **IR** (ATR): $\tilde{\nu}$ (cm^{-1}) = 3416.4 (vw), 2095.1 (m), 1600.6 (m), 1520.7 (m), 1418.2 (m), 1343.9 (m), 1237.8 (w), 1163.8 (w), 1098.6 (w), 927.2 (vw), 866.1 (w), 847.1 (w), 787.3 (m), 748.2 (w), 691.3 (w),

578.8 (w), 442.3 (vw), 406.2 (vw). – **EA**: calcd for $Eu(L)_3(H_2O)_2$, where L=4-(azidomethyl)-2-fluorobenzoate, % C 37.42, H 2.49, N 16.38, found C 37.53, H 2.66, N 16.56.

methyl 2-fluoro-4-((trimethylsilyl)ethynyl)benzoate (2.7-Me-TMS)

To a solution of 2.00 g of methyl 4-bromo-2-fluorobenzoate **2.10-Me** (8.58 mmol, 1.00 equiv.) in 15 mL 1,4-dioxane, 15 mL trimethylamine, 1.60 mL trimethylsilylacetylene (11.2 mmol, 1.30 equiv.), 301 mg of dichlorobis-(triphenylphosphine)palladium(II) (0.43 mmol, 0.05 equiv.) and 49.0 mg of copper(I) iodide (0.26 mmol, 0.03 equiv.) were added in the mentioned order and the mixture was further stirred at 60 °C for 1 h. The reaction mixture was allowed to cool down to room temperature, filtered through Celite® and washed with diethyl ether. The filtrate was concentrated under reduced pressure. EtOAc and water were added to the obtained residue, the layers were separated and the aqueous layer was extracted with EtOAc. The combined organic layers were washed with water and brine and dried over Na_2SO_4. After filtration, the filtrate was concentrated under reduced pressure and the obtained residue was purified by flash column chromatography (toluene:$CHCl_3$, 8:1) to afford 1.43 g (67%) of a brown oil.

^{1}H NMR (300 MHz, $CDCl_3$): δ (ppm) = 7.97 – 7.81 (m, 1H, H_6), 7.32 – 7.10 (m, 2H, $H_{5,3}$), 3.94 (s, 3H, CH_3), 0.26 (s, 9H, Si(CH_3)$_3$). – **^{13}C NMR** (75 MHz, $CDCl_3$): δ (ppm) = 164.3 ($C_{quart.}$, COOH), 160.4 ($C_{quart.}$, C_{Ar}), 133.8 (+, C_{Ar}H), 127.9 (+, C_{Ar}H), 121.6 ($C_{quart.}$, C_{Ar}), 116.0 ($C_{quart.}$, C_{Ar}, +, C_{Ar}H), 104.8 ($C_{quart.}$, C_{Ar}–C≡C), 96.1 ($C_{quart.}$, C≡CSi), 52.3 (+, OCH_3), -0.1 (+, Si(CH_3)$_3$).

4-ethynyl-2-fluorobenzoic acid (2.7-H)

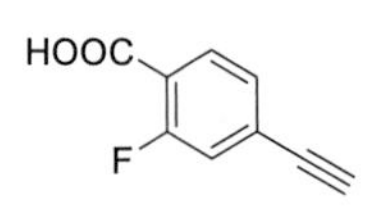

To a solution of 1.00 g of methyl 2-fluoro-4-((trimethylsilyl)ethynyl)-benzoate 2.7-Me-TMS (3.99 mmol, 1.00 equiv.) in 30 mL methanol, 320 mg of NaOH (7.99 mmol, 2.00 equiv.) were added and allowed to reflux overnight. The reaction was quenched with 6 M HCl to pH 2-3 (as indicated by pH paper) and extracted twice with EtOAc. The combined organic layers were dried over Na_2SO_4 and evaporated to afford 552 mg (85%) of a brown solid.

^{1}H NMR (300 MHz, $CDCl_3$): δ (ppm) = 7.92 (t, J = 7.8 Hz, 1H, H_6), 7.36 – 7.23 (m, 2H, $H_{5,3}$), 4.08 (s, 1H, CH). – **IR** (ATR): $\tilde{\nu}$ (cm^{-1}) = 3275.5 (w), 2827.6 (w), 1684.2 (s), 1614.2 (m), 1557.5 (m), 1493.4 (w), 1408.3 (m), 1291.9 (m), 1251.5 (m), 1152.3 (m), 1094.7 (m), 953.5 (m), 917.3 (w), 872.5 (m), 851.5 (m), 777.5 (m), 713.4 (w), 694.0 (w), 642.4 (m), 616.9 (m), 578.5 (w), 522.3 (w), 489.2 (w), 394.2 (w). – **EA**: calcd for $C_9H_5FO_2$,% C 65.86, H 3.07, found C 65.04, H 3.24.

methyl 4-bromo-2-fluorobenzoate (2.10-Me)

According to **GP3a**, 1.00 mL concentrated H_2SO_4 was added to a stirred solution of 2.36 g of 2-fluoro-4-bromobenzoic acid **2.10-H** (10.8 mmol, 1.00 equiv.) in 100 mL methanol and the mixture was refluxed for 48 h. After concentration in vacuo, the residue was triturated with water. The resulting precipitate was collected and dried in vacuo to afford 2.16 g (86%) of a yellow oil. Analytical data are consistent with the literature.[273]

^{1}H NMR (400 MHz, $CDCl_3$): δ (ppm) = 7.72 (dd, J = 2.1 Hz, J = 8.4 Hz, 1H, H_6), 7.65 (dd, J = 1.4 Hz, J = 8.1 Hz, 1H, H_5), 7.64 – 7.59 (m, 1H, H_3), 3.91 (s, 3H, CH_3). – **^{13}C NMR** (101 MHz, $CDCl_3$): δ (ppm) = 165.4 ($C_{quart.}$, COOMe), 159.0 ($C_{quart.}$, C_{Ar}), 138.8 ($C_{quart.}$, C_{Ar}), 131.5 (+, C_{Ar}H), 126.3 (+, C_{Ar}H), 117.5 ($C_{quart.}$, C_{Ar}), 114.9 (+, C_{Ar}H), 52.7 (+, CH_3).

methyl 2-fluoro-4-methylbenzoate (2.12-Me)

According to **GP3a**, 5.00 mL concentrated H_2SO_4 were added to a stirred solution of 2.36 g of 2-fluoro-4-methylbenzoic acid **2.12-H** (32.4 mmol, 1.00 equiv.) in 400 mL methanol and the mixture was refluxed for 48 h. After concentration in vacuo, the residue was triturated with water. The resulting precipitate was collected and dried in vacuo. Purification *via* flash column chromatography (CH/EtOAc 6:1) resulted in 4.53 g (83%) of a white solid.

^{1}H NMR (400 MHz, $CDCl_3$): δ (ppm) = 7.82 (t, J = 8.0 Hz, 1H, H_6), 6.98 (dd, J = 8.0 Hz, J = 1.0 Hz, 1H, H_5), 6.93 (d, J = 12.0 Hz, 1H, H_3), 3.91 (s, 3H, $COOCH_3$), 2.37 (s, 3H, CH_3).

methyl 4-amino-2,5-difluorobenzoate (2.14-Me)

According to **GP3a**, 1.00 mL concentrated H_2SO_4 was added to a stirred solution of 1.87 g of 2,5-difluoro-4-aminobenzoic acid **2.14-H** (10.8 mmol, 1.00 equiv.) in 100 mL methanol and the mixture was refluxed for 48 h. After concentration in vacuo, the residue was triturated with water. The resulting precipitate was collected and dried in vacuo to afford 1.75 g (86%) of a yellow solid.

^{1}H NMR (400 MHz, $CDCl_3$): δ (ppm) = 7.42 (dd, J = 11.5 Hz, J = 6.4 Hz, 1H, H_6), 6.45 (dd, J = 11.7 Hz, J = 7.1 Hz, 1H, H_3), 4.24 (bs, 2H, NH_2), 3.87 (s, 3H, CH_3). – **^{13}C NMR** (101 MHz, $CDCl_3$): δ (ppm) = 164.6 ($C_{quart.}$, COOMe), 161.3 ($C_{quart.}$, C_{Ar}), 146.8 ($C_{quart.}$, C_{Ar}), 141.0 ($C_{quart.}$, C_{Ar}), 118.2 (2C, $C_{quart.}$, +, C_{Ar}, C_{Ar}H), 103.6 (+, C_{Ar}H), 52.5 (+, CH_3). – **^{19}F NMR** (377 MHz, $CDCl_3$): δ (ppm) = -117.24 (s, 1F), -145.62 (s, 1F). – **IR** (ATR): $\tilde{\nu}$ (cm^{-1}) = 3398.8 (w), 3324.6 (w), 3216.8 (w),

3068.3 (vw), 2967.3 (vw), 1691.5 (m), 1643.1 (m), 1615.8 (m), 1522.4 (m), 1450.9 (m), 1431.7 (m), 1365.2 (w), 1315.9 (m), 1273.9 (m), 1201.5 (m), 1170.7 (m), 1095.8 (w), 1071.1 (m), 956.5 (m), 897.9 (w), 846.8 (w), 774.3 (m), 652.0 (w), 613.6 (m), 472.0 (w), 449.3 (w), 419.3 (w), 404.0 (w).

europium 4-amino-2,5-difluorobenzoate dihydrate (Eu-2.14)

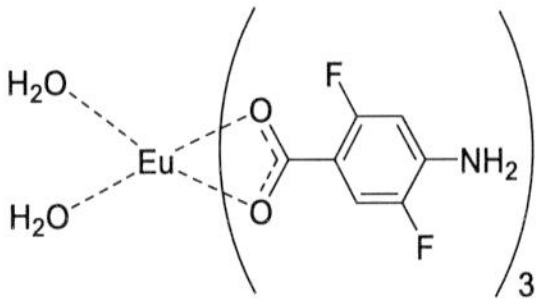

For the synthesis 366 mg of $EuCl_3 \cdot 6\,H_2O$ and 519 mg of 4-amino-2,5-difluorobenzoic acid **2.14-H** were used. 598 mg (85%) of an off-white solid of general formula $Eu(L_6)_3(H_2O)_2$ were obtained, where L=4-amino-2,5-difluorobenzoate.

¹H NMR (300 MHz, DMSO-d_6): δ (ppm) = 5.66 – 5.10 (m, 12H, 3 × 2 × H_{Ar}, 3 × NH_2). – **IR** (ATR): $\tilde{\nu}$ (cm^{-1}) = 3373.2 (vw), 1638.0 (w), 1602.4 (w), 1563.1 (w), 1528.5 (w), 1438.3 (w), 1379.4 (m), 1312.0 (w), 1231.7 (w), 1197.7 (w), 1171.5 (w), 910.7 (vw), 896.8 (vw), 854.5 (vw), 837.4 (w), 802.9 (w), 789.3 (w), 740.4 (w), 691.6 (vw), 609.3 (w), 391.9 (w). – **EA**: calcd for $Eu(L_6)_3(H_2O)_2$, where L=4-amino-2,5-difluorobenzoate, % C 38.81, H 2.29, N 5.97, found C 38.09, H 2.44, N 5.83

methyl 4-azido-2,5-difluorobenzoate (2.15-Me)

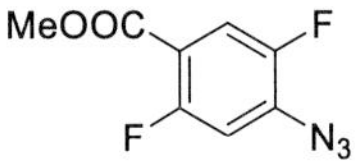

According to **GP7**, 1.08 g of methyl 4-amino-fluorobenzoate (5.92 mmol, 1.00 equiv.) were stirred in 10 mL of 50% HCl. The suspension was cooled to 0 °C and a solution of 531 mg of sodium nitrite (7.70 mmol, 1.30 equiv.) in 5 mLwater was added dropwise over a period of 15 min with stirring. After 15 min of further stirring at 0–5 °C, a solution of 578 mg of sodium azide (8.89 mmol, 1.50 equiv.) in 5 mL water was added at such a rate that the temperature of the reaction mixture did not exceed 15 °C towards the end of the addition. The solid product, that precipitated after 10–15 min was filtered, washed with ice water and dried in vacuo to afford 1.03 g (82%) of a pale-yellow solid.

¹H NMR (400 MHz, DMSO-d_6): δ (ppm) = 7.75 (dd, J = 11.4 Hz, J = 6.4 Hz, 1H, H_6), 7.45 (dd, J = 11.0 Hz, J = 6.7 Hz, 1H, H_3), 3.84 (s, 3H, CH_3). – **¹⁹F NMR** (377 MHz, DMSO-d_6): δ (ppm) = -113.08 (s, 1F), -132.66 (s, 1F).

4-azido-2,5-difluorobenzoic acid (2.15-H)

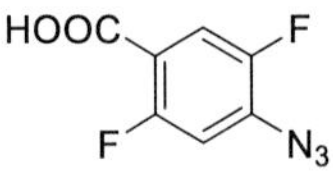

According to **GP4a**, 0.90 g of methyl 4-azido-2,5-difluorobenzoate **2.15-Me** (4.22 mmol, 1.00 equiv.) were dissolved in 60 mL methanol and treated with ~ 10 mL 1 M NaOH solution (9.29 mmol, 2.20 equiv.), then stirred for 2.5 h.

A solution of 1 M HCl was added to acidify the reaction mixture. The carboxylic acid was collected by filtration, then dried in vacuo to afford 722 mg (86%) of a yellow solid.

^{1}H NMR (400 MHz, DMSO-*d$_6$*): δ (ppm) = 13.47 (s, 1H, COO*H*), 7.68 (dd, J = 11.2 Hz, J = 6.4 Hz, 1H, *H$_6$*), 7.37 (dd, J = 10.8 Hz, J = 6.8 Hz, 1H, *H$_3$*). – **^{13}C NMR** (101 MHz, DMSO-*d$_6$*): δ (ppm) = 163.5 ($C_{quart.}$, *C*OOH), 158.9 ($C_{quart.}$, C_{Ar}), 156.3($C_{quart.}$, C_{Ar}), 150.5 (C_{quart}, C_{Ar}), 133.4 (C_{quart}, C_{Ar}), 118.9 (+, C_{Ar}H), 110.4 (+, C_{Ar}H). – **^{19}F NMR** (377 MHz, $CDCl_3$): δ (ppm) = -117.04 (s, 1F), -135.64 (s, 1F). – **IR** (ATR): $\tilde{\nu}$ (cm^{-1}) = 2840.3 (vw), 2133.6 (vw), 1693.7 (vw), 1621.1 (vw), 1506.5 (vw), 1455.4 (vw), 1394.1 (vw), 1333.0 (vw), 1297.4 (vw). – **EA**: calcd for $C_7H_3F_2N_3O_2$,% C 42.22, H 1.52, N 21.10, found C 42.59, H 1.49, N 20.90.

2-fluoro-5-nitrobenzoic acid (2.16-H)

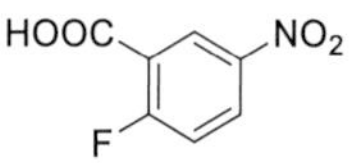

HOOC NO$_2$ F

Variant A. A stirred suspension of 1.70 g of 2-fluoro-5-nitrotoluene **2.22** (11.0 mmol, 1.00 equiv.) and 6.00 g of potassium dichromate (20.3 mmol, 1.85 equiv.) in 10 mL water was cooled down to 0 °C. Then 18 mL concentrated H_2SO_4 were added dropwise, so that the temperature did not exceed 50–60 °C. After the addition was complete (15 min), the reaction was stirred for additional 15 min and then heated to 65 °C for 3 h. The reaction mixture was allowed to cool to room temperature and poured onto 60 mL crushed ice. The resulting mixture was stirred for 15 min and filtered. The solid was added to 20 mL of a 2 M Na_2CO_3 solution and the resulting suspension was filtered. The filtrate was cooled to 0 °C and 14 mL of 25% aq. HCl was added dropwise. The mixture was extracted with EtOAc and washed with water and brine. The combined organic layers were dried over Na_2SO_4 and evaporated to dryness to afford 1.54 g (76%) of a white solid.

Variant B. A stirred solution of 4.00 g of 2-fluorobenzoic acid (28.5 mmol, 1.00 equiv.) in 12 mL concentrated H_2SO_4 was cooled down to 0 °C. Then 10 mL HNO_3 were added dropwise so that the temperature did not exceed 50–60 °C. The mixture was stirred for additional 4 h and then poured onto crushed ice. After filtration, the collected white solid was washed with a copious amount of water and dried under high vacuum to afford 4.60 g, (87%) of a white solid. Analytical data are consistent with the literature.[274]

^{1}H NMR (400 MHz, $CDCl_3$): δ (ppm) = 10.80 (bs, 1H, COO*H*), 8.77 (dd, J = 6.1 Hz, J = 2.9 Hz, 1H, *H$_6$*), 8.51 (dt, J = 9.1 Hz, J = 3.4 Hz, 1H, *H$_4$*), 7.40 (t, J = 9.2 Hz, *H$_3$*). – **IR** (ATR): $\tilde{\nu}$ (cm^{-1}) = 2921.0 (s), 2855.1 (s), 2736.6 (w), 2657,3 (w), 1694.3 (m), 1632.2 (m), 1585.8 (w), 1530.1 (m), 1456.2 (s), 1377.1 (m), 1354.5 (s), 1329.1 (m), 1290.6 (m), 1272.4 (m), 1237.6 (m), 1150.9 (m), 1124.2 (m), 1078.5 (m), 951.3 (w), 921.5 (m), 675.2 (m), 636.1 (m), 585.0 (w), 532.6 (w), 514.7 (w). – **EA**: calcd for $C_7H_4FNO_4$,% C 45.42, H 2.18, N 7.57, found C 45.28, H 2.17, N 7.64.

methyl 2-fluoro-5-nitrobenzoate (2.16-Me)

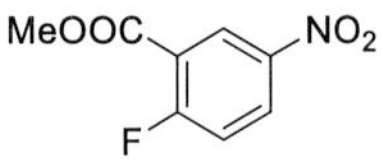

According to **GP3a**, 1.00 mL concentrated H_2SO_4 was added to a stirred solution of 2.00 g of 2-fluoro-5-nitrobenzoic acid **2.16-H** (10.8 mmol, 1.00 equiv.) in 100 mL methanol and the mixture was refluxed for 48 h. After concentration in vacuo, the residue was triturated with water. The resulting precipitate was collected and dried in vacuo to afford 1.80 g (83%) of a yellow solid.

^{1}H NMR (300 MHz, $CDCl_3$): δ (ppm) = 8.81 (dd, *J* = 6.1 Hz, *J* = 2.9 Hz, 1H, H_6), 8.38 (dd, *J* = 6.4 Hz, *J* = 2.6 Hz, 1H, H_4), 7.37 – 7.23 (m, 1H, H_3), 3.95 (s, 3H, CH_3). – **^{13}C NMR** (101 MHz, $CDCl_3$): δ (ppm) = 164.4 ($C_{quart.}$, COOMe), 162.4 ($C_{quart.}$, C_{Ar}), 145.6 ($C_{quart.}$, C_{Ar}), 129.6 (+, C_{Ar}H), 124.6 (+, C_{Ar}H), 117.4 ($C_{quart.}$, CCOOMe), 116.1 (+, C_{Ar}H), 52.3 (+, CH_3). – **IR** (ATR): $\tilde{\nu}$ (cm^{-1}) = 3100.1 (w), 2924.3 (s), 2852.2 (s), 1722.6 (s), 1600.5 (vw), 1591.2 (w), 1548.3 (m), 1458,1 (s), 1429.1 (s), 1383.8 (m), 1356.7 (s), 1299.3 (s), 1244.6 (m), 1190.2 (w), 1131.0 (m), 1048.0 (m), 978.5 (w), 947.8 (m), 891.3 (m), 857.6 (w), 825.1 (m), 782.2 (w), 759.3 (m), 741.9 (m), 664.5 (w), 586.9 (vw), 511.1 (w). – **EA**: calcd for $C_8H_6FNO_4$,% C 48.25, H 3.04, N 7.03, found C 47.99, H 3.10, N 7.08.

5-amino-2-fluorobenzoic acid (2.17-H)

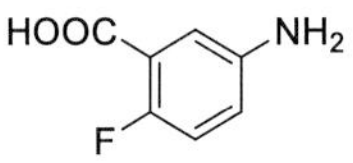

According to **GP4a**, 1.50 g of methyl 5-amino-2-fluorobenzoate **2.17-H** (8.87 mmol, 1.00 equiv.) were dissolved in 100 mL methanol and treated with ~ 20 mL 1 M NaOH solution (19.5 mmol, 2.20 equiv.), then stirred for 2.5 h. A solution of 1 M HCl was added to acidify the reaction mixture. The carboxylic acid was collected by filtration, then dried in vacuo to afford 1.21 g (88%) of a grey solid.

^{1}H NMR (400 MHz, DMSO-d_6): δ (ppm) = 12.91 (s, 1H, COO*H*), 7.68 (s, 2H, NH_2), 7.06 (dd, *J* = 6.2 Hz, *J* = 3.0 Hz, 1H, H_6), 6.94 (dd, *J* = 10.8 Hz, *J* = 8.8 Hz, 1H, H_3), 6.76 (dt, *J* = 8.7 Hz, *J* = 3.4 Hz, 1H, H_4). – **^{13}C NMR** (101 MHz, DMSO-d_6): δ (ppm) = 165.7 ($C_{quart.}$, COOH), 156.9 ($C_{quart.}$, C_{Ar}), 154.8 ($C_{quart.}$, C_{Ar}), 127.6 (+, C_{Ar}H), 119.1 (+, C_{Ar}H), 117.0 ($C_{quart.}$, C_{Ar}), 116.8 (+, C_{Ar}H). – **^{19}F NMR** (377 MHz, DMSO-d_6): δ (ppm) = -132.09 (s, 1F). – **EA**: calcd for $C_7H_6FNO_2$,% C 54.20, H 3.90, N 9.03, found C 55.09, H 4.08, N 9.24.

methyl 5-amino-2-fluorobenzoate (2.17-Me)

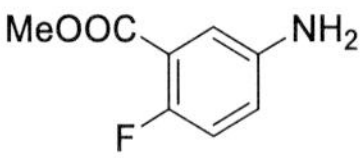

A solution of 1.75 g of methyl 2-fluoro-5-nitrobenzoate **2.1-Me** (9.00 mmol, 1.00 equiv.) in 15 mL methanol and 5 mL THF was hydrogenated (H_2, baloon) at rt overnight in the presence of 300 mg of Raney-Ni (89%, 50 mol%, 4.50 mmol, 0.50 equiv.). After filtration through Celite®, the mixture was concentrated in vacuo to afford 1.00 g (70%) of a yellow solid. Analytical data are consistent with the literature.[267]

^{1}H NMR (400 MHz, DMSO-*d_6*): δ (ppm) = 8.80 – 8.86 (m, 1H, *H_6*), 8.33 – 8.43 (m, 1H, *H_3*), 7.26 – 7.35 (m, 1H, *H_4*), 5.29 (bs, 2H, N*H_2*), 3.99 (s, 3H, C*H_3*). – **^{19}F NMR** (377 MHz, DMSO-*d_6*): δ (ppm) = -100.03 (s, 1F).

europium 5-amino-2-fluorobenzoate dihydrate (Eu-2.17)

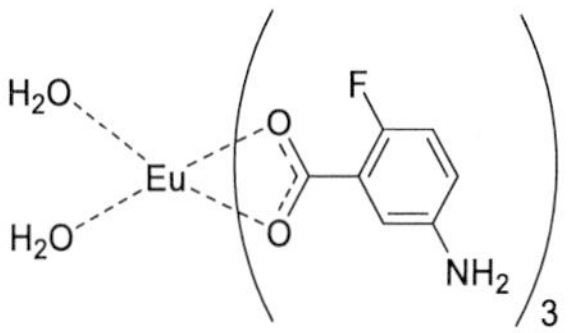

For the synthesis 366 mg of $EuCl_3 \cdot 6\ H_2O$ and 465 mg of 5-amino-2-fluorobenzoic acid **2.17-H** were used. 547 mg (84%) of an off-white solid of general formula $Eu(L_6)_3(H_2O)_2$ were obtained, where L=4-amino-2-fluorobenzoate.

^{1}H NMR (300 MHz, DMSO-*d_6*): δ (ppm) = 6.33 – 5.97 (m, 9H, 3 × 3 × *H_{Ar}*), 5.21 (bs, 6H, 3 × N*H_2*). – **^{19}F NMR** (377 MHz, DMSO-*d_6*): δ (ppm) = -111.36 (bs, 1F, *F*). – **IR** (ATR): $\tilde{\nu}$ (cm^{-1}) = 3317.4 (vw), 1655.1 (vw), 1596.6 (vw), 1580.9 (vw), 1529.9 (vw), 1492.4 (vw), 1443.0 (vw), 1392.6 (w), 1314.9 (vw), 1252.7 (vw), 1211.0 (vw), 1143.3 (vw), 1087.2 (vw), 891.2 (vw), 818.0 (vw), 804.0 (vw), 764.7 (vw), 672.2 (vw), 578.8 (vw), 523.5 (vw), 474.9 (vw), 410.5 (vw). – **EA**: calcd for $Eu(L)_3(H_2O)_2$, where L=4-amino-2-fluorobenzoate, % C 38.78, H 2.94, N 6.46, found C 39.01, H 3.04, N 6.36.

methyl 5-azido-2-fluorobenzoate (2.18)

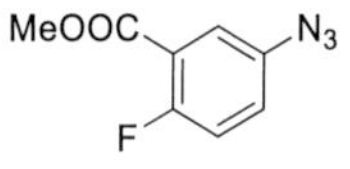

According to **GP7**, 800 mg of methyl 5-amino-2-fluorobenzoate **2.17-Me** (4.72 mmol, 1.00 equiv.) were stirred in 10 mL of 50% HCl. The suspension was cooled to 0 °C and a solution of 424 mg of sodium nitrite (6.14 mmol, 1.30 equiv.) in 5 mLwater was added dropwise over a period of 15 min with stirring. After 15 min of further stirring at 0–5 °C, a solution of 461 mg of sodium azide (7.09 mmol, 1.50 equiv.) in 5 mLwater was added at such a rate that the temperature of the reaction mixture did not exceed 15 °C towards the end of the addition. The reaction mixture was extracted with DCM three times. The combined organic layers were dried over Na_2SO_4, filtered and evaporated under reduced pressure to afford 0.414 g (45%) of a pale-yellow oil.

^{1}H NMR (400 MHz, $CDCl_3$): δ (ppm) = 7.54 – 7.58 (m, 1H, *H_6*), 7.10 – 7.18 (m, 2H, *$H_{3,4}$*), 3.93 (s, 3H, C*H_3*). – **^{19}F NMR** (377 MHz, DMSO-*d_6*): δ (ppm) = -118.44 (s, 1F).

ethyl pentafluorobenzoate (2.19-Et)

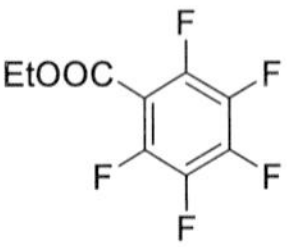

To a solution of 5.00 g of pentafluorobenzoic acid **2.19-H** (23.6 mmol, 1.00 equiv.) in 35 mL ethanol, 15 mL $SOCl_2$ (60.0 mmol, 2.54 equiv.) were added dropwise at 0 °C. The reaction mixture was allowed to warm to room temperature and refluxed for 12 h. All volatiles were removed under reduced pressure, the residue was dissolved in EtOAc, washed with water and an aq. $NaHCO_3$ solution and dried over Na_2SO_4. The solvent was removed under reduced pressure to afford 5.05 g (88%) of a colorless liquid.

^{1}H NMR (300 MHz, $CDCl_3$): δ (ppm) = 4.43 (q, J = 7.1 Hz, 2H, OCH_2), 1.38 (t, J = 6.4 Hz, 3H, CH_3). – **^{13}C NMR** (75 MHz, DMSO-d_6): δ (ppm) = 159.2 ($C_{quart.}$, COOEt), 147.5 ($C_{quart.}$, C_{Ar}), 146.8 ($C_{quart.}$, C_{Ar}), 139.3 ($C_{quart.}$, 2 × C_{Ar}), 108.9 ($C_{quart.}$, C_{Ar}COOEt), 63.1 (–, CH_2), 14.1 (+, CH_3). – **IR** (ATR): $\tilde{\nu}$ (cm^{-1}) = 2990.5 (m), 2947.1 (w), 1741.4 (s), 1655.3 (s), 1526.8 (s), 1500.5 (s), 1446.4 (m), 1424.9 (s), 1397.5 (m), 1370.4 (s), 1235.9 (s), 1103.5 (s), 997.1 (s), 936.3 (s), 865.5 (m), 809.5 (m), 767.0 (m), 743.8 (s), 703.2 (w), 612.4 (w). – **EA**: calcd for $C_9H_5F_5O_2$,% C 45.02, H 2.10, found C 45.14, H 2.12.

ethyl 4-azido-2,3,5,6-tetrafluorobenzoate (2.20-Et)

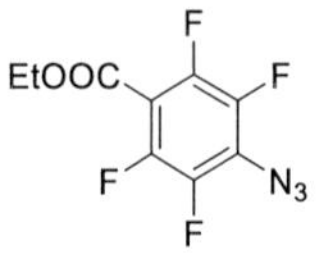

To a solution of 4.44 g of ethyl pentafluorobenzoate (**2.19-Et**) (18.5 mmol, 1.00 equiv.) in 45 mL acetone/water (2/1), 1.27 g of sodium azide (19.5 mmol, 1.05 equiv.) were added. The reaction mixture was refluxed for 8 h, allowed to cool to room temperature, partially evaporated, extracted with diethylether and dried over Na_2SO_4. All volatiles were removed under reduced pressure to afford 4.53 g (93%) of a yellow oil.

^{1}H NMR (300 MHz, $CDCl_3$): δ (ppm) = 4.40 (q, J = 7.2 Hz, 2H, OCH_2), 1.37 (t, J = 7.2 Hz, 3H, CH_3). – **^{13}C NMR** (75 MHz, $CDCl_3$): δ (ppm) = 160.4 ($C_{quart.}$, COOEt), 147.1 ($C_{quart.}$, C_{Ar}), 141.3 ($C_{quart.}$, C_{Ar}), 139.2 ($C_{quart.}$, C_{Ar}), 122.9 ($C_{quart.}$, C_{Ar}), 114.9 ($C_{quart.}$, C_{Ar}COOEt), 108.3 ($C_{quart.}$, C_{Ar}), 62.6 (–, CH_2), 14.0 (+, CH_3). – **^{19}F NMR** (377 MHz, $CDCl_3$): δ (ppm) = -139.13 – -139.22 (m, 2F), -151.17 – -151.25 (m, 2F). – **EA**: calcd,% C 41.08, H 1.92, N 15.97 found C 41.24, H 1.98, N 15.7.

4-azido-2,3,5,6-tetrafluorobenzoate (2.20-H)

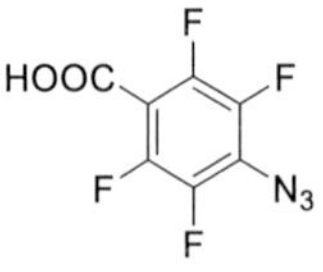

According to **GP4a**, 500 mg of ethyl 4-azido-2,3,5,6-tetrafluorobenzoate (**2.20-Et**) (1.90 mmol, 1.00 equiv.) was dissolved in 60 mL methanol, treated with ~ 4.2 mL NaOH 1 M solution (4.18 mmol, 2.20 equiv.), then stirred overnight. A solution of 1 M HCl was added to acidify the reaction mixture. The reaction mixture was extracted with $CHCl_3$ and the combined organic layers were dried over

Na_2SO_4. The solvent was evaporated under reduced pressure to afford 335 mg (75%) of a white solid. Single crystal suitable for SC-XRD analysis was grown from methanol solution.

^{13}C NMR (75 MHz, CD_3OD): δ (ppm) = 161.9 ($C_{quart.}$, *C*OOH), 146.9 ($C_{quart.}$, 2 × C_{Ar}F), 141.3 ($C_{quart.}$, 2 × C_{Ar}F), 122.9 ($C_{quart.}$, *C*COOH), 108.3 ($C_{quart.}$, $C_{Ar}N_3$). – **^{19}F NMR** (377 MHz, $CDCl_3$): δ (ppm) = -142.49 – -142.56 (m, 2F), -154.03 – -154.10 (m, 2F). – **EA**: calcd,% C 35.76, H 0.43, N 17.87 found C 35.82, H 1.08, N 17.87.

ethyl 2,3,5,6-tetrafluoro-4-(4-phenyl-1H-1,2,3-triazol-1-yl)benzoate (2.24-Et)

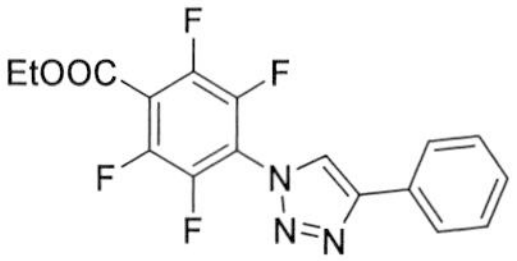

According to **GP10a**, to a solution of 1.00 g of ethyl 4-azido-2,3,5,6-tetrafluorobenzoate (**2.20-Et**) (3.80 mmol, 1.00 equiv.) in 35 mL dry DCM, 2.00 mL triethylamine (12.7 mmol, 3.60 equiv.), 28.0 mg of copper(I) bromide (0.190 mmol, 0.0500 equiv.) and 0.500 mL phenylacetylene (4.37 mmol, 1.15 equiv.) were added and stirred for 4 h. The reaction mixture was then filtered, the solvent was evaporated and the product was recrystallized from ethanol to afford 1.17 g (84%) of a yellow solid.

^{1}H NMR (400 MHz, $CDCl_3$): δ (ppm) = 8.12 (s, 1H, C*H*), 7.92 – 7.88 (m, 2H, 2 × H_{Ar}), 7.48 – 7.42 (m, 2H, 2 × H_{Ar}), 7.40 – 7.35 (m, 1H, H_{Ar}), 4.40 (q, J = 7.1 Hz, 2H, OCH_2), 1.42 (t, J = 7.1 Hz, 3H, CH_3). – **^{13}C NMR** (101 MHz, $CDCl_3$): δ (ppm) = 159.1 ($C_{quart.}$, *C*OOEt), 148.4 ($C_{quart.}$, Ph-C_{Ar}N), 146.1 ($C_{quart.}$, 2 × C_{Ar}F), 140.0 ($C_{quart.}$, FC*C*N), 138.9 ($C_{quart.}$, 2 × C_{Ar}F), 129.1 ($C_{quart.}$, C_{Ph}), 128.9 (+, 2 × C_{Ar}H), 127.8 (+, C_{Ar}H), 125.9 (+, 2 × C_{Ar}H), 121.8 (+, H*C*N), 114.0 ($C_{quart.}$, C_{Ar}COOEt), 63.3 (–, O*C*H_2), 13.0 (+, *C*H_3). – **^{19}F NMR** (377 MHz, $CDCl_3$): δ (ppm) = -137.07 – -137.21 (m, 2F), -145.04 – -145.12 (m, 2F). – **MS** (70 eV, EI): m/z (%) = 365 (1) $[M]^+$, 337 (100) $[M-N_2]^+$, 309 (58) $[M-N_2-C_2H_4]^+$, 292 (11) $[M-CO_2Et]^+$, 264 (19) $[M-CO_2Et-N_2]^+$, 176 (11) $[C_6F_4CO]^+$, 148 (15) $[C_6F_4]^+$, 116 (88) $[PhC_2HN]^+$, 89 (56), 63 (14), 29 (28). – **EA**: calcd for $C_{17}H_{11}F_4N_3O_2$,% C 55.90, H 3.04, N 11.50 found C 54.94, H 2.98, N 12.01.

2,3,5,6-tetrafluoro-4-(4-phenyl-1H-1,2,3-triazol-1-yl)benzoic acid (2.24-H)

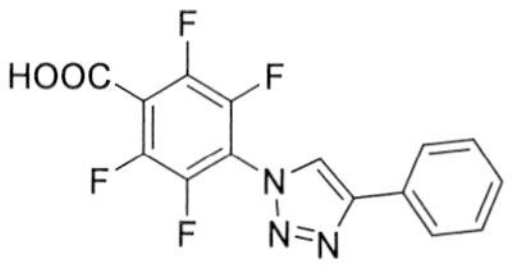

According to **GP4a**, 1.00 g of ethyl 2,3,5,6-tetrafluoro-4-(4-phenyl-1H-1,2,3-triazol-1-yl)benzoate (**2.24-Et**) (2.73 mmol, 1.00 equiv.) was dissolved in 30 mL methanol, treated with ~ 6 mL 1 M NaOH solution (6.02 mmol, 2.20 equiv.), then stirred overnight. A solution of 1 M HCl was added to acidify the reaction mixture. The carboxylic acid was collected by filtration, dried in vacuo and purified by recrystalization from methanol to afford 709 mg (77%) of a yellow solid.

^{1}H NMR (300 MHz, $CDCl_3$): δ (ppm) = 8.12 (s, 1H, NC*H*), 7.92 – 7.88 (m, 2H, 2 × H_{Ar}), 7.48 – 7.42 (m, 2H, 2 × H_{Ar}), 7.40 – 7.35 (m, 1H, H_{Ar}). – **^{13}C NMR** (75 MHz, $CDCl_3$): δ (ppm) = 159.1 ($C_{quart.}$, *C*OOH), 148.8 ($C_{quart.}$, Ph-C_{Ar}N), 146.1 ($C_{quart.}$, 2 × C_{Ar}F), 140.3 ($C_{quart.}$, FC*C*N), 138.6 ($C_{quart.}$, 2 × C_{Ar}F), 129.2 ($C_{quart.}$, C_{Ph}), 129.0 (+, 2 × C_{Ar}H), 127.9 (+, C_{Ar}H), 125.9 (+, 2 × C_{Ar}H), 121.7 (+, H*C*N), 114.0 ($C_{quart.}$, C_{Ar}COOH). – **^{19}F NMR** (377 MHz, $CDCl_3$): δ (ppm) = -137.07 – -137.21 (m, 2F), -145.04 – -145.12 (m, 2F). – **MS** (70 eV, EI): *m/z* (%) = 365 (1) $[M]^+$, 337 (100) $[M–N_2]^+$, 309 (58) $[M–N_2–C_2H_4]^+$, 292 (11) $[M–CO_2Et]^+$, 264 (19) $[M–CO_2Et–N_2]^+$, 176 (11) $[C_6F_4CO]^+$, 148 (15) $[C_6F_4]^+$, 116 (88) $[PhC_2HN]^+$, 89 (56), 63 (14), 29 (28). – **EA**: calcd for $C_{15}H_7F_4N_3O_2$,% C 53.42, H 2.09, N 12.46 found C 51.88, H 2.14, N 12.65.

((4-fluorophenyl)ethynyl)trimethylsilane (2.26-TMS)

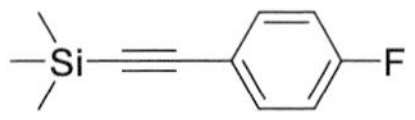

According to **GP8**, to a solution of 1.53 g of bromo-4-fluorobenzene (8.73 mmol, 1.00 equiv.) in 15 mL dry THF, 18 mL trimethylamine, 1.88 mL trimethylsilylacetylene (13.1 mmol, 1.50 equiv.), 504 mg of tetrakis(triphenylphosphine)palladium(0) (430 µmol, 0.0500 equiv.) and 50 mg of copper(I) iodide (260 µmol, 0.0300 equiv.) were added in the mentioned order and the mixture was further refluxed for 24 h. The reaction was allowed to cool to room temperature, the black solution was diluted with DCM and then treated with 0.1 M aq HCl until the aqueous layer reached a pH value of 1. The aqueous layer was extracted with DCM and filtered through Celite®. The combined organic layers were washed with water and brine and dried over Na_2SO_4. Then the solvent was removed under reduced pressure and the obtained crude product was purified by distillation to afford 1.51 g (90%) of a colorless liquid.

^{1}H NMR (300 MHz, $CDCl_3$): δ (ppm) = 7.44 – 7.36 (m, 2H, 2 × H_{Ar}), 6.98 – 7.11 (m, 2H, 2 × H_{Ar}), 0.26 (s, 9H, Si(CH_3)$_3$). – **^{13}C NMR** (75 MHz, $CDCl_3$): δ (ppm) = 160.4 ($C_{quart.}$, C_{Ar}F), 131.8 (+, 2 × C_{Ar}H), 114.3 (+, 2 × C_{Ar}H), 110.7 ($C_{quart.}$, C_{Ar}C), 103.6 ($C_{quart.}$, C_{Ar}*C*≡C), 97.7 ($C_{quart.}$, C≡*C*Si), -0.2 (+, Si(CH_3)$_3$).

1-ethynyl-4-fluorobenzene (2.26)

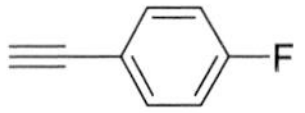

According to **GP9**, to a solution of 1.50 g of ((4-fluorophenyl)ethynyl)trimethylsilane (**2.26-TMS**) (7.85 mmol, 1.00 equiv.) in 60 mL methanol, 628 mg of NaOH (15.7 mmol, 2.00 equiv.) were added and allowed to reflux for 4 h. The reaction mixture was concentrated and the residual liquid was treated with 5% aq. $NaHCO_3$ and DCM. The organic layer was separated and the aqueous layer was extracted twice with DCM. Combined organic layers were dried over Na_2SO_4 and evaporated to afford 600 mg (84%) of a colorless liquid.

^{1}H NMR (300 MHz, $CDCl_3$): δ (ppm) = 7.58 – 7.51 (m, 2H, 2 × H_{Ar}), 7.29 – 7.20 (m, 2H, 2 × H_{Ar}), 4.17 (s, 1H, C≡C*H*). – **^{13}C NMR** (75 MHz, $CDCl_3$): δ (ppm) = 163.4 ($C_{quart.}$, $C_{Ar}F$), 134.0 (+, 2 × $C_{Ar}H$), 118.2 ($C_{quart.}$, $C_{Ar}C$), 113.7 (+, 2 × $C_{Ar}H$), 81.1 ($C_{quart.}$, C_{Ar}*C*≡C), 40.2 (+, C≡*C*H).

((2,4-difluorophenyl)ethynyl)trimethylsilane (2.27-TMS)

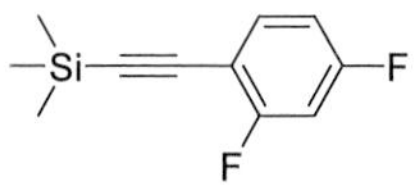

According to **GP8**, to a solution of 1.68 g of bromo-2,4-difluorobenzene (8.73 mmol, 1.00 equiv.) in 15 mL dry THF, 18 mL trimethylamine, 1.88 mL trimethylsilylacetylene (13.1 mmol, 1.50 equiv.), 504 mg of tetrakis(triphenylphosphine)palladium(0) (430 µmol, 0.0500 equiv.) and 50 mg of copper(I) iodide (260 µmol, 0.0300 equiv.) were added in the mentioned order and the mixture was further refluxed for 24 h. After allowing to cool to room temperature, the black solution was diluted with DCM and then treated with 0.1 M aq. HCl until the aqueous layer reached a pH value of 1. The aqueous layer was extracted with DCM and filtered through Celite®. The combined organic layers were washed with water and brine and dried over Na_2SO_4. Then the solvent was removed under reduced pressure and the obtained crude product was purified by distillation to afford 1.63 g (89%) of a colorless liquid.

^{1}H NMR (300 MHz, $CDCl_3$): δ (ppm) = 7.54 – 7.48 (m, 1H, H_{Ar}), 7.49 – 7.37 (m, 1H, H_{Ar}), 6.89 – 6.78 (m, 1H, H_{Ar}), 0.26 (s, 9H, Si(CH_3)$_3$). – **^{13}C NMR** (75 MHz, $CDCl_3$): δ (ppm) = 163.8 ($C_{quart.}$, C_{Ar}), 160.4 ($C_{quart.}$, C_{Ar}), 134.1 (+, $C_{Ar}H$), 112.4 (+, $C_{Ar}H$), 110.7 ($C_{quart.}$, $C_{Ar}C$), 104.3 (+, $C_{Ar}H$), 103.6 ($C_{quart.}$, C_{Ar}*C*≡C), 98.7 ($C_{quart.}$, C≡*C*Si), -0.2 (+, Si(*C*H_3)$_3$).

1-ethynyl-2,4-difluorobenzene (2.27)

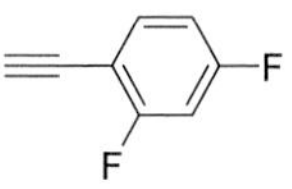

According to **GP9**, to a solution of 1.60 g of ((2,4-difluorophenyl)ethynyl)trimethylsilane (**2.27-TMS**) (7.61 mmol, 1.00 equiv.) in 40 mL methanol, 608 mg of NaOH (15.2 mmol, 2.00 equiv.) were added and allowed to reflux for 4 h. The reaction mixture was concentrated and the residual liquid was treated with 5% aq. $NaHCO_3$ and DCM. The organic layer was separated and the aqueous layer was extracted twice with DCM. The combined organic layers were dried over Na_2SO_4 and evaporated to afford 840 mg (80%) of a low-melting white solid.

^{1}H NMR (300 MHz, $CDCl_3$): δ (ppm) = 7.39 – 7.47 (m, 1H, H_{Ar}), 6.89 – 6.81 (m, 2H, 2 × H_{Ar}), 3.28 (s, 1H, C≡C*H*). – **^{13}C NMR** (75 MHz, $CDCl_3$): δ (ppm) = 164.2 ($C_{quart.}$, $C_{Ar}F$), 163.4 ($C_{quart.}$, $C_{Ar}F$), 135.3 (+, $C_{Ar}H$), 112.1 (+, $C_{Ar}H$), 108.3 ($C_{quart.}$, $C_{Ar}C$), 104.8 (+, $C_{Ar}H$), 82.7 ($C_{quart.}$, C_{Ar}*C*≡C), 77.8 (+, C≡*C*H).

ethyl 2,3,5,6-tetrafluoro-4-[4-(4-fluorophenyl)-1H-1,2,3-triazol-1-yl]benzoate (2.28-Et)

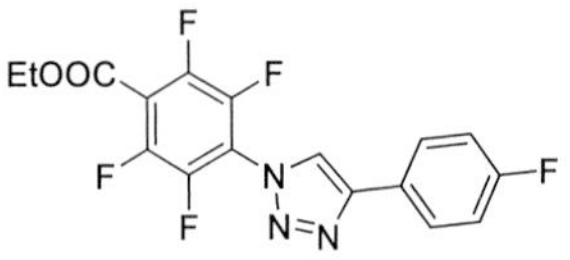

According to **GP10a**, to a solution of 1.00 g of ethyl 4-azido-2,3,5,6-tetrafluorobenzoate (**2.20-Et**) (3.80 mmol, 1.00 equiv.) in 35 mL dry DCM, 2.0 mL triethylamine (12.67 mmol, 3.60 equiv.), 28 mg of copper (I) bromide (0.19 mmol, 0.05 equiv.) and the 525 mg of 1-ethynyl-4-fluorobenzene (**2.26**) (4.37 mmol, 1.15 equiv.) were added and allowed to stir for 4 h. The reaction mixture was then filtered, the solvent was evaporated and the product was recrystallized from EtOAc/hexane (1/1) to afford 1.17 g (84%) of a white solid.

^{1}H NMR (400 MHz, $CDCl_3$): δ (ppm) = 8.02 (s, 1H, C*H*), 7.88 – 7.80 (m, 2H, 2 × H_{Ar}), 7.34 – 7.29 (m, 2H, 2 × H_{Ar}), 4.47 (q, *J* = 7.1 Hz, 2H, OCH_2), 1.41 (t, *J* = 7.1 Hz, 3H, CH_3). – **^{13}C NMR** (101 MHz, $CDCl_3$): δ (ppm) = 158.5 ($C_{quart.}$, *C*OOEt), 156.2 ($C_{quart.}$, HC*C*F), 146.2 ($C_{quart.}$, 2 × C_{Ar}F), 146.1 ($C_{quart.}$, N*C*CH), 140.0 ($C_{quart.}$, FC*C*N), 138.9 ($C_{quart.}$, 2 × C_{Ar}F), 129.1 ($C_{quart.}$, NC*C*CH), 130.8 (+, 2 × C_{Ar}H), 125.8 (+, N*C*H), 116.3 (+, 2 × C_{Ar}H), 113.4 ($C_{quart.}$, C_{Ar}COOEt), 30.4 (–, O*C*H_2CH_3), 14.0 (+, OCH_2*C*H_3). – **^{19}F NMR** (377 MHz, $CDCl_3$): δ (ppm) = -137.27 – -137.39 (m, 2F), -145.08 – -145.21 (m, 2F). – **EA**: calcd for $C_{17}H_{10}F_5N_3O_2$,% C 53.27, H 2.63, N 10.96 found C 52.94, H 2.67, N 10.75.

2,3,5,6-tetrafluoro-4-[4-(4-fluorophenyl)-1H-1,2,3-triazol-1-yl]benzoic acid (2.28-H)

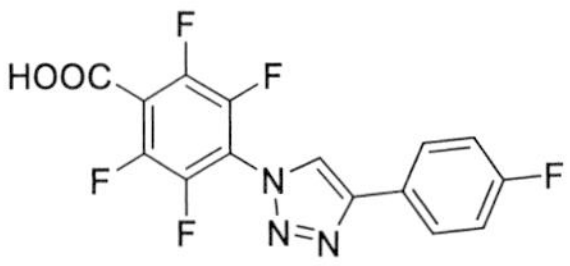

According to **GP4a**, 1.00 g of 2,3,5,6-tetrafluoro-4-[4-(4-fluorophenyl)-1H-1,2,3-triazol-1-yl] benzoate (**2.28-Et**) (2.61 mmol, 1.00 equiv.) was dissolved in 30 mL methanol, treated with ~6 mL NaOH 1 M solution (5.74 mmol, 2.20 equiv.), then stirred overnight. A solution of 1 M HCl was added to acidify the reaction mixture. The carboxylic acid was collected by filtration, dried in vacuo and purified by recrystalization from methanol to afford 927 mg (82%) of a yellow solid.

^{1}H NMR (300 MHz, $CDCl_3$): δ (ppm) = 8.06 (s, 1H, C*H*), 7.88 – 7.80 (m, 2H, 2 × H_{Ar}), 7.34 – 7.29 (m, 2H, 2 × H_{Ar}). – **^{13}C NMR** (101 MHz, $CDCl_3$): δ (ppm) = 158.5 ($C_{quart.}$, *C*OOH), 156.1 ($C_{quart.}$, HC*C*F), 146.4 ($C_{quart.}$, 2 × C_{Ar}F), 146.1 ($C_{quart.}$, N*C*CH), 139.8 ($C_{quart.}$, FC*C*N), 138.9 ($C_{quart.}$, 2 × C_{Ar}F), 129.1 ($C_{quart.}$, NC*C*CH), 130.7 (+, 2 × C_{Ar}H), 125.8 (+, N*C*H), 116.3 (+, 2 × C_{Ar}H), 112.6 ($C_{quart.}$, C_{Ar}COOH). – **^{19}F NMR** (377 MHz, $CDCl_3$): δ (ppm) = -137.26 – -137.39 (m, 2F), -145.09 – -145.20 (m, 2F). – **EA**: calcd,% C 53.27, H 2.63, N 10.96 found C 52.94, H 2.67, N 10.75.

ethyl 4-[4-(2,4-difluorophenyl)-1H-1,2,3-triazol-1-yl]-2,3,5,6-tetrafluorobenzoate (2.29-Et)

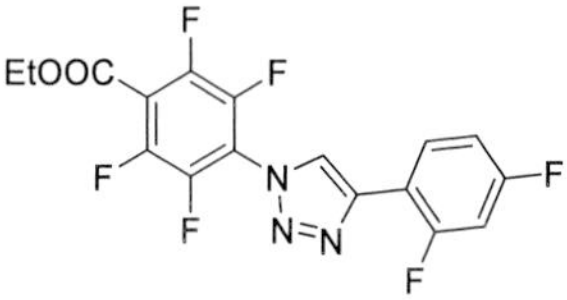

According to **GP10a**, to a solution of 1.00 g of ethyl 4-azido-2,3,5,6-tetrafluorobenzoate (**2.20-Et**) (3.80 mmol, 1.00 equiv.) in 35 mL dry DCM, 2.0 mL triethylamine (12.67 mmol, 3.60 equiv.), 28 mg of copper (I) bromide (5.00 mol%) and 604 mg of 1-ethynyl-2,4-difluorobenzene (**2.27**) (4.37 mmol, 1.15 equiv.) were added and allowed to stir for 4 h. The reaction mixture was then filtered, the solvent was evaporated and the product was recrystallized from toluene and subsequently from ethanol to afford 0.70 g (46%) of a white solid.

^{1}H NMR (400 MHz, $CDCl_3$): δ (ppm) = 8.02 (s, 1H, C*H*), 7.73 – 7.80 (m, 2H, 2 × H_{Ar}), 7.34 – 7.29 (m, 1H, H_{Ar}), 4.47 (q, *J* = 7.1 Hz, 2H, OCH_2), 1.41 (t, *J* = 7.1 Hz, 3H, CH_3). – **^{13}C NMR** (101 MHz, $CDCl_3$): δ (ppm) = 164.1 ($C_{quart.}$, *C*OOEt), 162.5 ($C_{quart.}$, HC*C*F), 160.4 ($C_{quart.}$, HC*C*F), 146.8 ($C_{quart.}$, 2 × C_{Ar}F), 145.5 ($C_{quart.}$, N*C*CH), 140.0 ($C_{quart.}$, FC*C*N), 138.9 ($C_{quart.}$, 2 × C_{Ar}F), 130.6 ($C_{quart.}$, NC*C*CH), 127.8 (+, C_{Ar}H), 125.8 (+, N*C*H), 115.1 (+, C_{Ar}H), 111.3 ($C_{quart.}$, C_{Ar}COOEt), 104.2 (+, C_{Ar}H), 61.2 (–, O*C*H_2), 14.1 (+, OCH$_2$*C*H$_3$).

4-[4-(2,4-difluorophenyl)-1H-1,2,3-triazol-1-yl]-2,3,5,6-tetrafluorobenzoate (2.29-H)

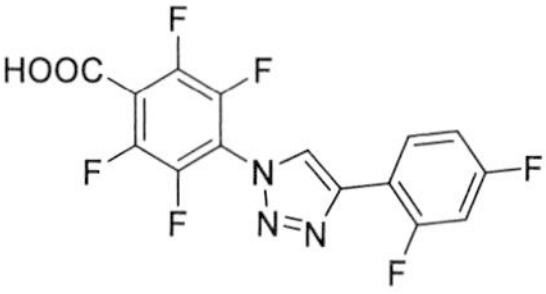

According to **GP4a**, 0.60 g of ethyl 4-[4-(2,4-difluorophenyl)-1H-1,2,3-triazol-1-yl]-2,3,5,6-tetrafluorobenzoat (**2.29-Et**) (1.60 mmol, 1.00 equiv.) was dissolved in 20 mL methanol, treated with ~ 4 mL NaOH 1 M solution (3.52 mmol, 2.20 equiv.), then stirred overnight. A solution of 1 M HCl was added to acidify the reaction mixture. The carboxylic acid was collected by filtration, dried in vacuo and purified by recrystalization from methanol to afford 483 mg (81%) of a yellow solid.

^{1}H NMR (400 MHz, $CDCl_3$): δ (ppm) = 8.05 (s, 1H, C*H*), 7.73 – 7.79 (m, 2H, 2 × H_{Ar}), 7.31 – 7.26 (m, 1H, H_{Ar}). – **^{13}C NMR** (101 MHz, $CDCl_3$): δ (ppm) = 164.1 ($C_{quart.}$, *C*OOH), 162.5 ($C_{quart.}$, HC*C*F), 160.4 ($C_{quart.}$, HC*C*F), 146.8 ($C_{quart.}$, 2 × C_{Ar}F), 145.5 ($C_{quart.}$, N*C*CH), 140.0 ($C_{quart.}$, FC*C*N), 138.9 ($C_{quart.}$, 2 × C_{Ar}F), 130.6 ($C_{quart.}$, NC*C*CH), 127.8 (+, C_{Ar}H), 125.8 (+, N*C*H), 115.1 (+, C_{Ar}H), 111.3 ($C_{quart.}$, C_{Ar}COOH), 104.2 (+, C_{Ar}H).

ethyl 4-(4-{[(tert-butoxycarbonyl)amino]methyl}-1H-1,2,3-triazol-1-yl)-2,3,5,6-tetra-fluorobenzoate (2.30^{Boc}-Et)

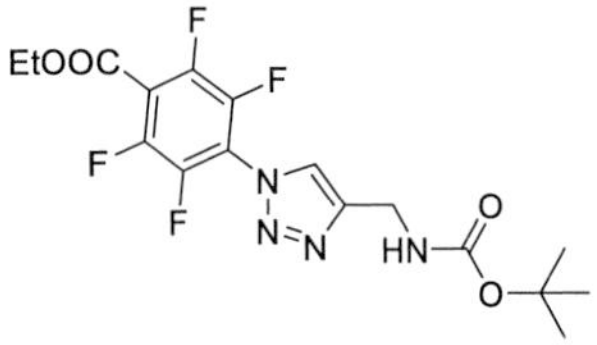

According to **GP10a**, to a solution of 1.00 g of ethyl 4-azido-2,3,5,6-tetrafluorobenzoate (**2.20-Et**) (3.80 mmol, 1.00 equiv.) in 35 mL dry DCM, 2.00 mL triethylamine (12.7 mmol, 3.60 equiv.), 28.0 mg of copper(I) bromide (5.00 mol%) and 678 mg of N-Boc-propargylamine (4.37 mmol, 1.15 equiv.) were added and stirred for 4 h. The reaction mixture was then filtered, the solvent was evaporated and the product was recrystallized from ethanol to afford 1.19 g (75%) of a yellow solid.

^{1}H NMR (300 MHz, DMSO-d_6): δ (ppm) = 8.41 (m, 1H, C*H*), 7.47 – 7.53 (m, 1H, N*H*), 4.40 (q, J = 7.1 Hz, 2H, OCH_2), 4.30 (s, 2H, CH_2), 1.39 – 1.32 (m, 12H, 4 × CH_3). – **^{13}C NMR** (75 MHz, DMSO-d_6): δ (ppm) = 165.9 ($C_{quart.}$, *C*OOEt), 156.1 ($C_{quart.}$, *C*OO*t*Bu), 146.8 ($C_{quart.}$, 2 × C_{Ar}F), 140.5 ($C_{quart.}$, FC*C*N), 138.1 ($C_{quart.}$, 2 × C_{Ar}F), 137.5 ($C_{quart.}$, *C*CH$_2$NH-Boc), 125.9 (+, *C*H), 112.3 ($C_{quart.}$, C_{Ar}COOEt), 80.5 ($C_{quart.}$, *C*(CH$_3$)$_3$), 61.6 (–, O*C*H$_2$), 40.6 (–, HN*C*H$_2$), 28.1 (+, C(*C*H$_3$)$_3$), 14.3 (+, CH$_2$*C*H$_3$). – **^{19}F NMR** (377 MHz, DMSO-d_6): δ (ppm) = -137.25 – -137.37 (m, 2F), -145.15 – -145.25 (m, 2F). – **EA**: calcd for $C_{17}H_{18}F_4N_4O_4$,% C 48.81, H 4.34, N 13.39 found C 48.98, H 4.51, N 13.44.

ethyl 4-[4-(aminohydrochloridemethyl)-1H-1,2,3-triazol-1-yl]-2,3,5,6-tetrafluorobenzoate (2.30-Et)

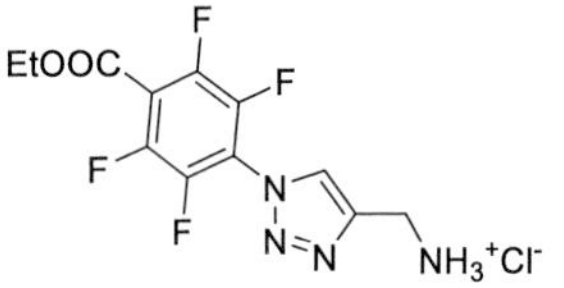

To a solution of 1.01 g of ethyl 4-(4-{[(tertbutoxycarbonyl)amino]methyl}-1H-1,2,3-triazol-1-yl)-2,3,5,6-tetrafluorobenzoate (**2.30^{Boc}-Et**) (2.42 mmol, 1.00 equiv.) in 25 mL 1,4-dioxane, 17.0 mL of 4 M HCl were added and the mixture was stirred for 5 h. The volatiles were removed under reduced pressure and 20 mL diethyl ether were added. The obtained solid was filtered through a medium porosity fritted glass funnel, additionally washed with another 20 mL diethyl ether and the solvent was removed under reduced pressure to afford 702 mg (82%) of a white solid.

^{1}H NMR (300 MHz, DMSO-d_6): δ (ppm) = 8.75 (s, 1H, C*H*), 8.70 (s, 3H, NH_3^+), 4.46 (q, J = 7.1 Hz, 2H, OCH_2), 4.27 (bs, 2H, CH_2NH$_2$), 1.34 (t, J = 7.1 Hz, 3H, CH_3).

4-[4-(aminohydrochloridemethyl)-1H-1,2,3-triazol-1-yl]-2,3,5,6-tetrafluorobenzoic acid (2.30-H)

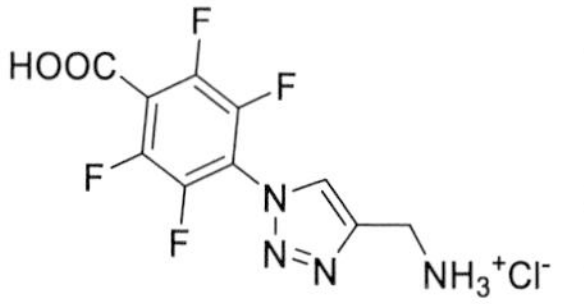

According to **GP4a**, 600 mg of ethyl 4-[4-(aminohydro-chloridemethyl)-1H-1,2,3-triazol-1-yl]-2,3,5,6-tetra-fluorobenzoate (**2.30-Et**) (1.69 mmol, 1.00 equiv.) were dissolved in 60 mL methanol, treated with ~ 4 mL 1 M NaOH solution (3.72 mmol, 2.20 equiv.), then stirred overnight. A solution of 1 M HCl was added to acidify the reaction mixture. The carboxylic acid was collected by filtration, dried in vacuo and purified by recrystalization from methanol to afford 331 mg (60%) of a white solid.

^{1}H NMR (300 MHz, DMSO-d_6): δ (ppm) = 12.01 (bs, 1H, COO*H*), 8.75 (s, 1H, C*H*), 8.70 (s, 3H, NH_3^+), 4.26 (bs, 2H, CH_2N). – **^{13}C NMR** (75 MHz, DMSO-d_6): δ (ppm) = 165.9 ($C_{quart.}$, *C*OOH), 146.8 ($C_{quart.}$, 2 × C_{Ar}F), 140.5 ($C_{quart.}$, FC*C*N), 138.8 ($C_{quart.}$, 2 × C_{Ar}F), 137.5 ($C_{quart.}$, *C*CH$_2$NH$_2$), 125.9 (+, *C*H), 112.3 ($C_{quart.}$, C_{Ar}COOH), 40.6 (HN*C*H$_2$). – **^{19}F NMR** (377 MHz, DMSO-d_6): δ (ppm) = -137.21 – -137.34 (m, 2F), -145.16 – -145.25 (m, 2F). – **EA**: calcd for $C_{10}H_7ClF_4N_4O_2$,% C 36.77, H 2.16, N 17.15 found C 38.70, H 2.31, N 16.98.

tris europium 4-[4-(amino)-1H-1,2,3-triazol-1-yl]-2,3,5,6-tetrafluorobenzoate trihydrate (Eu-2.30)

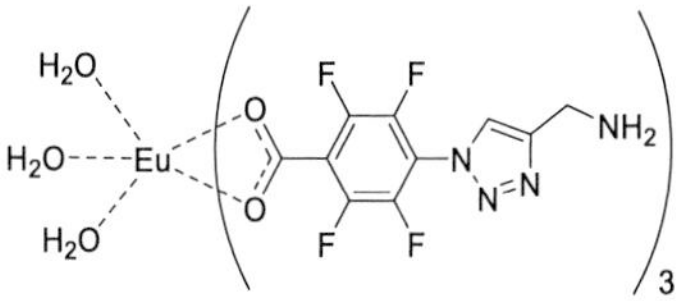

According to **GP1a**, a suspension of 163 mg of (1-(4-carboxy-2,3,5,6-tetrafluorophenyl)-1H-1,2,3-triazol-4-yl)methanaminium chloride **2.30-H** (0.50 mmol, 3.00 equiv.) in water was treated with 1 M NaOH at 60 °C until the pH was slightly acidic (pH = 5–6, approx. 6 mL) and the solution became clear. Then an aq. solution of 61 mg of $EuCl_3 \cdot x\ H_2O$ (0.17 mmol, 1.00 equiv.) was added following by an immediate precipitation of a product. The reaction mixture was stirred for additional 3 h at rt, the crude product was filtered, washed with ethanol and dried in vacuo to afford a product as a yellow solid.

^{1}H NMR (300 MHz, DMSO-d_6): δ (ppm) = 8.84 (s, 3H, 3 × H_{Ar}), 4.38 – 4.14 (m, 6H, 3 × CH_2). – **EA**: calcd for $Eu(L)_3(H_2O)_3$, where L=(1-(4-carboxy-2,3,5,6-tetrafluorophenyl)-1H-1,2,3-triazol-yl)methanaminium chloride, % C 33.56, H 1.97, N 15.66, found C 33.42, H 2.15, N 15.72.

methyl 2-fluoro-4-(4-phenyl-1H-1,2,3-triazol-1-yl)benzoate (2.31-Me)

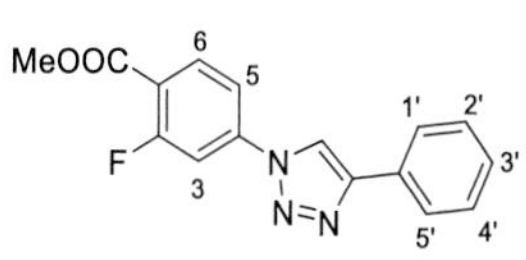

According to **GP10a**, to a solution of 1.00 g of methyl 4-azido-2-fluorobenzoate **2.3-Me** (5.52 mmol, 1.00 equiv.) in 50 mL dry DCM, 2.8 mL triethylamine (19.9 mmol, 3.60 equiv.), 40 mg of copper(I) bromide (5.00 mol%) and

0.70 mL phenylacetylene (6.35 mmol, 1.15 equiv.) were added and stirred for 4 h. The reaction mixture was then filtered, the solvent was evaporated and the product was recrystallized from ethanol to afford 1.11 g (68%) of a yellow solid.

^{1}H NMR (300 MHz, DMSO-d_6): δ (ppm) = 9.48 (s, 1H, NC*H*), 8.20 – 7.88 (m, 5H, $H_{3,6,5,1',5'}$), 7.55 – 7.46 (m, 2H, $H_{2',4'}$), 7.46 – 7.38 (m, 1H, $H_{3'}$), 3.90 (s, 3H, CH_3). – **^{13}C NMR** (75 MHz, DMSO-d_6): δ (ppm) = 164.1 ($C_{quart.}$, *C*OOMe), 160.4 ($C_{quart.}$, C_{Ar}), 147.3 ($C_{quart.}$, C_{Ar}), 139.4 (C_{quart}, C_{Ar}N), 137.5 (+, C_{Ar}H), 131.1 (+, C_{Ar}H), 129.0 (+, 2 × C_{Ar}H), 128.4 (+, C_{Ar}H), 127.8 (+, 2 × C_{Ar}H), 125.2 (+, C_{Ar}H), 117.2 ($C_{quart.}$, C_{Ar}) 105.6 (+, C_{Ar}H), 52.2 (+, CH_3).

5-nitro-1,10-phenanthroline (2.33)

N N NO2

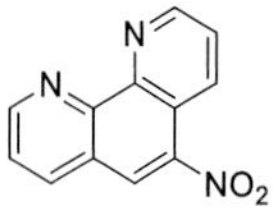

A stirred solution of 3.00 g of 1,10-phenanthroline **2.32** (16.7 mmol, 1.00 equiv.) in 30 mL H_2SO_4 was cooled down to at 0 °C. Then 24 mL HNO_3 was added dropwise so that the temperature did not exceed 50-60 °C. The mixture was then refluxed for 3 h, allowed to cool to room temperature and carefully poured onto crushed ice and the pH was adjusted to ~ 7 with 2 M NaOH. The precipitate was filtered, washed with cold water and dried in vacuo to afford 2.85 g (76%) of a yellow solid.

^{1}H NMR (400 MHz, $CDCl_3$): δ (ppm) = 9.37 (dd, *J* = 4.3 Hz, *J* = 1.7 Hz, 1H, H_{Ar}), 9.31 (dd, *J* = 4.3 Hz, *J* = 1.6 Hz, 1H, H_{Ar}), 9.01 (dd, *J* = 8.6 Hz, *J* = 1.6 Hz, 1H, H_{Ar}), 8.69 (s, 1H, H_{Ar}), 8.48 – 8.44 (m, 1H, H_{Ar}), 7.80 – 7.86 (m, 2H, 2 × H_{Ar}). – **IR** (ATR): $\tilde{\nu}$ (cm^{-1}) = 3414.8 (w), 3080.4 (w), 1618.7 (w), 1519.0 (m), 1419.3 (w), 1386.5 (vw), 1353.8 (w), 1202.1 (w), 1145.4 (w), 1045.1 (w), 986.9 (m), 906.6 (m), 809.8 (w), 734.4 (vw), 623.8 (w)$^{-}$. – **MS** (EI, 70 eV): m/z (%) = 225 (64) [M]$^{+}$, 179 (100) [M–NO_2], 167 (37), 152 (24), 125 (17), 99 (10), 75 (13).

5-amino-1,10-phenanthroline (2.34)

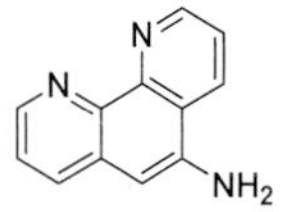

N N NH2

To a suspension of 3.15 g of 5-nitro-1,10-phenanthroline (**2.33**) (14.00 mmol, 1.00 equiv.) and 60 mg of Pd/C (10wt%, 0.56 mmol, 0.04 equiv.) in 100 mL abs. ethanol, 6.80 mL hydrazine monohydrate (140 mmol, 10.00 equiv.) was added dropwise and stirred under Ar. The mixture was refluxed for 4 h then filtered hot through a Celite® to remove the catalyst and washed with ethanol. The solvent was removed under reduced pressure, the residue was dissolved in 100 mL $CHCl_3$ and washed three times with 100 mL water to remove the excess of hydrazine. The organic phase was dried over Na_2SO_4 and the solvent was removed under reduced pressure to afford 2.88 g (95%) of a yellow solid.

^{1}H NMR (300 MHz, $CDCl_3$): δ (ppm) = 9.21 (dd, J = 4.4 Hz, J = 1.7 Hz, 1H, H_{Ar}), 8.96 (dd, J = 4.3 Hz, J = 1.7 Hz, 1H, H_{Ar}), 8.30 (dd, J = 8.3 Hz, J = 1.7 Hz, 1H, H_{Ar}), 8.01 (dd, J = 8.1 Hz, J = 1.7 Hz, 1H, H_{Ar}), 7.70 – 7.61 (m, 1H, H_{Ar}), 7.72 (dd, J = 8.1 Hz, J = 4.3 Hz, 1H, H_{Ar}), 6.97 (s, 1H, H_{Ar}).

Synthesis of Ln(III) complexes

According to **GP1a**, a suspension of the carboxylic acid (3.00 mmol, 3.00 equiv.) in water was treated with 1 M NaOH at 60 °C until the pH was slightly acidic (pH = 5–6, approx. 3 mL) and the solution became clear. Then an aq. solution of $EuCl_3 \cdot x\ H_2O$ (1.00 mmol, 1.00 equiv.) was added following by an immediate precipitation of a product. The reaction mixture was stirred for additional 3 h at rt, the crude product was filtered, washed with ethanol and dried in vacuo to afford a product as an off-white solid.

2.36, yellow solid, 421 mg (19%)

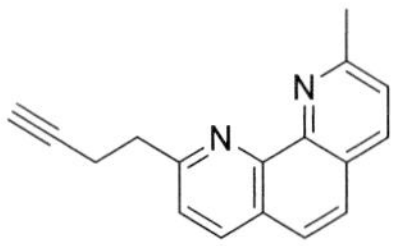

***R*f** (CH/EtOAc 1:1) = 0.29. – **^{1}H NMR** (500 MHz, $CDCl_3$): δ (ppm) = 8.18 (d, J = 8.2 Hz, 1H, H_{Ar}), 8.13 (d, J = 8.2 Hz, 1H, H_{Ar}), 7.72 (s, 2H, 2 × H_{Ar}), 7.63 (d, J = 8.3 Hz, 1H, H_{Ar}), 7.50 (d, J = 8.2 Hz, 1H, H_{Ar}), 3.46 (t, J = 7.6 Hz, 2H, CH_2CH_2), 2.94 (s, 3H, CH_3), 2.91 (t, J = 2.6 Hz, 1H, CH), 2.86 (td, J = 7.6 Hz, J = 2.6 Hz, 2H, CH_2CCH). – **^{13}C NMR** (126 MHz, $CDCl_3$): δ (ppm) = 160.7 ($C_{quart.}$, C_{Ar}), 159.6 ($C_{quart.}$, C_{Ar}), 145.6 ($C_{quart.}$, C_{Ar}), 136.6 (+, $C_{Ar}H$), 136.4 (+, $C_{Ar}H$), 127.5 ($C_{quart.}$, C_{Ar}), 126.9 ($C_{quart.}$, 2 × C_{Ar}), 125.9 (+, $C_{Ar}H$), 125.5 (+, $C_{Ar}H$), 123.7 (+, $C_{Ar}H$), 122.7 (+, $C_{Ar}H$), 84.1 ($C_{quart.}$, CCH), 69.1 (+, CH), 38.1 (–, CH_2), 26.1 (–, CH_2), 18.7 (+, CH_3). – **IR** (ATR): $\tilde{\nu}$ (cm^{-1}) = 3156.2 (w), 2914.1 (w), 1618.6 (w), 1586.3 (w), 1548.3 (w), 1494.2 (w), 1419.6 (w), 1365.8 (w), 1192.3 (w), 1141.4 (w), 1100.7 (w), 881.0 (w), 850.1 (m), 833.8 (w), 759.2 (w), 722.7 (m), 701.3 (w), 634.4 (w), 558.7 (w), 534.1 (w), 467.3 (vw), 411.8 (w). – **MS** (EI, 70 eV): m/z (%) = 246 (100) $[M]^+$, 193 (10) $[M–HCCCH_2CH_2]^+$.

2-(but-3-yn-1-yl)-9-methyl-1,10-phenanthroline (2.36) and 2,9-di(but-3-yn-1-yl)-1,10-phenanthroline (2.37)

Under Ar atmosphere 1.88 g of neocuproine **2.35** (9.00 mmol, 1.00 equiv.)were dissolved in 50 mL dry THF and stirred for 10 min. Then 5.40 mL of 2 M LDA solution in dry THF was carefully added to the reaction mixture at 0 °C , turning the solution from yellow to red. The mixture was stirred for 1 h at 0 °C. Subsequently, 13 mL propargyl chloride (19.8 mmol, 2.20 equiv.) was added so that the temperature of the reaction mixture did not exceed 0 °C. The reaction was stirred for 2 h at 0 °C , then it was slowly allowed to warm to room temperature and stirred for additional 16 h. The reaction mixture was then poured onto crushed ice and extracted with DCM. Purification of the crude product by means of flash column chromatography (CH/EtOAc 1:1) afforded mono and disubstituted products.

2.37 yellow solid, 153 mg (6%)

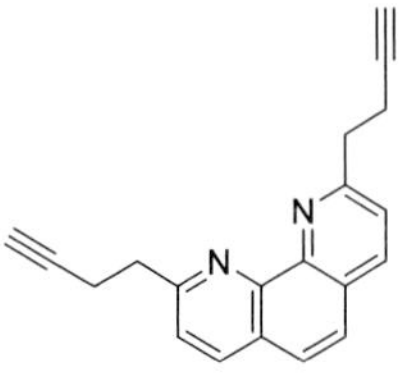

***R*f** (CH/EtOAc 1:1) = 0.25. – **^{1}H NMR** (500 MHz, $CDCl_3$): δ (ppm) = 8.17 (d, *J* = 8.2 Hz, 2H, 2 × H_{Ar}), 7.75 (s, 2H, 2 × H_{Ar}), 7.61 (d, *J* = 8.2 Hz, 2H, 2 × H_{Ar}), 3.44 (t, *J* = 7.6 Hz, 4H, 2 × CH_2CH_2), 2.88 (td, *J* = 8.8 Hz, *J* = 2.9 Hz, 4H, 2 × CH_2CCH), 2.01 (q, *J* = 2.9 Hz, 2H, 2 × *CH*). – **^{13}C NMR** (126 MHz, $CDCl_3$): δ (ppm) = 160.6 ($C_{quart.}$ 2 × C_{Ar}), 145.6 ($C_{quart.}$, 2 × C_{Ar}), 136.4 (+, 2 × $C_{Ar}H$), 127.5 (+, 2 × $C_{Ar}H$), 125.8 (+, 2 × $C_{Ar}H$), 122.8 (+, 2 × $C_{Ar}H$), 84.1 ($C_{quart.}$, 2 × *C*CH), 68.9 (+, 2 × *C*H), 37.9 (–, 4 × CH_2). – **IR** (ATR): $\tilde{\nu}$ (cm^{-1}) = 3299.1 (vw), 3184.0 (w), 2907.4 (vw), 1608.5 (vw), 1587.3 (w), 1548.8 (vw), 1495.2 (w), 1421.2 (w), 1362.4 (w), 1242.1 (vw), 1148.0 (w), 1092.5 (vw), 985.3 (vw), 953.6 (vw), 881.6 (vw), 852.4 (w), 808.2 (vw), 751.7 (w), 740.1 (w), 687.4 (w), 641.8 (w), 623.1 (w), 600.7 (w), 531.9 (w), 534.3 (w), 467.5 (vw), 445.2 (vw), 411.2 (w). – **MS** (EI, 70 eV): m/z (%) = 284 (10) $[M]^+$, 245 (100) $[M–CH_2CCH]^+$.

V.2.3.2 Synthesis and properties of europium conjugates with linear peptoids

***tert*-butyl (4-aminobutyl)carbamate (2.41)**

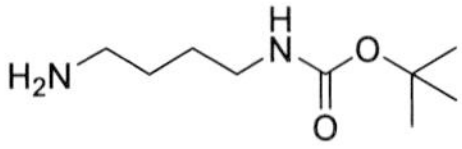

62.3 g (707 mmol, 1.00 equiv.) of 1,4-butanediamine **2.38** was dissolved in 400 mL THF and a solution of 23.1 g of Boc-anhydride **2.43** (106mmol, 0.15 equiv.) in 220 mL THF were added dropwise over 5 h. The mixture was stirred at room temperature overnight. After concentration in vacuo and resuspension in water, the product was extracted with DCM (3 × 100 mL). The combined organic phases were washed with water. The solvent was removed under reduced pressure to afford 13.7 g (68%) of a colorless oil.

^{1}H NMR (300 MHz, $CDCl_3$): δ (ppm) = 4.72 (bs, 1H, N*H*CO), 2.95 – 3.17 (m, 2H, CH_2NH), 2.67 (t, *J* = 6.6 Hz, 2H, CH_2NH_2), 1.13 – 1.56 (m, 15H, CH_3CO, CH_2CH_2N, NH_2)

***tert*-butyl (6-aminohexyl)carbamate (2.42)**

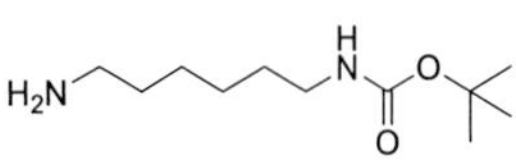

33.6 g (289 mmol, 1.00 equiv.) of 1,6-hexanediamine **2.39** were dissolved in 275 mL THF and a solution of 9.45 g of Boc-anhydride **2.43** (43.3 mmol, 0.15 equiv.) in 125 mL THF were added dropwise over 5 h. The mixture was stirred at room temperature overnight. After concentration in vacuo and resuspension in water, the product was extracted with DCM (3 × 100 mL). The combined organic phases were washed with water. The solvent was removed under reduced pressure to afford 5.80 g (62%) of a pale-yellow oil. Analytical data are consistent with the literature.[116]

***R*f** (EE/MeOH/TEA = 7:1:0.1) = 0.07. – **^{1}H NMR** (400 MHz, $CDCl_3$): δ (ppm) = 4.74 (bs, 1H, N*H*CO), 2.93 – 3.09 (m, 2H, C*H*₂NH), 2.58 (t, *J* = 6.9 Hz, 2H, CH_2NH_2), 1.71 (s, 2H, NH_2), 1.29 – 1.46 (m, 13H, $CH_2CH_2CH_2N$, CH_3CO), 1.18 – 1.28 (m, 4H, $CH_2CH_2CH_2N$). – **^{13}C NMR** (101 MHz, $CDCl_3$): δ (ppm) = 156.1 ($C_{quart.}$, NHCOO), 78.9 ($C_{quart.}$, CH_3CO), 41.9 (–, CH_2NH_2), 40.5 (–, CH_2NH), 33.4 (–, $CH_2CH_2CH_2NH_2$), 30.0 (–, $CH_2CH_2CH_2NH$), 28.4 (+, $C(CH_3)_3$), 26.6 (–, $CH_2CH_2CH_2NH_2$), 26.5 (– ,$CH_2CH_2CH_2NH$).

3-azidopropan-1-amine (2.45)

To a solution of 4.00 g of 3-chloropropylamine hydrochloride **2.44** (30.8 mmol, 1.00 equiv.) in 25 mL water 4.00 g of sodium azide (61.6 mmol, 2.00 equiv.) were added. The mixture was heated to 80 °C and stirred at this temperature overnight. Two-thirds of the solvent were removed under reduced pressure, diethyl ether was added and the mixture was cooled to 0 °C. Subsequently, 2.00 g of potassium hydroxide were added, the organic phase was separated and the aq. phase was extracted twice with 30 mL diethyl ether. The combined organic phases were dried over Na_2SO_4 and the solvent was removed under reduced pressure to afford 2.10 g (68%) of a pale-yellow oil.

^{1}H NMR (300 MHz, $CDCl_3$): δ (ppm) = 3.32 (t, *J* = 6.7 Hz, 2H, CH_2N_3), 2.76 (t, *J* = 6.8 Hz, 2H, CH_2NH_2), 1.67 (p, *J* = 6.8 Hz, 2H, $CH_2CH_2CH_2$), 1.33 (s, 2H, NH_2).

H-Eu-(N0b1)₂-(N0b1)-(N6am)₃-SP (Eu-2.47-SP)

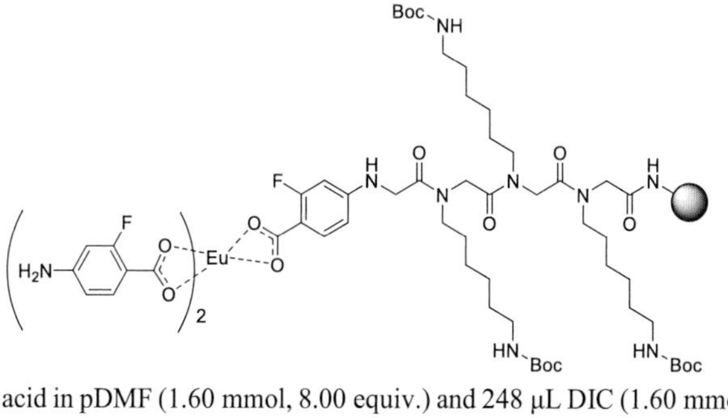

According to **GP11a**, 328 mg of Rink amide resin (0.20 mmol, 1.00 equiv.) were swollen and deprotected. The synthesis of the peptoid tetramer was carried out by acylation (**GP11a**, step 3) with 1.33 mL 1.20 M bromoacetic acid in pDMF (1.60 mmol, 8.00 equiv.) and 248 µL DIC (1.60 mmol, 8.00 equiv.). *N*6am^{Boc} group was attached, using 346 mg of *tert*-butyl(4-aminohexyl)carbamate (1.60 mmol, 8.00 equiv.)

(**GP11a**, step 4) and *Eu*-(*N*0b1)$_2$-(*N*0b1) group was attached, using the solution of 337 mg of **Eu-2.2** (0.54 mmol, 2.67 equiv.) in 1.60 mL pDMF (shaking overnight). The formation of the desired product on solid support was proven by the emission spectra measurement of the resin, showing the characteristic Eu-based luminescence. However, MALDI-TOF-MS only showed the mass of *H*-(*N*0b1)-(*N*6am)$_3$-fragment.

MS (MALDI-TOF, CHCA/DHB): m/z = 681.4 [M–×2 (**2.2-H**)+H–×3 Boc]$^+$; 704.4 [M–×2(**2.2-H**)+Na–×3 Boc]$^+$.

***H-Eu*-(*N*0b2)$_2$-(*N*0b2)-(*N*6am)$_3$-*SP* (Eu-2.48-SP)**

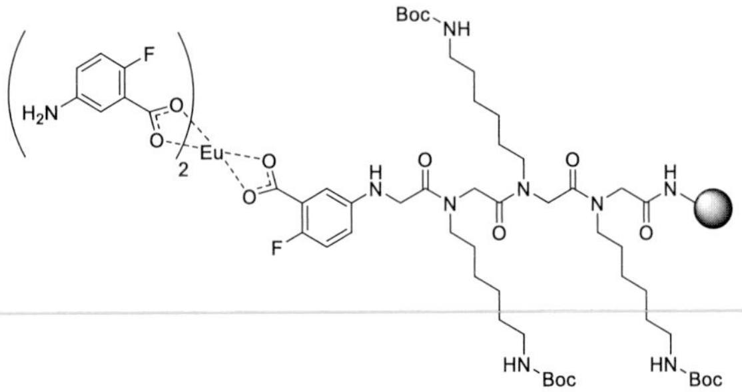

According to **GP11a**, 328 mg of Rink amide resin (0.20 mmol, 1.00 equiv.) were swollen and deprotected. The synthesis of the peptoid tetramer was carried out by acylation (**GP11a**, step 3) with 1.33 mL 1.20 M bromoacetic acid in pDMF (1.60 mmol, 8.00 equiv.) and 248 μL DIC (1.60 mmol, 8.00 equiv.). *N*6am^{Boc} group was attached, using 346 mg of *tert*-Butyl(4-aminohexyl)carbamate (1.60 mmol, 8.00 equiv.) (**GP11a**, step 4) and *Eu*-*N*0b2 group was attached, using the solution of 337 mg of **Eu-2.17** (0.54 mmol, 2.67 equiv.) in 1.60 mL pDMF (shaking overnight). The formation of the desired product on solid support was proven by the emission spectra measurement of the resin, showing the characteristic Eu-based luminescence. However, MALDI-TOF-MS only showed the mass of *H*-(*N*0fb)-(*N*6am)$_3$-fragment.

MS (MALDI-TOF, CHCA/DHB): m/z = 681.4 [M– ×2 (**2.17-H**)+H–×3 Boc]$^+$; 704.4 [M–×2 (**2.17-H**) +Na–×3 Boc]$^+$.

***H-Eu-*(*N*0b3)$_2$-(*N*0b3)-(*N*6am)$_3$-*SP* (Eu-2.49-SP)**

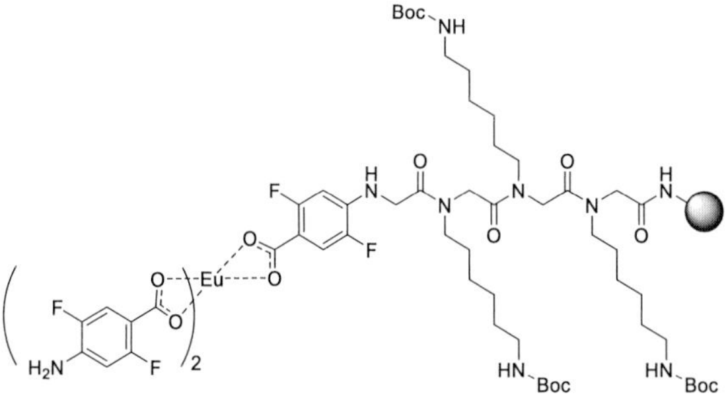

According to **GP11a**, 328 mg of Rink amide resin (0.20 mmol, 1.00 equiv.) were swollen and deprotected. The synthesis of the peptoid tetramer was carried out by acylation (**GP11a**, step 3) with 1.33 mL 1.20 M bromoacetic acid in pDMF (1.60 mmol, 8.00 equiv.) and 248 µL DIC (1.60 mmol, 8.00 equiv.). *N*6am^{Boc} group was attached, using 346 mg of *tert*-Butyl(4-aminohexyl)carbamate (1.60 mmol, 8.00 equiv.) (**GP11a**, step 4) and *Eu*-(*N*0b3)$_2$-(*N*0b3) group was attached, using the solution of 376 mg of **Eu-2.14** (0.54 mmol, 2.67 equiv.) in 1.60 mL pDMF (shaking overnight). The formation of the desired product on solid support was proven by the emission spectra measurement of the resin, showing the characteristic Eu-based luminescence. However, MALDI-TOF-MS only showed the mass of *H*-(*N*0fb)-(*N*6am)$_3$-fragment.

MS (MALDI-TOF, CHCA/DHB): m/z = 681.4 [M–×2 (**2.14**)+H–×3 Boc]$^+$; 704.4 [M–×2 (**2.14**)+Na–×3 Boc]$^+$.

***H*-(*N*0b1)-(*N*6am)$_3$-*NH*$_2$ (2.50)**

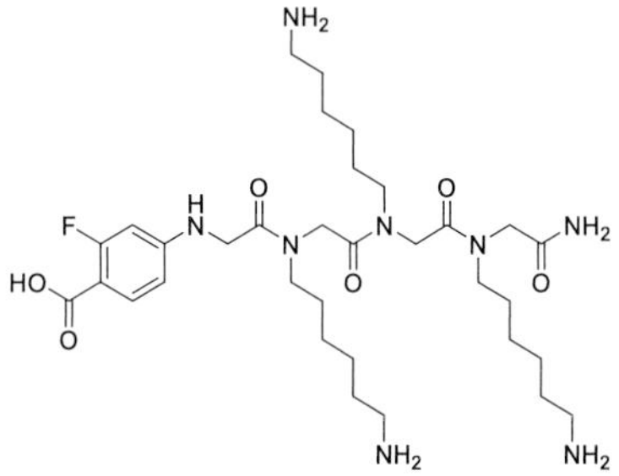

According to **GP12a**, **Eu-2.47-SP** was cleaved from the resin by treatment with 3 mL 95% TFA in DCM at room temperature for 2 h. The solution was filtered off and the next cleavage mixture was added to the resin and let shaken for additional 2 h. The resin was washed with 3 mL DCM (3 ×) and methanol (3 ×). The cleavage cocktail was evaporated under airflow, the rest was combined with washes and the solvent was removed under reduced pressure (**GP12a**, step 1). The crude product was then dissolved in 5 mL acetonitrile, passed through a syringe filter, purified by means of preparative HPLC (5-95% acetonitrile + 0.1% TFA in 30 min, 25 °C) and lyophilized (**GP12a**, step 2) to afford 26 mg (19% over 9 steps).

MS (MALDI-TOF, CHCA): m/z = 681.4 [M+H]$^+$; 704.4 [M+Na]$^+$. – **Analytical HPLC** (5–95% acetonitrile + 0.1% TFA in 20 min, detection at 218 nm): t_{Ret} = 8.49 min, 79% purity.

***Eu*-[*H*-(*N*0b1⁻)-(*N*6am)₃-NH₂]Cl₂ (Eu-2.50)**

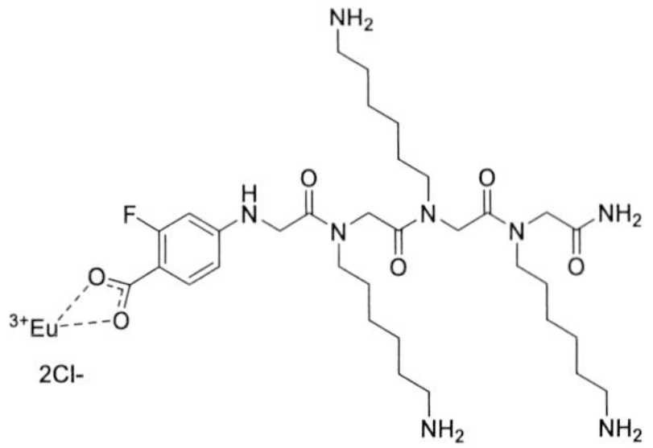

A solution of 20.0 mg of peptoid-based carboxylic acid **2.50-H** (30.0 µmol, 3.00 equiv.) in ethanol was treated with 30 µL of 1 M NaOH at 60 °C (30.0 µmol, 3.00 equiv.). Then 19.5 µL of 0.512 M aq. solution of $EuCl_3 \cdot 6\,H_2O$ (10.0 µmol, 1.00 equiv.) was added. The reaction mixture was refluxed for additional 3 h, ethanol was removed under reduced pressure and the remaining residue was lyophilized. The crude product was solubilized in acetonitrile/water (1/1), passed through a syringe filter, purified by preparative RP-HPLC and lyophilized. The final product possesses characteristic Eu luminescence, although three-coordinated "Eu(peptoid)₃" complex was not found.

MS (MALDI-TOF, CHCA): m/z = 1020.39 $(M - \times 2\,Cl + CHCA^-)^+$; (MALDI-TOF, THAP): m/z = 867.3 $[M - Cl]^+$. – **Analytical HPLC** (5–95% acetonitrile + 0.1% TFA in 20 min, detection at 218 nm): t_{Ret} = 9.16 min, 84% purity.

***H*-(*N*0b2)-(*N*6am)₃-*NH₂* (2.51-H)**

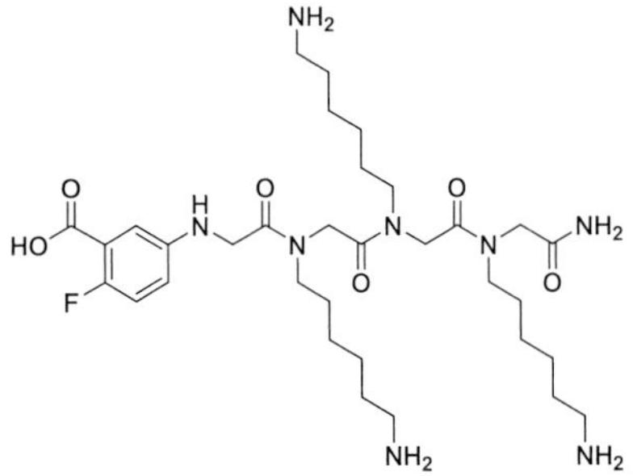

According to **GP12a**, **Eu-2.48-SP** was cleaved from the resin by treatment with 3 mL 95% TFA in DCM at room temperature for 2 h. The solution was filtered off and the next cleavage mixture was added to the resin and let shaken for additional 2 h. The resin was washed with 3 mL DCM (3 ×) and methanol (3 ×). The cleavage cocktail was evaporated under airflow, the rest was combined with washes and the solvent was removed under reduced pressure (**GP12a**, step 1). The crude product was then dissolved in 5 mL acetonitrile, passed through a syringe filter, purified by means of preparative HPLC (5-95% acetonitrile + 0.1% TFA in 30 min, 25 °C) and lyophilized (**GP12a**, step 2) to afford 23 mg (17% over 9 steps).

MS (MALDI-TOF, CHCA): m/z = 681.4 $[M+H]^+$; 704.4 $[M+Na]^+$. – **Analytical HPLC** (5–95% acetonitrile + 0.1% TFA in 20 min, detection at 218 nm): t_{Ret} = 8.93 min, 79% purity.

***Eu*-[*H*-(*N*0b2⁻)-(*N*6am)₃-NH₂]Cl₂ (Eu-2.51)**

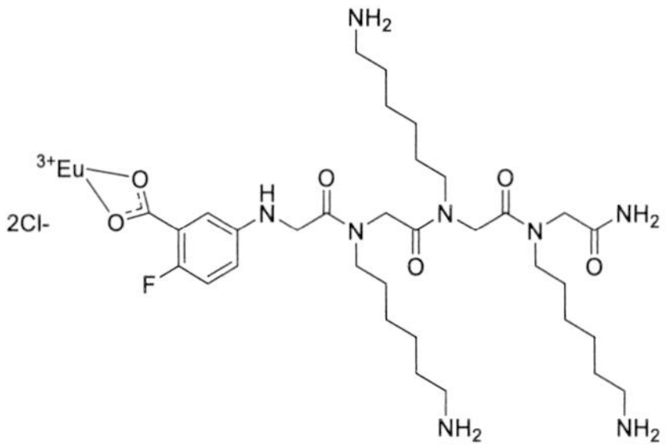

A solution of 20.0 mg of peptoid-based carboxylic acid **2.51-H** (30.0 µmol, 3.00 equiv.) in ethanol was treated with 30 µL of 1 M NaOH at 60 °C (30.0 µmol, 3.00 equiv.). Then 19.5 µL of 0.512 M aq. solution of $EuCl_3 \cdot 6\ H_2O$ (10.0 µmol, 1.00 equiv.) was added. The reaction mixture was refluxed for additional 3 h, ethanol was removed under reduced pressure and the remaining residue was lyophilized. The crude product was solubilized in acetonitrile/water (1/1), passed through a syringe filter, purified by preparative RP-HPLC and lyophilized. The final product possesses characteristic Eu luminescence, although three-coordinated "Eu(peptoid)₃" complex was not found.

MS (MALDI-TOF, THAP): m/z = 867.3 $[M-Cl]^+$. – **Analytical HPLC** (5–95% acetonitrile + 0.1% TFA in 20 min, detection at 218 nm): t_{Ret} = 8.93 min, 80% purity.

***H*-(*N*0b3)-(*N*6am)₃-*NH₂* (2.52-H)**

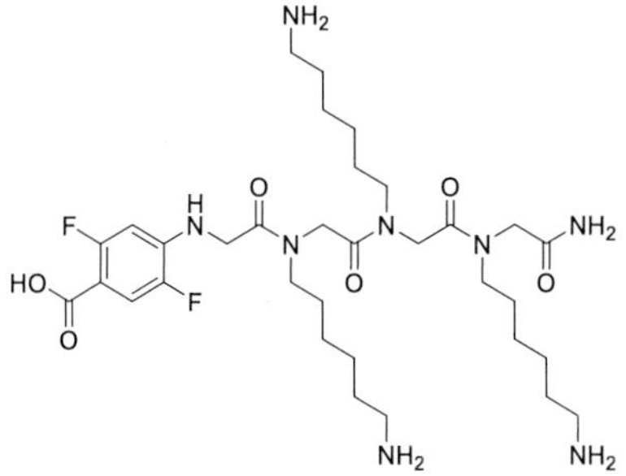

According to **GP12a**, **Eu-2.49-SP** was cleaved from the resin by treatment with 3 mL 95% TFA in DCM at room temperature for 2 h. The solution was filtered off and the next cleavage mixture was added to the resin and let shaken for additional 2 h. The resin was washed with 3 mL DCM (3 ×) and methanol (3 ×). The cleavage cocktail was evaporated under airflow, the rest was combined with washes and the solvent was removed under reduced pressure (**GP12a**, step 1). The crude product was then dissolved in 5 mL acetonitrile, passed through a syringe filter, purified by means of preparative HPLC (5-95% acetonitrile + 0.1% TFA in 30 min, 25 °C) and lyophilized (**GP12a**, step 2) to afford 14 mg (10% over 9 steps) of a white solid.

MS (MALDI-TOF, CHCA): m/z = 699.4 $[M+H]^+$; 721.4 $[M+Na]^+$. – **Analytical HPLC** (5–95% acetonitrile + 0.1% TFA in 20 min, detection at 218 nm): t_{Ret} = 9.65 min, 71% purity.

***Eu*-[*H*-(*N*0b3⁻)-(*N*6am)₃-NH₂]Cl₂ (Eu-2.52)**

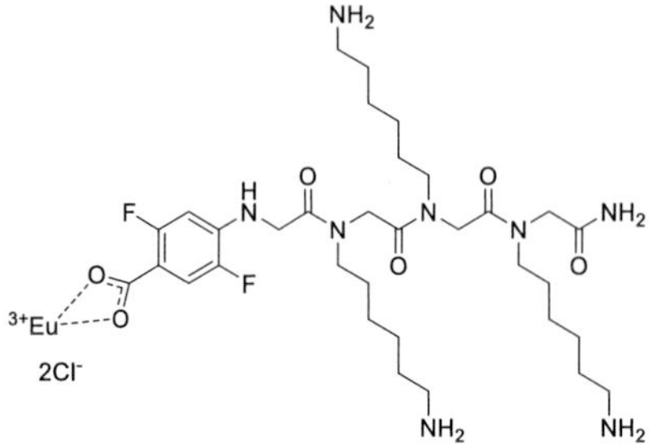

A solution of 20.0 mg of peptoid-based carboxylic acid (30.0 μmol, 3.00 equiv.) **2.51-H** in ethanol was treated with 30 μL of 1 M aq. NaOH at 60 °C (30.0 μmol, 3.00 equiv.). Then 19.5 μL of 0.512 M aq. solution of $EuCl_3 \cdot 6\ H_2O$ (10.0 μmol, 1.00 equiv.) was added. The reaction mixture was refluxed for additional 3 h, ethanol was removed under reduced pressure and the remaining residue was lyophilized. The crude product was solubilized in acetonitrile/water (1/1), passed through a syringe filter, purified by preparative RP-HPLC and lyophilized. The final product possesses characteristic Eu luminescence, although three-coordinated "Eu(peptoid)₃" complex was not found.

MS (MALDI-TOF, THAP): m/z = 885.3 $[M-Cl]^+$. – **Analytical HPLC** (5–95% acetonitrile + 0.1% TFA in 20 min, detection at 218 nm): t_{Ret} = 9.65 min, 78% purity.

***H*-(*N*Phe)-(*N*6am^{Boc})₃-*SP* (2.53-SP)**

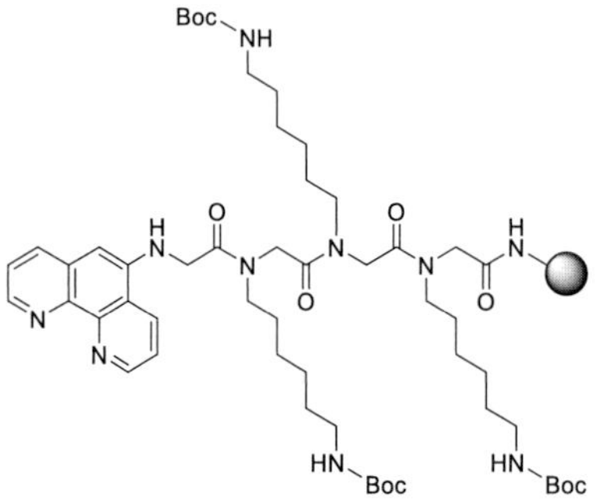

According to **GP11a**, 328 mg of Rink amide resin (0.20 mmol, 1.00 equiv.) were swollen and deprotected. The synthesis of the peptoid tetramer was carried out by acylation (**GP11a**, step 3) with 1.33 mL 1.20 M bromoacetic acid in pDMF (1.60 mmol, 8.00 equiv.) and 248 μL DIC (1.60 mmol, 8.00 equiv.). *N*6am^{Boc} group was attached, using 346 mg of *tert*-butyl(4-aminohexyl)carbamate (1.60 mmol, 8.00 equiv.) (**GP11a**, step 4) and *N*Phe group was attached, using the solution of 117 mg of 2.34 (0.60 mmol, 3.00 equiv.) in 1.60 mL pDMF (shaking overnight). A few beads were cleaved off and the formation of the desired product was detected by MALDI-TOF-MS.

MS (MALDI-TOF, CHCA/DHB): m/z = 721.5 $[M+H-\times 3\ Boc]^+$.

***H*-(*N*Phe)-(*N*6am^{Boc})$_3$-*NH$_2$* (2.53)**

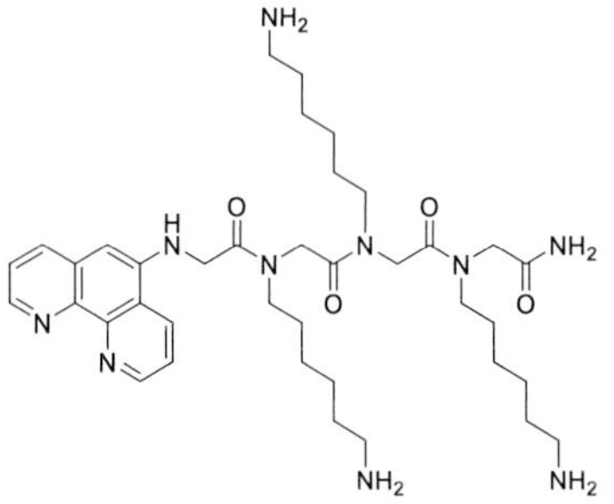

According to **GP12a**, **2.53-SP** was cleaved from the resin by treatment with 3 mL 95% TFA in DCM at room temperature for 2 h. The solution was filtered off and the next cleavage mixture was added to the resin and let shaken for additional 2 h. The resin was washed with 3 mL DCM (3 ×) and methanol (3 ×). The cleavage cocktail was evaporated under airflow, the rest was combined with washes and the solvent was removed under reduced pressure (**GP12a**, step 1). The crude product was then dissolved in 5 mL acetonitrile, passed through a syringe filter, purified by means of preparative HPLC (5-95% acetonitrile + 0.1% TFA in 30 min, 25 °C) and lyophilized (**GP12a**, step 2) to afford 23 mg (16% over 9 steps) of a red solid.

MS (MALDI-TOF, CHCA): m/z = 721.5 $[M + H]^+$. – **Analytical HPLC** (5–95% acetonitrile + 0.1% TFA in 20 min, detection at 218 nm): t_{Ret} = 8.42 min, 96% purity.

***Eu-2.8*-[*H*-(*N*Phe)-(*N*6am)$_3$-*NH$_2$*] (2.54)**

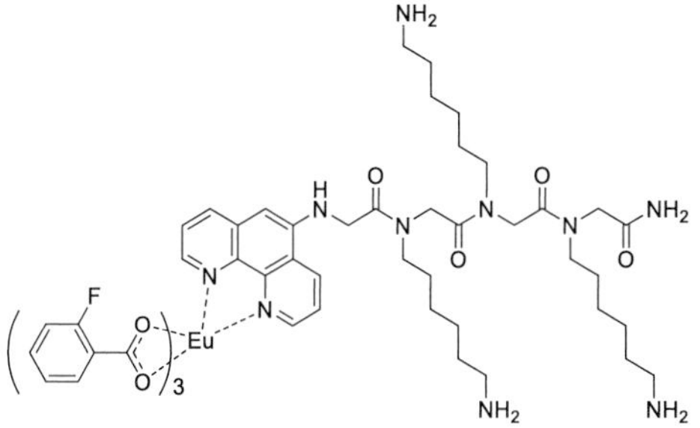

According to **GP2**, a solution of 20.0 mg of *H*-(*N*Phe)-(*N*6am)$_3$-NH$_2$ (**2.53**) (30.0 µmol, 1.00 equiv.) in ethanol was added to a ethanol/water solution of 18.0 mg of **Eu-2.8** (30.0 µmol, 1.00 equiv.) and the mixture was refluxed for 2 h. The reaction mixture was then allowed to cool to room temperature, ethanol was removed under reduced pressure and the remaining residue was lyophilized to afford 33.0 mg (87%) of red cristalline solid.

MS (MALDI-TOF, THAP): m/z = 1290.5 $[M]^+$; 1291.5 $[M+H]^+$. – **Analytical HPLC** (5–95% acetonitrile + 0.1% TFA in 20 min, detection at 218 nm): t_{Ret} = 9.35 min, 94% purity.

***H*-(*NPhe*)-(*N*1ph)-(*N*1ay)-(*N*1ph)-(*N*1Pc)-*SP* (2.56-SP)**

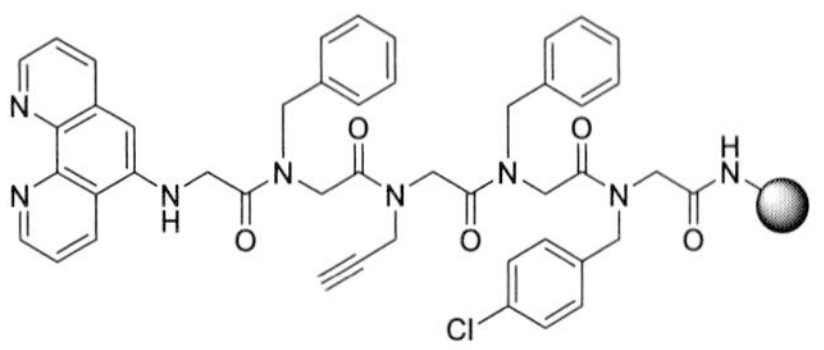

According to **GP11a**, 328 mg of Rink amide resin (0.20 mmol, 1.00 equiv.) were swollen and deprotected. The synthesis of the peptoid hexamer was carried out by acylation (**GP11a**, step 3) with 1.33 mL 1.20 M bromoacetic acid in pDMF (1.60 mmol, 8.00 equiv.) and 248 µL DIC (1.60 mmol, 8.00 equiv.). *N*1Pc group was attached, using 194 µL of 4-chlorobenzylamine (226 mg, 1.60 mmol, 8.00 equiv.), N1ph group was attached, using 175 µL of benzylamine (172 mg, 1.60 mmol, 8.00 equiv.) in 1.60 mL pDMF, *N*1ay group was attached, using 110 µL of propargylamine (90 mg, 1.60 mmol, 8.00 equiv.) and *N*Phe group was attached, using 156 mg of 4-amino-1,10-phenanthroline **2.34** (0.80 mmol, 4.00 equiv., stirring overnight) (**GP11a**, step 4). A few beads were cleaved off and the formation of the desired product was detected by MALDI-TOF-MS.

MS (MALDI-TOF, CHCA): m/z = 823.3 $[M+H]^+$; 845.5 $[M+Na]^+$.

***Eu*-2.8-[*H*-(*NPhe*)-(*N*1ph)-(*N*1ay)-(*N*1ph)-(*N*1Pc)]-*NH₂* (Eu-2.57)**

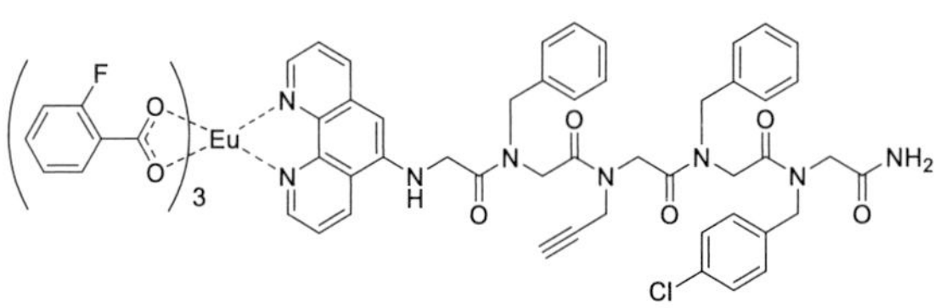

According to **GP2**, a solution of 16.0 mg of *H*-(*NPhe*)-(*N*1ph)-(*N*1ay)-(*N*1ph)-(*N*1Pc)]-NH_2 (**2.56**) (20.0 µmol, 1.00 equiv.) in ethanol was added to an ethanol/water solution of 12.0 mg of **Eu-2.8** (20.0 µmol, 1.00 equiv.) and the mixture was refluxed for 2 h. The reaction mixture was then allowed to cool to room temperature, ethanol was removed under reduced pressure and the remaining residue was lyophilized to afford 25.0 mg (92%) of a red cristalline solid.

MS (MALDI-TOF, THAP): m/z = 1392.5 $[M]^+$; 1415.5 $[M+Na]^+$. – **Analytical HPLC** (5–95% acetonitrile + 0.1% TFA in 20 min, detection at 218 nm): t_{Ret} = 9.36 min, 94% purity.

***H*-(*NPhe*)-(*N*1ph)-(*N*1ay)-(*N*1ph)-(*N*1Pc)-*NH₂* (2.56)**

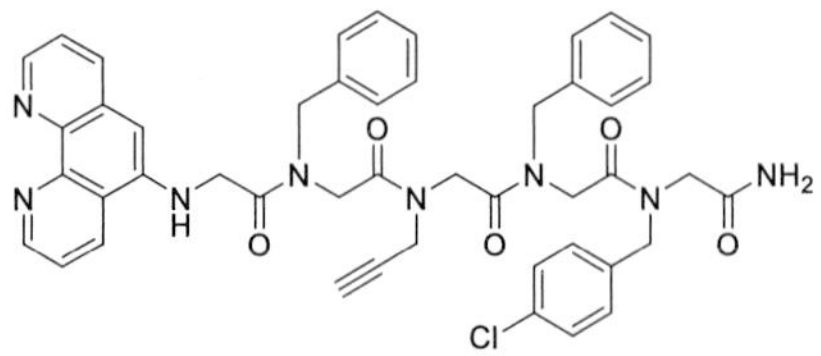

According to **GP12a**, **2.56-SP** was cleaved from the resin by treatment with 3 mL 95% TFA in DCM at room temperature for 2 h. The solution was filtered off and the next cleavage mixture was added to the resin and let shaken for additional 2 h. The resin was washed with 3 mL DCM (3 ×) and methanol (3 ×). The cleavage cocktail was evaporated under airflow, the rest was combined with washes and the solvent was removed under reduced pressure (**GP12a**, step 1). The crude product was then dissolved in 5 mL acetonitrile, passed through a syringe filter, purified by means of preparative HPLC (5-95% acetonitrile + 0.1% TFA in 30 min, 25 °C) and lyophilized (**GP12a**, step 2) to afford 24 mg (18% over 11 steps) of a red solid.

MS (MALDI-TOF, CHCA): m/z = 823.3 $[M+H]^+$; 845.5 $[M+Na]^+$.– **Analytical HPLC** (5–95% acetonitrile + 0.1% TFA in 20 min, detection at 218 nm): t_{Ret} = 10.09 min, 95% purity.

***H*-(*NPhe*)-(*N*1ph)-(*N*1Pc)-(*N*1ay)$_2$-*SP* (2.59-SP)**

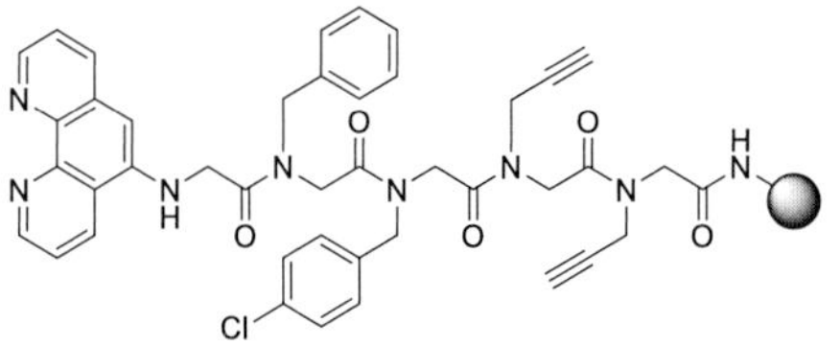

According to **GP11a**, 328 mg of Rink amide resin (0.20 mmol, 1.00 equiv.) were swollen and deprotected. The synthesis of the peptoid hexamer was carried out by acylation (**GP11a**, step 3) with 1.33 mL 1.20 M bromoacetic acid in pDMF (1.60 mmol, 8.00 equiv.) and 248 μL DIC (1.60 mmol, 8.00 equiv.). *N*1ay group was attached, using 110 μL of propargylamine (90 mg, 1.60 mmol, 8.00 equiv.), *N*1Pc group was attached, using 194 μL of 4-chlorobenzylamine (226 mg, 1.60 mmol, 8.00 equiv.), N1ph group was attached, using 175 μL of benzylamine (172 mg, 1.60 mmol, 8.00 equiv.) and *N*Phe group was attached, using 156 mg of 4-amino-1,10-phenanthroline (**2.34**) (0.80 mmol, 4.00 equiv., stirring overnight) in 1.60 mL pDMF (**GP11a**, step 4). A few beads were cleaved off and the formation of the desired product was detected by MALDI-TOF-MS.

MS (MALDI-TOF, CHCA): m/z = 771.3 $[M+H]^+$; 793.3 $[M+Na]^+$.

***H*-(*NPhe*)-(*N*1ph)-(*N*1Pc)-(*N*1ay)$_2$-*NH$_2$* (2.59)**

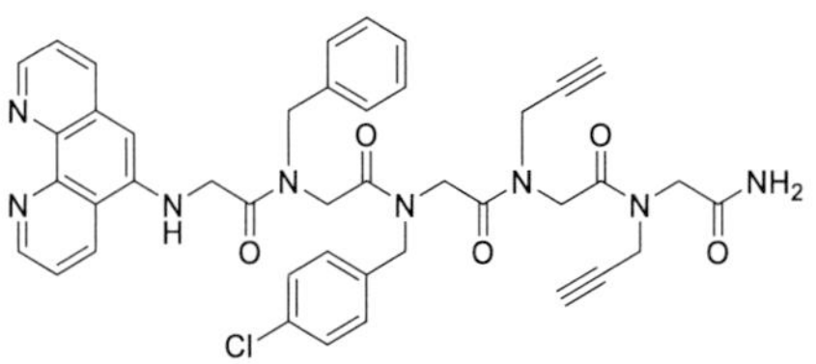

According to **GP12a**, **2.59-SP** was cleaved from the resin by treatment with 3 mL 95% TFA in DCM at room temperature for 2 h. The solution was filtered off and the next cleavage mixture was added to the resin and let shaken for additional 2 h. The resin was washed with 3 mL DCM (3 ×) and methanol (3 ×). The cleavage cocktail was evaporated under airflow, the rest was combined with washes and the solvent was removed under reduced pressure (**GP12a**, step 1). The crude product was then dissolved in 5 mL acetonitrile, passed through a syringe filter, purified by means of preparative

HPLC (5-95% acetonitrile + 0.1% TFA in 30 min, 25 °C) and lyophilized (**GP12a**, step 2) to afford 25 mg (16% over 11 steps) of a red solid.

MS (MALDI-TOF, CHCA): m/z = 771.3 $[M + H]^+$; 793.3 $[M + Na]^+$.– **Analytical HPLC** (5–95% acetonitrile + 0.1% TFA in 20 min, detection at 218 nm): t_{Ret} = 9.95 min, 89% purity.

***Eu-2.8*-[*H*-(*NPhe*)-(*N*1ph)-(*N*1ay)-(*N*1ph)-(*N*1Pc)]-*NH₂* (Eu-2.60)**

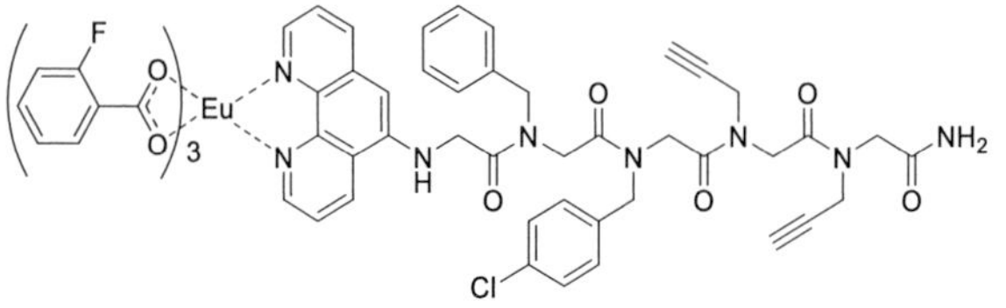

According to **GP2**, a solution of 14.8 mg of *H*-(*NPhe*)-(*N*1ph)-(*N*1ay)-(*N*1ph)-(*N*1Pc)]-NH_2 (**2.59**) (19.2 µmol, 1.00 equiv.) in ethanol was added to an ethanol/water solution of 11.2mg of **Eu-2.8** (19.2 µmol, 1.00 equiv.) and the mixture was refluxed for 2 h. The reaction mixture was then allowed to cool to room temperature, ethanol was removed under reduced pressure and the remaining residue was lyophilized to afford 25.0 mg (92%) of a red cristalline solid.

MS (MALDI-TOF, THAP): m/z = 1339.1 $[M]^+$. – **Analytical HPLC** (5–95% acetonitrile + 0.1% TFA in 20 min, detection at 218 nm): t_{Ret} = 9.24 min, 93% purity.

***H*-(*N*1ay)-(*N*6am^{Boc})$_3$-*SP* (2.61-SP)**

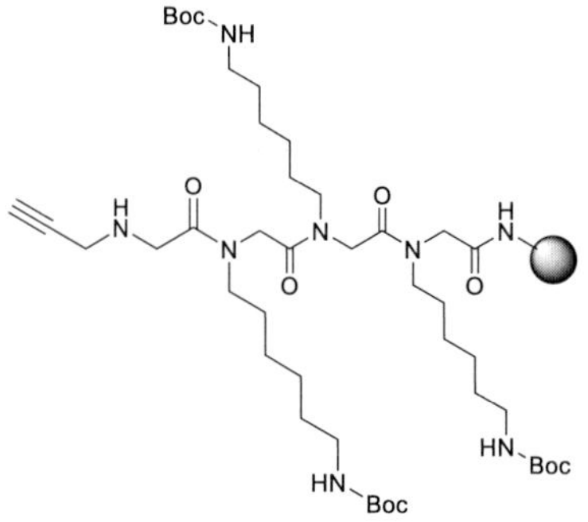

According to **GP11a**, 328 mg of Rink amide resin (0.20 mmol, 1.00 equiv.) were swollen and deprotected. The synthesis of the peptoid tetramer was carried out by acylation (**GP11a**, step 3) with 1.33 mL 1.20 M bromoacetic acid in pDMF (1.60 mmol, 8.00 equiv.) and 248 µL DIC (1.60 mmol, 8.00 equiv.). *N*6am^{Boc} group was attached, using 346 mg of *tert*-butyl(4-aminohexyl)carbamate (1.60 mmol, 8.00 equiv.) (**GP11a**, step 4) and *N*1ay group was attached by 102 µL of propargylamine (1.60 mmol, 8.00 equiv.) in 1.60 mL pDMF. A few beads were cleaved off and the formation of the desired product was detected by MALDI-TOF-MS.

MS (MALDI-TOF, CHCA): m/z = 581.4 $[M+H–\times3\ Boc]^+$.

***H*-(*N*1ay)-(*N*6am)$_3$-*NH*$_2$ (2.61)**

According to GP12a, **2.61-SP** was cleaved from solid support by treatment with 3.5 mL of 95% TFA in DCM solution (**GP12a**, step 1). Purification by preparative HPLC (**GP12a**, step 2) (5-95% acetonitrile + 0.1% TFA in 30 min, 25 °C) and subsequent lyophilization afforded 24 mg (21% over 9 steps) of the desired product.

MS (MALDI-TOF, CHCA): m/z = 581.4 $[M+H]^+$. – **Analytical HPLC** (5–95% acetonitrile + 0.1% TFA in 20 min, detection at 218 nm): t_{Ret} = 8.01 min, 89% purity.

***H*-(*N*3az)-(*N*6am^{Boc})$_3$-*SP* (2.62-SP)**

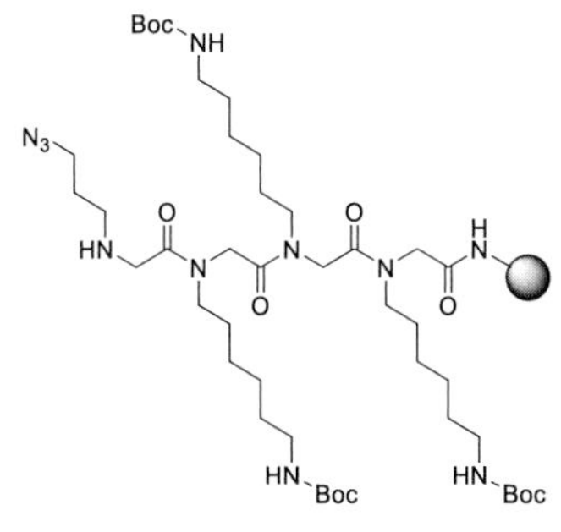

According to **GP11a**, 328 mg of Rink amide resin (0.20 mmol, 1.00 equiv.) were swollen and deprotected. The synthesis of the peptoid tetramer was carried out by acylation (**GP11a**, step 3) with 1.33 mL 1.20 M bromoacetic acid in pDMF (1.60 mmol, 8.00 equiv.) and 248 µL DIC (1.60 mmol, 8.00 equiv.). *N*6am^{Boc} group was attached, using 346 mg of *tert*-Butyl(4-aminohexyl)carbamate (1.60 mmol, 8.00 equiv.) (**GP11a**, step 4) and *N*3az group was attached, using the solution of 108 mg of 3-azidopropan-1-amine (**2.45**) (1.08 mmol, 5.40 equiv.) in 1.60 mL pDMF (shaking overnight). A few beads were cleaved from the resin and the formation of the desired product was detected by MALDI-TOF-MS.

MS (MALDI-TOF, CHCA): m/z = 626.5 $[M + H - \times 3\ Boc]^+$.

***H*-(*N*3az)-(*N*6am^{Boc})$_3$-*NH*$_2$ (2.62)**

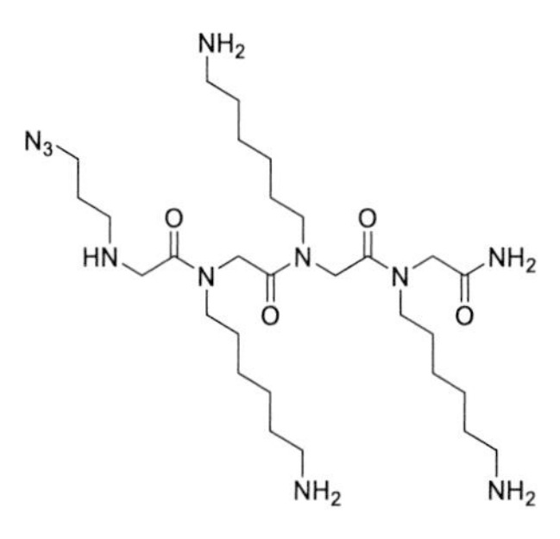

According to **GP12a**, **2.62-SP** was cleaved from the resin by treatment with 3 mL 95% TFA in DCM at room temperature for 2 h. The solution was filtered off and the next cleavage mixture was added to the resin and let shaken for additional 2 h. The resin was washed with 3 mL DCM (3 ×) and methanol (3 ×). The cleavage cocktail was evaporated under airflow, the rest was combined with washes and the solvent was removed under reduced pressure (**GP12a**, step 1). The crude product was then dissolved in 5 mL acetonitrile, passed through a syringe filter, purified by means of preparative HPLC (5-95% acetonitrile + 0.1% TFA in 30 min, 25 °C) and lyophilized (**GP12a**, step 2) to afford 25 mg (20% over 9 steps).

MS (MALDI-TOF, CHCA): m/z = 626.5 $[M + H]^+$. – **Analytical HPLC** (5–95% acetonitrile + 0.1% TFA in 20 min, detection at 218 nm): t_{Ret} = 7.96 min, 91% purity.

***H*-(*N*3az)-(*N*1ph)-(*N*1ay)-(*N*1Pc)-*SP* (2.63-SP)**

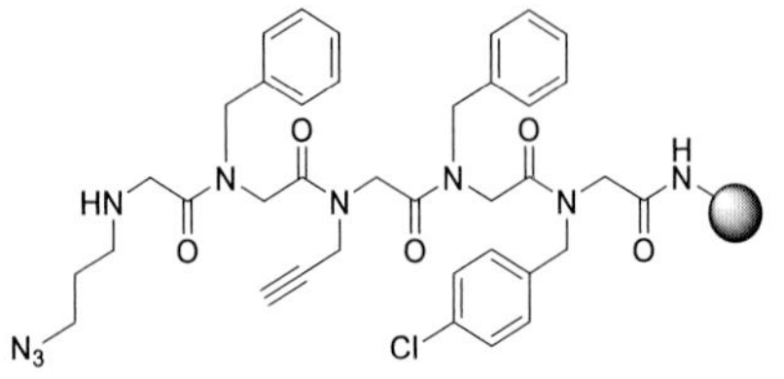

According to **GP11a**, 328 mg of Rink amide resin (0.20 mmol, 1.00 equiv.) were swollen and deprotected. The synthesis of the peptoid hexamer was carried out by acylation (**GP11a**, step 3) with 1.33 mL 1.20 M bromoacetic acid in pDMF (1.60 mmol, 8.00 equiv.) and 248 µL DIC (1.60 mmol, 8.00 equiv.). *N*1Pc group was attached, using 194 µL of 4-chlorobenzylamine (226 mg, 1.60 mmol, 8.00 equiv.), N1ph group was attached, using 175 µL of benzylamine (172 mg, 1.60 mmol, 8.00 equiv.) in 1.60 mL pDMF, *N*1ay group was attached, using 110 µL of propargylamine (90 mg, 1.60 mmol, 8.00 equiv.) and *N*3az group was attached, using 158 µL of 3-azidopropane-1-amine (**2.45**) (**GP11a**, step 4). A few beads were cleaved from the resin and the formation of the desired product was detected by MALDI-TOF-MS.

MS (MALDI-TOF): m/z = 727.3 $[M]^+$.

***H*-*0fb*^{Me}-(*N*1tz^{C})-(*N*6am^{Boc})$_3$-*SP* (2.64-Me-SP)**

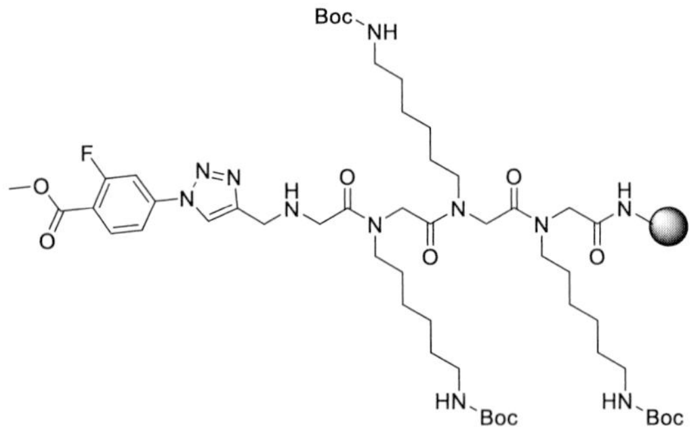

According to **GP10b**, 2.61^{Boc}-SP (0.20 mmol, 1.00 equiv.) was swollen in DMF for 2 h. Then CuI 23 mg (0.12 mmol, 0.60 equiv.) together with the pDMF solution of 41 mg of methyl 4-azido-2-fluorobenzoate (2.3-Me) (0.21 mmol, 1.05 equiv.) and 84 µL of DIPEA (0.48 mmol, 2.40 equiv.) in pDMF were added to the resin and the mixture was shaken at room temperature for 16 h. Afterwards the liquid was filtered off and the resin was washed with THF, methanol, DMF and DCM until the filtrate became colorless (3 × 3.00 mL for each washing step). A few beads were cleaved off and the formation of the desired product was detected by MALDI-TOF-MS.

MS (MALDI-TOF, CHCA): m/z = 776.5 $[M+H]^+$.

***H-0fb*-(*N*1tz^{C})-(*N*6am^{Boc})$_3$-*SP* (2.64-H-SP)**

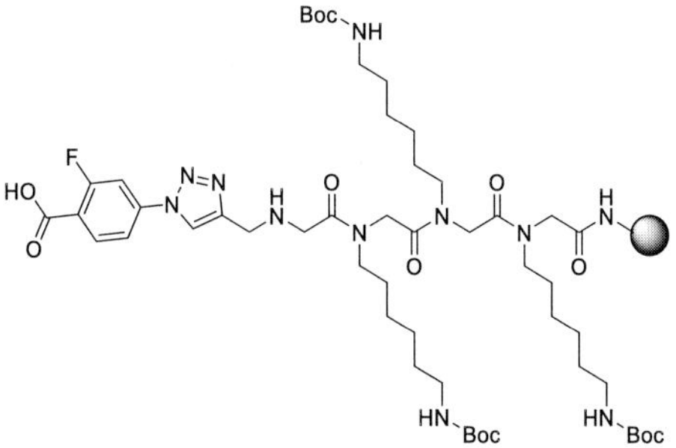

2.64-Me-SP (0.20 mmol, 1.00 equiv.) was swollen in DMF for 2 h and then the resin was washed with DMF, pDMF, methanol and twice with THF The resin was then treated with the solution of 15 mg of LiOH (0.63 mmol, 3.15 equiv.) in THF/methanol (1/1) and the mixture was shaken for 4 h. Afterwards the resin was washed with THF, methanol, water, methanol, DCM in this order (2 × 3.00 mL for each washing step). A few beads were cleaved off and the formation of the desired product was detected by MALDI-TOF-MS.

MS (MALDI-TOF): m/z = 783.5 $[M+Na+H-\times 3\ Boc]^+$.

***H-0fb*-(*N*1tz^{C})-(*N*6am)$_3$-*NH$_2$* (2.64-H)**

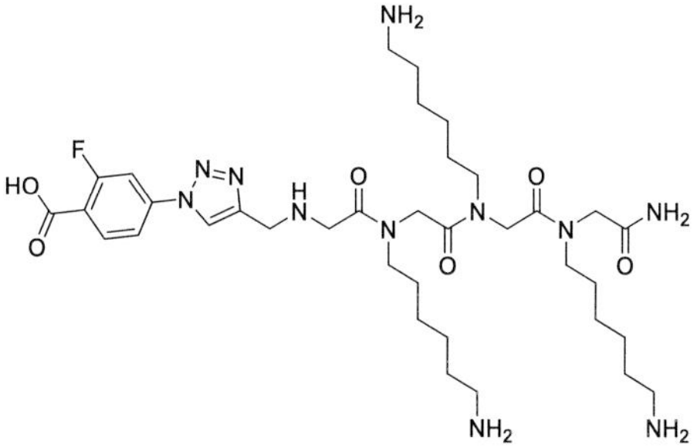

According to **GP12a**, **2.64-H-SP** was cleaved from the resin by treatment with 3 mL 95% TFA in DCM for 2 h at room temperature. The solution was filtered off and the next cleavage mixture was added to the resin and let shaken for additional 2 h. The resin was washed with 3 mL DCM (3 ×) and methanol (3 ×). The cleavage cocktail was evaporated under airflow, the rest was combined with washes and the solvent was removed under reduced pressure (**GP12a**, step 1). The crude product was then dissolved in 5 mL acetonitrile, passed through a syringe filter, purified by means of preparative HPLC (5-95% acetonitrile + 0.1% TFA in 30 min, 25 °C) and lyophilized (**GP12a**, step 2) to afford 17 mg (11% over 11 steps).

MS (MALDI-TOF, CHCA): m/z = 797.5 $[M+Na]^+$. – **Analytical HPLC** (5–95% Acetonitrile + 0.1% TFA in 20 min, detection at 218 nm): t_{Ret} = 9.98 min, 84% purity.

***H-1fb^Me^-(N1tz^C^)-(N6am^Boc^)~3~-SP* (2.65-Me-SP)**

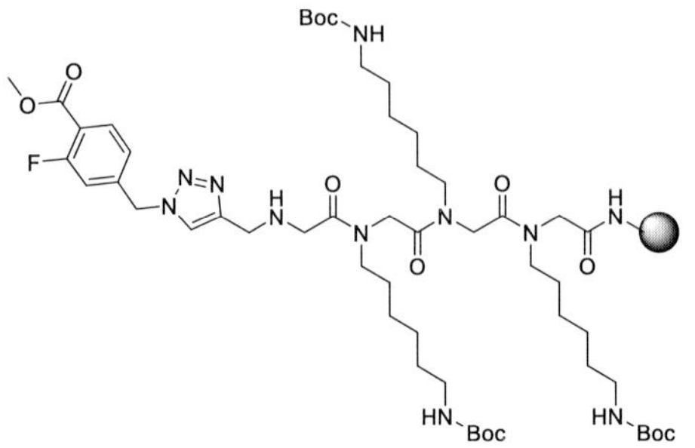

According to **GP10b**, **2.61-SP** (0.20 mmol, 1.00 equiv.) was swollen in DMF for 2 h. Then CuI 23 mg (0.12 mmol, 0.60 equiv.) together with the pDMF solution of 44 mg of methyl 4-(azidomethyl)-2-fluorobenzoate **(2.6-Me)** (0.21 mmol, 1.05 equiv.) and 84 µL of DIPEA (0.48 mmol, 2.40 equiv.) in pDMF was added to the resin and the mixture was shaken at room temperature for 16 h. Afterwards the liquid was filtered off and the resin was washed with THF, methanol, DMF and DCM until the filtrate became colorless (3 × 3.00 mL for each washing step). A few beads were cleaved off and the formation of the desired product was detected by MALDI-TOF-MS.

MS (MALDI-TOF, CHCA): m/z = 790.5 [M+H–×3 Boc]$^+$.

***H-1fb-(N1tz^C^)-(N6am^Boc^)~3~-SP* (2.65-H-SP)**

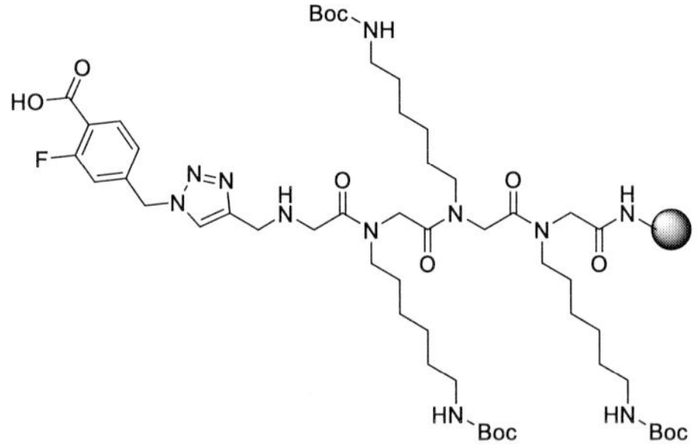

2.65-Me-SP (0.20 mmol, 1.00 equiv.) was swollen in DMF for 2 h and then the resin was washed with DMF, pDMF, methanol and twice with THF. The resin was then treated with the solution of 15 mg of LiOH (0.63 mmol, 3.15 equiv.) in THF/methanol (1/1) and the mixture was shaken for 4 h. Afterwards the resin was washed with THF, methanol, water, methanol, DCM in this order (2 × 3.00 mL for each washing step). A few beads were cleaved off and the formation of the desired product was detected by MALDI-TOF-MS.

MS (MALDI-TOF, CHCA): m/z = 797.5 [M+Na–×3 Boc]$^+$.

***H-1fb*-(*N*1tz^{C})-(*N*6am)$_3$-*NH$_2$* (2.65-H)**

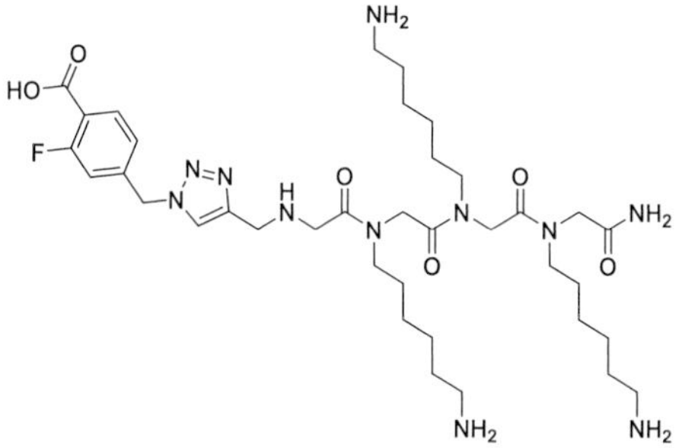

According to **GP12a**, **2.65-H-SP** was cleaved from the resin by treatment with 3 mL 95% TFA in DCM for 2 h at room temperature. The solution was filtered off and the next cleavage mixture was added to the resin and let shaken for additional 2 h. The resin was washed with 3 mL DCM (3 ×) and methanol (3 ×). The cleavage cocktail was evaporated under airflow, the rest was combined with washes and the solvent was removed under reduced pressure (**GP12a**, step 1). The crude product was then dissolved in 5 mL acetonitrile, passed through a syringe filter, purified by means of preparative HPLC (5-95% acetonitrile + 0.1% TFA in 30 min, 25 °C) and lyophilized (**GP12a**, step 2) to afford 14 mg (9% over 11 steps).

MS (MALDI-TOF, CHCA): m/z = 797.5 $[M+Na]^+$. – **Analytical HPLC** (5–95% Acetonitrile + 0.1% TFA in 20 min, detection at 218 nm): t_{Ret} = 9.87 min, 77% purity.

***H-Phen1*-(*N*3tz^{N})-(*N*6am^{Boc})$_3$-*SP* (2.66^{Boc}-SP)**

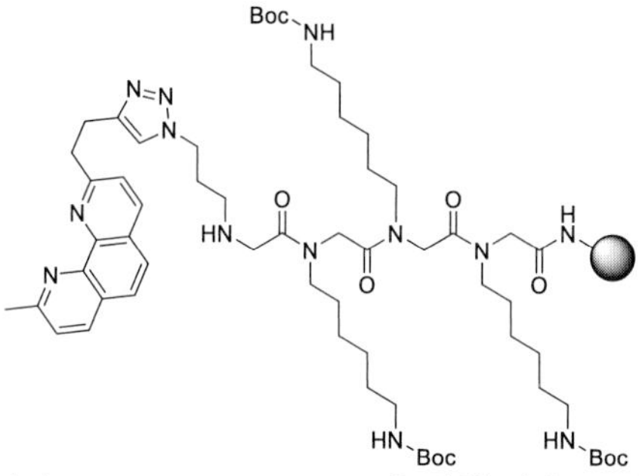

According to **GP10b**, **2.62^{Boc}-SP** (0.20 mmol, 1.00 equiv.) was swollen in DMF for 2 h. Then CuI 23 mg (0.12 mmol, 0.60 equiv.) together with the pDMF solution of 51 mg of 2-(but-3-yn-1-yl)-9-methyl-1,10-phenanthroline (**2.36**) (0.21 mmol, 1.05 equiv.) and 84 µL of DIPEA (0.48 mmol, 2.40 equiv.) in pDMF were added to the resin and the mixture was shaken at room temperature for 16 h. Afterwards the liquid was filtered off and the resin was washed with THF, methanol, DMF and DCM until the filtrate became colorless (3 × 3 mL for each washing step). A few beads were cleaved off and the formation of the desired product was detected by MALDI-TOF-MS.

MS (MALDI-TOF, CHCA): m/z = 873.2 $[M+H–\times 3\ Boc]^+$; 895.2 $[M+Na–\times 3\ Boc]^+$; 935.6 $[M+Cu–\times 3\ Boc]^+$.

H-Phen1-(*N*3tz^{N})-(*N*6am^{Boc})$_3$-*NH$_2$* (2.66)

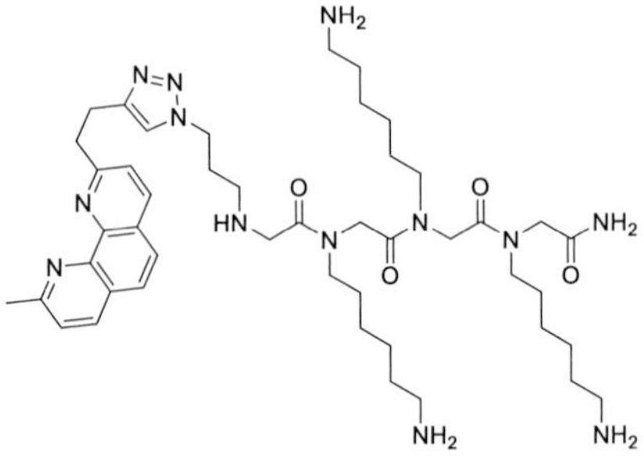

According to **GP12a**, **2.66^{Boc}-SP** was cleaved from the resin by treatment with 3 mL 95% TFA in DCM at room temperature for 2 h. The solution was filtered off and the next cleavage mixture was added to the resin and let shaken for additional 2 h. The resin was washed with 3 mL DCM (3 ×) and methanol (3 ×). The cleavage cocktail was evaporated under airflow, the rest was combined with washes and the solvent was removed under reduced pressure (**GP12a**, step 1). The crude product was then dissolved in 5 mL acetonitrile, passed through a syringe filter, purified by means of preparative HPLC (5-95% acetonitrile + 0.1% TFA in 30 min, 25 °C) and lyophilized (**GP12a**, step 2) to afford 14 mg (8% over 10 steps).

MS (MALDI-TOF, CHCA): m/z = 873.2 $[M+H]^+$; 895.2 $[M+Na]^+$. – **Analytical HPLC** (5–95% acetonitrile + 0.1% TFA in 20 min, detection at 218 nm): t_{Ret} = 9.02 min, 80% purity.

H-Phen2-(*N*3tz^{N})-(*N*6am^{Boc})$_3$-*SP* (2.67^{Boc}-SP)

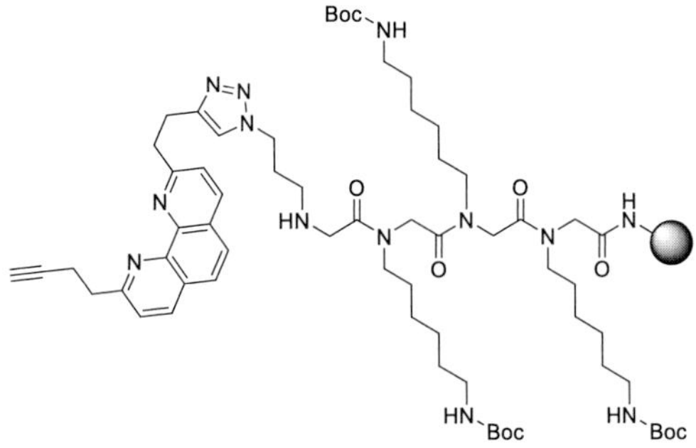

According to **GP10b**, **2.62^{Boc}-SP** (0.20 mmol, 1.00 equiv.) was swollen in DMF for 2 h. Then CuI 23 mg (0.12 mmol, 0.60 equiv.) together with the pDMF solution of 59 mg of 2,9-di(but-3-yn-1-yl)-1,10-phenanthroline (**2.37**) (0.21 mmol, 1.05 equiv.) and 84 µL of DIPEA (0.48 mmol, 2.40 equiv.) in pDMF were added to the resin and the mixture was shaken at room temperature for 16 h. Afterwards the liquid was filtered off and the resin was washed with THF, methanol, DMF and DCM until the filtrate became colorless (3 × 3.00 mL for each washing step). A few beads were cleaved off and the formation of the desired product was detected by MALDI-TOF-MS.

MS (MALDI-TOF, CHCA): m/z = 910.6 $[M + H - \times 3\ Boc]^+$; 933.7 $[M + Na - \times 3\ Boc]^+$; 974.1 $[M + Cu - \times 3\ Boc]^+$.

***H-Phen2*-(*N*3tz^{N})-(*N*6am^{Boc})$_3$-*NH$_2$* (2.67)**

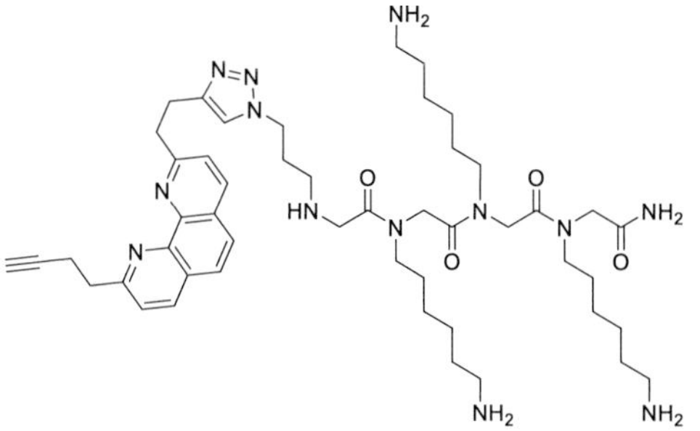

According to **GP12a**, **2.67^{Boc}-SP** was cleaved from the resin by treatment with 3 mL 95% TFA in DCM at room temperature for 2 h. The solution was filtered off and the next cleavage mixture was added to the resin and let shaken for additional 2 h. The resin was washed with 3 mL DCM (3 ×) and methanol (3 ×). The cleavage cocktail was evaporated under airflow, the rest was combined with washes and the solvent was removed under reduced pressure (**GP12a**, step 1). The crude product was then dissolved in 5 mL acetonitrile, passed through a syringe filter, purified by means of preparative HPLC (5-95% acetonitrile + 0.1% TFA in 30 min, 25 °C) and lyophilized (**GP12a**, step 2) to afford 16 mg (9% over 10 steps).

MS (MALDI-TOF, CHCA): m/z = 910.7 $[M + H]^+$; 933.7 $[M + Na]^+$. – **Analytical HPLC** (5–95% acetonitrile + 0.1% TFA in 20 min, detection at 218 nm): t_{Ret} = 9.24 min, 85% purity.

***H-Phen1*-(*N*3tz^{N})-(*N*1ph)-(*N*1ay)-(*N*1Pc)-*SP* (2.68-SP)**

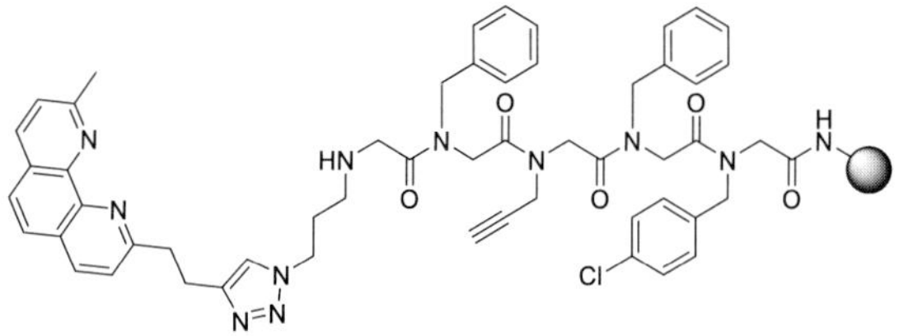

According to **GP10b**, **2.63-SP** (0.20 mmol, 1.00 equiv.) was swollen in DMF for 2 h. Then CuI 23 mg (0.12 mmol, 0.60 equiv.) together with the pDMF solution of 51 mg of 2-(but-3-yn-1-yl)-9-methyl-1,10-phenanthroline (**2.36**) (0.21 mmol, 1.05 equiv.) and 84 µL of DIPEA (0.48 mmol, 2.40 equiv.) in pDMF were added to the resin and the mixture was shaken at room temperature for 16 h. Afterwards the liquid was filtered off and the resin was washed with THF, methanol, DMF and DCM until the filtrate became colorless (3 × 3.00 mL for each washing step). A few beads were cleaved off and the formation of the desired product was detected by MALDI-TOF-MS.

MS (MALDI-TOF, CHCA): m/z = 974.5 $[M+H]^+$; 996.5 $[M+Na]^+$; 1037.1 $[M+Cu]^+$.

H-Phen1-(_N_3tz^{N})-(_N_1ph)-(_N_1ay)-(_N_1Pc)-_NH_$_2$ (2.68)

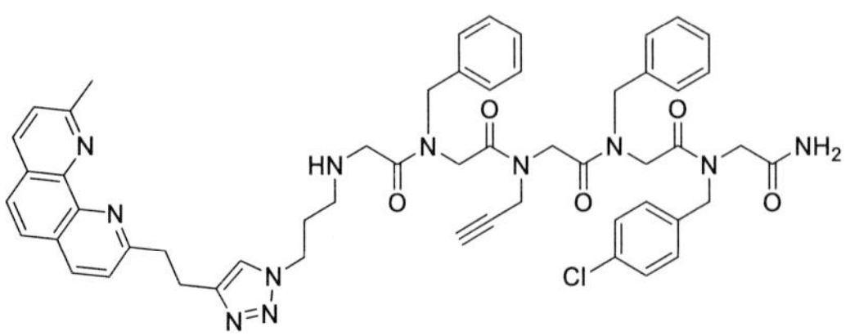

According to **GP12a**, **2.63-SP** was cleaved from the resin by treatment with 3 mL 95% TFA in DCM at room temperature for 2 h. The solution was filtered off and the next cleavage mixture was added to the resin and let shaken for additional 2 h. The resin was washed with 3 mL DCM (3 ×) and methanol (3 ×). The cleavage cocktail was evaporated under airflow, the rest was combined with washes and the solvent was removed under reduced pressure (**GP12a**, step 1). The crude product was then dissolved in 5 mL acetonitrile, passed through a syringe filter, purified by means of preparative HPLC (5-95% acetonitrile + 0.1% TFA in 30 min, 25 °C) and lyophilized (**GP12a**, step 2) to afford 8 mg (4% over 12 steps) of a white solid.

MS (MALDI-TOF, CHCA): m/z = 974.5 $[M+H]^+$; 996.5 $[M+Na]^+$. – **Analytical HPLC** (5–95% acetonitrile + 0.1% TFA in 20 min, detection at 218 nm): t_{Ret} = 11.95 min, 77% purity.

H-Phen2-(_N_3tz^{N})-(_N_1ph)-(_N_1ay)-(_N_1Pc)-_SP_ (2.69-SP)

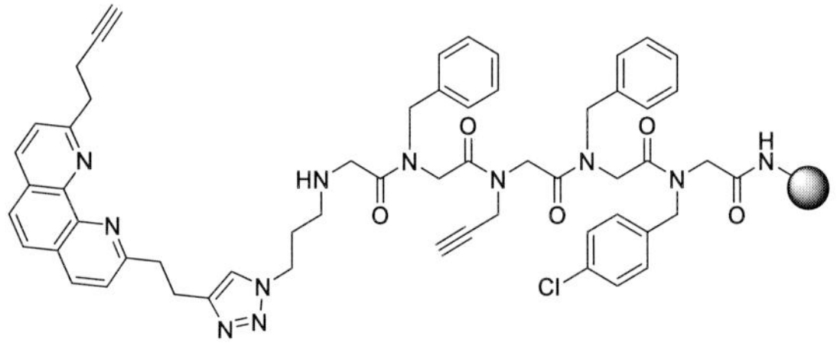

According to **GP10b**, **2.63-SP** (0.20 mmol, 1.00 equiv.) was swollen in DMF for 2 h. Then 23 mg of CuI (0.12 mmol, 0.60 equiv.) together with the pDMF solution of 59 mg of 2,9-di(but-3-yn-1-yl)-1,10-phenanthroline (**2.37**) (0.21 mmol, 1.05 equiv.) and 84 µL of DIPEA (0.48 mmol, 2.40 equiv.) in pDMF were added to the resin and the mixture was shaken at room temperature for 16 h. Afterwards the liquid was filtered off and the resin was washed with THF, methanol, DMF and DCM until the filtrate became colorless (3 × 3.00 mL for each washing step). A few beads were cleaved off and the formation of the desired product was detected by MALDI-TOF-MS.

MS (MALDI-TOF, CHCA): m/z = 1012.5 $[M+H]^+$; 1034.5 $[M+Na]^+$; 1076.1 $[M+Cu]^+$.

H-Phen2-(*N*3tz^{N})-(*N*1ph)-(*N*1ay)-(*N*1Pc)-*NH$_2$* (2.69-SP)

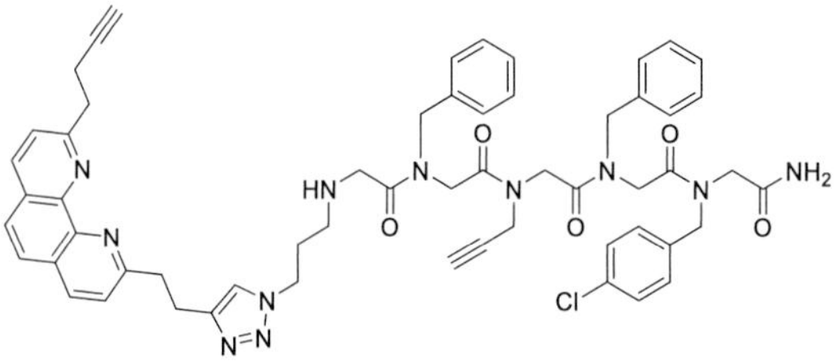

According to **GP12a**, **2.69-SP** was cleaved from the resin by treatment with 3 mL 95% TFA in DCM at room temperature for 2 h. The solution was filtered off and the next cleavage mixture was added to the resin and let shaken for additional 2 h. The resin was washed with 3 mL DCM (3 ×) and methanol (3 ×). The cleavage cocktail was evaporated under airflow, the rest was combined with washes and the solvent was removed under reduced pressure (**GP12a**, step 1). The crude product was then dissolved in 5 mL acetonitrile, passed through a syringe filter, purified by means of preparative HPLC (5-95% acetonitrile + 0.1% TFA in 30 min, 25 °C) and lyophilized (**GP12a**, step 2) to afford 8 mg (4% over 12 steps) of a white solid.

MS (MALDI-TOF, CHCA): m/z = 1012.5 $[M+H]^+$; 1034.5 $[M+Na]^+$. – **Analytical HPLC** (5–95% acetonitrile + 0.1% TFA in 20 min, detection at 218 nm): t_{Ret} = 12.04 min, 82% purity.

H-[(*N*1ph)-(*N*4am^{Boc})]-*SP* (2.70^{Boc}-SP)

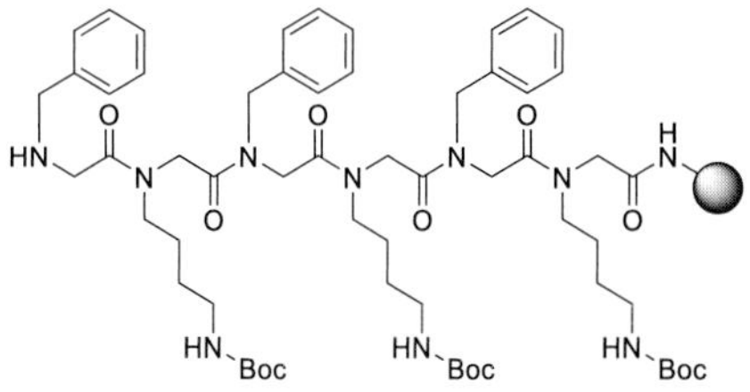

According to **GP11a**, 328 mg of Rink amide resin (0.20 mmol, 1.00 equiv.) were swollen and deprotected. The synthesis of the peptoid hexamer was carried out by acylation (**GP11a**, step 3) with 1.33 mL 1.20 M bromoacetic acid in pDMF (1.60 mmol, 8.00 equiv.) and 248 µL DIC (1.60 mmol, 8.00 equiv.). *N*6am^{Boc} group was attached, using 346 mg of *tert*-Butyl(4-aminobutyl)carbamate (1.60 mmol, 8.00 equiv.) (**GP11a**, step 4) and N1ph group was attached by 175 µL of benzylamine (1.60 mmol, 8.00 equiv.) in 1.60 mL pDMF. A few beads were cleaved off and the formation of the desired product was detected by MALDI-TOF-MS.

MS (MALDI-TOF): m/z = 842.5 [M–3 × Boc]$^+$.

***Cyo*-[(N1ph)-(N4am^{Boc})]$_3$-*SP* (2.71^{Boc}-SP)**

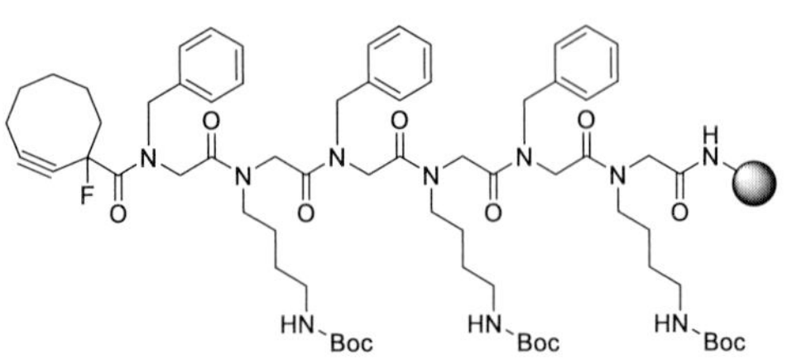

2.70^{Boc}-SP (0.20 mmol, 1.00 equiv.) was treated with 6 mL the solution of 68.0 mg 1 fluorocyclooct-2-yne carboxylic acid (0.40 mmol, 2.00 equiv.), 61.2 mg HOBt (0.40 mmol, 2.00 equiv.) and 62 µL DIC (50.4 mg, 0.40 mmol, 2.00 equiv.) in pDMF. The suspension was shaken overnight. Afterwards the solution was filtered off and the resin was washed three times with DMF and once with DCM. The product was detected by MALDI-TOF-MS and further reacted directly on the resin.

MS (MALDI-TOF): m/z = 994.6 [M–3 × Boc]$^+$.

***Terpy-Cyo*-[(N1ph)-(N4am^{Boc})]$_3$-*SP* (2.72^{Boc}-SP)**

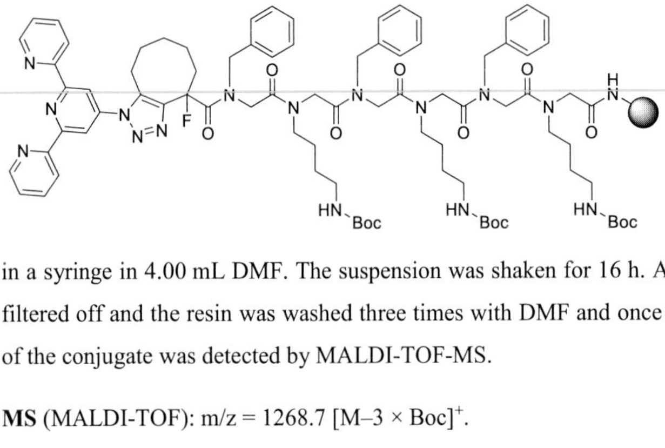

2.71^{Boc}-SP (0.20 mmol, 1.00 equiv.) was treated with 82.2 mg 4-azido-2,2:6,2-terpyridine (0.30 mmol, 1.50 equiv.) in a syringe in 4.00 mL DMF. The suspension was shaken for 16 h. Afterwards the solution was filtered off and the resin was washed three times with DMF and once with DCM. The formation of the conjugate was detected by MALDI-TOF-MS.

MS (MALDI-TOF): m/z = 1268.7 [M–3 × Boc]$^+$.

***Terpy-Cyo*-[(N1ph)-(N4am^{Boc})]$_3$-*NH$_2$* (2.72)**

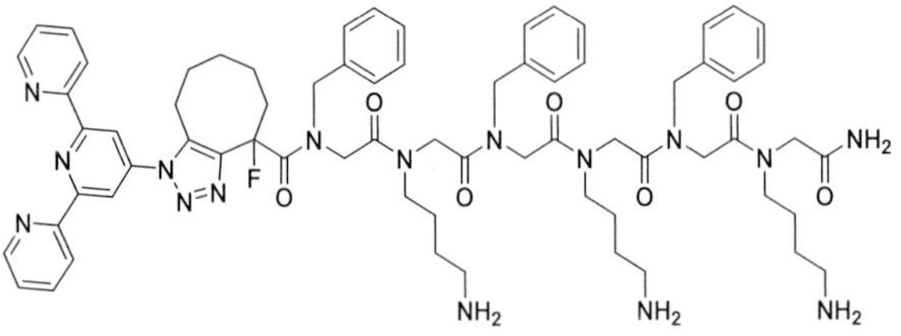

According to **GP12a**, **2.72^{Boc}-SP** was cleaved from the resin by treatment with 3 mL 95% TFA in DCM at room temperature for 2 h. The solution was filtered off and the next cleavage mixture was added to the resin and let shaken for additional 2 h. The resin was washed with 3 mL DCM (3 ×) and methanol (3 ×). The cleavage cocktail was evaporated under airflow, the rest was combined with washes and the solvent was removed under reduced pressure

(**GP12a**, step 1). The crude product was then dissolved in 5 mL acetonitrile, passed through a syringe filter, purified by means of preparative HPLC (5-95% acetonitrile + 0.1% TFA in 30 min, 25 °C) and lyophilized (**GP12a**, step 2) to afford 10 mg (4% over 15 steps).

MS (MALDI-TOF, CHCA): m/z = 1268.7 $[M]^+$. – **Analytical HPLC** (5–95% acetonitrile + 0.1% TFA in 20 min, detection at 218 nm): t_{Ret} = 8.79 min, 98% purity.

***Eu-2.8-Terpy-Cyo*-[(*N*1ph)-(*N*4am)]$_3$-*NH$_2$* (Eu-2.73)**

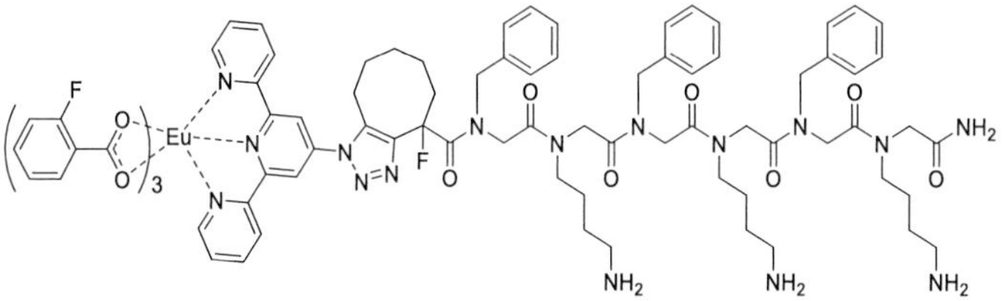

According to **GP2**, a solution of 20.0 mg of *Terpy-Cyo*-[(*N*1ph)-(*N*4am^{Boc})]$_3$-*NH$_2$* (30.0 µmol, 1.00 equiv.) in ethanol was added to a ethanol/water solution of 18.0 mg of **Eu-2.8** (30.0 µmol, 1.00 equiv.) and the mixture was refluxed for 2 h. The reaction mixture was then allowed to cool to room temperature, ethanol was removed under reduced pressure and the remaining residue was lyophilized to afford 21.0 mg (38%) of a white solid.

MS (MALDI-TOF, THAP): m/z = 1874.8 $[M + H]^+$. – **Analytical HPLC** (5–95% acetonitrile + 0.1% TFA in 20 min, detection at 218 nm): t_{Ret} = 9.02 min, 96% purity.

V.2.3.3 Synthesis and properties of lanthanide conjugates with cyclic peptoids

3-(((benzyloxy)carbonyl)amino)propanoic acid (2.77)

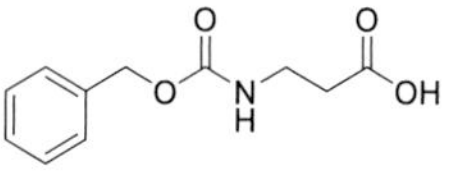

To a solution of 21.6 g of beta alanine **2.75** (242 mmol, 1.00 equiv.) in 150 mL 2 M NaOH, a suspension of 37 mL benzyl chloroformate **2.74** (262 mmol, 1.08 equiv) and 30 mL 2 M NaOH were added slowly at 0 °C. The reaction was stirred for 40 min at 0 °C and then overnight at room temperature. The crude product was washed with diethyl ether (3 × 50 mL), acidified with conc. HCl to pH 3 while cooling at 0 °C. The white precipitate was filtered, washed with cold 0.1 M HCl and finally dried to afford 23.3 g (43%) of a white solid. Analytical data are consistent with the literature.[265]

^{1}H NMR (400 MHz, $CDCl_3$): δ (ppm) = 7.35 – 7.21 (m, 5H, H_{Ar}), 5.31 (bs, 1H, N*H*), 5.13 (s, 2H, OCH_2), 3.47 (q, *J* = 6.1 Hz, 2H, $NHCH_2$), 2.61 (t, *J* = 5.8 Hz, 2H, CH_2CO). – **^{13}C NMR** (101 MHz, $CDCl_3$): δ (ppm) = 177.6 ($C_{quart.}$, COOH), 156.5 ($C_{quart.}$, NHCO), 136.4 ($C_{quart.}$, C_{Ar}), 128.7 (+, 2 × C_{Ar}H), 128.4 (+, C_{Ar}H), 128.3 (+, 2 × C_{Ar}H), 67.6 (–, OCH_2), 37.0 (–, $NHCH_2$), 34.6 (–, CH_2CO).

***tert*-butyl 3-(((benzyloxy)carbonyl)amino)propanoate (2.79)**

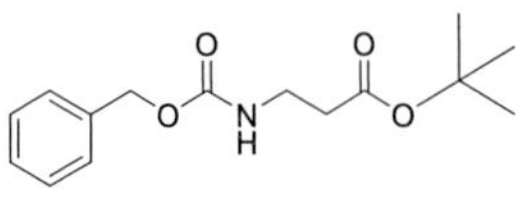

According to **GP3b**, 11.1 g of 3-(((benzyloxy)carbonyl)-amino)propanoic acid **2.77** (50.0 mmol, 1.00 equiv.) were dissolved in 20 mL dry DCM under Ar atmosphere. Then 14 mL dry *t*BuOH (150 mmol, 3.00 equiv.) with 1.22 g of DMAP (10.0 mmol, 0.20 equiv.) and 8.90 mL DIC (57.5 mmol, 1.15 equiv.) were added. The reaction mixture was stirred for 3 days under Ar atmosphere. The precipitate was filtered off, washed with DCM and the filtrate was washed with water (3 × 300 mL), 5% aq. $NaHCO_3$ solution (3 × 300 mL) and water again. The combined organic layers were dried over Na_2SO_4, the volatiles were removed under reduced pressure and the residue was purified *via* flash column chromatography (CH/EtOAc 3:1) to afford 9.36 g (67%) of a pale-yellow oil.

^{1}H NMR (300 MHz, $CDCl_3$): δ (ppm) = 7.35 – 7.21 (m, 5H, H_{Ar}), 5.21 (bs, 1H, N*H*), 5.02 (s, 2H, OCH_2), 3.36 (q, J = 6.1 Hz, 2H, NHCH_2), 2.38 (t, J = 6.0 Hz, 2H, CH_2CO), 1.37 (s, 9H, C(CH_3)$_3$). – **^{13}C NMR** (101 MHz, $CDCl_3$): δ (ppm) = 171.7 ($C_{quart.}$, *C*OO*t*Bu), 156.3 ($C_{quart.}$, NH*C*O), 136.6 ($C_{quart.}$, C_{Ar}), 128.6 (+., 2 × C_{Ar}H), 128.5 (+., C_{Ar}H), 128.1 (+., 2 × C_{Ar}H), 80.9 (C_{quart}, *C*(CH$_3$)$_3$), 65.5 (–, O*C*H$_2$), 36.6 (–, NH*C*H$_2$), 35.4 (–, *C*H$_2$CO), 27.9 (+, C(*C*H$_3$)$_3$).

4-(((benzyloxy)carbonyl)amino)butanoic acid (2.78)

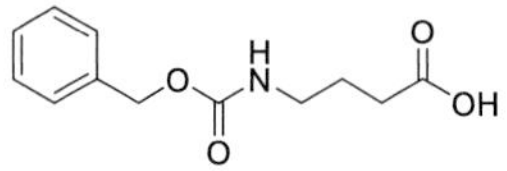

To a solution of 25.0 g of γ-aminobutyric acid (2.76) (242 mmol, 1.00 equiv.) in 150 mL 2 M NaOH, a suspension of 37 mL benzyl chloroformate **2.74** (262 mmol, 1.08 equiv) and 30 mL 2 M NaOH were added slowly at 0 °C. The reaction was stirred for 40 min at 0 °C and then overnight at room temperature. The crude product was washed with diethyl ether (3 × 50 mL), acidified with conc. HCl to pH 3 and cooled in an ice bath. The white precipitate was filtered, washed with cold 0.1 M HCl and finally dried to afford 26.5 g (46%) of a white solid. Analytical data are consistent with the literature.

^{1}H NMR (400 MHz, $CDCl_3$): δ (ppm) = 7.41 – 7.27 (m, 5H, H_{Ar}), 5.09 (s, 2H, OCH_2), 4.98 (bs, 1H, N*H*), 3.25 (q, J = 6.7 Hz, 2H, NHCH_2), 2.39 (t, J = 7.3 Hz, 2H, CH_2CO), 1.83 (p, J = 7.1 Hz, 2H, CH$_2$CH_2). – **^{13}C NMR** (101 MHz, $CDCl_3$): δ (ppm) = 178.3 ($C_{quart.}$, *C*OOH), 156.4 ($C_{quart.}$, NH*C*O), 136.2 ($C_{quart.}$, C_{Ar}), 128.3 (+, 2 × C_{Ar}H), 128.0 (+, C_{Ar}H), 127.9 (+, 2 × C_{Ar}H), 66.6 (–, O*C*H$_2$), 40.1 (–, NH*C*H$_2$), 30.9 (–, CH$_2$*C*H$_2$), 24.7 (–, *C*H$_2$CO).

tert-butyl 4-(((benzyloxy)carbonyl)amino)butanoate (2.80)

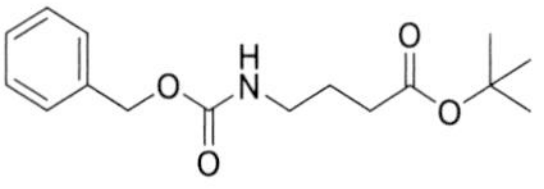

According to **GP3b**, 11.9 g of 4-(((benzyloxy)carbonyl)amino)-butanoic acid (**2.78)** (50.0 mmol, 1.00 equiv.) were dissolved in 20 mL dry DCM under Ar atmosphere. Then 14 mL dry *t*BuOH (150 mmol, 3.00 equiv.) with 1.22 g of DMAP (10.0 mmol, 0.20 equiv.) and 8.90 mL DIC (57.5 mmol, 1.15 equiv.) were added. The reaction mixture was stirred for 3 days under Ar atmosphere. The precipitate was filtered off, washed with DCM and the filtrate was washed with water (3 × 300 mL), 5% aq. $NaHCO_3$ solution (3 × 300 mL) and water again. The combined organic layers were dried over Na_2SO_4, the volatiles were removed under reduced pressure and the residue was purified *via* flash column chromatography (CH/EE 3:1) to afford 10.1 g (69%) of a pale-yellow oil.

^{1}H NMR (300 MHz, $CDCl_3$): δ (ppm) = 7.39 – 7.27 (m, 5H, H_{Ar}), 5.08 (s, 1H, OCH_2), 4.97 (s, 2H, N*H*), 3.22 (q, *J* = 6.6 Hz, 2H, $NHCH_2$), 2.26 (t, *J* = 7.3 Hz, 2H, CH_2CO), 1.79 (q, *J* = 7.1 Hz, 2H, CH_2CH_2), 1.43 (s, 9H, $C(CH_3)_3$). – **^{13}C NMR** (101 MHz, $CDCl_3$): δ (ppm) = 172.7 ($C_{quart.}$, *C*OO*t*Bu), 156.5 ($C_{quart.}$, NH*C*O), 136.7 ($C_{quart.}$, C_{Ar}), 128.6 (+, 2 × $C_{Ar}H$), 128.2 (+, 2 ×$C_{Ar}H$), 80.6 (C_{quart}, $C(CH_3)_3$), 66.7 (–, OCH_2), 40.6 (–, $NHCH_2$), 32.9 (–, CH_2CO), 28.2 (–, CH_2CH_2), 25.3 (+, $C(CH_3)_3$).

tert-butyl 3-aminopropanoate (2.82)

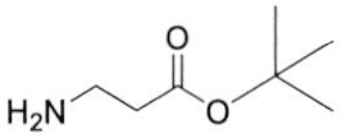

According to **GP6**, a solution of 9.30 g of *tert*-butyl 3-(((benzyloxy)carbonyl)amino)propanoate **2.79** (33.3 mmol, 1.00 equiv.) in 100 mL methanol was treated with 1.06 g of Pd/C (9.99 mmol, 0.03 equiv.). The reaction flask was flushed with H_2 and then fitted with a balloon filled with H_2. The mixture was stirred under balloon pressure of H_2 until TLC showed the reaction to be complete (usually 3–4 h). Afterwards, the catalyst was filtered through a Celite® pad and washed with methanol. The solvent was evaporated under reduced pressure to afford 4.40 g (89%) of a colorless oil.

^{1}H NMR (300 MHz, $CDCl_3$): δ (ppm) = 2.94 (t, *J* = 6.2 Hz, 2H, $NHCH_2$), 2.41 (t, *J* = 6.2 Hz, 2H, CH_2CO), 1.97 (bs, 2H, NH_2), 1.50 – 1.42 (m, 9H, $C(CH_3)_3$). – **^{13}C NMR** (75 MHz, $CDCl_3$): δ (ppm) = 171.7 ($C_{quart.}$, *C*OO*t*Bu), 81.9 (C_{quart}, $C(CH_3)_3$), 37.9 (–, $NHCH_2$), 37.7 (–, CH_2CO), 28.6 (+, $C(CH_3)_3$).

tert-butyl 4-aminobutanoate (2.83)

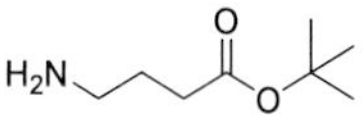

According to **GP6**, a solution of 9.77 g of *tert*-butyl 3-(((benzyloxy)-carbonyl)amino)propanoate (33.3 mmol, 1.00 equiv.) (**2.80)** in 100 mL methanol was treated with 1.06 g of Pd/C (9.99 mmol, 0.03 equiv.). The reaction flask was flushed with H_2 and then fitted with a balloon filled with H_2. The mixture was stirred under balloon pressure of H_2 until TLC showed the reaction to be complete (usually 3–4 h). Afterwards, the catalyst was filtered through a Celite® pad and washed with methanol. The solvent was then evaporated under reduced pressure to afford 5.04 g (95%) of a colorless oil.

^{1}H NMR (300 MHz, $CDCl_3$): δ (ppm) = 2.73 (t, J = 7.1 Hz, 2H, NHCH_2), 2.27 (t, J = 6.2 Hz, 2H, CH_2CO), 1.97 (bs, 2H, NH_2), 1.75 (q, J = 7.2 Hz, 2H, CH$_2$CH_2), 1.44 (s, 9H, C(CH_3)$_3$). – **^{13}C NMR** (75 MHz, $CDCl_3$): δ (ppm) = 173.4 ($C_{quart.}$, *COOt*Bu), 81.9 (C_{quart}, *C*(CH$_3$)$_3$), 41.4 (–, NHCH$_2$), 32.6 (–, CH$_2$CO), 26.1 (+, C(CH$_3$)$_3$), 25.1 (–, CH$_2$CH$_2$).

6,6'-bis(bromomethyl)-2,2'-bipyridine (2.85)

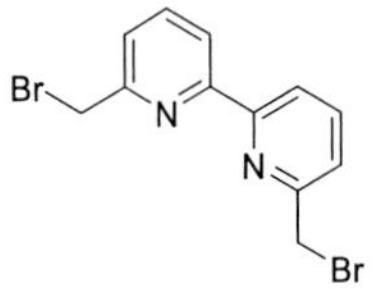

A mixture of 1.00 g of 6,6'-dimethyl 2,2'-bipyridine **2.89** (5.43 mmol, 1.00 equiv.) and 1.93 g of N-bromosuccinimide (10.9 mmol, 2.00 equiv.) in 150 mL CCl_4, was refluxed for 30 min. Then benzoyl peroxide (13.0 mg, 0.054 mmol, 0.01 equiv.) was added. The mixture was refluxed for another 2 h and the succinimide was filtered off. The solution was cooled to 0 °C and the solid which deposited was filtered and washed with methanol to give 427 mg of the product (23% yield) as a white crystalline solid which may be recrystallized from CCl_4. The filtrate was concentrated. Purification *via* flash column chromatography (DCM/MeOH 98:2) resulted in additional 335 mg (18%) of a white solid. Total yield 762 mg (41%). Analytical data are consistent with the literature[104].

^{1}H NMR (300 MHz, $CDCl_3$): δ (ppm) = 8.49 – 8.32 (m, 2H, 2 × H_{Ar}), 7.83 (t, J = 7.8 Hz, 2H, 2 × H_{Ar}), 7.47 (dd, J = 7.7 Hz, J = 1.0 Hz, 2H, 2 × H_{Ar}), 4.63 (s, 4H, 2 × CH_2).

2,9-bis(bromomethyl)-1,10-phenanthroline (2.86)

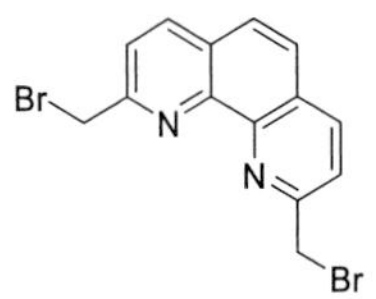

Variant A. To a solution of 1.00 g of neocuproine (4.80 mmol, 1.00 equiv.) in 200 mL CH_3CN, 2.56 g of *N*-bromosuccinimide (14.4 mmol, 3.00 equiv.) and 13.0 mg of benzoil peroxide (50.0 µmol, 0.01 equiv.) were added and the mixture was heated to reflux for 24 h under Ar. After cooling to room temperature, the solvent was evaporated under reduced pressure, the mixture was dissolved in diethyl ether and the succinimide was removed by filtration. The solid was washed with 50 mL aq. $NaHCO_3$ and purified by flash column chromatography, using DCM as eluent and recrzstalliyed from ethanol. The solvent was removed under reduced pressure to afford 597 mg (34%) of the product as pale-yellow solid. Analytical data are consistent with the literature.[242,275]

Variant B. A solution of 250 mg of (1,10-phenanthroline-2,9-diyl)dimethanol **2.88** (1.04 mmol, 1.00 equiv.) and 156 mL hydrobromic acid (47% in H_2O solution) was heated under reflux for an hour, then cooled in an ice bath and treated with solid sodium carbonate until the mixture reached a pH value of 10. The mixture was extracted with DCM and the organic phase was dried over Na_2SO_4. The solvent was evaporated under reduced pressure and the residue was purified by flash column chromatography (DCM/EtOAc 5:1) to afford 262 mg (69%) of the product as yellow needles.

^{1}H NMR (300 MHz, $CDCl_3$): δ (ppm) = 8.23 (dd, *J* = 8.6 Hz, *J* = 2.0 Hz, 2H, 2 × H_{Ar}), 7.86 (dd, *J* = 8.2 Hz, *J* = 1.6 Hz, 2H, 2 × H_{Ar}), 7.76 (d, *J* = 2.8 Hz, 2H, 2 × H_{Ar}), 4.91 (s, 4H, 2 × CH_2). – **MS** (70 eV, EI): *m/z* (%) = 368 (100) $[M+H]^+$, 287 (40) $[M-Br+2H]^+$. – **EA**: calcd for $C_{14}H_{10}Br_2N_2$,% C 45.94, H 2.75, N 7.65, found C 46.11, H 2.81, N 7.34.

1,10-phenanthroline-2,9-dicarbaldehyde (2.87)

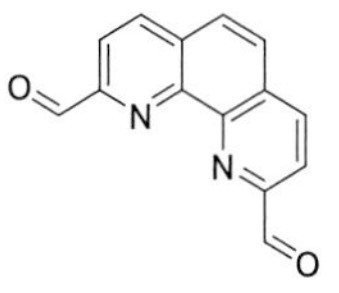

A mixture of 4.50 g of neocuproine **2.35** (21.6 mmol, 1.00 equiv.) and 9.59 g of selenium dioxide (86.4 mmol, 4.00 equiv.) in 1,4-dioxane containing 4% H_2O (500 mL) was heated under reflux for an hour and then filtered through Celite® while hot. The solid separated in the cold filtrate was recrystallized from 1,4-dioxane containing 4% H_2O to afford 3.06 g (60%) of the product as white needles. Analytical data are consistent with the literature.[241]

^{1}H NMR (300 MHz, $CDCl_3$): δ (ppm) =10.57 (s, 2H, 2 × C*H*O), 8.52 (dd, *J* = 8.2 Hz, *J* = 0.9 Hz, 2H, 2 × H_{Ar}), 8.39 (d, *J* = 8.3 Hz, 2H, 2 × H_{Ar}), 8.06 (s, 2H, 2 × H_{Ar}).

(1,10-phenanthroline-2,9-diyl)dimethanol (2.88)

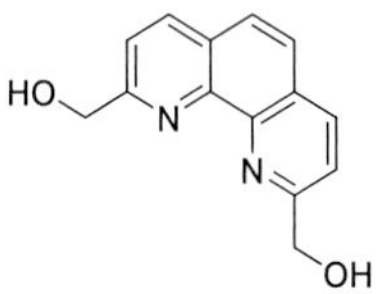

A solution of 1.00 g of 1,10-phenanthroline-2,9-dicarbaldehyde (**2.87**) (4.23 mmol, 1.00 equiv.) and 328 mg of sodium borohydride (8.47 mmol, 2.00 equiv.) in 100 mL ethanol was heated under reflux for half an hour. The mixture was extracted with ethyl acetate and the organic layers were dried over Na_2SO_4. The solvent was evaporated under reduced pressure to afford 629 mg (62%) of the product as yellow needles. Analytical data are consistent with the literature.[241]

^{1}H NMR (300 MHz, $CDCl_3$): δ (ppm) = 8.30 (d, *J* = 8.4 Hz, 2H, 2 × H_{Ar}), 7.84 (s, 2H, 2 × H_{Ar}), 7.63 (d, *J* = 7.9 Hz, 2H, 2 × H_{Ar}), 5.12 (s, 4H, 2 × CH_2).

***H*-(*N*1cx^{tBu})-Pyr-(*N*1cx^{tBu})$_2$-*SP* (2.90-*t*Bu-SP)**

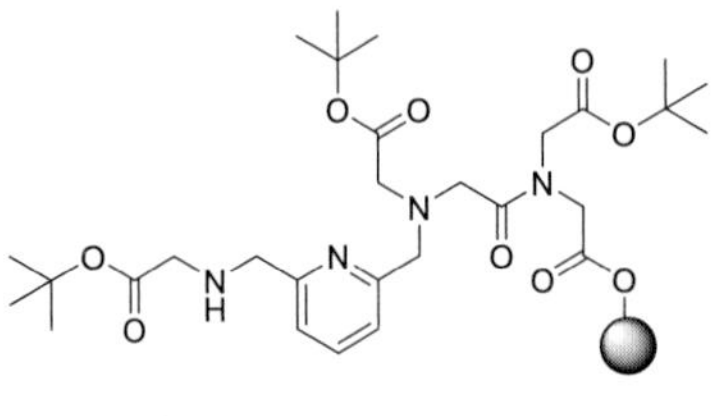

According to **GP11b**, 800 mg of 2-chlortritylchlorid resin (0.64 mmol, 1.00 equiv.) were swollen in DCM. The synthesis of the peptoid was carried out by acylation (**GP11b**, step 2) with 2.66 mL 1.20 M bromoacetic acid in DCM (3.20 mmol, 5.00 equiv.) and 558 μL DIPEA (3.20 mmol, 5.00 equiv.). *N*1cx^{tBu} group was attached, using 419 mg of *tert*-butyl glycinate (3.20 mmol, 5.00 equiv.) (**GP11b**, step 3). Subsequent acylation was carried out, using 2.66 mL 1.20 M bromoacetic acid in pDMF (3.20 mmol, 5.00 equiv.) and 496 μL DIC (3.20 mmol, 5.00 equiv.) (**GP11b**, step 4). After the second substitution step, the resin was treated with 509 mg of 2,6-Bis(bromomethylpyridine) (1.92 mmol, 3.00 equiv.) and 335 μL DIPEA (1.92 mmol, 3.00 equiv.) in 2.00 mL pDMF. The mixture was stirred for 1 h. The subsequent *N*1cx^{tBu} group was attached, using 419 mg of *tert*-butyl glycinate (3.20 mmol, 5.00 equiv.) (**GP11b**, step 3) in 2.00 mL pDMF. A few beads were cleaved off and the formation of the desired product was detected by MALDI-TOF-MS.

MS (MALDI-TOF, CHCA): m/z = 594.3 $[M]^+$.

***H*-(*N*1cx^{tBu})-Pyr-(*N*1cx^{tBu})$_2$-*OH* (2.90-*t*Bu)**

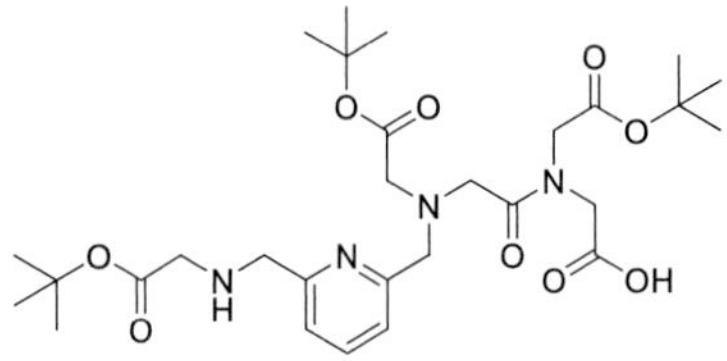

According to **GP12b**, **ASK381S1** was cleaved from the resin by treatment with 8.00 mL 20% HFIP solution in DCM at room temperature for 2 h. The solution was filtered off and the next cleavage mixture was added to the resin and let shaken for additional 2 h. The resin was washed with 3 mL DCM (3 ×) and methanol (3 ×). The cleavage cocktail was evaporated under airflow, the residue was combined with the washes and the solvent was removed under reduced pressure (**GP12b**, step 1). The crude product was then dissolved in 5 mL acetonitrile, passed through a syringe filter, purified by means of preparative RP-HPLC (5-95% acetonitrile + 0.1% TFA in 30 min, 25 °C) and lyophilized (**GP12b**, step 2) to afford 80.0 mg (21% over 7 steps) of an off-white solid.

^{1}H NMR (300 MHz, CD_3CN): δ (ppm) = 7.89 (m, 1H, H_{Ar}), 7.70 – 7.30 (m, 2H, 2 × H_{Ar}), 4.59 – 3.69 (m, 15H, 7 × CH_2, *NH*), 1.48 – 1.39 (m, 27H, 9 × CH_3). – **MS** (MALDI-TOF, CHCA): m/z = 594.3 $[M]^+$. – **Analytical HPLC** (5–95% acetonitrile + 0.1% TFA in 20 min, detection at 218 nm): t_{Ret} = 10.75 min, 80% purity.

c-(N1cx^{tBu})$_3$-Pyr (c-2.90-*t*Bu)

According to **GP13**, to a solution of 211 mg of the linear peptoid **2.90-*t*Bu** (360 µmol, 1.00 equiv.) in 200 mL pDMF, a diluted solution of 383 µL DIPEA (2.20 mmol, 6.20 equiv.) and 540 mg of HATU (1.42 mmol, 4.00 equiv.) in 250 mL pDMF was added in 12 h under Ar atmosphere, using a dropping funnel. The resulting mixture was stirred for 3 days under Ar at room temperature. Afterwards, the solvent was evaporated under reduced pressure and the residue was taken up in 100 mL EtOAc. The organic layer was washed with 100 mL water and the aqueous layer was extracted with EtOAc (3 ×). The combined organic layers were dried over Na_2SO_4, filtered and concentrated under reduced pressure. The crude product was used in the subsequent reaction without further purification.

^{1}H NMR (300 MHz, CD_3CN): δ (ppm) = 7.98 – 7.86 (m, 1H, H_{Ar}), 7.52 – 7.38 (m, 2H, 2 × H_{Ar}), 4.80 – 3.80 (m, 14H, 7 × CH_2), 1.52 – 1.32 (m, 27H, 9 × CH_3). – **MS** (MALDI-TOF, CHCA): m/z = 577.3 $[M+H]^+$, 465.2 $[M–2t\text{Bu}+H]^+$.

c-(N1cx)$_3$-Pyr (c-2.90-H)

According to **GP4b**, 61.0 mg of the crude peptoid **c-(N1cx^{tBu})$_3$-pyr** (c-2.90-*t*Bu**)** was treated with a mixture of TFA, TIPS and water (95/2.5/2.5, volumetric) for 4 h. The volatiles were then removed under airflow and reduced pressure, the crude product was solubilized in an CH_3CN/H_2O mixture, purified by means of preparative RP-HPLC (5-95% acetonitrile + 0.1% TFA in 30 min, 25 °C) and lyophilized to afford 8.00 mg (18% over 2 steps) of the product as a white powder.

MS (MALDI-TOF, CHCA): m/z = 409.3 $[M+H]^+$. – **Analytical HPLC** (5–95% acetonitrile + 0.1% TFA in 20 min, detection at 256 nm): t_{Ret} = 10.72 min, 82% purity.

***Eu*-c-(N1cx)$_3$-Pyr (Eu-c-2.90)**

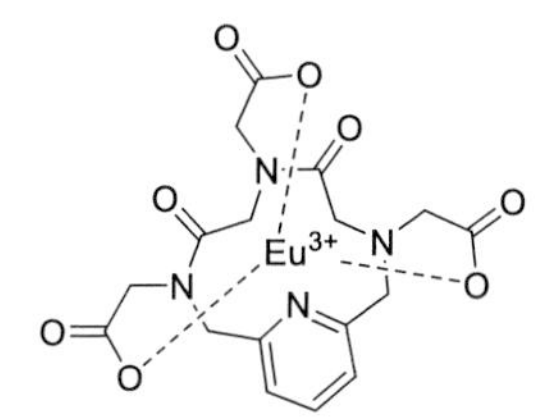

A suspension of 8.00 mg of **c-(N1cx)$_3$-pyr (c-2.90)** (19.6 µmol, 1.00 equiv.) in methanol was treated with 588 µL of 0.10 M NaOH (58.8 µmol, 3.00 equiv.). Then an aq. solution of 7.32 mg of $EuCl_3 \cdot x\, H_2O$ (19.6 µmol, 1.00 equiv.) was added. The reaction mixture was refluxed for an additional hour, then the solvent was removed under reduced pressure and the crude product was purified by means of preparative

RP-HPLC (5-95% acetonitrile + 0.1% TFA in 30 min, 25 °C) and lyophilized to afford 3.08 mg (27%) of a white solid.

MS (MALDI-TOF, THAP): m/z = 559.1 $[M+H]^+$, 500.1 $[M–Eu+4Na]^+$. – **Analytical HPLC** (5–95% acetonitrile + 0.1% TFA in 20 min, detection at 256 nm): t_{Ret} = 10.75 min, 77% purity.

***H*-(*N*2cx^{tBu})-Pyr-(*N*2cx^{tBu})$_2$-*SP* (2.91-*t*Bu-SP)**

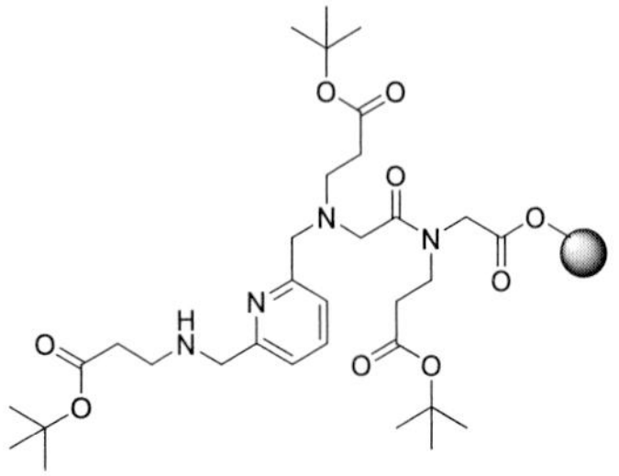

According to **GP11b**, 800 mg of 2-chlortritylchlorid resin (640 µmol, 1.00 equiv.) were swollen. The synthesis of the peptoid was carried out by acylation (**GP11b**, step 2) with 2.66 mL 1.20 M bromoacetic acid in DCM (3.20 mmol, 5.00 equiv.) and 558 µL DIPEA (3.20 mmol, 5.00 equiv.). *N*2cx^{tBu} group was attached, using 465 mg of *tert*-butyl 3-aminopropanoate (3.20 mmol, 5.00 equiv.) (**GP11b**, step 3). Subsequent acylation was carried out, using 2.66 mL 1.20 M bromoacetic acid in pDMF (3.20 mmol, 5.00 equiv.) and 496 µL DIC (3.20 mmol, 5.00 equiv.) (**GP11b**, step 4). After the second substitution step, the resin was treated with 509 mg of 2,6-bis(bromomethylpyridine (**2.84**) (1.92 mmol, 3.00 equiv.) and 335 µL DIPEA (1.92 mmol, 3.00 equiv.) in 2.00 mL pDMF. The mixture was stirred for 1 h. The subsequent *N*2cx^{tBu} group was attached, using 465 mg of *tert*-butyl 3-aminopropanoate (3.20 mmol, 5.00 equiv.) (**GP11b**, step 3) in 2.00 mL pDMF. A few beads were cleaved off and the formation of the desired product was detected by MALDI-TOF-MS.

MS (MALDI-TOF, CHCA): m/z = 636.4 $[M]^+$.

***H*-(*N*2cx^{tBu})-Pyr-(*N*2cx^{tBu})$_2$-*OH* (2.91-*t*Bu)**

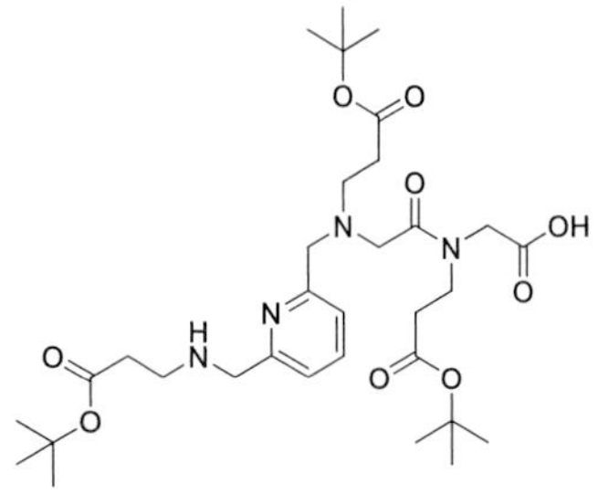

According to **GP12b, 2.91-*t*Bu-SP** was cleaved from the resin by treatment with 8.00 mL 20% HFIP solution in DCM at room temperature for 2 h. The solution was filtered off and the next cleavage mixture was added to the resin and let shaken for additional 2 h. The resin was washed with 3 mL DCM (3 ×) and methanol (3 ×). The cleavage cocktail was evaporated under airflow, the residue was combined with the washes and the solvent was removed under reduced pressure (**GP12b**, step 1). The crude product was then dissolved in 5 mL acetonitrile, passed through a syringe filter, purified by means of preparative RP-HPLC (5-95% acetonitrile + 0.1% TFA in

30 min, 25 °C) and lyophilized (**GP12b**, step 2) to afford 78.0 mg (12% over 7 steps) of a pale-yellow solid.

^{1}**H NMR** (300 MHz, D_2O): δ (ppm) = 8.06 –7.81 (m, 1H, H_{Ar}), 7.68 –7.33 (m, 2H, 2 × H_{Ar}), 4.88 – 2.28 (m, 20H, 10 × CH_2), 1.52 – 1.27 (m, 27H, 9 × CH_3). The compound contains 17% of starting material **2.82**: 2.20 – 2.13 (m, 0.21H, 2 × CH_2), 1.77 – 1.68 (m, 0.21H, 2 × CH_2), 1.47 (m, 2H, 9 × CH_3). – **MS** (MALDI-TOF, CHCA): m/z = 636.5 $[M]^+$.

c-(*N*2cx^{tBu})$_3$-Pyr (c-2.91-*t*Bu)

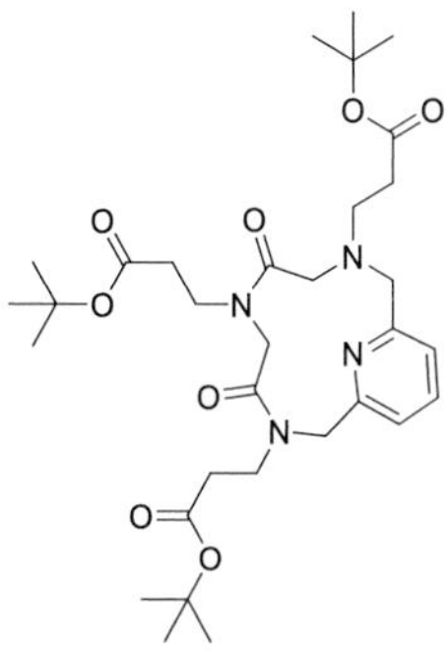

According to **GP13**, to a solution of 153 mg of the linear peptoid ***H*-(*N*2cx^{tBu})-Pyr-(*N*2cx^{tBu})$_2$-*OH* 2.91-*t*Bu** (0.24 mmol, 1.00 equiv.) in 200 mL pDMF, a diluted solution of 259 µL DIPEA (1.49 mmol, 6.20 equiv.) and 365 mg of HATU (0.96 mmol, 4.00 equiv.) in 250 mL pDMF was added in 12 h under Ar, using a dropping funnel. The resulting mixture was stirred for 3 days under Ar at room temperature. Afterwards, the solvent was evaporated under reduced pressure and the residue was taken up in 100 mL EtOAc. The organic layer was washed with 100 mL water and the aqueous layer was extracted with EtOAc (3 ×). The combined organic layers were dried over Na_2SO_4, filtered and concentrated under reduced pressure. The crude product was used in the subsequent reaction without further purification.

MS (MALDI-TOF, CHCA): m/z = 619.4 $[M+H]^+$, 642.5 $[M+Na]^+$.

c-(*N*2cx)$_3$-Pyr (c-2.91)

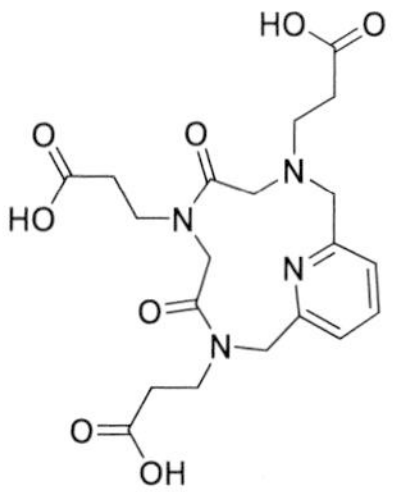

According to **GP4b**, 69.0 mg of the crude peptoid **c-(*N*2cx^{tBu})$_3$-Pyr (c-2.91-*t*Bu)** was treated with a mixture of TFA, TIPS and water (95/2.5/2.5, volumetric) for 4 h. The volatiles were then removed under airflow and reduced pressure, the crude product was solubilized in an CH_3CN/H_2O mixture, purified by means of preparative RP-HPLC (5-95% acetonitrile + 0.1% TFA in 30 min, 25 °C) and lyophilized to afford 11.0 mg (10% over 2 steps) of the product as a pale-yellow powder.

MS (MALDI-TOF, CHCA): m/z = 552.3 $[M+H]^+$. – **Analytical HPLC** (5–95% acetonitrile + 0.1% TFA in 20 min, detection at 256 nm): t_{Ret} = 10.10 min, 98% purity.

***Eu*-c-(*N*2cx)$_3$-Phen (Eu-c-2.91)**

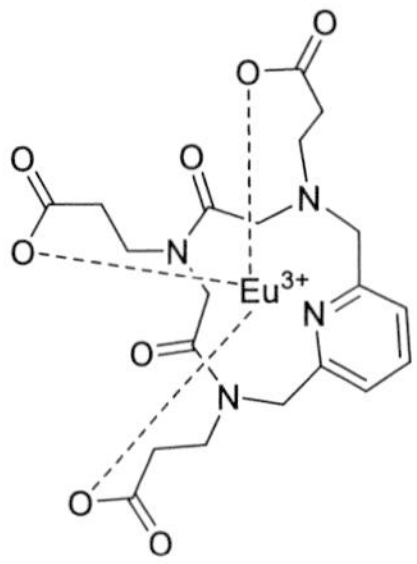

A suspension of 8.00 mg **c-(*N*2cx)$_3$-Phen (c-2.91)** (17.8 µmol, 1.00 equiv.) in methanol was treated with 534 µL of 0.10 M aqueous solution NaOH (53.3 µmol, 3.00 equiv.). Then an aqueous solution of 6.50 mg of $EuCl_3 \cdot x\ H_2O$ (17.8 µmol, 1.00 equiv.) was added. The reaction mixture was refluxed for additional 1 h, then the solvent was removed under reduced pressure and the crude product was purified by means of preparative RP-HPLC (5-95% acetonitrile + 0.1% TFA in 30 min, 25 °C) and lyophilized to afford 2.00 mg (19%) of a white solid.

MS (MALDI-TOF, THAP): m/z = 600.1 $[M]^+$, 598.2 $[M]^+$. – **Analytical HPLC** (5–95% acetonitrile + 0.1% TFA in 20 min, detection at 256 nm): t_{Ret} = 10.35 min, 90% purity.

***H*-(*N*2cx^{tBu})-Phen-(*N*2cx^{tBu})$_2$-*SP* (2.92-*t*Bu-SP)**

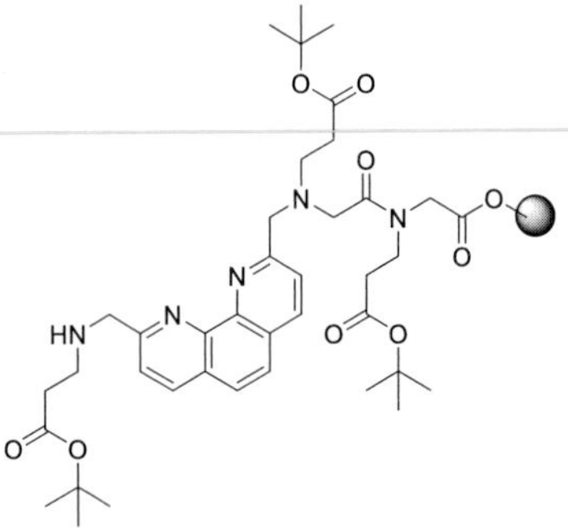

According to **GP11b**, 800 mg of 2-chlortritylchlorid resin (0.64 mmol, 1.00 equiv.) were swollen in DCM. The synthesis of the peptoid was carried out by acylation (**GP11b**, step 2) with 2.66 mL 1.20 M bromoacetic acid in DCM (3.20 mmol, 5.00 equiv.) and 558 µL DIPEA (3.20 mmol, 5.00 equiv.). *N*2cx^{tBu} group was attached, using 465 mg of *tert*-butyl 3-aminopropanoate (3.20 mmol, 5.00 equiv.) (**GP11b**, step 3). Subsequent acylation was carried out, using 2.66 mL 1.20 M bromoacetic acid in pDMF (3.20 mmol, 5.00 equiv.) and 496 µL DIC (3.20 mmol, 5.00 equiv.) (**GP11b**, step 4). After the second substitution step, the resin was treated with 703 mg of 2,9-bis(bromomethyl)-1,10-phenanthroline (**2.86**) (1.92 mmol, 3.00 equiv.) and 335 µL DIPEA (1.92 mmol, 3.00 equiv.) in 2.00 mL pDMF. The mixture was stirred for 1 h. The subsequent *N*2cx^{tBu} group was attached, using 465 mg of *tert*-butyl 3-aminopropanoate (3.20 mmol, 5.00 equiv.) (**GP11b**, step 3) in 2.00 mL pDMF. A few beads were cleaved off and the formation of the desired product was detected by MALDI-TOF-MS.

MS (MALDI-TOF, CHCA): m/z = 737.4 $[M]^+$.

***H*-(*N*2cx^{tBu})-Phen-(*N*2cx^{tBu})$_2$-*OH* and *OH*-(*N*2cx^{tBu})$_2$-Phen-(*N*2cx^{tBu})$_2$-*OH* (2.92-*t*Bu)**

According to **GP12b**, **2.92-*t*Bu-SP** was cleaved from the resin by treatment with 8.00 mL 20% HFIP solution in DCM at room temperature for 2 h. The solution was filtered off and the next cleavage mixture was added to the resin and let shaken for additional 2 h. The resin was washed with 3 mL DCM (3 ×) and methanol (3 ×). The cleavage cocktail was evaporated under airflow, the residue was combined with the washes and the solvent was removed under reduced pressure (**GP12b**, step 1). The crude product was then dissolved in 5 mL acetonitrile, passed through a syringe filter, purified by means of preparative RP-HPLC (5-95% acetonitrile + 0.1% TFA in 30 min, 25 °C) and lyophilized (**GP12b**, step 2) to afford 61.0 mg (8% over 7 steps) of a pale-yellow solid as well as 3.0 mg of a dimeric by-product as a yellow solid.

_H_-(_N_2cx^{tBu})-Phen-(_N_2cx^{tBu})$_2$-_OH_ (2.92-_t_Bu)

^{1}H NMR (300 MHz, D_2O): δ (ppm) = 8.61 (dd, J = 4.6 Hz, J = 1.5 Hz, 2H, 2 × H_{Ar}), 8.29 (dd, J = 8.5 Hz, J = 1.6 Hz, 2H, 2 × H_{Ar}), 7.57 – 7.48 (m, 2H, 2 × H_{Ar}), 3.67 – 3.50 (m, 12H, 6 × CH_2), 3.17 – 2.80 (m, 8H, 4 × CH_2), 1.27 – 1.10 (m, 27H, 9 × CH_3). – **MS** (MALDI-TOF, CHCA): m/z = 747.3 $[M]^+$. – **Analytical HPLC** (5–95% acetonitrile + 0.1% TFA in 20 min, detection at 256 nm): t_{Ret} = 10.00 min, 84% purity.

_OH_-(_N_2cx^{tBu})$_2$-Phen-(_N_2cx^{tBu})$_2$-_OH_ (2.92-_t_Bu-dimer)

MS (MALDI-TOF, CHCA): m/z = 980.6 $[M]^+$. – **Analytical HPLC** (5–95% acetonitrile + 0.1% TFA in 20 min, detection at 256 nm): t_{Ret} = 12.63 min, 67% purity.

c-(_N_2cx^{tBu})$_3$-Phen (c-2.92-_t_Bu)

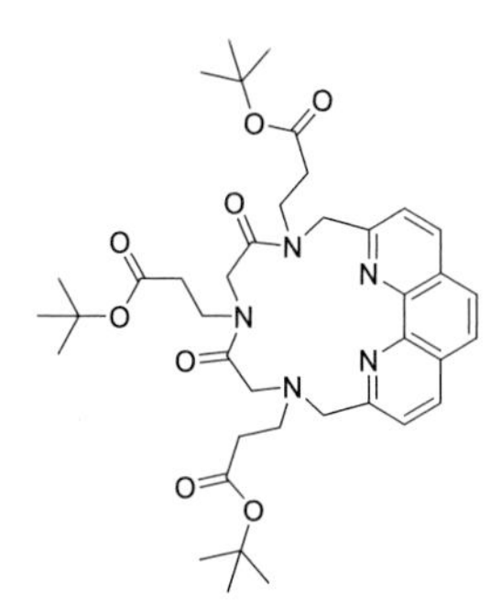

According to **GP13**, to a solution of 184 mg of the linear peptoid **2.92-_t_Bu** (0.25 mmol, 1.00 equiv.) in 200 mL pDMF, a diluted solution of 270 µL DIPEA (1.55 mmol, 6.20 equiv.) and 380 mg of HATU (1.00 mmol, 4.00 equiv.) in 250 mL pDMF was added in 12 h under Ar, using a dropping funnel. The resulting mixture was stirred for 3 days under Ar at room temperature. Afterwards, the solvent was evaporated under reduced pressure and the residue was taken up in 100 mL EtOAc. The organic layer was washed with

100 mLwater and the aqueous layer was extracted with EtOAc (3 ×). The combined organic layers were dried over Na_2SO_4, filtered and concentrated under reduced pressure. The crude product was used in the subsequent reaction without further purification.

MS (MALDI-TOF, CHCA): m/z = 720.4 $[M+H]^+$, 742.5 $[M+Na]^+$.

c-(*N*2cx)₃-Phen (c-2.92-H)

According to **GP4b**, 82.0 mg of the crude peptoid **c-(*N*2cx^{tBu})$_3$-Phen (c-2.92-*t*Bu)** was treated with a mixture of TFA, TIPS and water (95/2.5/2.5, volumetric) for 4 h. The volatiles were then removed under airflow and reduced pressure, the crude product was solubilized in an CH_3CN/H_2O mixture, purified by means of preparative RP-HPLC (5-95% acetonitrile + 0.1% TFA in 30 min, 25 °C) and lyophilized to afford 10.0 mg (16% over 2 steps) of the product as a pale-yellow powder.

MS (MALDI-TOF, CHCA): m/z = 552.3 $[M+H]^+$. – **Analytical HPLC** (5–95% acetonitrile + 0.1% TFA in 20 min, detection at 256 nm): t_{Ret} = 10.82 min, 83% purity.

***Eu*-c-(*N*2cx)₃-Phen (Eu-c-2.92)**

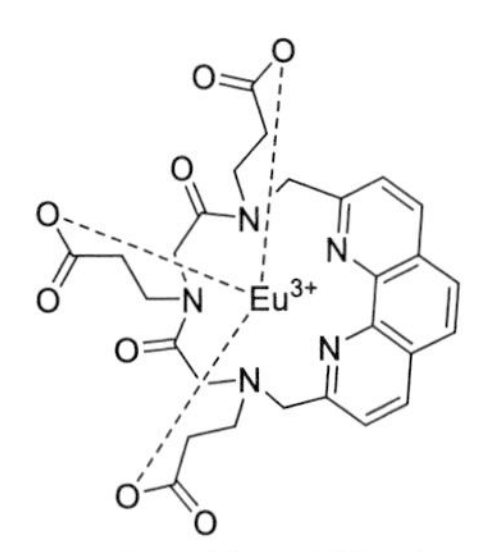

A suspension of 8.00 mg **c-(*N*2cx)₃-Phen (c-2.92)** (19.6 µmol, 1.00 equiv.) in methanol was treated with 588 µL of 0.10 M NaOH (58.8 µmol, 3.00 equiv.). Then an aq. solution of 7.32 mg of $EuCl_3 \cdot x\, H_2O$ (19.6 µmol, 1.00 equiv.) was added. The reaction mixture was refluxed for additional 1 h, then the solvent was removed under reduced pressure and the crude product was purified by means of preparative RP-HPLC (5-95% acetonitrile + 0.1% TFA in 30 min, 25 °C) and lyophilized to afford 2.50 mg (23%) of a white solid.

MS (MALDI-TOF, THAP): m/z = 701.5 $[M+H]^+$, 724.5 $[M+Na]^+$. – **Analytical HPLC** (5–95% acetonitrile + 0.1% TFA in 20 min, detection at 256 nm): t_{Ret} = 10.30 min, 81% purity.

***H*-(*N*3cx^{tBu})-Pyr-(*N*3cx^{tBu})-(*N*1ph)-(*N*3cx^{tBu})-*SP* (2.93-*t*Bu-SP)**

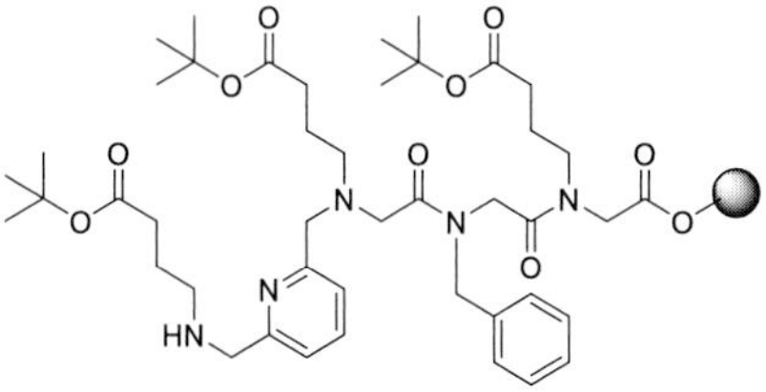

According to **GP11b**, 800 mg of 2-chlortritylchlorid resin (640 µmol, 1.00 equiv.) were swollen in DCM. The synthesis of the peptoid was carried out by acylation (**GP11b**, step 2) with 2.66 mL 1.20 M bromoacetic acid in DCM (3.20 mmol, 5.00 equiv.) and 558 µL DIPEA (3.20 mmol, 5.00 equiv.). *N*3cx^{tBu} group was attached, using 509 mg of *tert*-butyl 4-aminobutanoate (3.20 mmol, 5.00 equiv.) (**GP11b**, step 3). Subsequent acylation was carried out, using 2.66 mL 1.20 M bromoacetic acid in pDMF (3.20 mmol, 5.00 equiv.) and 496 µL DIC (3.20 mmol, 5.00 equiv.) (**GP11b**, step 4). *N*1ph group was attached, using 343 mg (353 µL, 3.20 mmol, 5.00 equiv.) of benzylamine (**GP11b**, step 3). After the third substitution step, the resin was treated with 509 mg of 2,6-bis(bromomethyl)pyridine **2.84** (1.92 mmol, 3.00 equiv.) and 335 µL DIPEA (1.92 mmol, 3.00 equiv.) in 2.00 mL pDMF. The mixture was stirred for 1 h. The subsequent *N*3cx^{tBu} group was attached, using 509 mg of *tert*-butyl 3-aminopropanoate (3.20 mmol, 5.00 equiv.) (**GP11b**, step 3) in 2.00 mL pDMF. A few beads were cleaved off and the formation of the desired product was detected by MALDI-TOF-MS.

MS (MALDI-TOF, CHCA): m/z = 826.6 $[M]^+$.

***H*-(*N*3cx^{tBu})-Pyr-(*N*3cx^{tBu})-(*N*1ph)-(*N*3cx^{tBu})-*OH* (2.93-*t*Bu)**

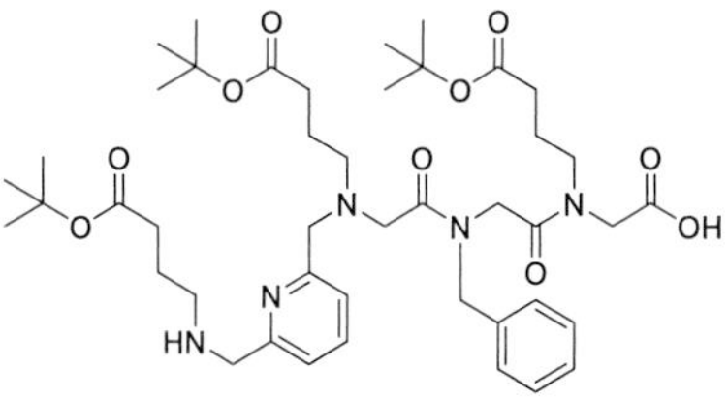

According to **GP12b**, **2.93-*t*Bu-SP** was cleaved from the resin by treatment with 8.00 mL 20% HFIP solution in DCM at room temperature for 2 h. The solution was filtered off and the next cleavage mixture was added to the resin and let shaken for additional 2 h. The resin was washed with 3 mL DCM (3 ×) and methanol (3 ×). The cleavage cocktail was evaporated under airflow, the residue was combined with the washes and the solvent was removed under reduced pressure (**GP12b**, step 1). The crude product was then dissolved in 5 mL acetonitrile, passed through a syringe filter, purified by means of preparative RP-HPLC (5-95% acetonitrile + 0.1% TFA in 30 min, 25 °C) and lyophilized (**GP12b**, step 2) to afford 89.0 mg (17% over 9 steps) of an off-white solid.

^{1}H NMR (300 MHz, CD_3CN): δ (ppm) = 7.29 – 7.16 (m, 8H, 8 × H_{Ar}), 3.97 – 2.10 (m, 30H, 15 × CH_2), 1.41 – 1.24 (m, 27H, 9 × CH_3). – **MS** (MALDI-TOF, CHCA): m/z = 826.5 $[M]^+$.

c-(*N*3cx^{tBu})-Pyr-(*N*3cx^{tBu})$_2$-(*N*1ph) (c-2.93-*t*Bu)

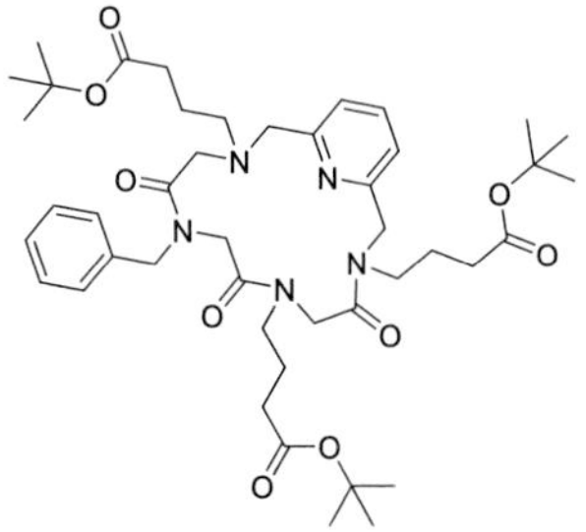

According to **GP13**, to a solution of 87.0 mg of the linear peptoid **2.93-*t*Bu** (110 µmol, 1.00 equiv.) in 100 mL pDMF, a diluted solution of 113 µL DIPEA (650 µmol, 6.20 equiv.) and 160 mg of HATU (420 µmol, 4.00 equiv.) in 150 mL pDMF was added in 12 h under Ar, using a dropping funnel. The resulting mixture was stirred for 3 days under Ar at room temperature. Afterwards, the solvent was evaporated under reduced pressure and the residue was taken up in 50 mL EtOAc. The organic layer was washed with 50 mLwater and the aqueous layer was extracted with EtOAc (3 ×). The combined organic layers were dried over Na_2SO_4, filtered and concentrated under reduced pressure. The crude product was used in the subsequent reaction without further purification.

MS (MALDI-TOF, CHCA): m/z = 807.5 $[M]^+$, 751.5 $[M–tBu+H]^+$.

c-(*N*3cx)-Pyr-(*N*3cx)$_2$-(*N*1ph) (2.93)

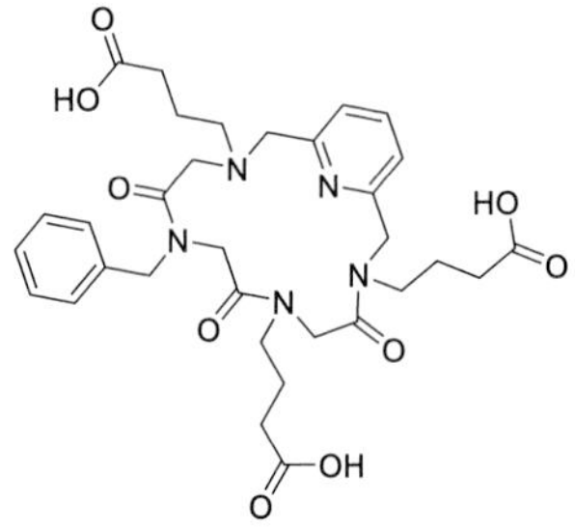

According to **GP4b**, 34.0 mg of the crude peptoid **c-(*N*3cx^{tBu})-Pyr-(*N*3cx^{tBu})$_2$-(*N*1ph) (c-2.93-*t*Bu)** was treated with a deprotection cocktail as a mixture of TFA, TIPS and water (95/2.5/2.5, volumetric) for 4 h. The volatiles were then removed under airflow and reduced pressure, the crude product was solubilized in an CH_3CN/H_2O mixture, purified by means of preparative RP-HPLC (5-95% acetonitrile + 0.1% TFA in 30 min, 25 °C) and lyophilized to afford 2.60 mg (4% over 2 steps) of the product as an off-white solid.

MS (MALDI-TOF, CHCA): m/z = 639.3 $[M]^+$. – **Analytical HPLC** (5–95% acetonitrile + 0.1% TFA in 20 min, detection at 256 nm): t_{Ret} = 10.21 min, 90% purity.

Eu-c-(*N*3cx)-Pyr-(*N*3cx)$_2$-(*N*1ph) (Eu-c-2.93)

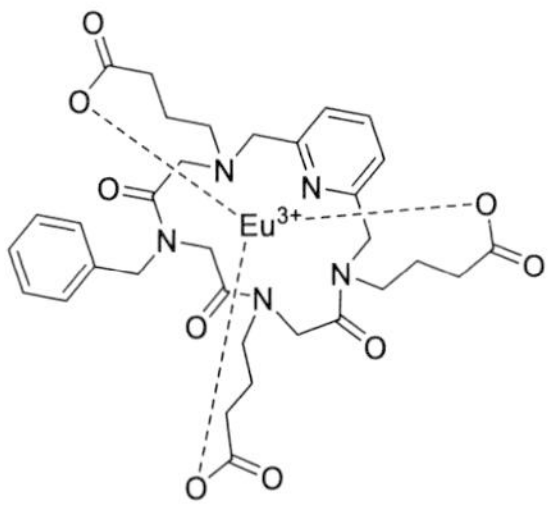

A solution of 2.6 mg of **c-(*N*3cx)-Pyr-(*N*3cx)$_2$-(*N*1ph)** (2.93) (4.06 µmol, 1.00 equiv.) in methanol was treated with 122 µL of 0.10 M aqueous solution of NaOH (12.2 µmol, 3.00 equiv.). Then an aqueous solution of 1.48 mg of $EuCl_3 \cdot x\,H_2O$ (4.06 µmol, 1.00 equiv.) was added. The reaction mixture was refluxed for additional 1 h, then the solvent was removed under reduced pressure and the crude product was purified by means of preparative HPLC (5-95% acetonitrile + 0.1% TFA in 30 min, 25 °C) and lyophilized to afford 2.3 mg (72%) of an off-white solid.

MS (MALDI-TOF, THAP): m/z = 789.2 $[M]^+$. – **Analytical HPLC** (5–95% acetonitrile + 0.1% TFA in 20 min, detection at 256 nm): t_{Ret} = 10.39 min, 87% purity.

H-(*N*2cx^{tBu})-BPyr-(*N*2cx^{tBu})$_2$-*SP* (2.94-*t*Bu-SP)

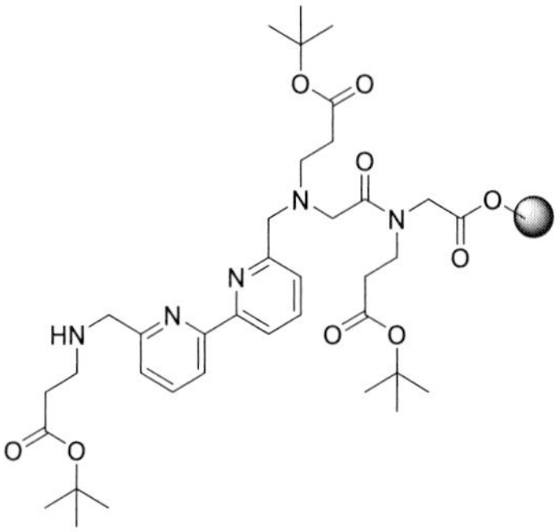

According to **GP11b**, 800 mg of 2-chlortritylchlorid resin (0.64 mmol, 1.00 equiv.) were swollen in DCM. The synthesis of the peptoid was carried out by acylation (**GP11b**, step 2) with 2.66 mL 1.20 M bromoacetic acid in DCM (3.20 mmol, 5.00 equiv.) and 558 µL DIPEA (3.20 mmol, 5.00 equiv.). *N*2cx^{tBu} group was attached, using 465 mg of *tert*-butyl 3-aminopropanoate (3.20 mmol, 5.00 equiv.) (**GP11b**, step 3). Subsequent acylation was carried out, using 2.66 mL 1.20 M bromoacetic acid in pDMF (3.20 mmol, 5.00 equiv.) and 496 µL DIC (3.20 mmol, 5.00 equiv.) (**GP11b**, step 4). After the second substitution step, the resin was treated with 657 mg of 6,6'-bis(bromomethyl)-2,2'-bipyridine **2.85** (1.92 mmol, 3.00 equiv.) and 335 µL DIPEA (1.92 mmol, 3.00 equiv.) in 2.00 mL pDMF. The mixture was stirred for 1 h. The subsequent *N*2cx^{tBu} group was attached, using 465 mg of *tert*-butyl 3-aminopropanoate (3.20 mmol, 5.00 equiv.) (**GP11b**, step 3) in 2.00 mL pDMF. A few beads were cleaved off and the formation of the desired product was detected by MALDI-TOF-MS.

MS (MALDI-TOF, CHCA): m/z = 714.4 $[M+H]^+$.

H-(N2cx^{tBu})-BPyr-(N2cx^{tBu})$_2$-OH (2.94-tBu)

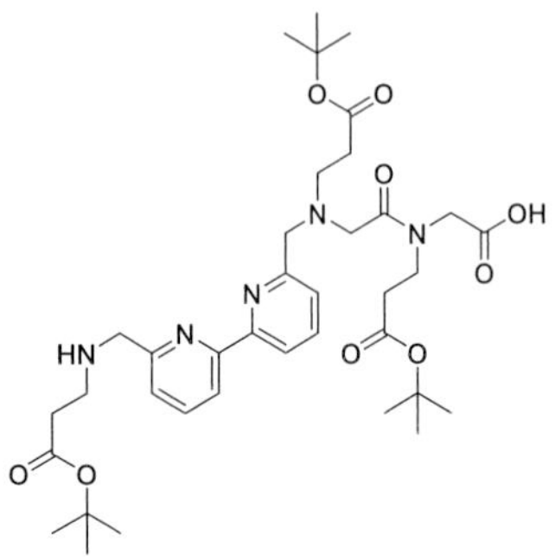

According to **GP12b**, **2.94-*t*Bu-SP** was cleaved from the resin by treatment with 8.00 mL 20% HFIP solution in DCM at room temperature for 2 h. The solution was filtered off and the next cleavage mixture was added to the resin and let shaken for additional 2 h. The resin was washed with 3 mL DCM (3 ×) and methanol (3 ×). The cleavage cocktail was evaporated under airflow, the residue was combined with the washes and the solvent was removed under reduced pressure (**GP12b**, step 1). The crude product was then dissolved in 5 mL acetonitrile, passed through a syringe filter, purified by means of preparative RP-HPLC (5-95% acetonitrile + 0.1% TFA in 30 min, 25 °C) and lyophilized (**GP12b**, step 2) to afford 51.0 mg (11% over 7 steps) of a pale-yellow solid.

MS (MALDI-TOF, CHCA): m/z = 714.4 $[M]^+$.

c-(N2cx^{tBu})$_3$-BPyr (c-2.94-tBu)

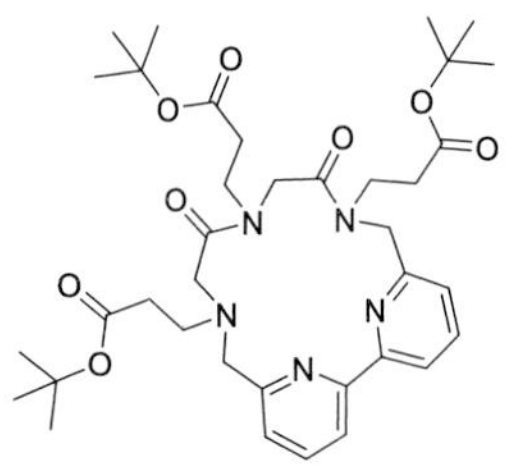

According to **GP13**, to a solution of 95.0 mg of the linear peptoid **2.94-*t*Bu** (160 µmol, 1.00 equiv.) in 150 mL pDMF, a diluted solution of 173 µL DIPEA (990 µmol, 6.20 equiv.) and 243 mg of HATU (640 µmol, 4.00 equiv.) in 200 mL pDMF was added in 12 h under Ar, using a dropping funnel. The resulting mixture was stirred for 3 days under Ar at room temperature. Afterwards, the solvent was evaporated under reduced pressure and the residue was taken up in 100 mL EtOAc. The organic layer was washed with 100 mL water and the aqueous layer was extracted with EtOAc (3 ×). The combined organic layers were dried over Na_2SO_4, filtered and concentrated under reduced pressure. The crude product was used in the subsequent reaction without further purification.

^{1}H NMR (300 MHz, D_2O): δ (ppm) = 8.68 – 8.59 (m, 2H, 2 × H_{Ar}), 8.29 (dt, J = 8.6 Hz, J = 1.6 Hz, 2H, 2 × H_{Ar}), 7.59 – 7.49 (m, 2H, 2 × H_{Ar}), 3.96 – 2.37 (m, 12H, 6 × CH_2 from the side chains), 1.47 – 0.92 (m, 27H, 9 × CH_3). The signals of the backbone protons (8H, 6 × CH_2) are not visible and are probably overlapped with the solvent signal (D_2O δ (ppm) = 5.27 –4.53). – **MS** (MALDI-TOF, CHCA): m/z = 695.9 $[M]^+$.

c-(*N*2cx)$_3$-BPyr (c-2.94-H)

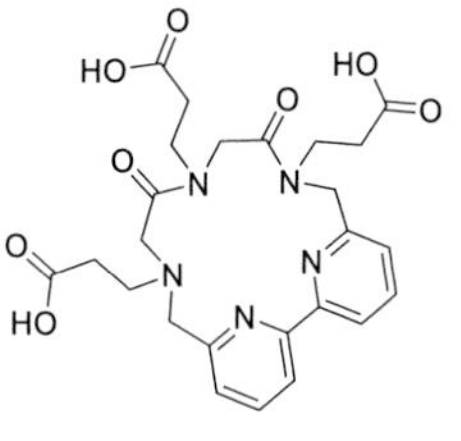

According to **GP4b**, 59.0 mg of the crude peptoid **c-(*N*2cx^{tBu})$_3$-BPyr c-2.94-*t*Bu** was treated with a mixture of TFA, TIPS and water (95/2.5/2.5, volumetric) for 4 h. The volatiles were then removed under airflow and reduced pressure, the crude product was solubilized in an CH_3CN/H_2O mixture, purified by means of preparative RP-HPLC (5-95% acetonitrile + 0.1% TFA in 30 min, 25 °C) and lyophilized to afford 12.0 mg (17% over 2 steps) of the product as an off-white solid.

MS (MALDI-TOF, CHCA): m/z = 528.2 [M+H]$^+$. – **Analytical HPLC** (5–95% acetonitrile + 0.1% TFA in 20 min, detection at 256 nm): t_{Ret} = 10.12 min, 97% purity.

***Eu*-c-(*N*2cx)$_3$-BPyr (Eu-c-2.94)**

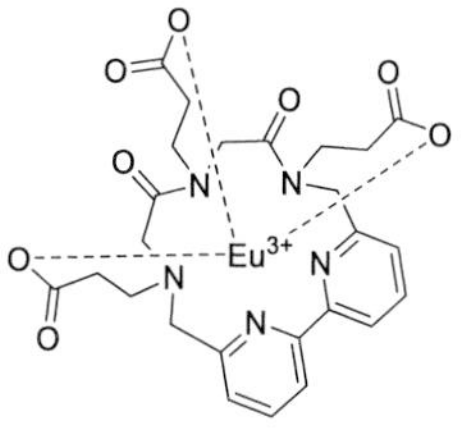

A suspension of 8.00 mg **c-(*N*2cx)$_3$-BPyr (c-2.94-H)** (15.2 µmol, 1.00 equiv.) in methanol was treated with 455 µL of 0.10 M aq. NaOH (45.5 µmol, 3.00 equiv.). Then an aq. solution of 5.56 mg of $EuCl_3 \cdot x\ H_2O$ (15.2 µmol, 1.00 equiv.) was added. The reaction mixture was refluxed for an additional 1 h, then the solvent was removed under reduced pressure and the crude product was purified by means of preparative RP-HPLC (5-95% acetonitrile + 0.1% TFA in 30 min, 25 °C) and lyophilized to afford 2.00 mg (20%) of a colorless solid.

MS (MALDI-TOF, THAP): m/z = 677.1 [M]$^+$. – **Analytical HPLC** (5–95% acetonitrile + 0.1% TFA in 20 min, detection at 256 nm): t_{Ret} = 10.10 min, 95% purity.

***H*-(*N*3cx^{tBu})-Phen-(*N*3cx^{tBu})$_2$-*SP* (2.95-*t*Bu-SP)**

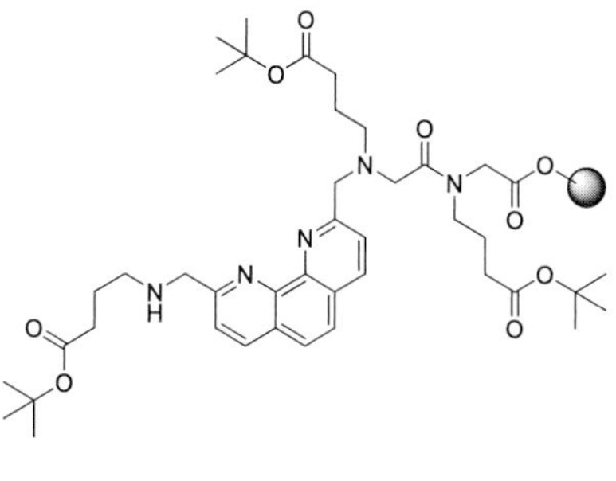

According to **GP11b**, 800 mg of 2-chlortritylchlorid resin (0.64 mmol, 1.00 equiv.) were swollen in DCM. The synthesis of the peptoid was carried out by acylation (**GP11b**, step 2) with 2.66 mL 1.20 M bromoacetic acid in DCM (3.20 mmol, 5.00 equiv.) and 558 µL DIPEA (3.20 mmol, 5.00 equiv.). *N*3cx^{tBu} group was attached, using 509 mg of *tert*-butyl 4-aminobutanoate (3.20 mmol, 5.00 equiv.) (**GP11b**, step 3). Subsequent acylation was carried out, using 2.66 mL 1.20 M bromoacetic acid in pDMF (3.20 mmol, 5.00 equiv.) and 496 µL DIC

(3.20 mmol, 5.00 equiv.) (**GP11b**, step 4). After the second substitution step, the resin was treated with 703 mg of 2,9-bis(bromomethyl)-1,10-phenanthroline (**2.86**) (1.92 mmol, 3.00 equiv.) and 335 µL DIPEA (1.92 mmol, 3.00 equiv.) in 2.00 mLpDMF. The mixture was stirred for 1 h. The subsequent *N*3cx^{tBu} group was attached, using 509 mg of *tert*-butyl 3-aminopropanoate (3.20 mmol, 5.00 equiv.) (**GP11b**, step 3) in 2.00 mL pDMF. A few beads were cleaved off and the formation of the desired product was detected by MALDI-TOF-MS.

MS (MALDI-TOF, CHCA): m/z = 780.8 $[M+H]^+$.

***H*-(*N*3cx$^{t\text{Bu}}$)-Phen-(*N*3cx$^{t\text{Bu}}$)$_2$-*OH* (2.95-*t*Bu)**

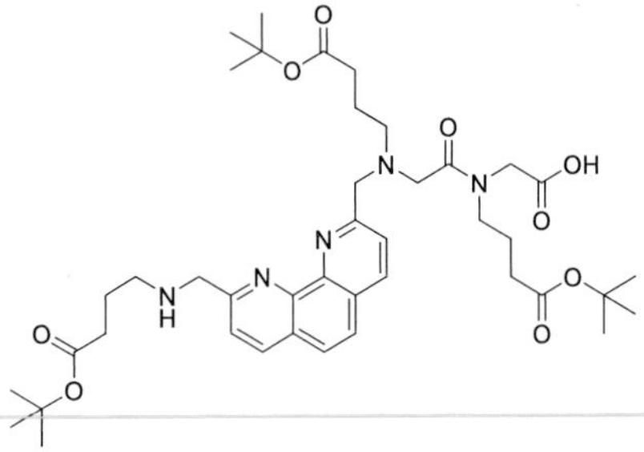

According to **GP12b**, **2.95-*t*Bu-SP** was cleaved from the resin by treatment with 8.00 mL 20% HFIP solution in DCM at room temperature for 2 h. The solution was filtered off and the next cleavage mixture was added to the resin and let shaken for additional 2 h. The resin was washed with 3 mL DCM (3 ×) and methanol (3 ×). The cleavage cocktail was evaporated under airflow, the residue was combined with the washes and the solvent was removed under reduced pressure (**GP12b**, step 1). The crude product was then dissolved in 5 mL acetonitrile, passed through a syringe filter, purified by means of preparative RP-HPLC (5-95% acetonitrile + 0.1% TFA in 30 min, 25 °C) and lyophilized (**GP12b**, step 2) to afford 59.0 mg (12% over 7 steps) of an off-white solid.

^{1}H NMR (300 MHz, CD_3CN): δ (ppm) = 8.58 – 7.15 (m, 6H, 6 × H_{Ar}), 5.06 – 2.80 (m, 26H, 13 × CH_2), 1.25 – 1.00 (m, 27H, 9 × CH_3). – **MS** (MALDI-TOF, CHCA): m/z = 780.8 $[M]^+$. – **Analytical HPLC** (5–95% acetonitrile + 0.1% TFA in 20 min, detection at 218 nm): t_{Ret} = 10.59 min, 82% purity.

c-(*N*3cx^{tBu})$_3$-Phen (c-2.95-*t*Bu)

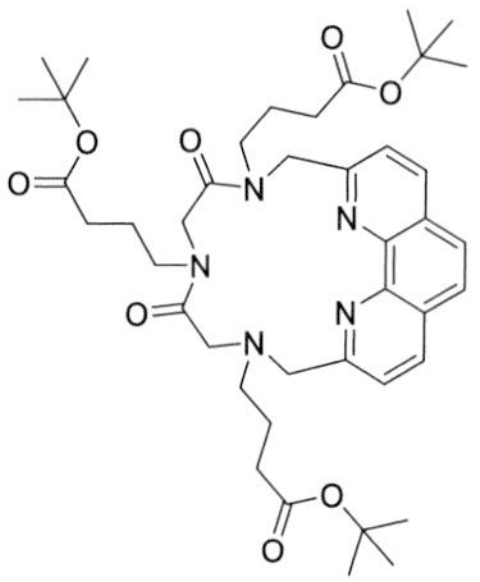

According to **GP13**, to a solution of 55.0 mg of the linear peptoid **2.95-*t*Bu** (70.0 µmol, 1.00 equiv.) in 100 mL pDMF, a diluted solution of 76.0 µL DIPEA (430 µmol, 6.20 equiv.) and 106 mg of HATU (280 µmol, 4.00 equiv.) in 150 mL pDMF was added in 12 h under Ar, using a dropping funnel. The resulting mixture was stirred for 3 days under Ar at room temperature. Afterwards, the solvent was evaporated under reduced pressure and the residue was taken up in 50 mL EtOAc. The organic layer was washed with 50 mLwater and the aqueous layer was extracted with EtOAc (3 ×). The combined organic layers were dried over Na_2SO_4, filtered and concentrated under reduced pressure. The crude product was used in the subsequent reaction without further purification.

MS (MALDI-TOF, CHCA): m/z = 762.4 $[M+H]^+$, 784.5 $[M+Na]^+$.

c-(*N*3cx)$_3$-Phen (c-2.95-H)

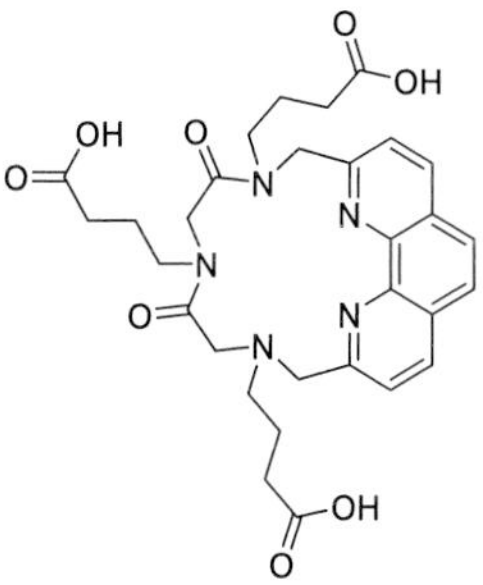

According to **GP4b**, 49.0 mg of the crude peptoid **c-(*N*3cx^{tBu})$_3$-Phen (c-2.95-*t*Bu)** was treated with a mixture of TFA, TIPS and water (95/2.5/2.5, volumetric) for 4 h. The volatiles were then removed under airflow and reduced pressure, the crude product was solubilized in an CH_3CN/H_2O mixture, purified by means of preparative RP-HPLC (5-95% acetonitrile + 0.1% TFA in 30 min, 25 °C) and lyophilized to afford 11.0 mg (24% over 2 steps) of the product as an off-white solid.

MS (MALDI-TOF, CHCA): m/z = 594.3 $[M+H]^+$. – **Analytical HPLC** (5–95% acetonitrile + 0.1% TFA in 20 min, detection at 256 nm): t_{Ret} = 10.18 min, 81% purity.

***Eu*-c-(*N*3cx)$_3$-Phen (Eu-c-2.95)**

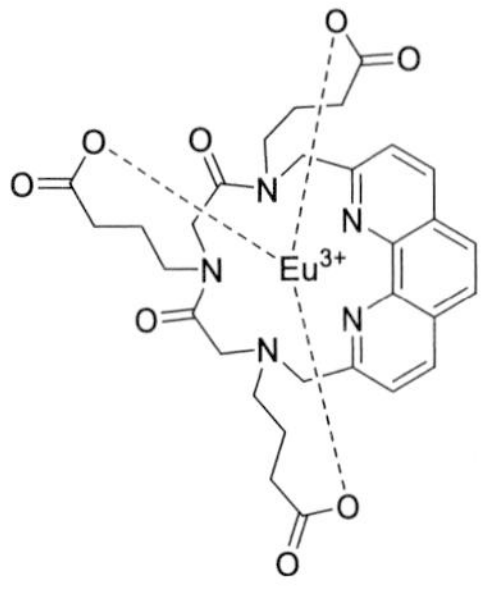

A suspension of 9.00 mg **c-(*N*3cx)$_3$-Phen (c-2.95-H)** (15.1 µmol, 1.00 equiv.) in methanol was treated with 455 µL of 0.10 M NaOH (45.4 µmol, 3.00 equiv.). Then an aq. solution of 5.54 mg of $EuCl_3 \cdot x H_2O$ (15.1 µmol, 1.00 equiv.) was added. The reaction mixture was refluxed for an additional hour, then the solvent was removed under reduced pressure and the crude product was purified by means of preparative RP-HPLC (5-95% acetonitrile + 0.1% TFA in 30 min, 25 °C) and lyophilized to afford 4.00 mg (36%) of an off-white solid.

MS (MALDI-TOF, THAP): m/z = 743.2 $[M]^+$. – **Analytical HPLC** (5–95% acetonitrile + 0.1% TFA in 20 min, detection at 256 nm): t_{Ret} = 10.21 min, 87% purity.

***H*-(*N*3cx^{tBu})-Pyr-[(*N*3cx^{tBu})-(*N*1ph)]$_2$-*SP* (2.96-*t*Bu-SP)**

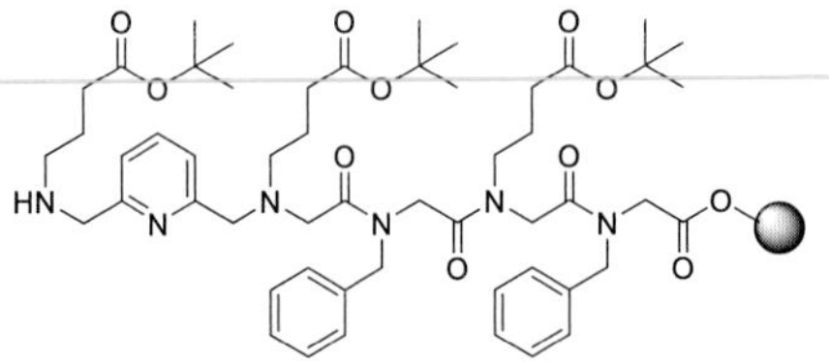

According to **GP11b**, 800 mg of 2-chlortritylchlorid resin (0.64 mmol, 1.00 equiv.) were swollen in DCM. The synthesis of the peptoid was carried out by acylation (**GP11b**, step 2) with 2.66 mL 1.20 M bromoacetic acid in DCM (3.20 mmol, 5.00 equiv.) and 558 µL DIPEA (3.20 mmol, 5.00 equiv.). *N*1Ph group was attached, using 343 mg (353 µL, 3.20 mmol, 5.00 equiv.) of benzylamine (**GP11b**, step 3). Subsequent acylation was carried out, using 2.66 mL 1.20 M bromoacetic acid in pDMF (3.20 mmol, 5.00 equiv.) and 496 µL DIC (3.20 mmol, 5.00 equiv.) (**GP11b**, step 4). *N*3cx^{tBu} group was attached, using 509 mg of *tert*-butyl 4-aminobutanoate (3.20 mmol, 5.00 equiv.) (**GP11b**, step 3). After the forth substitution step, the resin was treated with 509 mg of 2,6-bis(bromomethyl)pyridine (1.92 mmol, 3.00 equiv.) and 335 µL DIPEA (1.92 mmol, 3.00 equiv.) in 2.00 mL pDMF. The mixture was stirred for 1 h. The subsequent *N*3cx^{tBu} group was attached, using 509 mg of *tert*-butyl 3-aminopropanoate (3.20 mmol, 5.00 equiv.) (**GP11b**, step 3) in 2.00 mL pDMF. A few beads were cleaved from the resin and the formation of the desired product was detected by MALDI-TOF-MS.

MS (MALDI-TOF, CHCA): m/z = 972.6 $[M]^+$.

***H*-(*N*3cx^{tBu})-Pyr-[(*N*3cx^{tBu})-(*N*1ph)]$_2$-*OH* (2.96-*t*Bu)**

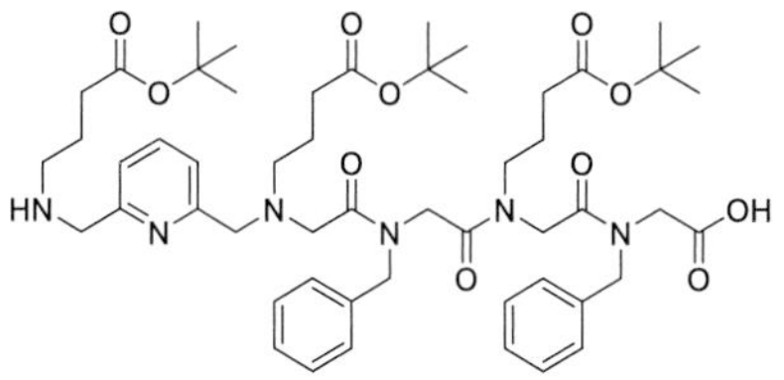

According to **GP12b**, **2.96-*t*Bu-SP** was cleaved from the resin by treatment with 8.00 mL 20% HFIP solution in DCM at room temperature for 2 h. The solution was filtered off and the next cleavage mixture was added to the resin and let shaken for additional 2 h. The resin was washed with 3 mL DCM (3 ×) and methanol (3 ×). The cleavage cocktail was evaporated under airflow, the residue was combined with the washes and the solvent was removed under reduced pressure (**GP12b**, step 1). The crude product was then dissolved in 5 mL acetonitrile, passed through a syringe filter, purified by means of preparative RP-HPLC (5-95% acetonitrile + 0.1% TFA in 30 min, 25 °C) and lyophilized (**GP12b**, step 2) to afford 82.0 mg (13% over 11 steps) of an off-white solid.

MS (MALDI-TOF, CHCA): m/z = 973.6 $[M+H]^+$. – **Analytical HPLC** (5–95% acetonitrile + 0.1% TFA in 20 min, detection at 256 nm): t_{Ret} = 10.77 min, 98% purity.

c-(*N*3cx^{tBu})-Pyr-[(*N*3cx^{tBu})-(*N*1ph)]$_2$ (c-2.96-*t*Bu)

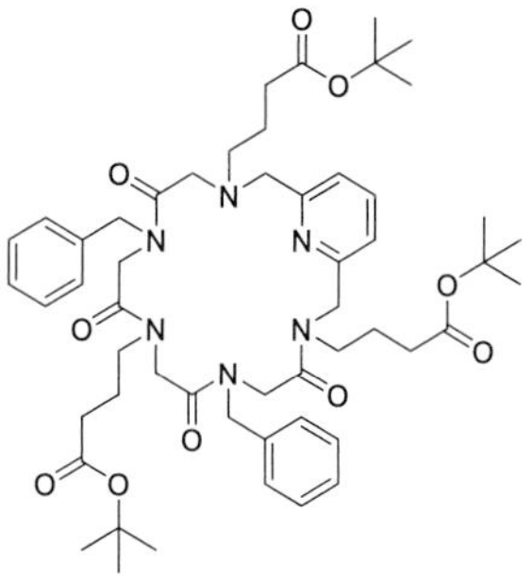

According to **GP13**, to a solution of 78.0 mg of the linear peptoid **2.96-*t*Bu** (80.0 μmol, 1.00 equiv.) in 100 mL pDMF, a diluted solution of 86.0 μL DIPEA (500 μmol, 6.20 equiv.) and 122 mg of HATU (280 μmol, 4.00 equiv.) in 150 mL pDMF was added in 12 h under Ar, using a dropping funnel. The resulting mixture was stirred for 3 days under Ar at room temperature. Afterwards, the solvent was evaporated under reduced pressure and the residue was taken up in 50 mL EtOAc. The organic layer was washed with 50 mLwater and the aqueous layer was extracted with EtOAc (3 ×). The combined organic layers were dried over Na_2SO_4, filtered and concentrated under reduced pressure. The crude product was used in the subsequent reaction without further purification.

MS (MALDI-TOF, CHCA): m/z = 956.2 $[M+H]^+$, 978.5 $[M+Na]^+$.

c-(*N*3cx)-Pyr-[(*N*3cx)-(*N*1ph)]$_2$ (c-2.96-H)

According to **GP4b**, 34.0 mg of the crude peptoid **c-(*N*3cx^{tBu})-Pyr-[(*N*3cx^{tBu})-(*N*1ph)]$_2$ (c-2.96-^{t}Bu)** was treated with a mixture of TFA, TIPS and water (95/2.5/2.5, volumetric) for 4 h. The volatiles are then removed under airflow and reduced pressure, the crude product was solubilized in an CH_3CN/H_2O mixture, purified by means of preparative RP-HPLC (5-95% acetonitrile + 0.1% TFA in 30 min, 25 °C) and lyophilized to afford 2.10 mg (3% over 2 steps) of the product as an off-white solid.

MS (MALDI-TOF, CHCA): m/z = 786.6 $[M]^+$. – **Analytical HPLC** (5–95% acetonitrile + 0.1% TFA in 20 min, detection at 256 nm): t_{Ret} = 9.91 min, 99% purity.

***Eu*-c-(*N*3cx)-Pyr-[(*N*3cx)-(*N*1ph)]$_2$ (Eu-c-2.96)**

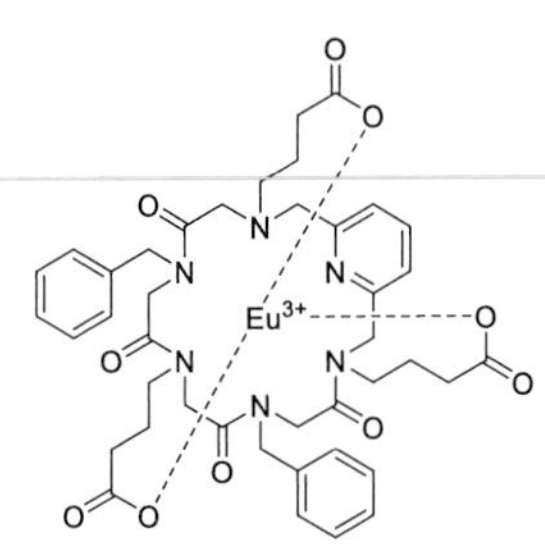

A solution of 2.00 mg of **c-(*N*3cx)-Pyr-[(*N*3cx)-(*N*1ph)]$_2$ (c-2.96-H)** (2.54 µmol, 1.00 equiv.) in methanol was treated with 76.0 µL of 0.10 M NaOH (7.62 µmol, 2.00 equiv.). Then an aq. solution of 0.93 mg of $EuCl_3 \cdot x\ H_2O$ (2.54 µmol, 1.00 equiv.) was added. The reaction mixture was refluxed for additional 1 h, then the solvent was removed under reduced pressure and the crude product was purifiedby means of preparative RP-HPLC (5-95% acetonitrile + 0.1% TFA in 30 min, 25 °C) and lyophilized to afford 1.50 mg (64%) of an off-white solid.

MS (MALDI-TOF, THAP): m/z = 936.3 $[M]^+$, 787.6 $[M–Eu^{3+}+4H]^+$. – **Analytical HPLC** (5–95% acetonitrile + 0.1% TFA in 20 min, detection at 256 nm): t_{Ret} = 10.40 min, 94% purity.

***H*-(*N*3cx^{tBu})-Phen-(*N*3cx^{tBu})-(*N*1ph)-(*N*3cx^{tBu})-*SP* (2.97-*t*Bu-SP)**

According to **GP11b**, 800 mg of 2-chlortritylchlorid resin (0.64 mmol, 1.00 equiv.) were swollen in DCM. The synthesis of the peptoid was carried out by acylation (**GP11b**, step 2) with 2.66 mL 1.20 M bromoacetic acid in DCM (3.20 mmol, 5.00 equiv.) and 558 µL DIPEA (3.20 mmol, 5.00 equiv.). *N*3cx^{tBu} group was attached, using 509 mg of *tert*-butyl

4-aminobutanoate (3.20 mmol, 5.00 equiv.) (**GP11b**, step 3). Subsequent acylation was carried out, using 2.66 mL 1.20 M bromoacetic acid in pDMF (3.20 mmol, 5.00 equiv.) and 496 µL DIC (3.20 mmol, 5.00 equiv.) (**GP11b**, step 4). *N*1ph group was attached, using 343 mg (353 µL, 3.20 mmol, 5.00 equiv.) of benzylamine (**GP11b**, step 3). After the third substitution step, the resin was treated with 703 mg of 2,9-bis(bromomethyl)-1,10-phenanthroline **2.86** (1.92 mmol, 3.00 equiv.) and 335 µL DIPEA (1.92 mmol, 3.00 equiv.) in 2.00 mL pDMF. The mixture was stirred for 1 h. The subsequent *N*3cx^{tBu} group was attached, using 509 mg of *tert*-butyl 3-aminopropanoate (3.20 mmol, 5.00 equiv.) (**GP11b**, step 3) in 2.00 mL pDMF. A few beads were cleaved off and the formation of the desired product was detected by MALDI-TOF-MS.

MS (MALDI-TOF, CHCA): m/z = 928.2 $[M+H]^+$.

***H*-(*N*3cx^{tBu})-Phen-(*N*3cx^{tBu})-(*N*1ph)-(*N*3cx^{tBu})-*OH* (2.97-*t*Bu) and *OH*-(*N*3cx^{tBu})-(*N*1ph)-(*N*3cx^{tBu})-Phen-(*N*3cx^{tBu})-(*N*1ph)-(*N*3cx^{tBu})-*OH* (2.97-*t*Bu-dimer)**

According to **GP12b**, **2.97-*t*Bu-SP** was cleaved from the resin by treatment with 8.00 mL 20% HFIP solution in DCM at room temperature for 2 h. The solution was filtered off and the next cleavage mixture was added to the resin and let shaken for additional 2 h. The resin was washed with 3 mL DCM (3 ×) and methanol (3 ×). The cleavage cocktail was evaporated under airflow, the residue was combined with the washes and the solvent was removed under reduced pressure (**GP12b**, step 1). The crude product was then dissolved in 5 mL acetonitrile, passed through a syringe filter, purified by means of preparative RP-HPLC (5-95% acetonitrile + 0.1% TFA in 30 min, 25 °C) and lyophilized (**GP12b**, step 2) to afford 60.0 mg (10% over 9 steps) of an off-white solid as well as 2.00 mg (2.3%) of a dimeric by-product as a yellow solid.

***H*-(*N*3cx^{tBu})-Phen-(*N*3cx^{tBu})-(*N*1ph)-(*N*3cx^{tBu})-*OH* (2.97-*t*Bu)**

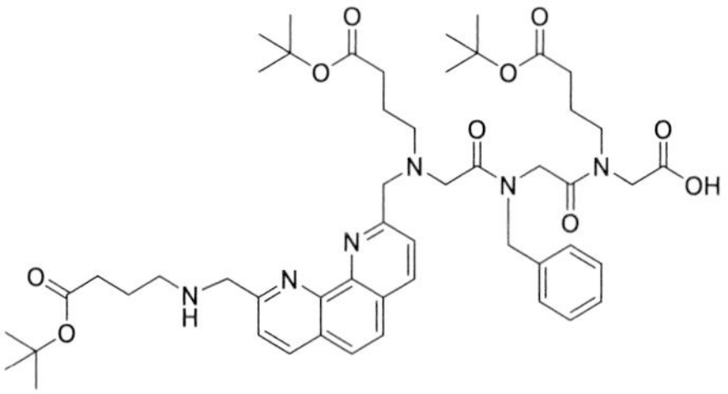

MS (MALDI-TOF, CHCA): m/z = 926.5 $[M]^+$. – **Analytical HPLC** (5–95% acetonitrile + 0.1% TFA in 20 min, detection at 256 nm): t_{Ret} = 11.75 min, 94% purity.

OH-(N3cx^{tBu})-(N1ph)-(N3cx^{tBu})-Phen-(N3cx^{tBu})-(N1ph)-(N3cx^{tBu})-OH (2.97-tBu-dimer)

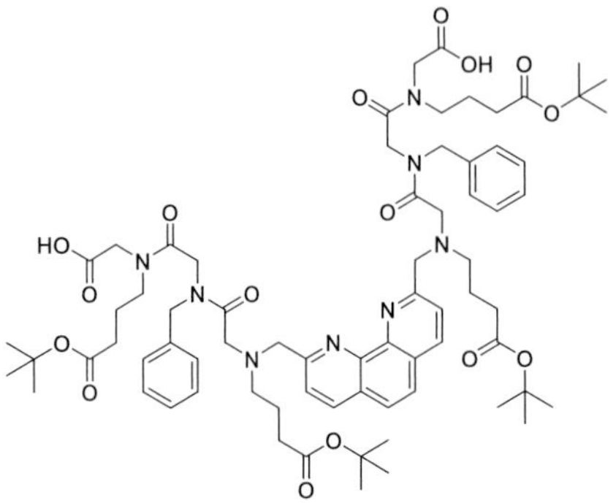

MS (MALDI-TOF, CHCA): m/z =1331.7 $[M]^+$. – **Analytical HPLC** (5–95% acetonitrile + 0.1% TFA in 20 min, detection at 256 nm): t_{Ret} = 13.48 min, 93% purity.

c-(N3cx^{tBu})$_2$-Phen-(N3cx^{tBu})-(N1ph) (c-2.97-tBu)

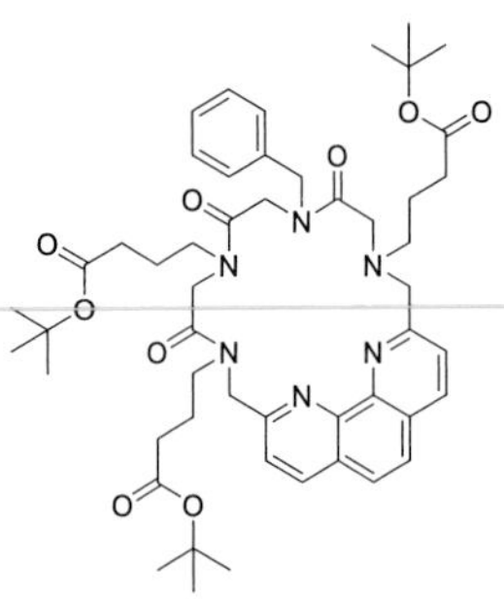

According to **GP13**, to a solution of 56.0 mg of the linear peptoid **3.97-tBu** (60.0 µmol, 1.00 equiv.) in 100 mL pDMF, a diluted solution of 65.0 µL DIPEA (370 µmol, 6.20 equiv.) and 91.0 mg of HATU (240 µmol, 4.00 equiv.) in 150 mL pDMF was added in 12 h under Ar, using a dropping funnel. The resulting mixture was stirred for 3 days under Ar at room temperature. Afterwards, the solvent was evaporated under reduced pressure and the residue was taken up in 50 mL EtOAc. The organic layer was washed with 50 mLwater and the aqueous layer was extracted with EtOAc (3 ×). The combined organic layers were dried over Na_2SO_4, filtered and concentrated under reduced pressure. The crude product was used in the subsequent reaction without further purification.

MS (MALDI-TOF, CHCA): m/z = 909.5 $[M+H]^+$.

c-(N3cx)$_2$-Phen-(N3cx)-(N1ph) (c-2.97-H)

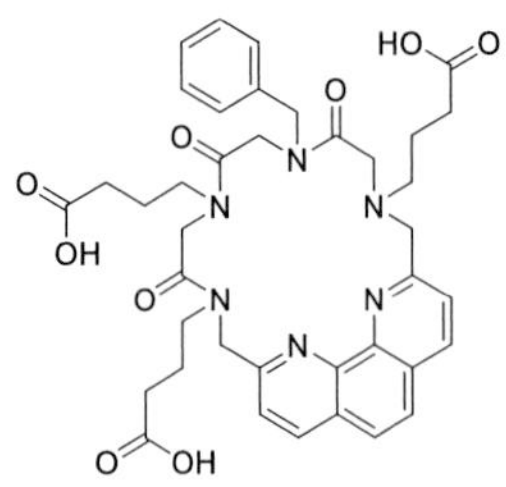

According to **GP4b**, 32.0 mg of the crude peptoid c-(N3cx^{tBu})$_2$-Phen-(N3cx^{tBu})-(N1ph) **(c-2.97-tBu)** was treated with a mixture of TFA, TIPS and water (95/2.5/2.5, volumetric) for 4 h. The volatiles were then removed under airflow and reduced pressure, the crude product was solubilized in an CH_3CN/H_2O mixture, purified by means of preparative RP-HPLC (5-95% acetonitrile + 0.1% TFA in 30 min, 25 °C) and lyophilized to afford 7.00 mg (16% over 2 steps) of the product as an off-white solid.

MS (MALDI-TOF, CHCA): m/z = 740.3 $[M]^+$. – **Analytical HPLC** (5–95% acetonitrile + 0.1% TFA in 20 min, detection at 256 nm): t_{Ret} = 10.22 min, 93% purity.

***Eu*-c-(*N*3cx)₂-Phen-(*N*3cx)-(*N*1ph) (Eu-c-2.97)**

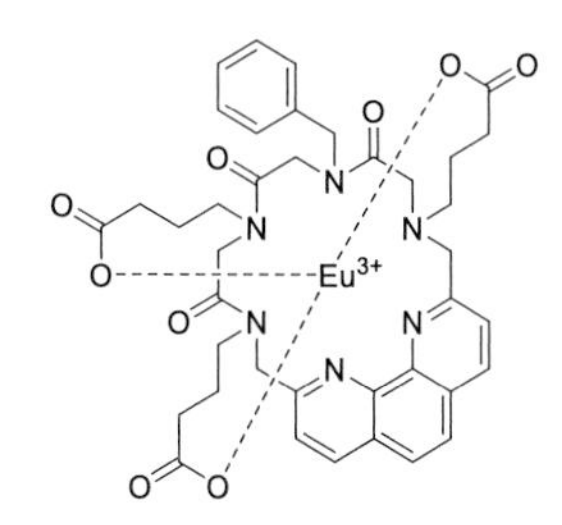

A suspension of 7.00 mg **c-(*N*3cx)$_2$-Phen-(*N*3cx)-(*N*1ph) (c-2.97-H)** (9.46 µmol, 1.00 equiv.) in methanol was treated with 284 µL of 0.10 M NaOH (28.4 µmol, 3.00 equiv.). Then an aq. solution of 3.46 mg of $EuCl_3 \cdot x\ H_2O$ (9.46 µmol, 1.00 equiv.) was added. The reaction mixture was refluxed for additional 1 h, then the solvent was removed under reduced pressure and the crude product was purified by means of preparative RP-HPLC (5-95% acetonitrile + 0.1% TFA in 30 min, 25 °C) and lyophilized to afford 2.00 mg (36%) of an off-white solid.

MS (MALDI-TOF, THAP): m/z = 891.2 $[M]^+$, 741.2 $[M–Eu^{3+}+4H]^+$. – **Analytical HPLC** (5–95% acetonitrile + 0.1% TFA in 20 min, detection at 256 nm): t_{Ret} = 10.00 min, 93% purity.

***H*-(*N*3cx^{tBu})-BPyr-(*N*3cx^{tBu})-(*N*1ph)-(*N*3cx^{tBu})-*SP* (2.98-*t*Bu-SP)**

According to **GP11b**, 800 mg of 2-chlortritylchlorid resin (640 µmol, 1.00 equiv.) were swollen in DCM. The synthesis of the peptoid was carried out by acylation (**GP11b**, step 2) with 2.66 mL 1.20 M bromoacetic acid in DCM (3.20 mmol, 5.00 equiv.) and 558 µL DIPEA (3.20 mmol, 5.00 equiv.). *N*3cx^{tBu} group was attached, using 509 mg of *tert*-butyl 4-aminobutanoate (3.20 mmol, 5.00 equiv.) (**GP11b**, step 3). Subsequent acylation was carried out, using 2.66 mL 1.20 M bromoacetic acid in pDMF (3.20 mmol, 5.00 equiv.) and 496 µL DIC (3.20 mmol, 5.00 equiv.) (**GP11b**, step 4). *N*1ph group was attached, using 343 mg (353 µL, 3.20 mmol, 5.00 equiv.) of benzylamine (**GP11b**, step 3). After the third substitution step, the resin was treated with 657 mg of 6,6'-bis(bromomethyl)-2,2'-bipyridine (ASK392) (1.92 mmol, 3.00 equiv.) and 335 µL DIPEA (1.92 mmol, 3.00 equiv.) in 2.00 mL pDMF. The mixture was stirred for 1 h. The subsequent *N*3cx^{tBu} group was attached, using 509 mg of *tert*-butyl 3-aminopropanoate (3.20 mmol, 5.00 equiv.) (**GP11b**, step 3) in 2.00 mLpDMF. A few beads were cleaved off and the formation of the desired product was detected by MALDI-TOF-MS.

MS (MALDI-TOF, CHCA): m/z = 902.7 $[M]^+$.

***H*-(*N*3cx^{tBu})-BPyr-(*N*3cx^{tBu})-(*N*1ph)-(*N*3cx^{tBu})-*OH* (2.98-*t*Bu)**

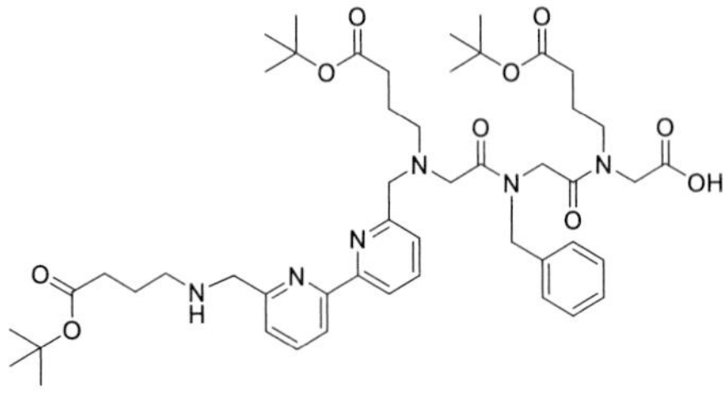

According to **GP12b**, **2.98-*t*Bu-SP** was cleaved from the resin by treatment with 8.00 mL 20% HFIP solution in DCM at room temperature for 2 h. The solution was filtered off and the next cleavage mixture was added to the resin and let shaken for additional 2 h. The resin was washed with 3 mL DCM (3 ×) and methanol (3 ×). The cleavage cocktail was evaporated under airflow, the residue was combined with the washes and the solvent was removed under reduced pressure (**GP12b**, step 1). The crude product was then dissolved in 5 mL acetonitrile, passed through a syringe filter, purified by means of preparative RP-HPLC (5-95% acetonitrile + 0.1% TFA in 30 min, 25 °C) and lyophilized (**GP12b**, step 2) to afford 59.0 mg (10% over 9 steps) of an off-white solid.

MS (MALDI-TOF, CHCA): m/z = 902.6 $[M]^+$. – **Analytical HPLC** (5–95% acetonitrile + 0.1% TFA in 20 min, detection at 256 nm): t_{Ret} = 11.80 min, 81% purity.

c-(*N*3cx^{tBu})$_2$-BPyr-(*N*3cx^{tBu})-(*N*1ph) (c-2.98-*t*Bu)

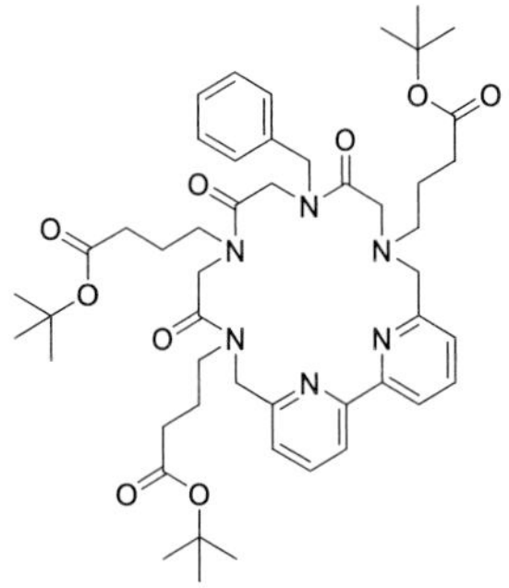

According to **GP13**, to a solution of 54.0 mg of the linear peptoid **2.98-*t*Bu** (60.0 µmol, 1.00 equiv.) in 100 mL pDMF, a diluted solution of 65.0 µL DIPEA (370 µmol, 6.20 equiv.) and 91.0 mg of HATU (240 µmol, 4.00 equiv.) in 150 mL pDMF was added in 12 h under Ar, using a dropping funnel. The resulting mixture was stirred for 3 days under Ar at room temperature. Afterwards, the solvent was evaporated under reduced pressure and the residue was taken up in 50 mL EtOAc. The organic layer was washed with 50 mLwater and the aqueous layer was extracted with EtOAc (3 ×). The combined organic layers were dried over Na_2SO_4, filtered and concentrated under reduced pressure. The crude product was used in the subsequent reaction without further purification.

MS (MALDI-TOF, CHCA): m/z = 884.5 $[M]^+$, m/z = 829.5 $[M–tBu+H]^+$.

c-(*N*3cx)$_2$-BPyr-(*N*3cx)-(*N*1ph) (c-2.98-H)

According to **GP4b**, 37.0 mg of the crude peptoid c-(*N*3cx^{tBu})$_2$-BPyr-(*N*3cx^{tBu})-(*N*1ph) **(c-2.98-*t*Bu)** was treated with a mixture of TFA, TIPS and water (95/2.5/2.5, volumetric) for 4 h. The volatiles were then removed under airflow and reduced pressure, the crude product was solubilized in an CH_3CN/H_2O mixture, purified by means of preparative RP-HPLC (5-95% acetonitrile + 0.1% TFA in 30 min, 25 °C) and lyophilized to afford 8.00 mg (19% over 2 steps) of the product as an off-white solid.

MS (MALDI-TOF, CHCA): m/z = 716.3 $[M]^+$. – **Analytical HPLC** (5–95% acetonitrile + 0.1% TFA in 20 min, detection at 256 nm): t_{Ret} = 10.42 min, 98% purity.

***Eu*-c-(*N*3cx)$_2$-BPyr-(*N*3cx)-(*N*1ph) (Eu-c-2.98)**

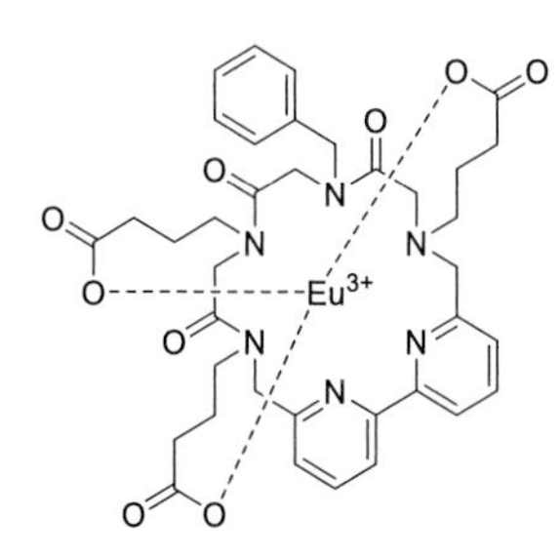

A solution of 6.00 mg **c-(*N*3cx)$_2$-BPyr-(*N*3cx)-(*N*1ph) (c-2.98)** (8.37 µmol, 1.00 equiv.) in methanol was treated with 252 µL of 0.10 M NaOH (25.1 µmol, 3.00 equiv.). Then an aqueous solution of 3.06 mg of $EuCl_3 \cdot 6\ H_2O$ (8.37 µmol, 1.00 equiv.) was added. The reaction mixture was refluxed for an additional 1 h, then the solvent was removed under reduced pressure and the crude product was purified by means of preparative RP-HPLC (5-95% acetonitrile + 0.1% TFA in 30 min, 25 °C) and lyophilized to afford 2.00 mg (28%) of an off-white solid.

MS (MALDI-TOF, THAP): m/z = 866.3 $[M]^+$, 717.4 $[M–Eu^{3+}+4H]^+$. – **Analytical HPLC** (5–95% acetonitrile + 0.1% TFA in 20 min, detection at 256 nm): t_{Ret} = 10.29 min, 97% purity.

V.3 Crystallographic data

Table V.3-1 Overview of $Ln(pfb)_3(H_2O)_x$ structural types

	α-Ln-1.1-sc	**β-Ln-1.1-sc**	**γ-Ln-1.1-sc**	**β′-Ln-1.1-sc**	**δ-Ln-1.1-sc**
found in	single crystal	single crystal	single crystal	single crystal	powder
composition	$[Ln_2(pfb)_6(H_2O)_8]$ $(H_2O)_2$	$[Ln(pfb)_2(H_2O)_6]^+$ $(pfb)^-$	$Ln(pfb)_3(H_2O)_5$ (H_2O)	$[Ln(pfb)_2(H_2O)_4]^+$ $(pfb)^-(H_2O)_2$	$[Ln(pfb)_3(H_2O)]_n$
CN	9	9	8	8	9
polyhedra	monocapped tetragonal antiprism	monocapped tetragonal antiprism	tetragonal antiprism	tetragonal antiprism	monocapped tetragonal antiprism
formula unit	$C_{42}H_{20}F_{30}O_{22}Ln_2$	$C_{21}H_{12}F_{15}O_{12}Ln$	$C_{21}H_{12}F_{15}O_{12}Ln$	$C_{21}H_{12}F_{15}O_{12}Ln$	$C_{21}H_2F_{15}O_7Ln$
ligand coordination	$[Ln_2(\mu_2\text{-}\kappa^1\text{:}\kappa^2\text{-}RCOO)_2(\kappa^1\text{-}RCOO)_4(H_2O)_8]$ $(H_2O)_2$	$[Ln(\kappa_2\text{-}RCOO)(\kappa_1\text{-}RCOO)(H_2O)_6]^+$ $(RCOO)^-$	$[Ln(\kappa^1\text{-}RCOO)_3(H_2O)_5]$ (H_2O)	$[Ln(\kappa^2\text{-}RCOO)_2(H_2O)_4]$ $(RCOO)^-(H_2O)_2$	$[Ln(\mu_2\text{-}\kappa_1\text{:}\kappa_2\text{-}RCOO)_2(\mu_2\text{-}\kappa^1\text{-}RCOO)(H_2O)]_n$
nuclearity	dimer	monomer	monomer	monomer	polymer
a, Å	7.6255(2)-7.6696(5)	17.3950(18)-17.4102(4)	7.503(2)-7.5116(7)	7.0698(7)	7.695(2)-7.873(2)
b, Å	9.2167(6)-9.2597(6)	20.4160(7)-20.4237(4)	7.696(2)-9.7184(3)	11.1908(11)	11.871(3)-11.980(3)
c, Å	18.1436(12)-18.1904(11)	7.2757(3)-7.2900(2)	18.642(1)-18.666(4)	17.8915(17)	13.057(3)-13.234(3)
α, °	80.912(2)-81.2310(10)	90	87.80(3)-87.893(2)	98.054(2)	91.71(5)-92.62(5)
β, °	84.3130(10)-84.3410(10)	90.232(2)-90.3380(10)	82.27(3)-82.53(3)	94.304(2)	83.77(2)-85.16(2)
γ, °	88.199(2)-88.996(2)	90	79.20(3)-79.45(1)	99.176(2)	104.26(2)-105.20(3)
V, Å³	1253.32(14)-1270.52(14)	2584.5(5)-2592.16(11)	1321.1(5)-1326.1(2)	1376.7(2)	1184.3(2)-1162.3(3)
observed for	Nd, Sm, Eu, Gd, Tb, Tb/Eu	Sm, Gd, Tb/Eu	Tb, Dy, Ho, Er, Tm	Lu	Sm, Eu, Gd, Tb, Dy, Ho, Er

Table V.3-2 Single-crystal X-ray diffraction data of lanthanide pentafluorobenzoates

	α-Nd-1.1-sc	α-Sm-1.1-sc	β-Sm-1.1-sc	α-Eu-1.1-sc	α-Gd-1.1-sc
Formula	$C_{42}H_{20}F_{30}Nd_2O_{22}$	$C_{42}H_{20}F_{30}O_{22}Sm_2$	$C_{21}H_{12}F_{15}O_{12}Sm$	$C_{42}H_{20}Eu_2F_{30}O_{22}$	$C_{42}H20F_{30}Gd_2O_{22}$
	1735.06	1747.28	891.66	1750.50	1761.08
Crystal system	triclinic	triclinic	monoclinic	triclinic	triclinic
Space group	P-1	P-1	P21/c	P-1	P-1
Z (Z')	1 (0.5)	1 (0.5)	4 (1)	1 (0.5)	1 (0.5)
a, b, c, Å	7.6696(5), 9.2597(6), 18.1904(11)	7.6434(2), 9.2356(2), 18.1635(4)	17.4102(4), 20.4237(4), 7.2900(2)	7.6263(5), 9.2167(6), 18.1436(12)	7.6255(2), 9.2186(3), 18.1601(5)
α, β, γ, °	81.2310(10), 84.3400(10), 88.996(2)	81.0070(10), 84.3130(10), 88.4690(10)	90, 90.232(2), 90	80.9940(10), 84.3410(10), 88.3480(10)	80.9340(10), 84.3240(10), 88.2070(10)
V, Å^3	1270.52(14)	1260.11(5)	2592.16(11)	1253.32(14)	1254.30(6)
d_{calc}, g·cm^{-3}	2.268	2.303	2.285	2.319	2.331
μ, cm^{-1}	22.16	191.58	186.74	26.77	28.19
F(000)	838	842	1724	844	846
Radiation	Mo Kα	Cu Kα	Cu Kα	Mo Kα	Mo Kα
2θ$_{max}$, °	50	135	135	50	50
Reflns measured	15587	19179	41509	5370	13440
Independent reflns	6739	4229	4399	5370	5465
Reflns with I>2σ(I)	5707	4094	4136	5262	4933
Number of refined parameters	433	433	442	434	433
R_1/R_{Bragg}	0.0345	0.0337	0.0230	0.0223	0.0239
wR_2/R_{wp}	0.0662	0.0885	0.0574	0.0564	0.0533
GOF	0.999	1.072	1.058	0.866	1.011
Residual electron density, e·Å^{-3} (d_{min}/d_{max})	0.966/-1.248	1.382/-0.955	0.661/-0.803	0.655/-0.758	1.080/-0.846

Table V.3-2 (cntd)

	β-Gd-1.1-sc	α-(TbEu) -1.1-sc	β-(TbEu) -1.1-sc	γ-Dy-1.1-sc	γ-Ho-1.1-sc
Formula	$C_{21}H_{12}F_{15}GdO_{12}$	$C_{42}H_{20}EuF_{30}O_{22}Tb$	$C_{21}H_{12}Eu_{0.5}F_{15}O_{12}Tb_{0.5}$	$C_{21}H_{12}DyF_{15}O_{12}$	$C_{21}H_{12}F_{15}HoO_{12}$
	898.56	1757.46	896.75	903.81	906.24
Crystal system	monoclinic	triclinic	monoclinic	triclinic	triclinic
Space group	P21/c	P-1	$P2_1/c$	P-1	P-1
Z (Z')	4 (1)	1 (0.5)	4 (1)	2 (1)	2 (1)
a, b, c, Å	17.4005(6), 20.4160(7), 7.2757(3)	7.6373(9), 9.2194(11), 18.165(2)	17.3950(18), 20.420(2), 7.2759(8)	7.5116(7), 9.7161(9), 18.6555(17)	7.5045(2), 9.7097(3), 18.6436(6)
α, β, γ, °	90, 90.3380(10), 90	80.912(2), 84.324(2), 88.199(2)	90, 90.311(2), 90	87.893(2), 82.345(2), 79.366(2)	87.8850(10), 82.3620(10), 79.3650(10)
V, Å^3	2584.64(16)	1256.6(3)	2584.5(5)	1326.1(2)	1323.21(7)
d_{calc}, g·cm^{-3}	2.309	2.322	2.305	2.263	2.275
μ, cm^{-1}	27.41	28.29	27.57	29.88	31.61
F(000)	1732	846	1732	870	872
Radiation	Mo Kα	Mo Kα	Mo Kα	Mo Kα	Mo Kα
2θ$_{max}$, °	50	50	50	50	50
Reflns measured	63651	5473	25294	13984	16559
Independent reflns	6853	5473	5640	5761	7032
Reflns with I>2σ(I)	5972	4411	4327	5125	6842
Number of refined parameters	442	433	442	442	448
R_1/R_{Bragg}	0.0224	0.0554	0.0353	0.0324	0.0173
wR_2/R_{wp}	0.0535	0.0820	0.0753	0.0817	0.0444
GOF	1.026	1.018	1.135	1.034	1.040
Residual electron density, e·Å^{-3} (d_{min}/d_{max})	1.626/-0.870	1.311/-2.197	2.418/-1.507	3.203/-1.285	0.881/-0.695

Table V.3-2 (cntd)

	γ-Er-1.1-sc	γ-Tm-1.1-sc	β′-Lu-1.1-sc	δ-Eu[a]-1.1
Formula	$C_{21}H_{12}ErF_{15}O_{12}$	$C_{21}H_{12}F_{15}O_{12}Tm$	$C_{21}H_{12}F_{15}LuO_{12}$	$C_{21}H_2EuF_{15}O_7$
	908.57	910.24	916.28	803.19
Crystal system	triclinic	triclinic	triclinic	triclinic
Space group	P-1	P-1	P-1	P-1
Z (Z')	2 (1)	2 (1)	2 (1)	2 (1)
a, b, c, Å	7.5050(15), 9.696(2), 18.666(4)	7.5027(15), 9.7020(19), 18.648(4)	7.0698(7), 11.1908(11), 17.8915(17)	7.85697(15), 11.9693(4), 13.1089(4)
α, β, γ, °	87.88(3), 82.53(3), 79.26(3)	87.80(3), 82.27(3), 79.20(3)	98.054(2), 94.304(2), 99.176(2)	92.355(2), 84.443(2), 105.1832(19)
V, Å^3	1323.1(5)	1321.1(5)	1376.7(2)	1184.01(6)
d_{calc}, g·cm^{-3}	2.281	2.288	2.21	2.253
μ, cm^{-1}	33.42	35.29	37.5	205.21
F(000)	874	876	880	764
Radiation	Mo Kα	Mo Kα	Mo Kα	Cu Kα1
$2\theta_{max}$, °	58	50	52	90
Reflns measured	7024	15235	13015	-
Independent reflns	7024	6365	5408	-
Reflns with I>2σ(I)	6514	5908	4616	-
Number of refined parameters	450	448	563	175
R_1/R_{Bragg}	0.0193	0.0222	0.0378	0.0050
wR_2/R_{wp}	0.0497	0.0605	0.0873	0.01696
GOF	1.040	0.882	1.04	2.664
Residual electron density, e·Å^{-3} (d_{min}/d_{max})	0.976/-1.376	1.130/-0.860	5.14/-0.89	-

V.3.1 Lanthanide fluorobenzoates with different fluorination degrees

Table V.3-3 Single-crystal X-ray diffraction data of lanthanide fluorobenzoates

	Er-1.2	Er-1.4	Er-1.5	Eu-1.2	Eu-1.4
Chemical formula	$0.5(C_{42}H_{14}Er_2F_{24}O_{16})$	$C_{25}H_{18}ErF_9O_8$	$C_{25}H_{18}ErF_9O_8$	$C_{21}H_7EuF_{12}O_8 \cdot H_2O$	$C_{21}H_8EuF_9O_7$
M_r	782.53	784.65	784.65	785.24	695.23
Crystal system, space group	monoclinic, *C*2/*c*	monoclinic, *C*2/*c*	monoclinic, *P*2$_1$/*n*	triclinic, *P*-1	triclinic, *P*-1
Temperature (K)	100	120	100	100	100
***a*, *b*, *c* (Å)**	18.732 (15), 14.452 (12), 9.569 (7)	14.3879 (11), 19.6338 (15), 19.2312 (14)	9.676 (6), 19.555 (14), 14.654 (10)	9.4209 (4), 10.8081 (5), 13.3175 (6)	7.7369 (14), 12.331 (2), 12.483 (2)
α, β, γ (°)	90, 99.147 (13), 90	90, 99.078 (2), 90	90, 103.63 (5), 90	69.257 (1), 71.408 (1), 85.339 (1)	112.563 (3), 102.773 (4), 94.392 (4)
***V* (Å^3)**	2558 (3)	5364.6 (7)	2695 (3)	1201.17 (9)	1055.2 (3)
Z	4	8	4	2	2
Radiation type	Mo *K*α	Mo *K*α	Cu *K*α	Mo *K*α	Mo *K*α
μ (mm^{-1})	3.41	3.24	6.81	2.76	3.10
T_{min}, T_{max}	0.373, 0.435	0.650, 0.746	0.070, 0.221	0.344, 0.438	0.191, 0.266
No. of measured, independent and observed [$I > 2\sigma(I)$] reflections	7185, 3365, 3034	27371, 5874, 4276	34738, 4742, 4155	12900, 5224, 4870	13669, 6227, 4738
R_{int}	0.026	0.089	0.078	0.026	0.067
(sin θ/λ)$_{max}$ (Å^{-1})	0.682	0.639	0.602	0.639	0.717
$R[F^2 > 2\sigma(F^2)]$, $wR(F^2)$, S	0.030, 0.071, 1.04	0.042, 0.086, 1.05	0.048, 0.130, 1.02	0.024, 0.060, 1.03	0.073, 0.209, 1.11
No. of reflections	3365	5874	4742	5224	6227
No. of parameters	214	390	398	402	351
No. of restraints	2	47	0	103	39
	$w = 1/[\sigma^2(F_o^2) + (0.0346P)^2 + 3.1623P]$ where $P = (F_o^2 + 2F_c^2)/3$	$w = 1/[\sigma^2(F_o^2) + (0.0279P)^2 + 10.5283P]$ where $P = (F_o^2 + 2F_c^2)/3$	$w = 1/[\sigma^2(F_o^2) + (0.0899P)^2 + 3.341P]$ where $P = (F_o^2 + 2F_c^2)/3$	$w = 1/[\sigma^2(F_o^2) + (0.0279P)^2 + 1.9116P]$ where $P = (F_o^2 + 2F_c^2)/3$	$w = 1/[\sigma^2(F_o^2) + (0.1098P)^2 + 6.6676P]$ where $P = (F_o^2 + 2F_c^2)/3$
$\Delta\rho_{max}$, $\Delta\rho_{min}$ (e Å^{-3})	1.28, −0.96	1.65, −2.00	2.23, −1.24	0.96, −0.66	7.53, −2.86

Table V.3-3 (cntd)

	Eu-1.5	Eu-1.6	Eu-1.7	Eu-1.8	Eu-1.9
Chemical formula	$C_{21}H_{10}EuF_9O_8 \cdot H_2O$	$C_{21}H_{13}EuF_6O_8 \cdot H_2O$	$C_{21}H_{13}EuF_6O_8 \cdot H_2O$	$0.5(C_{42}H_{26}Cl_6Eu_2F_6O_{16})$	$C_{21}H_{16}EuF_3O_8$
M_r	731.27	677.29	677.29	708.62	605.30
Crystal system, space group	monoclinic, $P2_1$	triclinic, P-1	triclinic, P-1	monoclinic, $C2/c$	triclinic, P-1
Temperature (K)	120	120	120	100	120
a, b, c (Å)	11.226 (2), 6.4448 (13), 15.717 (3)	6.3980 (3), 11.0491 (6), 15.9006 (9)	6.462 (2), 11.214 (4), 15.733 (5)	21.174 (2), 12.3789 (12), 9.0585 (9)	9.191 (5), 9.625 (5), 13.201 (7)
α, β, γ (°)	90, 91.49 (3), 90	92.545 (1), 90.570 (1), 93.066 (1)	90.146 (8), 90.993 (6), 93.823 (7)	90, 91.010 (2), 90	90.444 (16), 100.094 (16), 117.688 (11)
V (Å^3)	1136.8 (4)	1121.25 (10)	1137.4 (6)	2373.9 (4)	1012.8 (9)
Z	2	2	2	4	2
Radiation type	Mo $K\alpha$	Mo $K\alpha$	Mo $K\alpha$	Mo $K\alpha$	Mo $K\alpha$
μ (mm^{-1})	2.89	2.90	2.86	3.05	3.17
Diffractometer	Bruker *APEX*-II CCD	Bruker *APEX*-II DUO CCD	Bruker *APEX*-II CCD	Bruker *APEX*-II CCD	Bruker *APEX*-II DUO CCD
T_{min}, T_{max}	0.014, 0.039	0.282, 0.433	0.009, 0.033	0.337, 0.433	0.325, 0.438
No. of measured, independent and observed [$I > 2\sigma(I)$] reflections	11210, 4615, 3395	15722, 6926, 6447	13860, 4510, 3315	15004, 3523, 2309	24930, 9348, 8664
R_{int}	0.089	0.025	0.116	0.042	0.027
$(\sin \theta/\lambda)_{max}$ (Å^{-1})	0.625	0.718	0.625	0.708	0.835
$R[F^2 > 2\sigma(F^2)]$, $wR(F^2)$, S	0.053, 0.093, 0.92	0.022, 0.052, 1.03	0.082, 0.171, 1.01	0.038, 0.087, 1.03	0.024, 0.060, 1.05
No. of reflections	4615	6926	4510	3523	9348
No. of parameters	361	380	344	281	318
No. of restraints	1	24	3	187	2
	$w = 1/[\sigma^2(F_o^2) + (0.025P)^2]$ where $P = (F_o^2 + 2F_c^2)/3$	$w = 1/[\sigma^2(F_o^2) + (0.0272P)^2]$ where $P = (F_o^2 + 2F_c^2)/3$	$w = 1/[\sigma^2(F_o^2) + (0.0918P)^2]$ where $P = (F_o^2 + 2F_c^2)/3$	$w = 1/[\sigma^2(F_o^2) + (0.0303P)^2 + 5.8705P]$ where $P = (F_o^2 + 2F_c^2)/3$	$w = 1/[\sigma^2(F_o^2) + (0.0297P)^2 + 0.4684P]$ where $P = (F_o^2 + 2F_c^2)/3$
$\Delta\rho_{max}$, $\Delta\rho_{min}$ (e Å^{-3})	1.90, −0.94	1.40, −0.61	3.75, −1.54	0.81, −1.80	2.57, −0.79

V.3.2 Lanthanide ternary complexes

Table V.3-4 Selected Bond Lengths and Bond Angles of the ternary complexes **Eu-1.9-BPhen** and **Eu-1.7-Phen**

Bond	Coordination mode	Eu-1.9-BPhen	Eu-1.7-Phen
Eu(1)-O(1)	μ^2-κ^1:κ^2	2.356(3)	2.380(2)
Eu(1)-O(2)	μ^2-κ^1	2.355(3)	2.380(2)
Eu(1)-O(3)	κ^2	2.418(3)	2.470(3)
Eu(1)-O(4)	μ^2-κ^1:κ^2	2.406(3)	2.468(2)
Eu(1)-O(5)	κ^2	2.472(3)	2.457(2)
Eu(1)-O(6)	μ^2-κ^1	2.382(3)	2.361(2)
Eu(1)-O(7)	μ^2-κ^1:κ^2	3.158(4)*	2.708(2)
Eu(1)-N(1)	κ^1	2.557(4)	2.598(3)
Eu(1)-N(2)	κ^1	2.608(4)	2.601(3)

Table V.3-5 Single-crystal X-ray diffraction data for lanthanide ternary complexes.

	Eu-1.7-Phen	**Eu-1.9-BPhen**
Chemical formula	$C_{66}H_{34}Eu_2F_{12}N_4O_{12}$	$C_{97}H_{64}Eu_2F_6N_4O_{12}$
M_r	1606.89	1895.44
Crystal system, space group	monoclinic, $P2_1/n$	monoclinic, $P2_1/c$
Temperature (K)	120	120
a, b, c **(Å)**	14.583(2), 13.1391(17), 15.364(2)	11.765(7), 22.391(12), 16.151(9)
α, β, γ (°)	90, 103.713(4), 90	90, 110.732(13), 90
V **(Å**^{3}**)**	2859.9(7)	3979(4)
Z	2	2
Radiation type	Mo $K\alpha$ (λ = 0.71073)	Mo $K\alpha$ (λ = 0.71073)
μ (mm^{-1}**)**	2.28	1.645
Diffractometer	Bruker *APEX*-II CCD	Bruker *APEX*-II CCD
T_{min}, T_{max}	0.411, 0.493	0.631, 0.746
No. of measured and observed $[I > 2\sigma(I)]$ **reflections**	47998, 5647	46604, 9873
R_{int}	0.1461	0.087
$R[F^2 > 2\sigma(F^2)]$, $wR(F^2)$, S	0.0443, 0.0695, 0.958	0.0439, 0.0743, 1.016
No. of reflections	5647	9873
No. of parameters	433	542
No. of restraints	0	7
	$w = 1/[\sigma^2(F_o^2) + (0.0304P)^2]$, where $P = (F_o^2 + 2F_c^2)/3$	$w = 1/[\sigma^2(F_o^2) + (0.0428P)^2 + 3.637P]$, where $P = (F_o^2 + 2F_c^2)/3$
$\Delta\rho_{max}$, $\Delta\rho_{min}$ **(e Å**$^{-3}$**)**	1.65, −2.00	1.75, −1.28

V.3.3 Functionalized ligands and lanthanide complexes with them

Table V.3-6 Single-crystal X-ray diffraction data for *p*-modified ligands

	2.20-H	**2.31-Me**
Empirical formula	$C_{28}H_4F_{16}N_{12}O_8$	$C_{16}H_{12}FN_3O_2$
Formula weight	940.43	297.29
Temperature/K	120(2)	120
Crystal system	triclinic	monoclinic
Space group	P-1	$P2_1/c$
a/Å	5.9896(3)	16.123(9)
b/Å	8.2426(4)	11.607(6)
c/Å	16.0927(8)	7.091(4)
α/°	82.9650(10)	90
β/°	88.9500(10)	100.478(14)
γ/°	86.2700(10)	90
Volume/Å^3	786.80(7)	1304.8(13)
Z	1	4
ρ_{calc}**g/cm**^{3}	1.985	1.513
μ/mm^{-1}	0.209	0.112
F(000)	464.0	616.0
Crystal size/mm^3	0.260 × 0.220 × 0.180	0.1 × 0.1 × 0.02
Radiation	MoKα (λ = 0.71073)	MoKα (λ = 0.71073)
2Θ range for data collection/°	2.55 to 66.216	4.35 to 52.738
Index ranges	$-8 \leq h \leq 9$, $-12 \leq k \leq 12$, $-24 \leq l \leq 23$	$-20 \leq h \leq 19$, $0 \leq k \leq 14$, $0 \leq l \leq 8$
Reflections collected	11363	2563
Independent reflections	5496 [R_{int} = 0.0176, R_{sigma} = 0.0279]	2563 [R_{int} = 0.0174, R_{sigma} = 0.0743]
Data/restraints/parameters	5496/3/297	2563/0/200
Goodness-of-fit on F^2	1.022	1.176
Final R indexes [I>=2σ (I)]	R_1 = 0.0377, wR_2 = 0.0992	R_1 = 0.0829, wR_2 = 0.1594
Final R indexes [all data]	R_1 = 0.0581, wR_2 = 0.1101	R_1 = 0.1236, wR_2 = 0.1777
Largest diff. peak/hole / e Å^{-3}	0.54/-0.24	0.37/-0.39

Table V.3-7 Single-crystal X-ray diffraction data of *p*-modified lanthanide fluorobenzoates

	Eu-2.2	**Eu-2.6**	**Eu-2.9**	**Eu-2.10**	**Eu-2.13**
Empirical formula	$C_{21}H_{19}EuF_3N_3O_8$	$C_{48}H_{38}Eu_2F_6N_{18}O_{16}$	$C_{46}H_{34}Eu_2F_6I_6O_{16}$	$C_{44}H_{29}Br_6Eu_2F_6O_{16}$	$C_{26}H_{21}EuF_3N_3O_{10}$
Formula weight	650.35	1540.88	2022.05	1711.05	744.42
Temperature/ K	120	123(2)	120	120	120
Crystal system	triclinic	triclinic	monoclinic	monoclinic	triclinic
Space group	P-1	P-1	$P2_1/c$	$P2_1/n$	P-1
a/Å	9.188(4)	9.0797(4)	9.6765(15)	9.618(7)	9.9028(16)
b/Å	10.078(7)	9.3544(5)	25.431(4)	24.926(19)	11.208(2)
c/Å	13.425(8)	18.3812(8)	23.450(4)	23.027(17)	13.451(3)
α/°	87.298(14)	83.151(2)	90	90	87.871(4)
β/°	89.560(9)	75.998(2)	101.807(4)	101.845(17)	72.842(4)
γ/°	64.903(10)	63.780(2)	90	90	76.931(5)
Volume/Å³	1124.4(11)	1358.82(11)	5648.6(15)	5403(7)	1388.9(5)
Z	2	1	4	4	2
ρ_{calc}g/cm³	1.921	1.883	2.378	2.104	1.780
μ/mm⁻¹	2.869	2.396	5.563	6.823	2.340
F(000)	640.0	760.0	3744.0	3244.0	736.0
Crystal size/mm³	0.06 × 0.05 × 0.04	0.16 × 0.12 × 0.04	0.16 × 0.15 × 0.1	0.15 × 0.12 × 0.07	0.06 × 0.05 × 0.04
Radiation	MoKα (λ = 0.71073)	MoKα (λ = 0.71073)	MoKα (λ = 0.71073)	MoKα (λ = 0.71073)	MoKα (λ = 0.71073)
2Θ range for data collection/°	4.468 to 57.34	4.568 to 55.098	2.39 to 54	1.634 to 52.044	3.732 to 56.564
Index ranges	-12 ≤ h ≤ 12, -13 ≤ k ≤ 13, -17 ≤ l ≤ 18	-11 ≤ h ≤ 11, -12 ≤ k ≤ 12, -23 ≤ l ≤ 23	-12 ≤ h ≤ 12, -32 ≤ k ≤ 32, -29 ≤ l ≤ 29	-11 ≤ h ≤ 11, -30 ≤ k ≤ 30, -28 ≤ l ≤ 28	-13 ≤ h ≤ 13, -14 ≤ k ≤ 14, -17 ≤ l ≤ 17
Reflections collected	13138	51998	98781	34664	21995
Independent reflections	5659 [R_{int} = 0.1021, R_{sigma} = 0.1379]	6254 [R_{int} = 0.0282, R_{sigma} = 0.0157]	12298 [R_{int} = 0.0814, R_{sigma} = 0.0429]	10755 [R_{int} = 0.3373, R_{sigma} = 0.3331]	6891 [R_{int} = 0.0571, R_{sigma} = 0.0575]
Data/restraints /parameters	5659/56/378	6254/23/439	12298/146/751	10755/672/677	6891/6/414
Goodness-of-fit on F²	0.997	1.269	1.229	0.988	0.987
Final R indexes [I>=2σ (I)]	R_1 = 0.0642, wR_2 = 0.1355	R_1 = 0.0183, wR_2 = 0.0438	R_1 = 0.0653, wR_2 = 0.1408	R_1 = 0.1430, wR_2 = 0.3344	R_1 = 0.0340, wR_2 = 0.0706
Final R indexes [all data]	R_1 = 0.1041, wR_2 = 0.1548	R_1 = 0.0204, wR_2 = 0.0443	R_1 = 0.0794, wR_2 = 0.1479	R_1 = 0.3090, wR_2 = 0.4358	R_1 = 0.0463, wR_2 = 0.0753
Largest diff. peak/hole/ e Å⁻³	2.91/-1.53		3.44/-1.90	3.87/-1.83	1.11/-0.47

V.4 EXAFS/XANES spectroscopic data of lanthanide pentafluorobenzoates

EXAFS (Extended X-ray Absorption Fine Structure) spectra, measured at the lanthanide L_{III}-edge, were used to determine the nearest coordination environment of the lanthanides in lanthanide pentafluorobenzoates **Ln-1.1** in powders and in aqueous and methanol solutions (Ln = Eu, Gd, Tb, Dy, Ho, Er, Tm, Yb, Lu). For light lanthanides (Pr, Nd, Sm) the EXAFS method is not suitable, since L_{III}- and L_{II}-edges are very close to each other, limiting the available data range.[276]

The extent of information available from EXAFS for these objects was determined on the example of **δ-Ln-1.1**, for which X-ray diffraction data and DFT calculations were obtained. The first coordination sphere of central ions in these complexes consists of nine atoms, belonging to either water or $(pfb)^-$ COO^- groups. EXAFS spectroscopy allows to distinguish between these oxygen atoms by the appearance of the Ln···C scattering path in the second shell. Moreover, the observed Ln–C distance also allows to distinguish between κ^1 (R = 3.43 – 3.61 Å) and κ^2 (R = 2.68 – 2.80 Å) ligand coordination modes (Table V.4-1, Figure III.1-4).

The structural model of **δ-Ln-1.1** agreed well with the X-ray diffraction data and contained in total five different photoelectron scattering paths (Table V.4-2). Nine oxygen atoms at the distances of 2.24 – 2.55 Å are present in the first shell. Two carbon atoms of κ^2 carboxylic group (R = 2.68 – 2.73 Å) and four carbon atoms of κ^1 carboxylic group (R = 3.43 – 3.61 Å) are observed in the second shell (Table V.4-1). The contribution of the Ln···Ln contacts is apparent for all the Ln complexes in powder form and corresponds to the polymeric structure of **δ-Ln-1.1** (Figure III.1-3a). The decrease of the Ln···Ln distances from **δ-Eu-1.1** to **δ-Er-1.1** is in a good agreement with lanthanide contraction. The best-fit values for the local structure parameters are summarized in (Table V.4-2).

Next, the EXAFS spectroscopic study of the unknown phases **Tm-1.1, Yb-1.1** and **Lu-1.1** was carried out. The first lanthanide coordination sphere in all these samples consists of eight oxygen atoms (R = 2.25 – 2.29 Å), unlike **δ-Ln-1.1**. In case of **Tm-1.1** and **Yb-1.1** the second coordination sphere consists of three carbon atoms in κ^1 coordination mode (R = 3.50 – 3.66 Å). The second coordination shell of **«ε2»-Lu** contains two κ^2-carbon atoms (R = 2.76 Å) and one κ^1-carbon atom (R = 3.41 Å) (Table V.4-2).

Therefore, EXAFS spectroscopy data are completely consistent with the XRD results and confirm (i) the composition similarity of the complexes among **δ-Ln-1.1** group; (ii) the composition similarity of **Tm-1.1** and **Yb-1.1** and (iii) different ligand coordination in **Lu-1.1.**

Table V.4-1 The distances determined by EXAFS-spectroscopy for **δ-Eu-1.1**, **Eu-1.1** in H_2O and **$EuCl_3$** in H_2O.

	1st coordination sphere			2nd coordination sphere				3rd coordination sphere		
	path	**N**	**R(Å)**	**Path**	**N**	**coord. mode**	**R(Å)**	**path**	**N**	**R(Å)**
δ-Eu-1.1	Eu···O	9	2.38	Eu···C	2	κ^2 carboxylic	2.68	Eu···Eu	2	4.10
				Eu···C	4	κ^1 carboxylic	3.43			
Eu-1.1 in H_2O	Eu···O	10	2.46	Eu···C	2	κ^2 carboxylic	2.76			
				Eu···F	1		3.59			
$EuCl_3$ in H_2O	Eu···O	10	2.45	Eu···Cl	1.5		3.28			

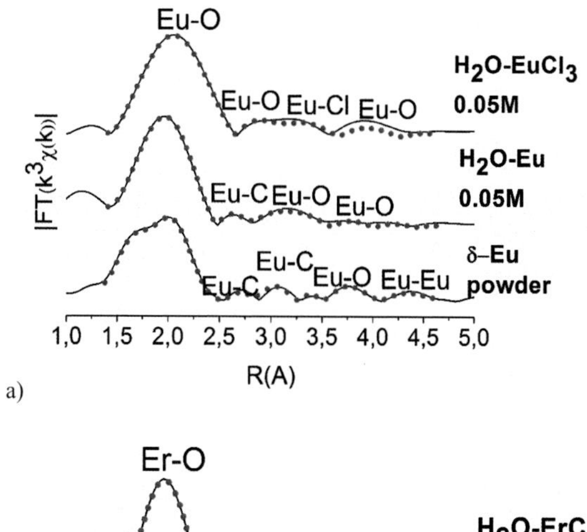

a)

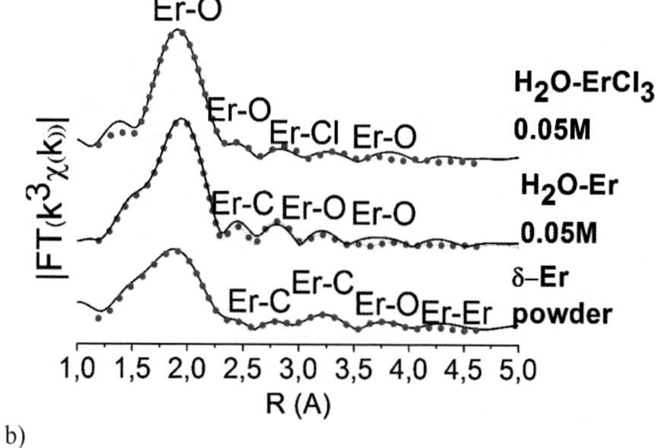

b)

Figure V.4-1 EXAFS spectra of **δ-Ln-1.1**, 0.05 M H_2O-Eu and 0.05 M H_2O-$EuCl_3$: a) Ln=Eu, b) Ln=Er, solid lines indicate experimental data, dashed lines indicate best fits.

To study the local structures of aqueous and methanol solutions (**Ln-1.1** in H_2O (0.05 M) and **Ln-1.1** in methanol (0.1 M), respectively*) their EXAFS spectra were compared with those of the corresponding powder samples and of the **$LnCl_3$** aqueous solutions (**$LnCl_3$** in H_2O), where the lanthanide is present as an aqua-ion (Table V.4-3, Figure I.2-3). The comparison between solution and powder data allows not only estimating the number of linked ligands, but also evaluating their coordination modes. Thus, in EXAFS spectra of complexes in aqueous solutions, the presence and the quantity of the Ln···C scattering paths should characterize the number and type of carboxylic groups and therefore, the linked ligands. The FTs of the experimental EXAFS spectra and fitting curves for **δ-Ln-1.1**, **Ln-1.1**, **Ln** and **$LnCl_3$** in H_2O are shown in Figure V.4-1 for Ln = Eu, Er. Table V.4-1 contains the structural parameters obtained for Eu complexes from the EXAFS optimization. The EXAFS spectra of **$LnCl_3$** in H_2O reveal two coordination spheres with the radii of 2.45 Å (oxygen atoms of the water molecules) and 3.28 Å (chloride atoms). It is in good agreement with the results obtained by MD simulation[277], which also shows that the second shell of water molecules appears at adistance of more than 4.50 Å.

Unlike **$EuCl_3$**, in H_2O the EXAFS spectrum of **Eu-1.1** in H_2O shows the presence of the coordination sphere with the radius of 2.80 Å, corresponding to two κ^2 carboxylic carbon atoms. Therefore, the EXAFS spectrum of **Eu-1.1** in H_2O shows a reduction of the number of carboxylate ligands in the coordination sphere when compared to **δ-Eu-1.1**, since only two ligands are coordinated and Eu^{3+} coordination sphere is further filled with water molecules. Similar results were obtained for **Gd-1.1**, **Tb-1.1** and **Dy-1.1** in H_2O (Table V.4-3). In contrast to them, in **Er-1.1** in H_2O three κ^1 carboxylic carbon atoms are present, according to the EXAFS spectrum (Figure V.4-1b), similarly to the powder form. It means that erbium pentafluorobenzoate does not dissociate in water at the given concentration. The same behaviour was found for **Ho-1.1**, **Tm-1.1** and **Yb-1.1** in H_2O, while the second coordination sphere of **Lu** in H_2O also contains two κ^2- and one κ^1-carbon atoms as in **Lu-1.1** powder (Table V.4-3). Therefore, aqueous solutions of the complexes with lanthanides from Eu to Dy partially dissociate, while those with lanthanides from Ho to Lu do not undergo dissociation at the given concentration (0.05 M). Unlike aqueous solutions, complex solutions in methanol **Ln-1.1** in CH_3OH do not undergo any dissociation for the whole set of lanthanides which is in line with the ^{19}F NMR data. The coordination number of the central ion remains CN = 5+3, corresponding to $Ln(pfb)_3(CH_3OH)_3$ coordination.

* The values of the solution concentrations suitable for EXAFS measurements were chosen with regard to method requirements.

Table V.4-2 EXAFS data of **δ-Ln-1.1** and **Ln-1.1**

Shell number	Atom type	**δ-Eu-1.1**		**δ-Gd-1.1**		**δ-Tb-1.1**		**δ-Dy-1.1**		**δ-Ho-1.1**	
		N	R, Å	N	R, Å	N	R, Å	N	R, Å	N	R, Å
1	O	9	2.38	9	2.40	9	2.39	9	2.38	9	2.36
2	κ^2 carbox.C	2	2.68	2	2.73	2	2.70	2	2.68	2	2.70
	κ^1 carbox.C	4	3.43	4	3.55	4	3.44	4	3.47	4	3.61
3	O	2	3.75	2	3.76	2	3.93	2	3.73	2	3.71
4	Ln	2	4.10	2	4.07	2	4.04	2	3.84	2	3.99
R-factor, %		1.56		0.92		0.47		0.23		0.36	
k-range, $Å^{-1}$		2.0÷12.0		2.0÷11.5		2.0÷12.0		2.0÷12.0		2.0÷12.0	
R-range, Å		1.6÷4.9		1.6÷4.9		1.6÷4.9		1.6÷4.9		1.6÷4.9	

Table V.4-2 (cntd)

Shell number	Atom type	**δ-Er-1.1**		**Tm-1.1**		**Yb-1.1**		**Lu-1.1**	
		N	R, Å	N	R, Å	N	R, Å	N	R, Å
1	O	9	2.34	8	2.29	8	2.25	8	2.25
2	κ^2 carbox.C	2	2.67					2	2.76
	κ^1 carbox.C	4	3.56	3	3.66	3	3.50	1	3.41
3	O	2	3.89	5	3.86	5	3.99	5	3.82
4	Ln	2	3.94						
R-factor, %		1.11		1.05		0.94		0.42	
k-range, $Å^{-1}$		2.0÷11.0		2.0÷11.3		2.0÷13.0		2.0÷11.0	
R-range, Å		1.6÷4.9		1.5÷5.5		1.8÷3.7		1.6÷3.9	

Table V.4-3 EXAFS data of **Ln-1.1** in H_2O

Shell number	Atom type	**Eu-1.1**		**Gd-1.1**		**Tb-1.1**		**Dy-1.1**		**Ho-1.1**	
		N	R, Å	N	R, Å	N	R, Å	N	R, Å	N	R, Å
1	O	10	2.46	12	2.43	11	2.43	11	2.39	11	2.45
2	κ^2 carbox.C	2	2.76	2	2.96	2	2.86	2	2.77		
	κ^1 carbox.C									3	3.54
3	F	1	3.59	3	3.88	2	3.98	2	3.48		
	O									1	3.70
R-factor, %		0.12		0.39		0.39		0.79		1.34	
k-range, $Å^{-1}$		2.0÷10.0		2.0÷9.0		2.0÷8.5		2.0÷10.5		2.0÷12.0	
R-range, Å		1.6÷3.8		1.6÷3.5		1.7÷4.0		1.6÷4.0		1.7÷4.0	

Table V.4-3 (cntd)

Shell number	Atom type	**Er-1.1**		**Tm-1.1**		**Yb-1.1**		**Lu-1.1**	
		N	R, Å	N	R, Å	N	R, Å	N	R, Å
1	O	11	2.38	12	2.36	10	2.36	10	2.32
2	κ^2 carbox.C							2	2.81
	κ^1 carbox.C	3	3.36	3	3.45	3	3.51	3	3.54
3	F			2	3.90	3	3.89		
	O	1	3.91					2	3.69
R-factor, %		1.30		0.23		0.78		0.55	
k-range, $Å^{-1}$		2.0÷11.2		2.0÷11.0		2.0÷10.0		2.0÷12.0	
R-range, Å		1.5÷4.0		1.5÷4.2		1.7÷4.0		1.6÷3.7	

Table V.4-4 EXAFS data of **Ln-1.1** in **CH_3OH**

Atom	N	R, Å			
		Eu-1.1	**Gd-1.1**	**Ho-1.1**	**Tm-1.1**
O	8	2.49	2.45	2.40	2.29
C	6	3.68	3.69	3.56	3.42
O	1	3.90	3.80	3.88	3.76
C	2	4.06	3.82	4.15	4.02
R-factor, %		0.42	0.31	0.65	0.81
k-range, $Å^{-1}$		2.0÷12.0	2.0÷11.5	2.0÷12.0	2.0÷12.0
R-range, Å		1.8÷4.2	1.8÷4.2	1.8÷4.2	1.8÷4.2

V.5 Analysis of single-crystal data of lanthanide pentafluorobenzoates

Single-crystal X-ray diffraction investigations of the Ln pentafluorobenzoate series revealed that the whole studied row of complexes can be divided into two groups: from the beginning of the lanthanide row (**α-Ln-1.1-sc** where Ln = Nd, Sm, Eu, Gd, Tb, Tb/Eu) and from its end (**γ-Ln-1.1-sc** where Ln = Tb, Dy, Ho, Er, Tm). The coordination environment of the lanthanide ion and the hydrate composition were found to be the main differences (Table III.1-1).

Prismatic crystals of **γ-Ln-sc** are isostructural with monomeric species in the unit cell corresponding to the $[Ln(\kappa^1\text{-}RCOO)_3(H_2O)_5](H_2O)$ composition ($R = C_6F_5^-$, Table V.3-1, Figure V.5-1a); in this composition the lanthanide contraction influences their structure only negligible. The metal ion polyhedron can be ascribed to a distorted tetragonal antiprism (Figure V.5-1b) which is composed of three $(pfb)^-$ anions possessing a monodentate coordination mode (κ^1 type) and five oxygen atoms of coordinated water molecules. The maximum distortion of the antiprism bases is meaningfully different and does not depend on the lanthanide radii: values of O(5)O(5W)O(2W)O(3W) and O(3)O(1W)O(4W)O(1) dihedral angles are the same within the standard uncertainty (0.03°) and on average equal to 30.2° and 10.8°, respectively. At the same time, the angle between the mean-square plane defined by atoms of the antiprism's bases is less than 9° for the whole group of complexes. The benzene rings of three pfb^- anions in the molecule are nearly parallel: the angles between corresponding mean-square planes are not more than 6.2°. Such a distance between two rings is small enough (C(4)...C(10) ~ 3.34 Å) to suppose the presence of an intermolecular stacking interaction between the aromatic fragments. The latter together with intramolecular hydrogen bonds between coordinated water molecules and donor atoms of pfb^- ligands (O(3W)-H(3WB)...F(6) O...F 2.928(4) – 2.933(3) and O(1W)-H(1WA)...O(4) O...O 2.668(4) – 2.655(3) Å correspondingly for **γ-Tb-1.1-sc** – **γ-Tm-1.1-sc**) can serve as additional factors stabilizing the complex structure in solids (Table V.3-1).

Similarly, non-covalent interactions were also observed analyzing crystal packing of **γ-Ln-1.1-sc** complexes. Intermolecular stacking interaction of nearly the same strength (C(7)...C(20) ~ 3.37 Å) and hydrogen bonds between complex molecules (O...O 2.734(5), 2.793(3), 2.904(3) and 2.732(3), 2.803(3), 2.913(3) Å for **γ-Tb-1.1-sc** and **γ-Tm-1.1-sc**, respectively) bind the units into chains (Figure V.5-2a), while hydrogen bonds between complex molecules and solvate water molecules and between complex molecules themselves bind these chains into layers (Figure V.5-2b, O...O 2.723(2)–3.188(5) and 2.737(3)–3.187(4) Å for **γ-Tb-1.1-sc** and **γ-Tm-1.1-sc**, respectively). Also, the F...F contacts (approximately 2.66 – 2.88 Å in all cases) between the layers are present in crystals; corresponding hydrophobic interactions are in turn responsible for the 3D structure formation.

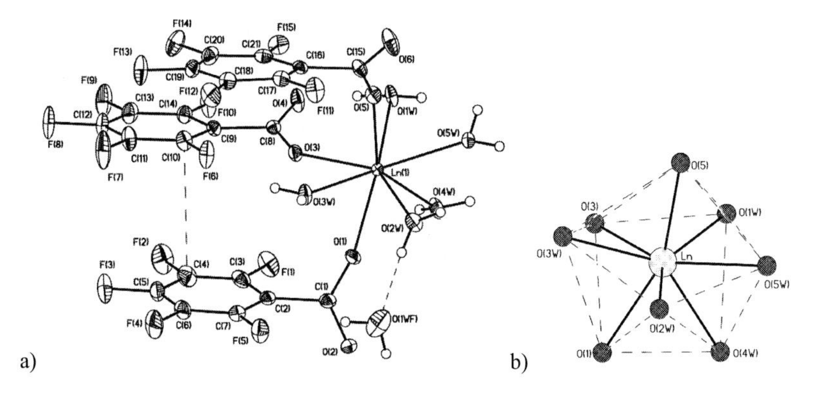

Figure V.5-1 a) General view of **γ-Ln-1.1-sc** complexes in representation of non-hydrogen atoms by probability ellipsoids of atomic displacement (p = 50%). Dash lines show the intramolecular stacking interaction and the hydrogen bonds between coordinated and solvate water molecules; b) the lanthanide ion polyhedron in **γ-Ln-1.1-sc** complexes.

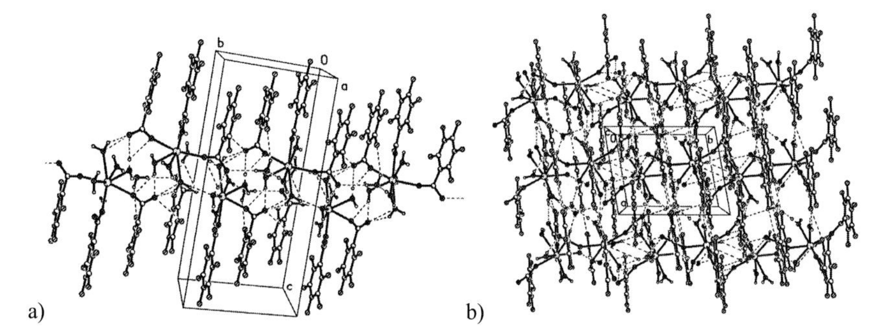

Figure V.5-2 a) Fragment of a hydrogen-bonded chain of **γ-Ln-1.1-sc** complexes in the crystal along the *b* axis; b) a fragment of hydrophobic layers in crystals of **γ-Ln-1.1-sc** complexes.

The increase of the lanthanide ionic radii leads to a formation of two types of crystals with different shapes and structures. Prismatic triclinic crystals of **α-Ln-1.1-sc** were obtained for the whole

Nd – Gd group (**α-Nd-1.1-sc, α-Sm-1.1-sc, α-Eu-1.1-sc, α-Gd-1.1-sc, α-(TbEu)-1.1-sc**) and were found to be isostructural with neodymium and terbium complexes. At the same time, needle monoclinic crystals of **β-Ln-1.1-sc** were successfully isolated only for Sm and Gd, as well as for heterometallic (TbEu) system (**β-Sm-1.1-sc, β-Gd-1.1-sc, β-(TbEu)-1.1-sc**); both types of crystals were found to be isostructural and the influence of the lanthanide ion contraction was again meaningless. In two crystal types, despite the concordance of the coordination number of a metal atom and the similarity of the metal polyhedron type which can be regarded as a distorted monocapped tetragonal antiprism, the nuclearity of the single molecules differ significantly.

The centrosymmetric dimer corresponding to the $[Ln_2(\mu_2\text{-}\kappa^1{:}\kappa^2\text{-}RCOO)_2(\kappa^1\text{-}RCOO)_4(H_2O)_8](H_2O)_2$ composition was found in the cell of **α-Ln-1.1-sc** crystals (Figure VIII.1-3a): each metal cation is surrounded by only four water molecules, two terminal $(pfb)^-$ anions and two semi-bridging $(pfb)^-$ ligands possessing different (monodentate and bidentate) coordination modes. The O(2) atom of the $\mu_2\text{-}\kappa^1{:}\kappa^2\text{–}C_6F_5COO^-$ ligand plays the cap role as it was expected from the CSD search (Figure V.5-3b). The distortion of the antiprism bases in this case are also similar for all studied complexes: the O(1W)O(2C)O(2W)O(1) and O(1A)O(3W)O(1B)O(4W) dihedral angles are on average 19.3° and 11.4° and retain these values to less than 1° upon lanthanide variation; the angle between base mean-square planes is not more than 5°. The angles between aromatic cycles of pfb^- anions are also small (less than 6°), however, shortened interatomic distances corresponding to intramolecular staking interactions were not observed. On the other hand, as in **γ-Ln-1.1-sc** series, intramolecular hydrogen bonds with close strength were observed for **α-Ln-1.1-sc** complexes. However, in this case their number is larger and they are formed not only by coordinated water molecules and benzoate ligands but also by coordinated water molecules themselves (O(2W)-H(2WB)-O(5WA), O…O 2.697(3) – 2.692(3) Å from **α-Nd-1.1-sc** to **α-Gd-1.1-sc**) that stabilizes the dimeric structure of the complexes.

Crystal packing analysis of **α-Ln-1.1-sc** complexes revealed the same types of intermolecular interaction as in **γ-Ln-1.1-sc** series: hydrogen bonds (O…O 2.808(3), 2.850(3) and 2.814(3), 2.867(3) Å for **α-Nd-1.1-sc** and **α-Gd-1.1-sc** correspondingly) and stacking interaction between complex molecules leads to chain formation (Figure V.5-4a), chains are stuck together by hydrogen bonds between complex molecules and solvate water molecules and between complex molecules themselves to form layers (Figure VIII.1-4b, O…O 2.667(3), 2.760(3), 2.830(3) and 2.661(3), 2.753(3), 2.870(3) Å for **α-Nd-1.1-sc** and **α-Gd-1.1-sc** correspondingly), the latter are bind by F…F contacts (approximately 2.83–2.94 Å in all cases). Nevertheless, two main differences between **α-Ln-1.1-sc** and **γ-Ln-1.1-sc** crystal packing were observed: there are more shortened π…π contacts in **α-Ln-1.1-sc** (C(6B)…C(3A) and C(3B)…C(6) distances varied from

~ 3.30 and ~ 3.22 Å in **α-Nd-1.1-sc** to ~ 3.29 and ~ 3.20 Å in **α-Gd-sc**). Besides the role of the F...F interactions, the three-dimensional structure is achieved by F...π interactions (F...C ~ 3.15 Å).

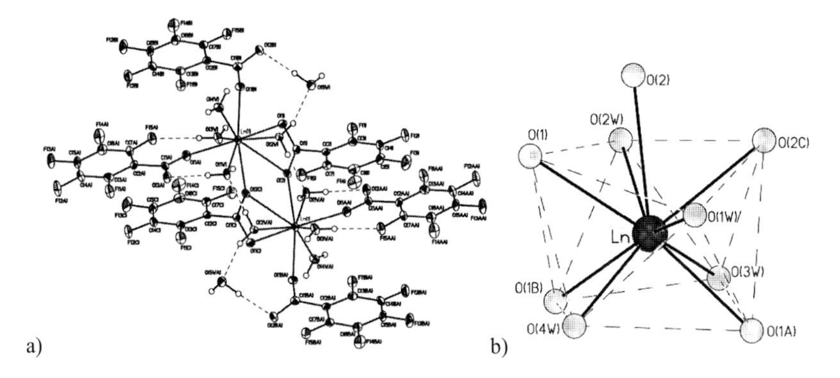

Figure V.5-3 a) General view of **α-Ln-1.1-sc** complexes in representation of non-hydrogen atoms by probability ellipsoids of atomic displacement (p = 50%). Dashed lines show the hydrogen bonds within the formula unit; b) lanthanide ion coordination polyhedron in **α-Ln-1.1-sc** complexes.

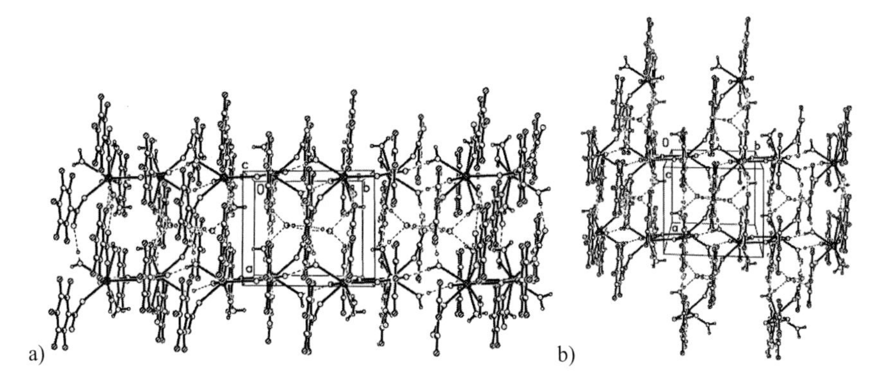

Figure V.5-4 a) Fragment of a chain of **α-Ln-1.1-sc** complexes in the crystal along the *b* axis; b) fragment of hydrogen-bonded layers in crystals of **γ-Ln-1.1-sc** complexes.

At the same time, the monomeric $Ln(\kappa^2\text{-RCOO})(\kappa^1\text{-RCOO})(H_2O)_6]^+(RCOO)^-$ species with the outer-sphere $(pfb)^-$ anion are presented in the unit cell of **β-Ln-1.1-sc** crystals (Figure VIII.1-5a). The coordination environment of the lanthanide ion is composed by six water molecules and only two $(pfb)^-$ ligands possessing monodentate and bidentate coordination modes. The antiprism distortion is negligible in **β-Ln-1.1-sc**: maximum dihedral angles are not more than 6°, the angle between base mean-square planes is less than 8°.

It should be noted that the low quality of crystals did not allow to unambiguously localize hydrogen atoms in the **β-Ln-1.1-sc** structures. Corresponding atomic positions were presupposed, using geometrical criteria. From this point of view, the short C(1F)-O(1F) and C(1F)-O(2F) bond lengths (1.242(4) and 1.260(4) Å) were used as an indirect but clear indication that the non-coordinated organic moiety is of an anionic nature. Even using such incomplete data on atomic positions and accounting for the interionic staking interaction between benzene rings (C(5A)...C(4F) ~ 3.26 Å), the hydrogen bonds should be considered as the most probable force bounding the counterions. Both, the O(1F) and O(2F) oxygen atoms of non-coordinated anion, form contacts with oxygen atoms of water molecules coordinated to the metal cation: only two of them are short enough (O(2F)...O(4W) 2.683(3) Å and 2.685(2) Å, O(1F)...O(6W) 2.717(3) Å and 2.722(2) Å, in **β-Sm-1.1-sc** and **β-Gd-1.1-sc**, respectively) while O...O separation for the other four contacts (2.800(3) – 2.995(3) Å and 2.800(3) – 2.988(3) Å for **β-Sm-1.1-sc** and **β-Gd-1.1-sc**, respectively) can serve as an indication on a possible disorder of hydrogen atoms.

The crystal packing motive in **β-Ln-1.1-sc** is similar to those observed for **α-Ln-1.1-sc** and **γ-Ln-1.1-sc** series with the only exception of the stacking interaction, which bound the complex moiety and the non-coordinated (pfb)$^-$ anion into the contact ion pair, being unique in the independent part of the unit cell. In turn, cation...cation and anion...anion hydrogen bonding leads to the formation of layers (Figure V.5-5b) which are stuck together by F...F interactions (~ 2.73 – 2.89 Å).

Based on the similarity of the crystal packing for all three series of single crystals one can exclude supramolecular forces as a factor influencing particular crystal structures of aqua complexes of lanthanide pentafluorobenzoates. In other words, the intramolecular factors govern structural diversity and they need to be divided and also analyzed. From this point of view, the lanthanide ion radii variation is not exclusive: since **α-Ln-1.1-sc** and **β-Ln-1.1-sc** series can both crystallize in an equilibrium between different forms of a complexes in water solutions should be accounted for that corresponds to a competition between (pfb)$^-$ ligand and water molecules for the metal atom and makes particularly interesting structure peculiarities and dynamics investigations in more dilute medium such as solution or isolated state. From the other hand some demonstrations of structure variations and an assumption of the isomerization process mechanism can be done analyzing more thoroughly trends of metal-ligand bond lengths within the whole row of the investigated single crystals.

In general, the Ln–O bond length variation is in line with the lanthanide ion contraction, the changes in the hydrate composition and the coordination mode of the (pfb)$^-$ ligands. For instance, the Ln–O distances are the shortest ones at the end of the lanthanide row (in **γ-Ln-1.1-sc** series) and with some exceptions the largest in complexes with a high content of water molecules within the first coordination sphere and a (pfb)$^-$ ligand possessing a κ^2 chelate coordination mode (**β-Ln-1.1-sc** series).

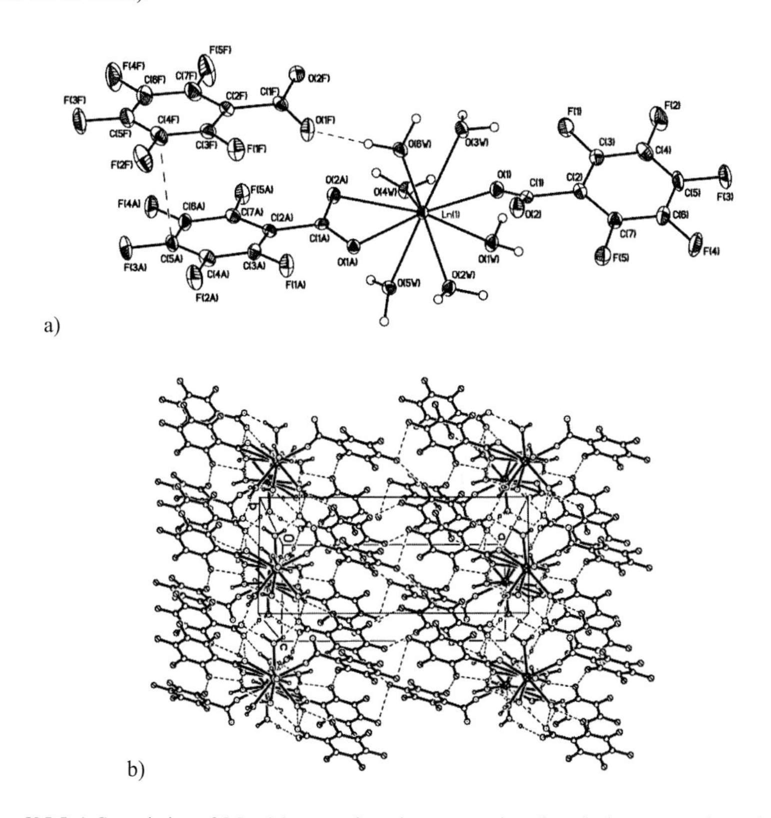

Figure V.5-5 a) General view of **β-Ln-1.1-sc** complexes in representation of non-hydrogen atoms by probability ellipsoids of atomic displacement (p = 50%). Dashed lines show the interionic stacking interaction and one of the interionic hydrogen bonds; b) fragment of hydrophobic layers in crystals of **β-Ln-1.1-sc** complexes.

At the same time, in dimer complexes, Ln–O bond lengths for different lanthanides change in a meaningfully different way. Thus, from **α-Nd-1.1-sc** to **α-Gd-1.1-sc**, the most pronounced shortening is observed for the Ln(1)-O(2C) bond with the oxygen atom of the semi-bridging ligand possessing a κ^1 coordination mode to the metal atom (on ~ 0.08 Å). This implies that its strength become comparable with Ln-O(κ^2-RCOO) bonds in **β-Ln-1.1-sc** complexes. The significant decrease is also observed for Ln–O bonds with terminal κ^1 (pfb)$^-$ ligands (on ~ 0.05 Å): their values become close to those for dysprosium complexes. On contrary, the Ln(1)–O(2) bond with the one of oxygen atoms of μ^2-κ^1:κ^2 ligand shortens only on 0.01 Å. Taking the close energy of the first coordination sphere stabilization into account, one can conclude that upon to lanthanide ion contraction the semi-bridging coordination mode of the (pfb)$^-$ ligand and the corresponding dimer structure becomes relatively less stable: as the strength of the Ln(1)–O(κ^1-RCOO) bonds becomes similar to that at the end of the row, the Ln(1)–O(2) bond does not become stronger. The Ln(1)–O(2C) bond indeed becomes much stronger than the Ln(1)–O(1) bond and together with small water molecule "saturates" the first coordination sphere better than the chelate bidentate structure. However, if the Ln(1)–O(2C) bond is not strong enough, the unfavorableness of the semi-bridging coordination mode can give structure with the outer-sphere anion if only the number of water molecule in the second coordination sphere is enough to take the vacant places.

Additional experimental details

For crystals of **α-Sm-1.1-sc** and **β-Sm-1.1-sc** Cu-Kα-radiation was used, while for all other crystals a Mo-Kα source was used. For the complexes of heavier Ln ions (Dy, Ho, Er, Tm), hydrogen atoms were found from difference FOURIER synthesis of the electron density, while for earlier Ln ions, H atoms were calculated, using a H-bonding network analysis both, in monomers and in dimers, due to the low quality of the crystals. For heterometallic complexes, coordinates and anisotropic displacement parameters of metal atoms were constrained on each other while their populations were fixed and equal to 0.5. Crystal of **α-Eu-1.1-sc**, **α-TbEu-1.1-sc** and **γ-Er-1.1-sc** were twins: indexing was performed, using CELL_NOW routine, structures were solved and initially refined, using only one domain, HKLF5 refinement was successful only for the **α-Eu-1.1-sc** crystal with populations of a second domain equal to 0.28. The model of the **γ-Dy-1.1-sc** complex has two, rather high, electron peaks near the metal atom (~ 3e^-) most likely due to the high absorption of the crystals. However, these results were achieved after several attempts to account for absorption by different models. CCDC numbers 1061319-1061332, 1061651 contain all crystallographic information and can be obtained free of charge *via* www.ccdc.ac.uk.

The **β′-Lu-1.1-sc** structure demonstrates a high (5.1 e^-) difference FOURIER density peak away from the Lu atom. The environment of the peak is not consistent with any metal polyhedron and its position corresponds to coordinates (X_{Lu1}-0.5, Y_{Lu1}-0.5, Z_{Lu1}). Thus, we consider this peak as a case of unresolved twinning or full-molecule disorder; it would amount to ≈ 2% of the Lu atom if the disorder could be modeled. To check whether every charge in the structure is accounted for, all hydrogen atoms were found in difference FOURIER maps and refined isotropically, without restraints on U_{iso} values and only with DFIX restraints on the O-H distances. All H atoms found this way formed hydrogen bonds and resulting U_{iso} values and H–O–H angles were in the acceptable range, indicating the correctness of the hydrogen positions. One of the pentafluorophenyl groups was disordered over two positions with occupancies of 0.542(10) and 0.458(10) and refined anisotropically, using SIMU restraints.

VI. List of abbreviations

AFM	atomic force microscopy
ATR	attenuated total reflection
BOC	*tert*-butyloxycarbonyl
BP	byproduct
BPhen	bathophenanthroline
Bpyr	bipyridine
^{13}C NMR	carbon nuclear magnetic resonance
cc pVDZ	correlation-consistent polarized valence-only Double-Zeta sets
CCDC	Cambridge Crystallographic Data Centre
CFP	cyan cluorescent protein
CHCA	α-cyano-4-hydroxycinnamic acid
CN	coordination number
CSD	Cambridge Crystallography Database
CuAAC	copper(I)-catalyzed alkyne-azide cycloaddition
Cyo	cyclooctine
DBU	1,8-diazabicyclo[5.4.0]undec-7-ene
DCM	dichloromethane
DEAD	diethyl azodicarboxylate
DELFIA	dissociation enhanced lanthanide fluoroimmuno assay
DEPT	distortionless enhancement by polarization transfer
DFT	density funtional theory
DHB	2,5 dihydroxybenzoic
DIAD	diisopropyl azodicarboxylate
DIC	*N,N'*-diisopropylcarbodiimide
DIPEA	*N,N*-diisopropylethylamine
DMAP	4-dimethylaminopyridine
DMEM	dulbecco's modified eagle's medium
DMF	dimethylformamide
DMSO	dimethyl sulfoxide
DOTA	1,4,7,10- tetracarboxymethyl-1,4,7,10-tetraazacyclododecane
DRS	diffuse reflection spectra
DSC	differential scanning calorimetry
DTPA	diethylenetriaminepentaacetic acid
EA	elemental analysis
EBL	electron blocking layer
ED	electric dipole

EDA	ethylene diamine	**FTIR**	micro fourier transform intefferometer
EI MS	electron ionization - mass spectrometry	**FWHM**	full width at half maximum
EL	electroluminescence	**GC MS**	gas chromatography - mass spectrometry
EML	emission layer	**GFP**	green fluorescent protein
EQE	electroluminescence quantum efficiency	**HATU**	1-[bis(dimethylamino)methylene]-1H-1,2,3-triazolo[4,5-b]pyridinium 3-oxid hexafluorophosphate, hexafluorophosphate azabenzotriazole tetramethyl uronium
ESI-MS	electrospray ionisation - mass spectrometry	**HBL**	hole blocking layer
ET	energy transfer	**HeLa cells**	human cervix carcinoma cells
ET_B	back energy transfer	**HFIP**	hexafluoroisopropanol
ETL	electron transport layer	**^{1}H NMR**	proton nuclear magnetic resonance
EtOAc	ethyl acetate	**HPLC**	high-performance liquid chromatography
EXAFS	extended X-ray absorption fine structure spectroscopy	**Hpfb**	pentafluorobenzoic acid
^{19}F NMR	fluorine nuclear magnetic resonance	**HR-MS**	high resolution-mass spectra
FAB MS	fast atom bombardment - mass spectrometry	**HTL**	hole transport layer
FCS	fetal calf serum	**HTRF**	homogeneous time-resolved fluorescence
FG	functional group	**HUW**	half uncertainty window
Fmoc	fluorenylmethyloxycarbonyl	**IAC**	2-methoxyisophtalate
FRET	Förster resonance energy transfer	**IAM**	2-hydroxyisophthalamide
FT	fourier-transformed	**ID**	identification number

ILCT	intraligand charge-transfer	**NIR**	near infra-red
IR	infra-red	**NMR**	nuclear magnetic resonance
ISC	inter-system crossing	**OLED**	organic light-emitting diode
ITO	indium tin oxide	**ORTEP**	drawing software
J-SD	snub disphenoid	**ovrn**	over night
J-TCTPR	distorted tricapped trigonal prism	**PCS**	pseudocontact shift
LCC	lanthanide coordination compounds	**pDMF**	peptidegrade dimethylformamide
LC-MS	liquid chromatography–mass spectrometry	**PEDOT:PSS**	poly(3,4-ethylenedioxythiophene) polystyrene sulfonate
LDA	lithium diisopropylamide	**$(pfb)^-$**	pentafluorobenzoate
LMCT	ligand-to-metal charge transfer	**PG**	protecting group
Ln	lanthanide	**Phen**	1,10-phenanthroline
LUMO	lowest unoccupied molecular orbital	**PHOLED**	phosphorescent organic light-emitting diode
MALDI-TOF-MS	matrix-assisted laser desorption/ionisation - time of flight	**PLQY**	photoluminescence quantum yield
MCT	mercury-cadmium-telluride	**PMA**	photonic multichannel analyzer
MD	molecular dynamics or magnetic dipole	**PT**	pentaethylene tetraamine
MRI	magnetic resonance imaging	**PVK**	poly(9-vinylcarbazole)
MTT	3-(4,5-dimethylthiazol-2-Yl)-2,5-diphenyltetrazolium bromide	**PW PBE**	plane-wave Perdew-Burke-Ernzerhof
NBS	*N*-bromosuccinimide	**PXRD**	powder X-ray diffraction
NHS	*N*-hydroxysuccinimide	**PyBOP**	(benzotriazol-1-yl-oxytripyrrolidinophosphonium hexafluorophosphate

Pyr	pyridine	**TG/TGA**	thermogravimetry
QY	quantum yield	**TGD**	time-gated detection
RP-HPLC	reverse phase high-performance liquid chromatogram	**THAP**	2,4,6 trihydroxyacetophenone
rt	room temperature	**THF**	tetrahydrofuran
SAR	structure-activity relationship	**TIPS**	triisopropylsilyl ether
SC-XRD	single-crystal X-ray diffraction	**TLC**	thin layer chromatography
SM	starting material	**TMS**	trimethylsilyl
SPAAC	strain-promoted azide-alkyne cycloddition	**TPBi**	2,2′,2"-(1,3,5-benzinetriyl)-tris(1-phenyl-1-H-benzimidazole)
SPR	structure-property relationship	**TPP**	triphenylphosphine
SVD	singular value decomposition	**TRF**	time-resolved fluorometry
TAZ	3-(biphenyl-4-yl)-5-(4-tert-butylphenyl)-4-phenyl-4H-1,2,4-triazole	**trFTIR**	time-resolved micro fourier transform intefferometer
TBS	tris-buffered saline	**UV**	ultraviolet
TC	ternary complexes	**UV/Vis**	ultraviolet/visible
TCSPC	time correlated single photon counting	**VASP**	vienna *ab initio* simulation package
TD-DFT	time dependent density funtional theory	**XANES**	X-ray absorption near-edge structure
TEM	transmission electron microscopy	**XAS**	X ray absorption spectroscopy
Terpy	terpyridine	**XRD**	X-ray diffraction
TFA	rifluoroacetic acid	**YFP**	yellow fluorescent protein

VII. References

[1] Thibon, A., Pierre, V. C. Principles of responsive lanthanide-based luminescent probes for cellular imaging. *Anal. Bioanal. Chem.* **394,** 107–20 (2009).

[2] Cockerill, A. F., Davies, G. L. O., Harden, R. C., Rackham, D. M. Lanthanide shift reagents for nuclear magnetic resonance spectroscopy. *Chem. Rev.* **73,** 553–588 (1973).

[3] Martin, L. J. *et al.* Double-Lanthanide-Binding Tags: Design, Photophysical Properties and NMR Applications. *J. Am. Chem. Soc* **129,** 7106–7113 (2007).

[4] Silvaggi, N. R., Martin, L. J., Schwalbe, H., Imperiali, B., Allen, K. N. Double-Lanthanide-Binding Tags for Macromolecular Crystallographic Structure Determination. *J. Am. Chem. Soc.* **129,** 7114–7120 (2007).

[5] And, K. N. R., Pierre, V. C. Next Generation, High Relaxivity Gadolinium MRI Agents. *Bioconjugate Chem.* **16,** 3–8 (2005).

[6] Rashid, H. U., Yu, K., Zhou, J. Lanthanide(III) chelates as MRI contrast agents: A brief description. *J. Struct. Chem.* **54,** 223–249 (2013).

[7] Allen, M. J., Meade, T. J. Synthesis and visualization of a membrane-permeable MRI contrast agent. *JBIC J. Biol. Inorg. Chem.* **8,** 746–750 (2003).

[8] And, W. J. E., Davis, B. L. Chemistry of Tris(pentamethylcyclopentadienyl) f-Element Complexes, (C5Me5)3M. *Chem. Rev.* **102,** 2119–2136 (2002).

[9] Bürgstein, M. R., Berberich, H., Roesky, P. W. Homoleptic Lanthanide Amides as Homogeneous Catalysts for Alkyne Hydroamination and the Tishchenko Reaction. *Chem. - A Eur. J.* **7,** 3078–3085 (2001).

[10] David J. Edmonds, Derek Johnston, A., Procter, D. J. Samarium(II)-Iodide-Mediated Cyclizations in Natural Product Synthesis. *Chem. Rev.* **104,** 3371–3404 (2004).

[11] Eliseeva, S. V., Bünzli, J.-C. G. Lanthanide luminescence for functional materials and bio-sciences. *Chem. Soc. Rev.* **39,** 189–227 (2010).

[12] Bünzli, J.-C. G., Piguet, C. Taking advantage of luminescent lanthanide ions. *Chem. Soc. Rev.* **34,** 1048–77 (2005).

[13] Judd, B. R. Optical absorption intensities of rare-earth ions. *Phys. Rev.* **127,** 750–761 (1962).

[14] Ofelt, G. Intensities of crystal spectra of rare earth ions. *J. Chem. Phys.* **37,** 511–520 (1962).

[15] Latva, M. *et al.* Correlation between the lowest triplet state energy level of the ligand and lanthanide(III) luminescence quantum yield. *J. Lumin.* **75,** 149–169 (1997).

[16] Monteiro, J. H. S. K., De Bettencourt-Dias, A., Mazali, I. O., Sigoli, F. A. The effect of 4-halogenobenzoate ligands on luminescent and structural properties of lanthanide complexes: Experimental and theoretical approaches. *New J. Chem.* **39,** 1883–1891 (2015).

[17] Pasatoiu, T. D., Madalan, A. M., Kumke, M. U., Tiseanu, C., Andruh, M. Temperature Switch of LMCT Role: From Quenching to Sensitization of Europium Emission in a Zn^{II} –Eu^{III} Binuclear Complex. *Inorg. Chem.* **49,** 2310–2315 (2010).

[18] Eliseeva, S. V & Bünzli, J.-C. G. Lanthanide luminescence for functional materials and bio-sciences. *Chem. Soc. Rev.* **39,** 189–227 (2010).

[19] Roh, S. G. *et al.* Er(III)-chelated prototype complexes based on benzoate and pentafluorobenzoate ligands: Synthesis and key parameters for near IR emission

enhancement. *Bull. Korean Chem. Soc.* **25,** 1503–1507 (2004).

[20] Ye, H.-Q. *et al.* Effect of Fluorination on the Radiative Properties of Er^{3+} Organic Complexes: An Opto-Structural Correlation Study. *J. Phys. Chem. C* **117,** 23970–23975 (2013).

[21] Doffek, C., Seitz, M. The Radiative Lifetime in Near-IR-Luminescent Ytterbium Cryptates: The Key to Extremely High Quantum Yields. *Angew. Chem. Int. Ed. Engl.* **54,** 9719–21 (2015).

[22] Scholten, J. *et al.* Anomalous Reversal of C-H and C–D Quenching Efficiencies in Luminescent Praseodymium Cryptates. *J. Am. Chem. Soc.* **134,** 13915–13917 (2012).

[23] Doffek, C. *et al.* Understanding the Quenching Effects of Aromatic C–H- and C–D-Oscillators in Near-IR Lanthanoid Luminescence. *J. Am. Chem. Soc.* **134,** 16413–16423 (2012).

[24] Doffek, C., Wahsner, J., Kreidt, E. & Seitz, M. Breakdown of the Energy Gap Law in Molecular Lanthanoid Luminescence: The Smallest Energy Gap Is Not Universally Relevant for Nonradiative Deactivation. *Inorg. Chem.* **53,** 3263–3265 (2014).

[25] Butler, S. J., Lamarque, L., Pal, R., Parker, D. EuroTracker dyes: highly emissive europium complexes as alternative organelle stains for live cell imaging. *Chem. Sci.* **5,** 1750 (2014).

[26] Faustino, W. M., Malta, O. L., de Sá, G. F. Intramolecular energy transfer through charge transfer state in lanthanide compounds: A theoretical approach. *J. Chem. Phys.* **122,** 054109 (2005).

[27] Binnemans, K. Lanthanide-Based Luminescent Hybrid Materials. *Chem. Rev.* **109,** 4283–4374 (2009).

[28] Főrster, T. 10th Spiers Memorial Lecture. Transfer mechanisms of electronic excitation. *Discuss. Faraday Soc.* **27,** 7–17 (1959).

[29] Dexter, D. L. A Theory of Sensitized Luminescence in Solids. *J. Chem. Phys.* **21,** 836–850 (1953).

[30] Ward, M. D. Mechanisms of sensitization of lanthanide(III)-based luminescence in transition metal/lanthanide and anthracene/lanthanide dyads. *Coord. Chem. Rev.* **254,** 2634–2642 (2010).

[31] Zaïm, A. *et al.* Lanthanide-to-Lanthanide Energy-Transfer Processes Operating in Discrete Polynuclear Complexes: Can Trivalent Europium Be Used as a Local Structural Probe? *Chem. - A Eur. J.* **20,** 12172–12182 (2014).

[32] Zaïm, A. *et al.* N-Heterocyclic Tridentate Aromatic Ligands Bound to [Ln(hexafluoroacetylacetonate)$_3$] Units: Thermodynamic, Structural and Luminescent Properties. *Chem. - A Eur. J.* **18,** 7155–7168 (2012).

[33] Kalyakina, A. S. *et al.* Lanthanide Fluorobenzoates as Bio-Probes: a Quest for the Optimal Ligand Fluorination Degree. *Chem. - A Eur. J.* **23,** (2017).

[34] Shavaleev, N. M., Eliseeva, S. V., Scopelliti, R., Bünzli, J. C. G. Influence of Symmetry on the Luminescence and Radiative Lifetime of Nine-Coordinate Europium Complexes. *Inorg. Chem.* **54,** 9166–9173 (2015).

[35] Malta, O. L. *et al.* Experimental and theoretical emission quantum yield in the compound Eu(thenoyltrifluoroacetonate)3.2(dibenzyl sulfoxide). *Chem. Phys. Lett.* **282,** 233–238 (1998).

[36] Ahmed, Z., Iftikhar, K. Efficient Layers of Emitting Ternary Lanthanide Complexes for Fabricating Red, Green, and Yellow OLEDs. *Inorg. Chem.* **54,** 11209–11225 (2015).

[37] Mal, S., Pietraszkiewicz, M. & Pietraszkiewicz, O. Luminescent studies of binuclear ternary europium(III) pyridineoxide tetrazolate complexes containing bis-phosphine oxide as auxiliary co-ligands. *Luminescence* (2017). doi:10.1002/bio.3423

[38] Bünzli, J. C. G. Lanthanide luminescence for biomedical analyses and imaging. *Chem. Rev.* **110,** 2729–2755 (2010).

[39] D'Vries, R. F., Gomez, G. E., Hodak, J. H., Soler-Illia, G. J. A. A., Ellena, J. Tuning the structure, dimensionality and luminescent properties of lanthanide metal–organic frameworks under ancillary ligand influence. *Dalt. Trans.* **45,** 646–656 (2016).

[40] Gillon, A. L., Feeder, N., Roger J. Davey, A., Storey, R. Hydration in Molecular CrystalsA Cambridge Structural Database Analysis. *Cryst. Growth Des.* **3,** 663–673 (2003).

[41] Alexandrov, E. V., Blatov, V. A., Kochetkov, A. V., Proserpio, D. M. Underlying nets in three-periodic coordination polymers: topology, taxonomy and prediction from a computer-aided analysis of the Cambridge Structural Database. *CrystEngComm* **13,** 3947 (2011).

[42] Blatov, V. A., Carlucci, L., Ciani, G., Proserpio, D. M. Interpenetrating metal–organic and inorganic 3D networks: a computer-aided systematic investigation. Part I. Analysis of the Cambridge structural database. *CrystEngComm* **6,** 377–395 (2004).

[43] Fábián, L. Cambridge Structural Database Analysis of Molecular Complementarity in Cocrystals. *Cryst. Growth Des.* **9,** 1436–1443 (2009).

[44] Ouchi, A., Sato, Y., Yukawa, Y. ,Takeuchi, T. Syntheses, Structure and Properties of Several Bis{(alkylthio)acetato}copper(II)–Amine Adducts. *Bull. Chem. Soc. Jpn.* **56,** 2241–2249 (1983).

[45] Bußkamp, H. *et al.* Structural variations in rare earth benzoate complexes. *CrystEngComm* **9,** 394–411 (2007).

[46] Ouchi, A., Suzuki, Y., Ohki, Y., Koizumi, Y. Structure of rare earth carboxylates in dimeric and polymeric forms. *Coord. Chem. Rev.* **92,** 29–43 (1988).

[47] Roitershtein, D. M. *et al.* Di- and Triphenylacetates of Lanthanum and Neodymium. Synthesis, Structural Diversity and Application in Diene Polymerization. *Organometallics* **32,** 1272–1286 (2013).

[48] Pecharsky, Vitalij K., Zavalij, P. Y. *Fundamentals of Powder Diffraction and Structural Characterization of Materials*. (Springer US, 2009). doi:10.1007/978-0-387-09579-0

[49] Kwon, S. *et al.* Self-Assembled Peptide Architecture with a Tooth Shape: Folding into Shape. *J. Am. Chem. Soc.* **133,** 17618–17621 (2011).

[50] Vilela, S. M. F. *et al.* Multifunctional micro- and nanosized metal–organic frameworks assembled from bisphosphonates and lanthanides. *J. Mater. Chem. C* **2,** 3311 (2014).

[51] Smrčok, Ľ., Jorík, V., Scholtzová, E., Milata, V. *Ab initio* structure determination of 5-anilinomethylene-2,2-dimethyl-1,3-dioxane-4,6-dione from laboratory powder data – a combined use of X-ray, molecular and solid-state DFT study. *Acta Crystallogr. Sect. B Struct. Sci.* **63,** 477–484 (2007).

[52] Balasubramanian, M. *et al.* Fine structure and chemical shifts in nonresonant inelastic X-ray scattering from Li-intercalated graphite. *Appl. Phys. Lett.* **91,** 031904 (2007).

[53] Caswell, N., Solin, S. A., Hayes, T. M., Hunter, S. J. Variations in the classical model of staging in graphite intercalates: EXAFS results. *Phys. B+C* **99,** 463–468 (1980).

[54] Pickering, I. J., Prince, R. C., Divers, T., George, G. N. Sulfur K-edge X-ray absorption spectroscopy for determining the chemical speciation of sulfur in biological systems. *FEBS Lett.* **441,** 11–14 (1998).

[55] Yano, J. *et al.* Where water is oxidized to dioxygen: structure of the photosynthetic Mn4Ca cluster. *Science* **314,** 821–5 (2006).

[56] Gschneidner, K. A., Eyring, L. & Hufner, S. *High energy spectroscopy*. (North-Holland, 1987).

[57] Terrier, C., Vitorge, P., Gaigeot, M.-P., Spezia, R. & Vuilleumier, R. Density functional theory based molecular dynamics study of hydration and electronic properties of aqueous La^{3+}. *J. Chem. Phys.* **133,** 044509 (2010).

[58] Antonio, M. R., McAlister, D. R., Horwitz, E. P. An europium(III) diglycolamide complex: insights into the coordination chemistry of lanthanides in solvent extraction. *Dalt. Trans.* **44,** 515–521 (2015).

[59] Hinckley, C. C. Paramagnetic shifts in solutions of cholesterol and the dipyridine adduct of trisdipivalomethanatoeuropium(III). A shift reagent. *J. Am. Chem. Soc.* **91,** 5160–5162 (1969).

[60] Kagawa, M., Machida, Y., Nishi, H., Haginaka, J. Enantiomeric purity determination of acetyl-l-carnitine by NMR with chiral lanthanide shift reagents. *J. Pharm. Biomed. Anal.* **38,** 918–923 (2005).

[61] Felitsky, D. J., Lietzow, M. A., Dyson, H. J., Wright, P. E. Modeling transient collapsed states of an unfolded protein to provide insights into early folding events. *Proc. Natl. Acad. Sci. U. S. A.* **105,** 6278–83 (2008).

[62] Otting, G. Protein NMR, using Paramagnetic Ions. *Annu. Rev. Biophys.* **39,** 387–405 (2010).

[63] Sastri, V. S., Bünzli, J.-C., Rao, V. R., Rayudu, G. V. S., Perumareddi, J. R. Lanthanide Nmr Shift Reagents. in *Modern Aspects of Rare Earths and Their Complexes* 779–843 (Elsevier, 2003).

[64] Bleaney, B. Nuclear magnetic resonance shifts in solution due to lanthanide ions. *J. Magn. Reson.* **8,** 91–100 (1972).

[65] Otting, G. Prospects for lanthanides in structural biology by NMR. *J. Biomol. NMR* **42,** 1–9 (2008).

[66] Bruno, J., Herr, B. R., Horrocks, W. D. Laser-induced luminescence studies of europium hexaazamacrocycle Eu(HAM)3+. An evaluation of complex stability and the use of the gadolinium analog as a relaxation agent in yttrium-89 NMR. *Inorg. Chem.* **32,** 756–762 (1993).

[67] Holz, R. C., Chang, C. A., Horrocks, W. D. Spectroscopic characterization of the europium(III) complexes of a series of N,N'-bis(carboxymethyl) macrocyclic ether bis(lactones). *Inorg. Chem.* **30,** 3270–3275 (1991).

[68] Holz, R. C., Meister, G. E., Horrocks, W. D. Spectroscopic characterization of a series of europium(III) amino phosphonate complexes in solution. *Inorg. Chem.* **29,** 5183–5189 (1990).

[69] Holz, R. C., Klakamp, S. L., Chang, C. A., Horrocks, W. D. Laser-induced europium(III) luminescence and NMR spectroscopic characterization of macrocyclic diaza crown ether complexes containing carboxylate ligating groups. *Inorg. Chem.* **29,** 2651–2658 (1990).

[70] Albin, M., Horrocks, W. D. Europium(III) luminescence excitation spectroscopy. Quantitive correlation between the total charge on the ligands and the $^7F_0$-$^5D_0$ transition frequency in europium(III) complexes. *Inorg. Chem.* **24,** 895–900 (1985).

[71] McNemar, C. W. & Horrocks, W. D. The determination of the Mg^{2+}·ATP dissociation constant by competition with Eu^{3+} ion, using laser-induced Eu^{3+} ion luminescence

spectroscopy. *Anal. Biochem.* **184,** 35–38 (1990).

[72] Albin, M., Farber, G. K., Horrocks, W. D. Europium(III) luminescence excitation spectroscopy. A species-specific method for the quantitation of lanthanide ion binding to chelating agents. Complexes of (1,2-ethanediyldioxy)diacetate. *Inorg. Chem.* **23,** 1648–1651 (1984).

[73] Horrocks, W. D. & Sudnick, D. R. Lanthanide ion luminescence probes of the structure of biological macromolecules. *Acc. Chem. Res.* **14,** 384–392 (1981).

[74] Horrocks, W. D., Rhee, M.-J., Snyder, A. P. & Sudnick, D. R. Laser-induced metal ion luminescence: interlanthanide ion energy transfer distance measurements in the calcium-binding proteins, parvalbumin and thermolysin. Metalloprotein models address a photophysical problem. *J. Am. Chem. Soc.* **102,** 3650–3652 (1980).

[75] Horrocks, W. D., Arkle, V. K., Liotta, F. J., Sudnick, D. R. Kinetic parameters for a system at equilibrium from the time course of luminescence emission: a new probe of equilibrium dynamics. Excited-state europium(III) as a species label. *J. Am. Chem. Soc.* **105,** 3455–3459 (1983).

[76] Werts, M. H. V., Jukes, R. T. F., Verhoeven, J. W. The emission spectrum and the radiative lifetime of Eu^{3+} in luminescent lanthanide complexes. *Phys. Chem. Chem. Phys.* **4,** 1542–1548 (2002).

[77] Kuke, S., Marmodée, B., Eidner, S., Schilde, U. & Kumke, M. U. Intramolecular deactivation processes in complexes of salicylic acid or glycolic acid with Eu(III). *Spectrochim. Acta Part A Mol. Biomol. Spectrosc.* **75,** 1333–1340 (2010).

[78] Werts, M. H. V., Jukes, R. T. F. & Verhoeven, J. W. The emission spectrum and the radiative lifetime of Eu^{3+} in luminescent lanthanide complexes. *Phys. Chem. Chem. Phys.* **4,** 1542–1548 (2002).

[79] Miniscalco, W. J. Erbium-doped glasses for fiber amplifiers at 1500 nm. *J. Light. Technol.* **9,** 234–250 (1991).

[80] Kropp, J. L., Dawson, W. R. Temperature-Dependent Quenching of Fluorescence of Europic-Ion Solutions. *J. Chem. Phys.* **45,** 2419–2420 (1966).

[81] Kropp, J. L., Windsor, M. W. Luminescence and Energy Transfer in Solutions of Rare-Earth Complexes. I. Enhancement of Fluorescence by Deuterium Substitution. *J. Chem. Phys.* **42,** 1599–1608 (1965).

[82] Kropp, J. L., Windsor, M. W. Enhancement of Fluorescence Yield of Rare-Earth Ions by Heavy Water. *J. Chem. Phys.* **39,** 2769–2770 (1963).

[83] Izzo, I. *et al.* Structural Effects of Proline Substitution and Metal Binding on Hexameric Cyclic Peptoids. *Org. Lett.* **15,** 598–601 (2013).

[84] Haas, Y., Stein, G. Radiative and nonradiative pathways in solutions. Excited states of the europium(III) ion. *J. Phys. Chem.* **76,** 1093–1104 (1972).

[85] Haas, Y. & Stein, G. Pathways of radiative and radiationless transitions in europium(III) solutions. The role of high energy vibrations. *J. Phys. Chem.* **75,** 3677–3681 (1971).

[86] Kropp, J. L., Windsor, M. W. Luminescence and energy transfer in solutions of rare earth complexes. II. Studies of the solvation shell in europium(III) and terbium(III) as a function of acetate concentration. *J. Phys. Chem.* **71,** 477–482 (1967).

[87] Kuke, S., Marmodée, B., Eidner, S., Schilde, U., Kumke, M. U. Intramolecular deactivation processes in complexes of salicylic acid or glycolic acid with Eu(III). *Spectrochim. Acta Part A Mol. Biomol. Spectrosc.* **75,** 1333–1340 (2010).

[88] Choppin, G. R., Wang, Z. M. Correlation between Ligand Coordination Number and the Shift of the $^7F_0$–$^5D_0$ Transition Frequency in Europium(III) Complexes. *Inorg. Chem.* **36,** 249–252 (1997).

[89] Bünzli, J. C. G., Comby, S., Chauvin, A. S., Vandevyver, C. D. B. New Opportunities for Lanthanide Luminescence. *J. Rare Earths* **25,** 257–274 (2007).

[90] Kalyakina, A. S. *et al.* OLED thin film fabrication from poorly soluble terbium o-phenoxybenzoate through soluble mixed-ligand complexes. *Org. Electron.* **28,** 319–329 (2016).

[91] Winkler, B. *et al.* New transport layers for highly efficient organic electroluminescence devices. *Synth. Met.* **102,** 1083–1084 (1999).

[92] Forrest, S. R. *et al.* Highly efficient phosphorescent emission from organic electroluminescent devices. *Nature* **395,** 151–154 (1998).

[93] Kido, J., Okamoto, Y. Organo Lanthanide Metal Complexes for Electroluminescent Materials. *Chem. Rev.* **102,** 2357–2368 (2002).

[94] Bünzli, J. C. G., Eliseeva, S. V. Lanthanide NIR luminescence for telecommunications, bioanalyses and solar energy conversion. *J. Rare Earths* **28,** 824–842 (2010).

[95] Kim, S. *et al.* Near-infrared fluorescent type II quantum dots for sentinel lymph node mapping. *Nat. Biotechnol.* **22,** 93–97 (2004).

[96] Ye, H. *et al.* Organo-erbium systems for optical amplification at telecommunications wavelengths. *Nat. Mater.* **13,** 382–6 (2014).

[97] Wei, H. *et al.* Constructing lanthanide [Nd(III), Er(III) and Yb(III)] complexes, using a tridentate N,N,O-ligand for near-infrared organic light-emitting diodes. *Dalton Trans.* **42,** 8951–60 (2013).

[98] Coya, C. *et al.* Novel Er-doped organic complexes for application in 1.5 μm emitting solution-processed organic light emitting diodes (OLEDs). *MRS Proc.* **1286,** 1286 (2011).

[99] D'Aléo, A., Pointillart, F., Ouahab, L., Andraud, C., Maury, O. Charge transfer excited states sensitization of lanthanide emitting from the visible to the near-infra-red. *Coord. Chem. Rev.* **256,** 1604–1620 (2012).

[100] Chen, Z. Q., Ding, F., Bian, Z. Q., Huang, C. H. Efficient near-infrared organic light-emitting diodes based on multimetallic assemblies of lanthanides and iridium complexes. *Org. Electron. physics, Mater. Appl.* **11,** 369–376 (2010).

[101] Katkova, M. A. *et al.* Near-infrared electroluminescent lanthanide [Pr(III), Nd(III), Ho(III), Er(III), Tm(III) and Yb(III)] N,O-chelated complexes for organic light-emitting devices. *J. Mater. Chem.* **21,** 16611 (2011).

[102] Hemmilä, I., Webb, S. Time-resolved fluorometry: an overview of the labels and core technologies for drug screening applications. *Drug Discov. Today* **2,** 373–381 (1997).

[103] Moore, E. G., Samuel, A. P. S., Raymond, K. N. From Antenna to Assay: Lessons Learned in Lanthanide Luminescence. *Acc. Chem. Res.* **42,** 542–552 (2009).

[104] Rodriguz-Ubis, J.-C., Alpha, B., Plancherel, D., Lehn, J.-M. Photoactive cryptands. Synthesis of the sodium cryptates of macrobicyclic ligands containing bipyridine and phenoanthroline groups. *Helv. Chim. Acta* **67,** 2264–2269 (1984).

[105] Sabbatini, N., Guardigli, M., Lehn, J.-M. Luminescent lanthanide complexes as photochemical supramolecular devices. *Coord. Chem. Rev.* **123,** 201–228 (1993).

[106] Xu, J. *et al.* Octadentate Cages of Tb(III) 2-Hydroxyisophthalamides: A New Standard for

Luminescent Lanthanide Labels. *J. Am. Chem. Soc.* **133,** 19900–19910 (2011).

[107] Butler, S. J., Lamarque, L., Pal, R., Parker, D. EuroTracker dyes: highly emissive europium complexes as alternative organelle stains for live cell imaging. *Chem. Sci.* **5,** 1750 (2014).

[108] Petoud, S., Cohen, S. M., Jean-Claude G. Bünzli, Raymond, K. N. Stable Lanthanide Luminescence Agents Highly Emissive in Aqueous Solution: Multidentate 2-Hydroxyisophthalamide Complexes of Sm^{3+}, Eu^{3+}, Tb^{3+}, Dy^{3+}. *J. Am. Chem. Soc.* **125,** 13324–13325 (2003).

[109] Delbianco, M. *et al.* Bright, Highly Water-Soluble Triazacyclononane Europium Complexes To Detect Ligand Binding with Time-Resolved FRET Microscopy. *Angew. Chemie Int. Ed.* **53,** 10718–10722 (2014).

[110] Delbianco, M., Lamarque, L., Parker, D. Synthesis of meta and para-substituted aromatic sulfonate derivatives of polydentate phenylazaphosphinate ligands: enhancement of the water solubility of emissive europium(III) EuroTracker® dyes. *Org. Biomol. Chem.* **12,** 8061–8071 (2014).

[111] Cheal, S. M. *et al.* Mapping protein-protein interactions by localized oxidation: consequences of the reach of hydroxyl radical. *Biochemistry* **48,** 4577–86 (2009).

[112] Gali, H. *et al.* Synthesis and in vitro evaluation of an ^{111}In-labeled ST-peptide enterotoxin (ST) analogue for specific targeting of guanylin receptors on human colonic cancers. *Anticancer Res.* **21,** 2785–92 (2001).

[113] Kielar, F. *et al.* Two-photon microscopy study of the intracellular compartmentalisation of emissive terbium complexes and their oligo-arginine and oligo-guanidinium conjugates. *Chem. Commun.*, 2435 (2008).

[114] Mohandessi, S., Rajendran, M., Magda, D., Miller, L. W. Cell-Penetrating Peptides as Delivery Vehicles for a Protein-Targeted Terbium Complex. *Chem. - A Eur. J.* **18,** 10825–10829 (2012).

[115] Kölmel, D. K., Rudat, B., Schepers, U., Bräse, S. Peptoid-Based Rare-Earth (Group 3 and Lanthanide) Transporters. *European J. Org. Chem.* **2013,** 2761–2765 (2013).

[116] Schröder, T. *et al.* Solid-Phase Synthesis, Bioconjugation and Toxicology of Novel Cationic Oligopeptoids for Cellular Drug Delivery. *Bioconjug. Chem.* **18,** 342–354 (2007).

[117] Vollrath, S. B. L., Fürniss, D., Schepers, U., Bräse, S. Amphiphilic peptoid transporters – synthesis and evaluation. *Org. Biomol. Chem.* **11,** 8197 (2013).

[118] Zuckermann, R. N., Kerr, J. M., Kent, S. B. H., Moos, W. H. Efficient method for the preparation of peptoids [oligo(N-substituted glycines)] by submonomer solid-phase synthesis. *J. Am. Chem. Soc.* **114,** 10646–10647 (1992).

[119] De Cola, C. *et al.* Gadolinium-binding cyclic hexapeptoids: synthesis and relaxometric properties. *Org. Biomol. Chem.* **12,** 424–431 (2014).

[120] De Cola, C. *et al.* Size-dependent cation transport by cyclic α-peptoid ion carriers. *Org. Biomol. Chem.* **7,** 2851 (2009).

[121] Morishetti, K. K., Russell, S. C., Zhao, X., Robinson, D. B., Ren, J. Tandem mass spectrometry studies of protonated and alkali metalated peptoids: Enhanced sequence coverage by metal cation addition. *Int. J. Mass Spectrom.* **308,** 98–108 (2011).

[122] Maayan, G., Ward, M. D., Kirshenbaum, K. Metallopeptoids. *Chem. Commun.*, 56–58 (2009).

[123] Pirrung, M. C., Park, K. Discovery of selective metal-binding peptoids, using 19F encoded combinatorial libraries. *Bioorg. Med. Chem. Lett.* **10,** 2115–2118 (2000).

[124] Kölmel, D. K., Rudat, B., Schepers, U., Bräse, S. Peptoid-Based Rare-Earth (Group 3 and Lanthanide) Transporters. *European J. Org. Chem.* **2013,** 2761–2765 (2013).

[125] Thielemann, D. T. *et al.* Peptoid-Ligated Pentadecanuclear Yttrium and Dysprosium Hydroxy Clusters. *Chem. - A Eur. J.* **21,** 2813–2820 (2015).

[126] Baranyai, Z. *et al.* Equilibrium and Kinetic Properties of the Lanthanoids(III) and Various Divalent Metal Complexes of the Heptadentate Ligand AAZTA. *Chem. - A Eur. J.* **15,** 1696–1705 (2009).

[127] Uha, H., Petoud, S. Novel antennae for the sensitization of near infrared luminescent lanthanide cations. *Comptes Rendus Chim.* **13,** 668–680 (2010).

[128] Utochnikova, V. V. *et al.* Lanthanide 9-anthracenate: solution processable emitters for efficient purely NIR emitting host-free OLEDs. *J. Mater. Chem. C* **4,** 9848–9855 (2016).

[129] de Silva, A. P., Gunaratne, H. Q. N., Rice, T. E. Proton-Controlled Switching of Luminescence in Lanthanide Complexes in Aqueous Solution: pH Sensors Based on Long-Lived Emission. *Angew. Chemie Int. Ed. English* **35,** 2116–2118 (1996).

[130] Rajendran, M., Miller, L. W. Evaluating the performance of time-gated live-cell microscopy with lanthanide probes. *Biophys. J.* **109,** 240–8 (2015).

[131] Law, G. L. *et al.* Emissive terbium probe for multiphoton in vitro cell imaging. *J. Am. Chem. Soc.* **130,** 3714–3715 (2008).

[132] Parker, D. Luminescent lanthanide sensors for pH, pO_2 and selected anions. *Coord. Chem. Rev.* **205,** 109–130 (2000).

[133] Aita, K., Temma, T., Kuge, Y., Saji, H. Development of a novel neodymium compound for in vivo fluorescence imaging. *Luminescence* **22,** 455–461 (2007).

[134] Berezin, M. Y., Achilefu, S. Fluorescence lifetime measurements and biological imaging. *Chem. Rev.* **110,** 2641–84 (2010).

[135] Parker, D. Excitement in f block: structure, dynamics and function of nine-coordinate chiral lanthanide complexes in aqueous media. *Chem. Soc. Rev.* **33,** 156–65 (2004).

[136] Prodi, L. Luminescent chemosensors: from molecules to nanoparticles. *New J. Chem.* **29,** 20 (2005).

[137] Meggers, E. Targeting proteins with metal complexes. *Chem. Commun. (Camb).* **0,** 1001–10 (2009).

[138] Hemmilä, I., Laitala, V. Progress in lanthanides as luminescent probes. *J. Fluoresc.* **15,** 529–42 (2005).

[139] Kalyakina, A. S. *et al.* Highly Luminescent, Water-Soluble Lanthanide Fluorobenzoates: Syntheses, Structures and Photophysics, Part I: Lanthanide Pentafluorobenzoates. *Chemistry* **21,** 17921–32 (2015).

[140] Utochnikova, V. V. *et al.* Luminescence Enhancement by *p* -Substituent Variation. *Eur. J. Inorg. Chem.* **2017,** 107–114 (2017).

[141] Mancino, G., Ferguson, A. J., Beeby, A., Long, N. J., Jones, T. S. Dramatic increases in the lifetime of the Er3+ ion in a molecular complex, using a perfluorinated imidodiphosphinate sensitizing ligand. *J. Am. Chem. Soc.* **127,** 524–5 (2005).

[142] Utochnikova, V. V., Kalyakina, A. S., Lepnev, L. S., Kuzmina, N. P. Luminescence enhancement of nanosized ytterbium and europium fluorides by surface complex formation with aromatic carboxylates. *J. Lumin.* **170,** 633–640 (2016).

[143] Roh, S.-G. *et al.* Er(III)-chelated Prototype Complexes Based on Benzoate and

Pentafluorobenzoate Ligands : Synthesis and Key Parameters for Near IR Emission Enhancement. *Bull. Korean Chem. Soc.* **25,** 1503–1507 (2004).

[144] Taylor, R., Allen, F. H. Statistical and Numerical Methods of Data Analysis. in *Chem. Commun.* (eds. Bürgi, H.-B. & Dunitz, J. D.) 111–161 (Wiley-VCH Verlag GmbH, 1994).

[145] Larionov, S. V. *et al.* Luminescence properties of complexes Ln(Phen)$(C_6F_5COO)_3$ (Ln = Tb, Eu) and $Ln(C_6F_5COO)_3 \cdot nH_2O$ (Ln = Tb, n = 2; Ln = Eu, n = 1). Structures of the $[Tb_2(H_2O)_8(C_6F_5COO)_6]$ complex and its isomer in the supramolecular compound $[Tb_2(H_2O)_8(C_6F_5COO)_6] \cdot 2C_6F_5COOH$. *Russ. J. Coord. Chem.* **35,** 798–806 (2009).

[146] Morgenthaler, M. *et al.* Predicting and Tuning Physicochemical Properties in Lead Optimization: Amine Basicities. *ChemMedChem* **2,** 1100–1115 (2007).

[147] Graton, J. *et al.* Influence of Alcohol β-Fluorination on Hydrogen-Bond Acidity of Conformationally Flexible Substrates. *Chem. - A Eur. J.* **23,** 2811–2819 (2017).

[148] Smart, B. E. Fluorine substituent effects (on bioactivity). *J. Fluor. Chem.* **109,** 3–11 (2001).

[149] Böhm, H.-J. *et al.* Fluorine in Medicinal Chemistry. *ChemBioChem* **5,** 637–643 (2004).

[150] Müller, K., Faeh, C., Diederich, F. Fluorine in Pharmaceuticals: Looking Beyond Intuition. *Science (80-.).* **317,** 1881–1886 (2007).

[151] Pálinkás, Z. *et al.* Stability, Water Exchange and Anion Binding Studies on Lanthanide(III) Complexes with a Macrocyclic Ligand Based on 1,7-Diaza-12-crown-4: Extremely Fast Water Exchange on the Gd3+ Complex. *Inorg. Chem.* **48,** 8878–8889 (2009).

[152] Burger, K. *Solvation, ionic and complex formation reactions in non-aqueous solvents : experimental methods for their investigation*. (Elsevier Scientific Pub. Co., 1983).

[153] Staemmler, V. The Donor-Acceptor Approach to Molecular Interactions. Von V. Gutmann. Plenum Press, New York 1978. XVI, 279 S., *Angew. Chemie* **91,** 595–595 (1979).

[154] Wang, G.-Y. *et al.* A luminescent 2D coordination polymer for selective sensing of nitrobenzene. *Dalt. Trans.* **42,** (2013).

[155] Zhou, X. *et al.* A microporous luminescent europium metal–organic framework for nitro explosive sensing. *Dalt. Trans.* **42,** 5718–5723 (2013).

[156] Lis, S. & Choppin, G. R. Luminescence lifetimes of aqueous europium perchlorate, chloride and nitrate solutions. *Mater. Chem. Phys.* **31,** 159–161 (1992).

[157] Ishida, H. *et al.* Guidelines for measurement of luminescence spectra and quantum yields of inorganic and organometallic compounds in solution and solid state (IUPAC Technical Report). *Pure Appl. Chem.* **88,** 1535 (2016).

[158] Uitert, L. G. Van, Johnson, L. F. Energy Transfer Between Rare-Earth Ions. *J. Chem. Phys.* **44,** 3514–3522 (1966).

[159] Gao, C. *et al.* Self-Assembly Synthesis, Structural Features and Photophysical Properties of Dilanthanide Complexes Derived from a Novel Amide Type Ligand: Energy Transfer from Tb(III) to Eu(III) in a Heterodinuclear Derivative. *Inorg. Chem.* **53,** 935–942 (2014).

[160] Meyer, L. V., Schönfeld, F., Müller-Buschbaum, K. Lanthanide based tuning of luminescence in MOFs and dense frameworks – from mono- and multimetal systems to sensors and films. *Chem. Commun.* **50,** 8093 (2014).

[161] Charbonnière, L. J., Hildebrandt, N., Ziessel, R. F. & Löhmannsröben, H.-G. Lanthanides to Quantum Dots Resonance Energy Transfer in Time-Resolved Fluoro-Immunoassays and Luminescence Microscopy. *J. Am. Chem. Soc.* **128,** 12800–12809 (2006).

[162] Charbonnière, L. J., Hildebrandt, N. Lanthanide Complexes and Quantum Dots: A Bright

Wedding for Resonance Energy Transfer. *Eur. J. Inorg. Chem.* **2008,** 3241–3251 (2008).

[163] Lill, D. T. de, Bettencourt-Dias, A. de, Cahill, C. L. Exploring Lanthanide Luminescence in Metal-Organic Frameworks: Synthesis, Structure and Guest-Sensitized Luminescence of a Mixed Europium/Terbium-Adipate Framework and a Terbium-Adipate Framework. *Inorg. Chem.* **46,** 3960–3965 (2007).

[164] Imbert, D., Cantuel, M., Bünzli, J.-C. G., Gérald Bernardinelli, Piguet, C. Extending Lifetimes of Lanthanide-Based Near-Infrared Emitters (Nd, Yb) in the Millisecond Range through Cr(III) Sensitization in Discrete Bimetallic Edifices. *J. Am. Chem. Soc.* **125,** 15698–15699 (2003).

[165] Aebischer, A., Gumy, F., Bünzli, J.-C. G. Intrinsic quantum yields and radiative lifetimes of lanthanide tris(dipicolinates). *Phys. Chem. Chem. Phys.* **11,** 1346 (2009).

[166] Khudoleeva, V. Y. *et al.* Surface modified $Eu_xLa_{1-x}F_3$ nanoparticles as luminescent biomarkers: Still plenty of room at the bottom. *Dye. Pigment.* **143,** 348–355 (2017).

[167] Bünzli, J.-C. G. On the design of highly luminescent lanthanide complexes. *Coord. Chem. Rev.* **293–294,** 19–47 (2015).

[168] Ansari, S. A. *et al.* Complexation of Lanthanides with Glutaroimide-dioxime: Binding Strength and Coordination Modes. *Inorg. Chem.* **55,** 1315–1323 (2016).

[169] Morris, R. H. Estimating the Acidity of Transition Metal Hydride and Dihydrogen Complexes by Adding Ligand Acidity Constants. *J. Am. Chem. Soc.* **136,** 1948–1959 (2014).

[170] Blasse, G., Bril, A. Hypersensitivity of the $^5D_0$ — $^7F_2$ Transition of Trivalent Europium in the Garnet Structure. *J. Chem. Phys.* **47,** 5442–5443 (1967).

[171] Li, H. *et al.* A Highly Transparent and Luminescent Hybrid Based on the Copolymerization of Surfactant-Encapsulated Polyoxometalate and Methyl Methacrylate. *Adv. Mater.* **17,** 2688–2692 (2005).

[172] Verlan, V. I. *et al.* Effective Transfer of UV Energy to Red Luminescence in the Nanocomposites Polymer/Eu Coordination Compounds. in *3rd International Conference on Nanotechnologies and Biomedical Engineering* 17–20 (Springer, Singapore, 2016).

[173] Kalyakina, A. S. *et al.* Remarkable high efficiency of red emitters, using Eu(iii) ternary complexes. *Chem. Commun.* **54,** 5221–5224 (2018).

[174] Shavaleev, N. M., Eliseeva, S. V., Scopelliti, R. & Bünzli, J.-C. G. Designing Simple Tridentate Ligands for Highly Luminescent Europium Complexes. *Chem. - A Eur. J.* **15,** 10790–10802 (2009).

[175] Sevéus, L. *et al.* Use of fluorescent europium chelates as labels in microscopy allows glutaraldehyde fixation and permanent mounting and leads to reduced autofluorescence and good long-term stability. *Microsc. Res. Tech.* **28,** 149–54 (1994).

[176] Bünzli, J.-C. G. Rising Stars in Science and Technology: Luminescent Lanthanide Materials. *Eur. J. Inorg. Chem.* **2017,** 5058–5063 (2017).

[177] Cuan, J. & Yan, B. Luminescent lanthanide-polyoxometalates assembling zirconia–alumina–titania hybrid xerogels through task-specified ionic liquid linkage. *RSC Adv.* **4,** 1735–1743 (2014).

[178] Moudam, O. *et al.* Europium complexes with high total photoluminescence quantum yields in solution and in PMMA. *Chem. Commun.* 6649 (2009).

[179] Glover, P. B. *et al.* Fully Fluorinated Imidodiphosphinate Shells for Visible- and NIR-Emitting Lanthanides: Hitherto Unexpected Effects of Sensitizer Fluorination on

Lanthanide Emission Properties. *Chem. - A Eur. J.* **13,** 6308–6320 (2007).

[180] Larionov, S. V., Rakhmanova, M. I., Karpov, V. M., Platonov, V. E., Fadeeva, V. P. Synthesis and photoluminescence of Eu(III) complex compounds containing anions of perfluoro-3-methylbenzoic acid. *Russ. J. Gen. Chem.* **86,** 2677–2681 (2016).

[181] Li, J.-J., Fan, T.-T., Qu, X.-L., Han, H.-L., Li, X. Temperature-induced 1D lanthanide polymeric frameworks based on Ln n (n = 2, 2, 4, 6) cores: synthesis, crystal structures and luminescence properties. *Dalt. Trans.* **45,** 2924–2935 (2016).

[182] Li, X., Zhang, Z.-Y. Synthesis, structure and luminescence properties of two novel lanthanide complexes with 2-fluorobenzoic acid and 1,10-phenanthroline. *J. Coord. Chem.* **59,** 1873–1882 (2006).

[183] Li, X., Zhang, Z.-Y., Zou, Y.-Q. Synthesis, Structure and Luminescence Properties of Four Novel Terbium 2-Fluorobenzoate Complexes. *Eur. J. Inorg. Chem.* **2005,** 2909–2918 (2005).

[184] Utochnikova, V. V. *et al.* Lanthanide tetrafluorobenzoates as emitters for OLEDs: New approach for host selection. *Org. Electron.* **44,** 85–93 (2017).

[185] González-Pérez, S. *et al.* Luminescent polymeric film containing an Eu(III) complex acting as UV protector and down-converter for Si-based solar cells and modules. *Surf. Coatings Technol.* **271,** 106–111 (2015).

[186] Han, L.-J., Kong, Y.-J., Sheng, N. & Jiang, X.-L. A new europium fluorous metal–organic framework with pentafluorobenzoate and 1,10-phenanthroline ligands: Synthesis, structure and luminescent properties. *J. Fluor. Chem.* **166,** 122–126 (2014).

[187] Ahmed, Z., Iftikhar, K. Sensitization of Visible and NIR Emitting Lanthanide(III) Ions in Noncentrosymmetric Complexes of Hexafluoroacetylacetone and Unsubstituted Monodentate Pyrazole. *J. Phys. Chem. A* **117,** 11183–11201 (2013).

[188] Reisfeld, R., Zigansky, E., Gaft, M. Europium probe for estimation of site symmetry in glass films, glasses and crystals. *Mol. Phys.* **102,** 1319–1330 (2004).

[189] Casanova, D., Llunell, M., Alemany, P., Alvarez, S. The Rich Stereochemistry of Eight-Vertex Polyhedra: A Continuous Shape Measures Study. *Chem. - A Eur. J.* **11,** 1479–1494 (2005).

[190] Ruiz-Martínez, A., Casanova, D., Alvarez, S. Polyhedral Structures with an Odd Number of Vertices: Nine-Coordinate Metal Compounds. *Chem. - A Eur. J.* **14,** 1291–1303 (2008).

[191] Hasegawa, Y. *et al.* Luminescent Polymer Containing the Eu(III) Complex Having Fast Radiation Rate and High Emission Quantum Efficiency. *J. Phys. Chem.* **107,** 1697–1702 (2003).

[192] Zimmer, M., Dietrich, F., Volz, D., Bräse, S., Gerhards, M. Solid-State Step-Scan FTIR Spectroscopy of Binuclear Copper(I) Complexes. *ChemPhysChem* **18,** 3023–3029 (2017).

[193] Zimmer, M. *et al.* Time-resolved IR spectroscopy of a trinuclear palladium complex in solution. *Phys. Chem. Chem. Phys.* **17,** 14138–14144 (2015).

[194] Sun, L. N. *et al.* Syntheses, structures and near-IR luminescent studies on ternary lanthanide (ErIII, HoIII, YbIII, NdIII) complexes containing 4,4,5,5,6,6,6-heptafluoro-1-(2-thienyl)hexane-1,3-dionate. *Eur. J. Inorg. Chem.* 3962–3973 (2006).

[195] Zheng, Y., Pearson, J., Tan, R. H. C., Gillin, W. P., Wyatt, P. B. Erbium bis(pentafluorophenyl) phosphinate: a new hybrid material with unusually long-lived infrared luminescence. *J. Mater. Sci. Electron.* **20,** 430–434 (2009).

[196] Gawryszewska, P., Moroz, O. V., Trush, V. a., Kulesza, D., Amirkhanov, V. M. Structure

and sensitized near-infrared luminescence of Yb(III) complexes with sulfonylamidophosphate type ligand. *J. Photochem. Photobiol. A Chem.* **217,** 1–9 (2011).

[197] Tan, R. H. C., Motevalli, M., Abrahams, I., Wyatt, P. B., Gillin, W. P. Quenching of IR luminescence of erbium, neodymium and ytterbium beta-diketonate complexes by ligand C-H and C-D bonds. *J. Phys. Chem. B* **110,** 24476–24479 (2006).

[198] Shavaleev, N. M., Scopelliti, R., Gumy, F., Bünzli, J. C. G. Surprisingly bright near-infrared luminescence and short radiative lifetimes of ytterbium in hetero-binuclear Yb-Na chelates. *Inorg. Chem.* **48,** 7937–7946 (2009).

[199] Li, W. *et al.* NIR luminescence of 2-(2,2,2-trifluoroethyl)-1-indone (TFI) neodymium and ytterbium complexes. *J. Lumin.* **146,** 205–210 (2014).

[200] Lo, W. S., Li, H., Law, G. L., Wong, W. T., Wong, K. L. Efficient and selective singlet oxygen sensitized NIR luminescence of a neodymium(III) complex and its application in biological imaging. *J. Lumin.* **169,** 549–552 (2016).

[201] Dmitrienko, A. O., Bushmarinov, I. S. Reliable structural data from Rietveld refinements via restraint consistency. *J. Appl. Crystallogr.* **48,** 1777–1784 (2015).

[202] Allen, F. The Cambridge Structural Database: a quarter of a million crystal structures and rising. *Acta Crystallogr. Sect. B* **58,** 380–388 (2002).

[203] Zhu, L., Al-Kaysi, R. O., Dillon, R. J., Tham, F. S., Bardeen, C. J. Crystal structures and photophysical properties of 9-anthracene carboxylic acid derivatives for photomechanical applications. *Cryst. Growth Des.* **11,** 4975–4983 (2011).

[204] Voigt-function model in diffraction line-broadening analysis. in *Defect and microstructure analysis by diffraction* (eds. Snyder, R. L., Fiala, J., Bunge, H. J.) (Oxford University Press, 1999).

[205] Werts, M. H. V., Verhoeven, J. W., Hofstraat, J. W. Efficient visible light sensitisation of water-soluble near-infrared luminescent lanthanide complexes. *J. Chem. Soc. Perkin Trans.* **2,** 433–439 (2000).

[206] Shuvaev, S. *et al.* Lanthanide complexes with aromatic o-phosphorylated ligands: synthesis, structure elucidation and photophysical properties. *Dalt. Trans.* **43,** 3121–3136 (2014).

[207] Utochnikova, V. V *et al.* Lanthanide complexes with 2-(tosylamino)benzylidene-N-benzoylhydrazone, which exhibit high NIR emission. *Dalton Trans.* **44,** 12660–9 (2015).

[208] Rajendran, M., Yapici, E., Miller, L. W. Lanthanide-Based Imaging of Protein–Protein Interactions in Live Cells. *Inorg. Chem.* **53,** 1839–1853 (2014).

[209] Hovinen, J. & Guy, P. M. Bioconjugation with Stable Luminescent Lanthanide(III) Chelates Comprising Pyridine Subunits. *Bioconjug. Chem.* **20,** 404–421 (2009).

[210] Wagnon, B. K., Jackels, S. C. Synthesis, characterization and aqueous proton relaxation enhancement of a manganese(II) heptaaza macrocyclic complex having pendant arms. *Inorg. Chem.* **28,** 1923–1927 (1989).

[211] Gauss, W., Moser, P., Schwarzenbach, G. N,N,N',N'-Tetrakis-(8-amino-thyl)-thylendiamin. *Helv. Chim. Acta* **35,** 2359–2363 (1952).

[212] Deng, X., Mani, N. S. A facile, environmentally benign sulfonamide synthesis in water. *Green Chem.* **8,** 835 (2006).

[213] Bird, R., Knipe, A. C., Stirling, C. J. M. Intramolecular reactions. Part X. Transition states in the cyclisation of N-ω-halogeno-alkylamines and sulphonamides. *J. Chem. Soc., Perkin Trans. 2* **0,** 1215–1220 (1973).

[214] Raymond, K.N. *et al.* Luminescent macrocyclic lanthanide complexes. *Pat. WO 2008063721* (2007).

[215] Hata, K., Doh, M.-K., Kashiwabara, K., Fujita, J. Preparation and Absorption and Circular Dichroism Spectra of Cobalt(III) Complexes with *N* , *N* , *N* ', *N* '-Tetrakis(2-aminoethyl)-1,2-ethanediamine, -1,3-propanediamine, -1,4-butanediamine and -(*R* , *R*)-and -(*R* , *S*)-2,4-pentanediamine. *Bull. Chem. Soc. Jpn.* **54,** 190–195 (1981).

[216] Searles, S., Nukina, S. Cleavage And Rearrangement Of Sulfonamides. *Chem. Rev.* **59,** 1077–1103 (1959).

[217] Weisblat, D. I., Magerlein, B. J., Myers, D. R. The Cleavage of Sulfonamides. *J. Am. Chem. Soc.* **75,** 3630–3632 (1953).

[218] Daumann, L. J., Werther, P., Ziegler, M. J. & Raymond, K. N. Siderophore inspired tetra- and octadentate antenna ligands for luminescent Eu(III) and Tb(III) complexes. *J. Inorg. Biochem.* **162,** 263–273 (2016).

[219] Zhu, J. *et al.* Hydrogen-Bonding-Induced Planar, Rigid and Zigzag Oligoanthranilamides. Synthesis, Characterization and Self-Assembly of a Metallocyclophane. *J. Org. Chem.* **69,** 6221–6227 (2004).

[220] Bernard, A. M., Ghiani, M. R., Piras, P. P. & Rivoldini, A. Dealkylation of Activated Alkyl Aryl Ethers, using Lithium Chloride in Dimethylformamide. *Synthesis (Stuttg).* **1989,** 287–289 (1989).

[221] Niwayama, S., Cho, H. Practical large scale synthesis of half-esters of malonic acid. *Chem. Pharm. Bull. (Tokyo).* **57,** 508–10 (2009).

[222] Seth M. Cohen, Stéphane Petoud, Raymond, K. N. A Novel Salicylate-Based Macrobicycle with a "Split Personality". *Inorg. Chem.* **38,** 4522–4529 (1999).

[223] Zhu, Y.-Y., Wang, G.-T., Li, Z.-T. A click chemistry approach for the synthesis of macrocycles from aryl amide-based precursors directed by hydrogen bonding. *Org. Biomol. Chem.* **7,** 3243 (2009).

[224] Aiello, F. *et al.* Design and Synthesis of 3-Carbamoylbenzoic Acid Derivatives as Inhibitors of Human Apurinic/Apyrimidinic Endonuclease 1 (APE1). *ChemMedChem* **7,** 1825–1839 (2012).

[225] Reis, S. A. *et al.* Light-controlled modulation of gene expression by chemical optoepigenetic probes. *Nat. Chem. Biol.* **12,** 317–323 (2016).

[226] Huisgen, R. 1,3-Dipolar Cycloadditions. Past and Future. *Angew. Chemie Int. Ed. English* **2,** 565–598 (1963).

[227] Rostovtsev, V. V., Green, L. G., Fokin, V. V., Sharpless, K. B. A Stepwise Huisgen Cycloaddition Process: Copper(I)-Catalyzed Regioselective "Ligation" of Azides and Terminal Alkynes. *Angew. Chemie Int. Ed.* **41,** 2596–2599 (2002).

[228] Huisgen, R., Szeimies, G., Möbius, L. 1.3-Dipolare Cycloadditionen, XXXII. Kinetik der Additionen organischer Azide an CC-Mehrfachbindungen. *Chem. Ber.* **100,** 2494–2507 (1967).

[229] Kolb, H. C., Finn, M. G., Sharpless, K. B. Click Chemistry: Diverse Chemical Function from a Few Good Reactions. *Angew. Chemie Int. Ed.* **40,** 2004–2021 (2001).

[230] Hein, J. E., Fokin, V. V. Copper-catalyzed azide-alkyne cycloaddition (CuAAC) and beyond: new reactivity of copper(I) acetylides. *Chem. Soc. Rev.* **39,** 1302–15 (2010).

[231] Meldal, M., Tornøe, C. W. Cu-Catalyzed Azide−Alkyne Cycloaddition. *Chem. Rev.* **108,** 2952–3015 (2008).

[232] Su, L. *et al.* Copper Catalysis for Selective Heterocoupling of Terminal Alkynes. *J. Am. Chem. Soc.* **138,** 12348–12351 (2016).

[233] Balaraman, K., Kesavan, V. Efficient Copper(II) Acetate Catalyzed Homo- and Heterocoupling of Terminal Alkynes at Ambient Conditions. *Synthesis (Stuttg).* **2010,** 3461–3466 (2010).

[234] Chadwick, J. *et al.* Design, synthesis and antimalarial/anticancer evaluation of spermidine linked artemisinin conjugates designed to exploit polyamine transporters in Plasmodium falciparum and HL-60 cancer cell lines. *Bioorg. Med. Chem.* **18,** 2586–2597 (2010).

[235] Schröder, T. *et al.* Solid-Phase Synthesis, Bioconjugation and Toxicology of Novel Cationic Oligopeptoids for Cellular Drug Delivery. *Bioconjug. Chem.* **18,** 342–354 (2007).

[236] Yoo, B., Shin, S. B. Y., Huang, M. L., Kirshenbaum, K. Peptoid Macrocycles: Making the Rounds with Peptidomimetic Oligomers. *Chem. - A Eur. J.* **16,** 5528–5537 (2010).

[237] Puntus, L. & Lyssenko, K. Influence of non-covalent interactions on properties of lanthanide systems with aromatic ligands. *J. Rare Earths* **26,** 146–152 (2008).

[238] Ma, Q., Zheng, Y., Armaroli, N., Bolognesi, M. & Accorsi, G. Synthesis and photoluminescence properties of asymmetrical europium(III) complexes involving carbazole, phenanthroline and bathophenanthroline units. *Inorganica Chim. Acta* **362,** 3181–3186 (2009).

[239] Salvadó, I. *et al.* Membrane-disrupting iridium(iii) oligocationic organometallopeptides. *Chem. Commun.* **52,** 11008–11011 (2016).

[240] Wright, S. W., Hageman, D. L., Wright, A. S., McClure, L. D. Convenient preparations of t-butyl esters and ethers from t-butanol. *Tetrahedron Lett.* **38,** 7345–7348 (1997).

[241] Higashi, T., Inami, K., Mochizuki, M. Synthesis and DNA-binding properties of 1,10-phenanthroline analogues as intercalating-crosslinkers. *J. Heterocycl. Chem.* **45,** 1889–1892 (2008).

[242] Hernández, D. J., Vázquez-Lima, H., Guadarrama, P., Martínez-Otero, D., Castillo, I. Solution and solid-state conformations of 1,5-pyridine and 1,5-phenanthroline-bridged p-tert-butylcalix[8]arene derivatives. *Tetrahedron Lett.* **54,** 4930–4933 (2013).

[243] Zhao, S., Ito, S., Ohba, Y., Katagiri, H. *Determination of the absolute configuration and identity of chiral carboxylic acids, using a Cu(II) complex of pyridine–benzimidazole-based ligand. Tetrahedron Letters* **55,** (2014).

[244] Becker, H. G. O. 1922-. *Organikum : organisch-chemisches Grundpraktikum*. (Wiley-VCH Verlag GmbH & Co. KGaA, 2015).

[245] Still, W. C., Kahn, M., Mitra, A. Rapid chromatographic technique for preparative separations with moderate resolution. *J. Org. Chem.* **43,** 2923–2925 (1978).

[246] Gottlieb, H. E., Vadim Kotlyar, Nudelman, A. NMR Chemical Shifts of Common Laboratory Solvents as Trace Impurities. *J. Org. Chem.* **62,** 7512–7515 (1997).

[247] Corpillo, D., Cabella, C., Crich, S. G., Barge, A., Aime, S. Detection and Quantification of Lanthanide Complexes in Cell Lysates by Matrix-Assisted Laser Desorption/Ionization Time-of-Flight Mass Spectrometry. *Anal. Chem.* **76,** 6012–6016 (2004).

[248] Coelho, A. A. Indexing of powder diffraction patterns by iterative use of singular value decomposition. *J. Appl. Crystallogr.* **36,** 86–95 (2003).

[249] *TOPAS 4.2 User Manual*. (Bruker AXS GmbH, 2009).

[250] Favre-Nicolin, V., Černý, R. *FOX*, `free objects for crystallography': a modular approach

to *ab initio* structure determination from powder diffraction. *J. Appl. Crystallogr.* **35,** 734–743 (2002).

[251] Bushmarinov, I. S., Dmitrienko, A. O., Korlyukov, A. A., Antipin, M. Y. Rietveld refinement and structure verification, using `Morse' restraints. *J. Appl. Crystallogr.* **45,** 1187–1197 (2012).

[252] Grimme, S. Semiempirical GGA-type density functional constructed with a long-range dispersion correction. *J. Comput. Chem.* **27,** 1787–1799 (2006).

[253] Kresse, G., Hafner, J. *Ab initio* molecular-dynamics simulation of the liquid-metal–amorphous-semiconductor transition in germanium. *Phys. Rev. B* **49,** 14251–14269 (1994).

[254] Kresse, G., Furthmüller, J. Efficiency of ab-initio total energy calculations for metals and semiconductors, using a plane-wave basis set. *Comput. Mater. Sci.* **6,** 15–50 (1996).

[255] Van de Streek, J., Neumann, M. A. Validation of experimental molecular crystal structures with dispersion-corrected density functional theory calculations. *Acta Crystallogr. Sect. B Struct. Sci.* **66,** 544–558 (2010).

[256] Ravel, B., Newville, M. *ATHENA* , *ARTEMIS* , *HEPHAESTUS*: data analysis for X-ray absorption spectroscopy, using *IFEFFIT*. *J. Synchrotron Radiat.* **12,** 537–541 (2005).

[257] Newville, M. *et al.* Analysis of multiple-scattering XAFS data, using theoretical standards. *Phys. B Condens. Matter* **208–209,** 154–156 (1995).

[258] Hilder, M. *et al.* Spectroscopic properties of lanthanoid benzene carboxylates in the solid state: Part 2. Polar substituted benzoates. *J. Photochem. Photobiol. A Chem.* **217,** 76–86 (2011).

[259] Sun, M. *et al.* Bright and monochromic red light-emitting electroluminescence devices based on a new multifunctional europium ternary complex. *Chem. Commun.* **0,** 702–703 (2003).

[260] Takahashi, K. *et al.* The design and optimization of a series of 2-(pyridin-2-yl)-1H-benzimidazole compounds as allosteric glucokinase activators. *Bioorg. Med. Chem.* **17,** 7042–7051 (2009).

[261] Mustafa, D., Ma, D., Zhou, W., Meisenheimer, P., Cali, J. J. Novel No-Wash Luminogenic Probes for the Detection of Transporter Uptake Activity. *Bioconjug. Chem.* **27,** 87–101 (2016).

[262] Bertrand, H. C. *et al.* Design, Synthesis and Evaluation of Triazole Derivatives That Induce Nrf2 Dependent Gene Products and Inhibit the Keap1–Nrf2 Protein–Protein Interaction. *J. Med. Chem.* **58,** 7186–7194 (2015).

[263] Knör, S. *et al.* Synthesis of Novel 1,4,7,10-Tetraazacyclodecane-1,4,7,10-Tetraacetic Acid (DOTA) Derivatives for Chemoselective Attachment to Unprotected Polyfunctionalized Compounds. *Chem. - A Eur. J.* **13,** 6082–6090 (2007).

[264] Pearson, D. A., Blanchette, M., Baker, M. Lou, Guindon, C. A. Trialkylsilanes as scavengers for the trifluoroacetic acid deblocking of protecting groups in peptide synthesis. *Tetrahedron Lett.* **30,** 2739–2742 (1989).

[265] Castonguay, R., Lherbet, C., Keillor, J. W. Mapping of the active site of rat kidney γ-glutamyl transpeptidase, using activated esters and their amide derivatives. *Bioorg. Med. Chem.* **10,** 4185–4191 (2002).

[266] Riva, R., Banfi, L., Basso, A., Zito, P. A new diversity oriented and metal-free approach to highly functionalized 3H-pyrimidin-4-ones. *Org. Biomol. Chem.* **9,** 2107 (2011).

[267] Soundararajan, N., Platz, M. S. Descriptive photochemistry of polyfluorinated azide

derivatives of methyl benzoate. *J. Org. Chem.* **55,** 2034–2044 (1990).

[268] Solodukhin, N. N., Borisova, N. E., Churakov, A. V. & Zaitsev, K. V. Substituted 4-(1H-1,2,3-triazol-1-yl)-tetrafluorobenzoates: Selective synthesis and structure. *J. Fluor. Chem.* **187,** 15–23 (2016).

[269] Das, B., Krishnaiah, M., Laxminarayana, K. Zirconium tetrachloride mediated regioselective transformation of N-tosylaziridines into β-chlorosulfonamides. *J. Chem. Res.* **2007,** 82–83 (2007).

[270] Smith, K., Musson, A., DeBoos, G. A. A Novel Method for the Nitration of Simple Aromatic Compounds. *J. Org. Chem.* **63,** 8448–8454 (1998).

[271] Mackman, R. L. *et al.* Exploiting Subsite S1 of Trypsin-Like Serine Proteases for Selectivity: Potent and Selective Inhibitors of Urokinase-Type Plasminogen Activator. *J. Med. Chem.* **44,** 3856–3871 (2001).

[272] Scharow, A. *et al.* Optimized Plk1 PBD Inhibitors Based on Poloxin Induce Mitotic Arrest and Apoptosis in Tumor Cells. *ACS Chem. Biol.* **10,** 2570–2579 (2015).

[273] Zimmerman, S. S. *et al.* Design, Synthesis and Structure–Activity Relationship of a Novel Series of GluN2C-Selective Potentiators. *J. Med. Chem.* **57,** 2334–2356 (2014).

[274] Fang, Y.-Q., Lautens, M. A Highly Selective Tandem Cross-Coupling of gem-Dihaloolefins for a Modular, Efficient Synthesis of Highly Functionalized Indoles. *J. Org. Chem.,* **73,** 538–549 (2008).

[275] Chandler, C. J., Deady, L. W., Reiss, J. A. Synthesis of some 2,9-disubstituted-1,10-phenanthrolines. *J. Heterocycl. Chem.* **18,** 599–601 (1981).

[276] Malet, P., Capitan, M. J., Centeno, M. A., Odriozola, J. A., Carrizosa, I. EXAFS data analysis for lanthanide sesquioxides. *J. Chem. Soc. Faraday Trans.* **90,** 2783 (1994).

[277] Beuchat, C., Hagberg, D., Spezia, R., Gagliardi, L. Hydration of Lanthanide Chloride Salts: A Quantum Chemical and Classical Molecular Dynamics Simulation Study. *J. Phys. Chem. B* **114,** 15590–15597 (2010).

VIII. Appendix

VIII.1 Curriculum Vitae

Chemist with physics and materials science background. Social, internationally oriented and enjoy being a productive part of a team as well as encouraging collaborative working.

Email: a.s.kalyakina@gmail.com
Mobil: +49 1577 645 0 645
linkedin.com/in/alena-kalyakina
Born 3.7.1992 in Obninsk, Russia

WORK EXPERIENCE

05/2018–02/2018 **Visiting scholar,** Harvard University/Massachusetts General Hospital, Boston, USA, Prof. Dr. R. Mazitschek
Multi-step synthesis of Tb(III) complex with a macrotricyclic ligand for a potential use in FRET-bioassays

07/2015–11/2013 **Junior project manager and researcher** at the startup company evOLED, Riga, Latvia
- Synthesis of emitting layers for OLEDs, optimization of OLED structure, OLED prototyping.
- Project management and coordination, business plan drafting, preparation of the reports and presentations.

12/2014–09/2014 **Intern**: National Research Centre «Kurchatov Institute», Moscow, Russia, Prof. Dr. Y.V. Zubavichus
EXAFS/XANES spectroscopy for the characterization of metal bonding in metal complexes.

06/2014–03/2014 **Intern:** Nesmeyanov Institute of Organoelement Compounds, Moscow, Russia, Prof. Dr. K.A. Lyssenko
Determination of the structures of organic and metalorganic compounds *via* single-crystal and powder X-ray diffraction.

11/2013–08/2013 **Intern:** Karlsruhe Institute of Technology, Karlsruhe, Germany, Prof. Dr. S. Bräse
Synthesis, characterization and photophysical properties of the metal complexes with fluorinated benzoic acid derivatives.

08/2013–09/2010 **Research assistant:** Lebedev Physical Institute of Russian Academy of Sciences, Moscow, Russia, Dr. L.S. Lepnev
Luminescence spectroscopy (measurements and data analysis).

EDUCATION

07/2018–08/2015 **Ph.D. in natural sciences**, specialization: Chemistry, Supervisor: Prof. Dr. S. Bräse, Karlsruhe Institute of Technology, Institute of Organic Chemistry, Karlsruhe, Germany
Thesis title: Novel lanthanide-based luminescent probes for biological and lighting applications.
- Organic and metal-organic synthesis (cross-coupling reactions, click chemistry (CuAAC and SPAAC), solid-phase synthesis, synthesis of peptides and peptoids), photophysical and structural characterization.
- Testing of the synthesized compounds: OLED prototyping, cellular experiments.
- Additional: supervision of intern and Master students, CCDC Database search and data analysis, DFT calculations.

11/2017–10/2015 **MBA** Fundamentals Program 18 ECTS (GPA 3.5), Karlsruhe Institute of Technology HECTOR School of Engineering and Management, Karlsruhe, Germany

07/2015–09/2013 **M.Sc. in Chemistry**, specialization: Solid State Chemistry (GPA 4.0), Supervisor Prof. Dr. Kuzmina N.P., Lomonosov Moscow State University, Department of Materials Science, Moscow, Russia.
Thesis title: Synthesis, structure and photophysical properties of lanthanide fluorobenzoates with different ligand fluorination degree – performed in the firm evOLED
- Synthesis, characterization and photophysical study of lanthanide fluorobenzoates, OLED prototyping.
- Structure elucidation of lanthanide complexes in solids and solutions (NMR, EXAFS and X-ray techniques).

06/2014–09/2011 **B.Sc. in Economics**, specialization: Finance (GPA 3.7) – 2nd B.Sc. degree, Supervisor: Dr. Chalova A.Yu. Plekhanov Russian University of Economics, Institute of finance and credit, Moscow, Russia
Thesis title: The foreign experience of government budgetary policy under crisis and economic growth and its implementation to Russia.

06/2013–09/2009 **B.Sc. in Materials Science**, specialization: Materials Chemistry, Physics and Mechanics (GPA 3.5) Lomonosov Moscow State University, Department of Materials Science, Moscow, Russia.
Thesis title: Chemical solution deposition of Eu and Tb aromatic carboxylate thin films and their use in OLEDs.
- Synthesis, characterization and photophysical study of lanthanide aromatic carboxylates.
- Thin film deposition techniques development and OLED fabrication

06/2009–09/2007 **Secondary school:** The Advanced Educational Scientific Center (faculty) – Kolmogorov's boarding school of Moscow State University (AESC MSU), Moscow, Russia – free attendance scholarship.

PROFESSIONAL SKILLS

Chemical syntheses
- Organic
- Inorganic
- Metal-organic

Cellular experiments
- Culturing
- MTT assay
- Cell microscopy

Analytics
- UV-Vis/ Luminescence
- X-ray diffraction
- NMR spectroscopy
- HPLC (analytic and prep.)
- Raman, IR spectroscopy
- Mass-spectrometry (MALDI-TOF, GC-, LC-MS)

Other
- Microscopy AFM, SEM/TEM, Confocal
- OLED prototyping (incl. Spin-coating/PVD/CVD)
- Synchrotron (EXAFS&XANES)
- DFT calculations

LEADERSHIP EXPERIENCE

- supervision of the practical research work of 2 bachelor students (8 semesters, materials science)
- supervision of the practical research work of 1 intern master student (3 months)
- supervision of the theoretical course of 3 KSOP master students (2 months)
- organic chemistry practical course assistance (13.8 ECTS, 10 students)

LANGUAGES

- **Russian** (mother tongue) ■■■■■
- **German** (fluent) ■■■■□
- **English** (fluent) ■■■■□

SELECTED AWARDS

- KHYS KIT Research Travel Grant for the internship to Harvard Medical School (**2017**)
- KHYS KIT Internship Grant for the supervision of a foreign master student (**2016**)
- DAAD PhD scholarship Graduate School Scholarship Programm (**2015–2018**)
- Winner of the LG Chem competition (**2015–2016**)
- Award at the ICFE9 (International conference on f-elements), Oxford, UK (**2015**)
- Award for the best master thesis in Inorganic Chemistry, Moscow, Russia
- Saint-Gobain Challenge award for master students (**2014**)
- 2-year funding of commercially oriented scientific and technical projects of young scientists (**2012-2014**), awarded by *Russian Global Innovation Foundation*

MEMBERSHIP

- KSOP – Karlsruhe School of Optics and Photonics (**2015–2018**)
- GRK2039 – Research Training Group – PhD graduate school (**2015–2018**)

IT-SKILLS

- Origin Pro, MATLAB (expert) ■■■■■
- ChemBioOffice, MSOffice (expert) ■■■■■
- ACDLabs, MestRe-C (expert) ■■■■■
- SHELXL/XS, OLEX2 (expert) ■■■■■
- Mercury, Diamond (expert) ■■■■■
- ConQuest (expert) ■■■■■
- IFFEFITT (intermediate) ■■■■■
- TmoleX (intermediate) ■■■□□
- LAS XF (intermediate) ■■■□□
- CSD Python API (beginner) ■■□□□

VIII.2 Publications and conference contributions

1st-author papers (peer-reviewed)

1. Kalyakina A.S., Utochnikova V.V., Zimmer M., Dietrich F., Kaczmarek A.M, Van Deun R., Vashchenko A.A., Goloveshkin A.S., Nieger M., Gerhards M., Schepers U., Bräse S. *Remarkable high efficiency of red emitters, using Eu(III) ternary complexes, Chemistry Communications* **2018,** 54 (41), 5221-5224, DOI: 10.1039/c8cc02930j.

2. Kalyakina A.S., Utochnikova V.V., Bushmarinov I.S., Le-Deygen I.M., Volz D., Schepers U., Kuzmina N.P., Bräse S. *Lanthanide fluorobenzoates as bio-probes: a quest for the optimal ligand fluorination degree, Chemistry – A European Journal* **2017**, *23*, 14944-14953, DOI: 10.1002/chem.201703543.

3. Kalyakina A.S., Utochnikova V.V., Sokolova E.Yu., Vashchenko A.A., Lepnev L.S., Van Deun R., Trigub A.L., Zubavichus Y.V., Hoffmann M., Mühl S., Kuzmina N.P. *OLED thin film fabrication from poorly soluble terbium o-phenoxybenzoate through soluble mixed-ligand complexes, Organic Electronics* **2016**, *28*, 319-329, DOI:10.1016/j.orgel.2015.11.006

4. Kalyakina A.S., Utochnikova V.V., Trigub A.L, Zubavichus Y.V., Kuzmina N.P., Bräse S. EXAFS characterization of metal bonding in highly luminescent, UV stable, water-soluble and biocompatible lanthanide complexes, *Journal of Physics: Conference Series* **2016**, *712*, 012137, DOI: 10.1088/1742-6596/712/1/012137.

5. Kalyakina A.S., Utochnikova V.V., Bushmarinov I.S., Ananyev I.V., Eremenko I.L., Volz D., Rönicke F., Schepers U., Van Deun R., Trigub A.L., Zubavichus Y.V., Kuzmina N.P., Bräse S. *Highly Luminescent, Water-Soluble Lanthanide Fluorobenzoates: Syntheses, Structures and Photophysics, Part I: Lanthanide Pentafluorobenzoates, Chemistry – A European Journal* **2015**, *21*, 17921-17932, DOI: 10.1002/chem.201501816.

Othet papers (peer-reviewed)

6. Utochnikova V.V., Abramvich M.S., Latipov E.V., Dalinger A.I., Goloveshkin A.S., Vashchenko A.A., Kalyakina A.S., Vatsadze S.Z., Schepers U., Bräse S., Kuzmina N.P. *Brightly luminescent lanthanide pyrazolecarboxylates: synthesis, luminescent properties and influence of ligand isomerism, Journal of Luminescence* **2019,** *205*, 429-439, DOI: 10.1016/j.jlumin.2018.09.027.

7. Utochnikova V.V., Latipov E.V., Dalinger A.I., Nelyubina Yu.V., Vashchenko A.A., Hoffman M., Kalyakina A.S., Vatsadze S.Z., Schepers U., Bräse S., Kuzmina N.P. *Lanthanide Pyrazolecarboxylates for OLEDs and Bioimaging, Journal of Luminescence* **2018**, *202*, 38-46, DOI: 10.1016/j.jlumin.2018.05.022.

8. Khudoleeva V.Yu., Kovalenko A.D., Kalyakina A.S., Goloveshkin A.S., Lepnev L.S., Utochnikova V.V. *Terbium-europium fluorides surface modified with benzoate and terephthalate anions for temperature sensing: does sensitivity depend on the ligand? Journal of Luminescence* **2018**, *201*, 500-508, DOI: 10.1016/j.jlumin.2018.05.002.

9. Khudoleeva V.Yu., Utochnikova V.V., Kalyakina A.S., Deygen I.M., Shiryaev A.A., Lebedev V.A., Roslyakov I.V., Garshev A.V., Lepnev L.S., Schepers U., Bräse S., Kuzmina N.P. *Surface modified $Eu_xLa_{1-x}F_3$ nanoparticles as luminescent biomarkers: Still plenty of room at the bottom, Dyes and Pigments* **2017**, *143*, 348-355, DOI: 10.1016/j.dyepig.2017.04.058.

10. Utochnikova V.V., Solodukhin N.N., Aslandukov A.N., Zaitsev K.V., Kalyakina A.S., Averin A.A., Ananyev I.V., Churakov A.V., Kuzmina N.P. *Luminescence enhancement by p-substituent variation, European Journal of Inorganic Chemistry* **2017**, *1*, 107-114, DOI: 10.1002/ejic.201600843.

11. Utochnikova V.V., Kalyakina A.S., Bushmarinov I.S., Vashchenko A.A., Marciniak L., Kaczmarek A.M., Van Deun R., Bräse S., Kuzmina N.P. *Lanthanide 9-anthracenate: solution processable emitters for efficient purely NIR emitting host-free OLEDs, Journal of Materials Chemistry C* **2016**, *4*, 41, 9848-9855, DOI: 10.1039/c6tc03586h.

12. Utochnikova V.V., Koshelev D.S., Medvedko A.V., Kalyakina A.S., Bushmarinov I.S., Grishko A.Yu, Schepers U. Bräse S., Vatsadze S.Z. *Europium 2-benzofuranoate: Synthesis and use for bioimaging, Optical Materials* **2017**, *74*, 191-196, DOI: 10.1016/j.optmat.2017.05.038.

13. Utochnikova V.V., Kalyakina A.S., Lepnev L.S., Kuzmina N.P. *Luminescence enhancement of nanosized ytterbium fluorides by surface complex formation with aromatic carboxylates, Journal of Luminescence* **2016**, *170*, 633-640, DOI: 10.1016/j.jlumin.2015.03.033.

14. Utochnikova V.V., Kovalenko A.D., Burlov A.S., Marciniak L., Ananyev I.V., Kalyakina A.S., Kurchavov N.A., Kuzmina N.P. *Lanthanide complexes with 2-(tosylamino)-benzylidene-N-benzoylhydrazone, which exhibit high NIR emission, Dalton Transactions* **2015**, *44*, 12660-12669, DOI: 10.1039/c5dt01161b.

15. Utochnikova V.V., Kalyakina A.S., Kuzmina N.P. *New approach to deposition of thin luminescent films of lanthanide aromatic carboxylates, Inorganic Chemistry Communications* **2012**, *16*, 4-7, DOI: 10.1016/j.inoche.2011.11.009.

Patents

16. Patent WO2015/030627 Utochnikova V., Kalyakina A., Sokolova E., Vashchenko A., Lepnev L., Kuzmina N., *Emission layers for organic light emitting diodes and methods for their preparation*, **2013.**

Proceedings (non-peer-reviewed)

17. Utochnikova V., Kalyakina A., Grishko A., Kovalenko A., Marciniak L., Van Deun R., Bräse S., Lepnev L., Kuzmina N. *Lanthanide complexes with fluorinated aromatic carboxylates for visible and near-infrared luminescence*, Proceedings to SPIE Photonics, **2014,** USA.

18. Kalyakina A.S., Sokolova E.Yu., Vaschenko A.A., Lepnev L.S., Utochnikova V.V., Kuzmina N.P. *High efficiency organic light-emitting diodes based on UV stable terbium aromatic carboxylates,* SID Mid-Europe Chapter Spring Meeting, **2013**, p.19-20.

Abstracts in conference abstract books

19. Kalyakina A.S., Schneider A.C., Bräse S. *Novel lanthanide-based luminescent probes for bioimaging*, Book of Abstracts, Symposium on molecular architectures for fluorescent imaging of cells. 4.10 – 6.10, **2017**, Karlsruhe, Germany.

20. Kalyakina A.S. *New highly luminescent probes for biological and lighting applications*, Book of Abstracts, Karlsruhe Days of Optics and Photonics, 07.11 – 08.11, **2017**, Karlsruhe, Germany.

21. Kalyakina A.S. *Lanthanide-based systems for multimodal optical and MRI cellular imaging*, Book of Abstracts, 5th Heidelberg forum for young life scientists, 08.06 – 09.06, **2017**, Heidelberg, Germany.

22. Kalyakina A.S., Utochnikova V.V., Kuzmina N.P., Bräse S., *Lanthanide fluorobenzoates for bioimaging and lighting applications*, Abstract Book, p.177, 42nd International Conference on Coordination Chemistry ICCC2016, 03.07 – 08.07, **2016**, Brest, France.

23. Kalyakina A.S., Utochnikova V.V., Kuzmina N.P., Bräse S., *Lanthanide fluorobenzoates as promising materials for cellular bio-imaging*, Book of Abstracts, Frontiers in Medicinal Chemistry, 13.03 – 16.03, **2016**, Bonn, Germany.

24. Utochnikova V., Kalyakina A., Grishko A., Kovalenko A., Marciniak L., Van Deun R., Bräse S., Lepnev L., Kuzmina N. *NIR emitting lanthanide coordination compounds for use as biomarkers*, Book of Abstracts, 17th International Conference on Luminescence and Optical Spectroscopy of condensed matter, 13.07 – 18.07, **2014**, Wrocław, Poland.

25. Utochnikova V., Kalyakina A., Sokolova E., Vaschenko A., Lepnev L., Kuzmina N. *Solution processed OLEDs with luminescent lanthanide aromatic carboxylate thin films as active layers*, Book of Abstracts, p. 3, E-MRS Spring Meeting 26.05 – 30.05, **2014**, Lille, France.

26. Kalyakina A., Sokolova E., Utochnikova V., Bräse S. *Toward the efficient NIR- and visible emitters based on luminescent lanthanide coordination compounds with aromatic ligands*, Book of abstracts, p.26, 2nd International conference on Bimetallic Complexes, 23.09 – 25.09, **2013**, Karlsruhe, Germany.

27. Kalyakina A., Vaschenko A., Utochnikova V. *Thin film deposition of luminescent terbium o-phenoxybenzoate for optoelectronic applications,* Book of Abstracts, XII International Krutyń Summer School «Solving the World's Energy Demands with Molecules and Nanostructures in Sunlight» 30.09 – 06.10, **2012**, Krutyń, Poland.

28. Utochnikova V., Kalyakina A., Vaschenko A., Pietraszkiewicz O., Pietraszkiewicz M., Kuzmina N. *Mixed-ligand compounds of lanthanide luminescent aromatic carboxylates and their role in thin film deposition of homo-ligand aromatic carboxylates*, Book of Abstracts, p. 17, 8th international conference on f-elements ICFE8, 26.08 – 31.08, **2012**, Udine, Italy

Oral presentations

1. **Alena S. Kalyakina**, Stefan Bräse. Symposium on molecular architectures for fluorescent imaging of cells. 4-6. October, 2017, Karlsruhe, Germany. *Lanthanide-based luminescent probes for cellular imaging.*

2. **Alena S. Kalyakina**, Stefan Bräse. Research training group GRK2039 retreat and 7th Seminar day, 11. April 2017, Speyer, Germany. *Lanthanide-based luminescent probes for cellular imaging.*

3. **Alena S. Kalyakina**, Stefan Bräse. Research training group GRK2039 retreat, Annweiler am Trifels, Germany, 15. February 2016 *Novel lanthanide-based probes for bioapplications.*

4. **Alena S. Kalyakina**, V. Utochnikova, Stefan Bräse. International conference on f-elements ICFE9, 6-9. September, 2015, Oxford, UK. *Synthesis, structure and photophysical properties of lanthanide fluorobenzoates with different ligand fluorination degrees.*

Posters

1. **Alena S. Kalyakina**, Stefan Bräse, Karlsruhe Days of Optics and Photonics, 7 – 8. November, **2017**, Karlsruhe, Germany. *New highly luminescent probes for biological and lighting applications.*

2. **Alena S. Kalyakina**, Stefan Bräse. Symposium on molecular architectures for fluorescent imaging of cells. 4-6. October, **2017**, Karlsruhe, Germany. *Lanthanide-based luminescent probes for cellular imaging.*

3. Jasmin Busch, **Alena S. Kalyakina**, Martin Nieger, Stefan Bräse 3MET PhD Meetiung, 27. September **2017**, Heidelberg, Germany. *Synthesis of metal complexes with unusual spectroscopic properties.*

4. **Alena S. Kalyakina**, Stefan Bräse, 5th Heidelberg forum for young life scientists.8 – 9 June **2017**, Heidelberg, Germany. *Ln-based systems for multimodal optical/MRI cellular imaging.*

5. **Alena S. Kalyakina**, Stefan Bräse. Research training group GRK2039 retreat, 11. April **2017**, Speyer, Germany. *Lanthanide-based luminescent probes for cellular imaging.*

6. **Alena S. Kalyakina**, Stefan Bräse, Research training group GRK2039 retreat. 04 – 06. April, **2016**, Lauterbad, Germany. *Lanthanide fluorobenzoates as promising materials for cellular bio-imaging.*

7. **Alena S. Kalyakina**, Valentina V. Utochnikova, Natalia P. Kuzmina, Stefan Bräse. 42nd International Conference on Coordination Chemistry, 3 – 8. July **2016**, Brest, France. *Lanthanide fluorobenzoates for bioimaging and lighting applications.*

8. **Alena S. Kalyakina**, Valentina V. Utochnikova, Natalia P. Kuzmina, Stefan Bräse, Frontiers in Medicinal Chemistry, 13 – 16. March **2016**, Bonn, Germany. *Lanthanide fluorobenzoates as promising materials for cellular bio-imaging.*

9. Alexander L. Trigub, **Alena S. Kalyakina**, Valentina V. Utochnikova, Yan V. Zubavichus, Stefan Bräse, 16th International Conference on X-ray Absorption Fine Structure, 23 – 28. August **2015**, Karlsruhe, Germany. *EXAFS characterization of metal bonding in highly luminescent stable, water-soluble and biocompatible lanthanide complexes.*

IX. Acknowledgements

I would like to start with a heartfelt thank to Prof. Dr. **Stefan Bräse** for giving me the opportunity to work in his team. I just think he is a very brave man, letting me work in his lab ☺. Stefan, you are a great chemist and mentor and I appreciate your true believe in me and my research. You supported all my ideas (including the craziest ones) and I guess these three years under your supervision made me an independent scientist.

I thank Prof. Dr. **Ute Schepers** for letting me work in her team. I bet I spent a quarter of my PhD there. Oh my God, how excited I was, when I first saw the cells through the ocular of the microscope and how even more excited I became when I first saw my cells emitting red light. Ute, thank you for your kindness and valuable advices.

I am also very grateful to Prof. Dr. **Ralph Mazitschek** for welcoming me in his amazing team in MGH/Harvard Medical School, Boston. Working in his lab was incredible and our project broadened both, my knowledge frontiers and my adaptability. I've experienced what does it mean to be a great chemist, not just a good one. I really appreciate that you gave me the opportunity to have this experience. It taught and shaped me a lot.

I am very grateful to Prof. Dr. **Peter Roesky**, who agreed to be my coreferent.

I am utterly thankful to Prof. Dr. **Natalia Kuzmina**, who is my "scientific grandmother" from my studentship. Despite the distance, I always felt your support and, honestly, when I was puzzled with some problems, sometimes I thought: "what would she think about it? How would she do it?" The experience I gained working in your group is just unforgettable. Thank you for being a family to me.

Huge thanks goes to Dr. **Valentina Utochnikova**. You are truly the person, who showed me, how cool the science can be. How many evenings we were sitting together, struggling on something and how proud we were getting the great results. You aroused my curiosity and brought up the ability to work hard. I am happy to be your friend and really hope we can see each other more often.

I cannot help but mention all my collaborators, who altogether made this research possible. Dr. **Anna Kaczmarek** and Prof. Dr. **Rik Van Deun** from Ghent University, Belgium, who were always ready to help with the measurements and their interpretation. You made a huge contribution and it was a real pleasure to work with you.

I am very thankful to Dr. **Patrick Weis** and **Katrina Brendle** for their timely response and their kind willingness to help. I would also like to thank Prof. Dr. **Markus Gerhards** and his group (Dr. **MANUEL ZIMMER** and Dr. **FABIAN DIETRICH**) for the opportunity to conduct such a fruitful collaboration. I also want to thank Dr. **MARTIN NIEGER** for his help with the crystal data analysis and expertise. Another "Thank you" goes to Dr. **ALEXANDER TRIGUB** and Prof. Dr. **YAN ZUBAVICHUS** for helping me with the EXAFS data analysis.

I would like to thank X-ray collaboration team. Dr. **IVAN ANANYEV**, you are a great talented person and a remarkable mentor. I am happy to learn so many cool things from you and I always will be grateful for your great support. Another "Thanks" goes to Dr. **IVAN BUSHMARINOV**. I really appreciate your willingness to help anytime and your efforts to teach me. You are truly one of the smartest person on this planet.

I am very thankful to Dr. **ANDREY VASHCHENKO** and Dr. **LEONID LEPNEV**. You are great physicists and specialists, thank you for always-timely response and your valuable expertise.

An invaluable thanks goes to my "correction team": **CLAUDINE HERLAN**, **NICOLAI WIPPERT** and Dr. **KSENIA KUTONOVA**. Thank you for helping me to make my thesis better even in such a limited time. Claudine, thank you for your kindness, you made a huge job and I really appreciate your readiness to read my thesis so carefully and so quickly. Nico, huge thanks to you as well, you helped me a lot. Ksenia, what you did was amazing. You were able read the whole main part of my thesis in one weekend day and give me unbelievably valuable comments. Beyond the thesis, you're just a great person and a top-friend. I am very happy to know you.

I would like to thank the whole **AK BRÄSE** team for each and every minute. Particular thanks goes to the "weekend workers team": **ANNE SCHNEIDER**, **TOBIAS BANTLE** and Dr. **YULING HU**. Anne, huge thanks to you, you helped me a lot, not only with lab stuff, but just personally. Yuling, you are so responsive, you are a great lab mate and my kind friend. Additionally, thanks to the whole 306-team, **ROBIN BÄR**, **JANINA BECK**, **MAREEN STAHLBERGER**, **NIKOLAI ROSENBAUM**, Dr. **LARISSA GEIGER**, Dr. **MIRELLA WAWRYSZYN**, **CHRISTINA RETICH**, **QAIS AHMAD PARSA**, **RIEKE SCHULTE**, thank you, guys, it was my pleasure to work with you together. My particular thanks goes to Dr. **STEPHAN MÜNCH**, thanks for your help and for being a great person. I guess I still owe you a beer. ☺ A huge thanks to **SELIN SAMUR** and **CHRISTIANE LAMPERT**, your help cannot be overestimated.

Dr. **ANGELA WANDLER**, I wish you all the best, you were very kind to me and I wish we could meet each other more often to continue the tradition of our amazing coffee times. Dr. **VANESSA KOCH**, thank you for your kindness and your great recipes of cookies, I hope we will repeat this great Christmas experience.

A huge thanks to Dr. **Thomas Hurrle**, you are just a cool person and I sincerely wish you and your great family all the best. I will always remember our bouldering activities, it was truly awesome.

I want to acknowledge my GRK2039-family, particularly Dr. **Mathilde Bichelberger**, **Mikhail Khorenko**, **Katharina Fanselau**, **Carolin Heiler**.

Thanks to Prof. Dr. **Hans-Achim Wagenknecht** for letting me be the part of this family.

Along the same lines, I would like to thank **AK Schepers**, particularly **Anna Meschkov**, Dr. **Bettina Olshausen** and **Bettina Fleck**. I was happy to work with you girls. Anna, it's even hard to express my gratitude to you, being always so supportive and responsive. I am happy to have such a great friend.

I am very grateful to my labmates, whom I was working with in Boston: **Mark Tye**, **Connor Payne**, **Alexa Jackson**, **Kritika Singh**, **Lola Fagbami**. Guys, you are amazing. I will always remember our late dinners in the lab and cool conversations. Mark, thanks for being such a kind and interesting person, I enjoyed working together with you and learned a great deal from you. Connor, you are really cool. Honestly, I have never seen such a clean fume hood! Your working style is unique. Alexa, a huge thanks to you, you are a great person. Thanks for showing me Boston Ballet and for your great character. I enjoyed our writing time in the conference room.

I would like also to show my gratitude to the **technical staff** of the Institute of Organic Chemistry, KIT, who has always been of tremendous help for both, science and organizational issues.

I am very grateful to my **family**. Thanks to my mom **Tatiana Kalyakina** for her kind support and for her care, I would not be able to handle everything without it. Thanks to my dad Dr. **Sergey Kalyakin** for his support and his great sense of humor, it really helped me a lot. Thanks to my Bro Dr. **Dmitriy Kalyakin**, his charming wife **Oxana Kalyakina** and their lovely son **Sergey Kalyakin**. They were together with me at my writing time and brightened up these hard times ☺. Thank you all, guys, I wouldn't make it without you.

Last, but not least, I would like to thank "meine bessere Hälfte" Dr. **Vladimir Korzinov** for everything: for his endless love and his boundless patience, for listening, talking and always being near. Everything makes much more sense with you. Thank you for making me a better person.

If you want to overcome the whole world,
overcome yourself

– Fyodor Dostoevsky, *Demons*

Если хочешь победить весь мир,
победи себя

– Федор Достоевский, *Бесы*